Lamprophyres

N.M.S. ROCK
MA, Ph.D. (Cantab), D.Sc. (UWA)

Key Centre for Strategic Mineral Deposits
Department of Geology
University of Western Australia

With additional invited contributions by

D.R. BOWES
Department of Geology and Applied Geology
University of Glasgow

and

A.E. WRIGHT
Department of Geological Sciences
University of Birmingham

Blackie
Glasgow and London

Van Nostrand Reinhold
New York

Blackie and Son Ltd
Bishopbriggs, Glasgow G64 2NZ
and
7 Leicester Place, London WC2H 7BP

Published in the United States of America by
Van Nostrand Reinhold
115 Fifth Avenue
New York, New York 10003

Distributed in Canada by
Nelson Canada
1120 Birchmount Road
Scarborough, Ontario M1K 5G4, Canada

16 15 14 13 12 11 10 9 8 7 6 5 4 3 2 1

First published 1991

British Library Cataloguing in Publication Data

Rock, N. M. S.
Lamprophyres.
1. Igneous rocks
I. Title
552

ISBN 0-216-92897-4

Library of Congress Cataloging-in-Publication Data

Rock, N.M.S.
Lamprophyres / by N.M.S. Rock : with additional invited contributions from D.R. Bowes and A.E. Wright.
p. cm.
ISBN 0-442-30396-3 (Van Nostrand Reinhold)
1. Lamprophyres. I. Title
QE461.R593 1990
552'.1—dc20

90-360 12
CIP

Printed in Great Britain by Thomson Litho Ltd, East Kilbride, Glasgow

Acknowledgements
I am indebted to UWA and to all my immediate colleagues, for providing such a stimulating atmosphere in which to write this work. I also thank Blackie and Son for commissioning it. I have greatly benefited over the years from communication with the following, some of whom have generously provided unpublished information: Zhang Andi, Peder Aspen, Alistair Baxter, Alan Cooper, Barry Dawson, Jean Bédard, Steve Bergman, Walter Boyd, Felix Chayes, Ashoka Dessai, Alan Edgar, Ron Elsdon, Eric Finlayson, Steve Foley, Bill Gaskarth, Brendon Griffin, Sue Golding, Charlie Harper, Bram Janse, Lynton Jaques, Roger Hamilton, Ji Haizhang, Ali Abba Kaka, Roger Le Maitre, Ray Macdonald, Greg McHone, Henry Meyer, Roger Mitchell, Dusian Nemec, Dalim Paul, Colin Price, Nick Ratcliffe, Chris Rundle, Henri Sabatier, Paul Scott, John Sheraton, Barbara Scott Smith, Chris Smith, Joe Smith, Lyn Sutherland, Wayne Taylor, Matthew Thirlwall, Bob Thompson, Brian Upton, Danielle Velde, Gianpero Venturelli, Alan Woolley, Derek Wyman, Hatton Yoder, Xue-Hui Yu, the invited contributors, and the late Sharon Bachinski. Research students Graeme Cameron, Damon Edwards, Paul Henney, Peter Hwang, Philip Mather, Maureen Muggeridge, Caroline Perring, Robert Ramsay, Paul Shand, Steve Sheppard and Penny Wallace-Bell have also contributed greatly. For computing assistance, I am indebted to Gary Carroll and Mike Wheatley. Financial support has been provided by ARC, CTEC, MERIWA and UWA.

Preface

Following their recognition by Gümbel (1874), lamprophyres were treated for an entire century as little more than obscure curiosities. Although this situation has changed recently, with a flowering of publications and active workers, lamprophyres remain almost the only group of igneous rocks which have not yet received attention in a dedicated monograph.

In five exploratory reviews (1977–1987), the writer aimed to set out what was known about these rocks. The IUGS Subcommission on igneous rock systematics had meanwhile presented its nomenclatural framework (Streckeisen 1979). All this has now been overtaken by a recent explosion of interest, epitomized not least by lamprophyres' greater prominence in the *4th International Kimberlite Conference Proceedings.* More data have become available since 1985 than over the entire previous century, and it is obviously impossible for such an extraordinary outpouring to be fully reviewed in this first, preliminary book.

At the risk of dissatisfying some readers, therefore, this book concentrates on factual matters, and on a broad overview rather than minutiae. Because not even a world map of known lamprophyres was previously available, almost half the book is deliberately taken up by the first global lamprophyre compilation, and its commensurately extensive Bibliography. Such a compendium of largely objective information is believed to be of more immediate interest and lasting value than a premature pottage of petrogenetic polemic.

Chapters 1-7 bring previous studies up to date, and concentrate on factual information. Chapter 8 shows why more workers are coming to see lamprophyres as nearer the centre than the periphery of modern petrology. Chapter 9 briefly outlines reasons for the growing economic interest in lamprophyres. The Bibliography offers some 1,500 references. Appendix A compiles the many abbreviations and symbols used. Appendix B aims to help non-specialist readers to steer their way through the terminological morasse formerly surrounding lamprophyres; geological terms cited in the text that do *not* agree with the usage in Appendix B are placed in "quotes". Appendix C, a primary *raison d'être* of this book, documents and (with Figs.2.1–2.13) locates ≈ 500 lamprophyre occurrences for the first time. Appendix D details the computerized database LAMPDA on which the book is largely based. Tables/Figures in the Appendices are referred to in the form 'Table C1' (i.e. Table C1, Appendix C), whilst those in the text receive decimal numbers, e.g.'Table 3.1'.

The entire text and figures of this book were generated on a Macintosh® IIx microcomputer and LaserWriter® IINTX, for photographic reproduction by the publishers.

Constructive comments from interested readers, helping towards the eventual preparation of a more comprehensive, seasoned future edition, would be most welcome.

N.M.S.R.

Contents

1 What are Lamprophyres? — History, Definitions, Classification

1.1 A brief history of lamprophyre research

To an uncomfortably large number of geologists, the term 'lamprophyre' conjures up predominantly negative images: of rotten dykes forming miniscule, impersistent outcrops; of legion obscure rock-types named after equally obscure European villages. To many, indeed, the term connotes little more than 'horribly altered', or 'impossible to classify'. This is all a pity, for it has severely delayed the recognition until recently that lamprophyres have an importance out of all proportion to their absolute abundance among igneous rocks: as windows into the deep mantle and mantle processes, as possible indicators for precious mineral deposits (notably diamond and gold), as parental magmas to an extraordinarily wide range of other igneous rock-types, and, in their own right, as sources of innumerable exotic mineral varieties and geochemical compositions. The reasons for this delayed appreciation of lamprophyres are salutory, and deserve a brief historical exposition.

1.1.1 The 19th and early 20th centuries: lamprophyres are characterized

The term 'lamprophyre' was coined by Gümbel (1874), making it one of petrography's oldest; in fact it predates such fashionable *wunderkinden* as komatiite and ophiolite by a full century. The name derives from the Greek *lampros porphyros* (λαμπροσ πορπηψροσ) meaning 'glistening porphyry' (purple rock), and was used because of the large, lustrous biotite phenocrysts which occur in the type-rocks from the Fichtelgebirge (Germany). Today, notwithstanding the blank look it often incites on petrologists' faces, lamprophyre is in fact one of few geological terms (along with *nahcolite* and *bazirite*) which really tells the reader something about what it stands for: 'glistening porphyry' actually describes the field appearance of many lamprophyres rather accurately. In stark contrast, most rock-names merely reflect their type-locality (e.g. *hawaiite, mugearite*), or have origins lost in the mists of antiquity (e.g. *syenite*), and tell the student absolutely nothing about the rock itself. Furthermore, lamprophyre is one of few rock-names ending in *-phyre* which retains its correct connotation of porphyritic character — contrast *granophyre, keratophyre*.

Because lamprophyres are abundant in Germany and Switzerland, where they were first recognized (Chapter 2), they received much attention in these countries during the closing decades of the 19th century, culminating in an influential exposition by Rosenbusch (1897). Parallel, though less numerous, field and petrographical studies were carried out over the same period into the equally abundant lamprophyres of northern Britain (e.g. Bonney & Houghton 1879; Harker 1892). By World War I, most of the macroscopic and microscopic idiosyncracies of lamprophyres described in Chapter 3 had been amply documented. Indeed, descriptive papers by H.G. Smith (1916–1946) and others on the crustal xenoliths in these rocks have scarcely yet been bettered. Over 130 'good quality' chemical analyses of lamprophyres had become available by the time Washington (1917) compiled his celebrated tables. It had also been recognized by then that certain lamprophyres are good indicators for the presence of gold deposits (e.g. McLennan 1915) — although this was soon afterwards forgotten, and neglected for fully 70 years (Section 9.2).

During this early period, when petrographers and petrologists were heavily concerned with classification and nomenclature, it was quickly recognized that lamprophyres show peculiarities which require them to be treated apart from other igneous rocks. For example: (1) conventional classifications based on feldspar compositions and modal ratios are inapplicable, because feldspars may commonly be decomposed and potash carried largely by mica rather than by feldspar; (2) lamprophyres' textures, which came to be described as panidiomorphic (Appendix B) and with a sugary (i.e. aplitic) matrix (Grout 1932) but no feldspar phenocrysts, are unique; (3) the intensity of autometasomatism leads to a superabundance of minerals normally considered secondary (carbonates, chlorite, epidote, zeolites, etc.); (4) lamprophyres' confinement to minor hypabyssal intrusions (dykes, etc.) required separate classification because (at this time), plutonic, hypabyssal and volcanic rocks tended to be treated as three quite distinct groups of igneous rocks. The result was that practically every petrological monograph of the period contained a new, idiosyncratic classification of lamprophyres, often conflicting violently with previous schemes.

These first 50–60 years of research were in fact marked by four milestone monographs. The first, Rosenbusch's (1897) detailed exposition of the lamprophyre concept, recognized several distinct but broad groups of lamprophyres: he realised that minettes (Appendix B) are typically associated with granites, and spessartites with diorites, but that camptonites and monchiquites usually occur with syenites. In this author's opinion, Rosenbusch's views are far nearer the 'truth' than many subsequent reassessments. One exception is his concept of *diaschistic dykes* — the notion that late-stage differentiates from granitoid bodies split spontaneously into two complementary magmas (lamprophyre and aplite). This idea was meant to explain the paradox that normal igneous differentiation moves from basic to acidic compositions, whereas lamprophyres in granitoid complexes (Section 2.4.1) are often late-stage, representing a return to basic compositions. The diaschistic concept held sway for several decades (e.g. Grout 1932, p.121), and has still not been buried in some countries, although it does contain some elements of truth.

The second monograph, Beger's (1923) monumental work on the lamprophyres of the Lausitz, Schwarzwald and St.Gotthard (Fig.2.5), presented a wealth of analyses, attesting very clearly to the interest lamprophyres had aroused among petrologists by this time. The third, accompanying work by Niggli (1923) marked the introduction of the term *lamproite* for extrusive rocks of lamprophyric aspect ('lamprophyre' itself had hitherto been applied only to minor intrusions). The fourth monograph was Bowen's (1928) classic work, containing a tightly argued chapter on lamprophyres which maintained, among other things, that lamprophyres did not correspond to fully liquid magmas (because of their persistently porphyritic condition), and that resorption of complex phases such as hornblende and biotite into evolved (e.g. granitic) liquids was a plausible model for their origin.

1.1.2 1940–1960: the wilderness period

Bowen's chapter was tantamout at the time to a papal encyclical *ex cathedra*, and comments in a fifth monograph (Tröger 1935), expanding the lamproite concept but adding little else, must have reinforced the impression that there was now little more to be said on the subject.

A further death-knell was dealt immediately afterwards by Knopf's (1936) reassessment of the lamprophyre concept in the Spanish Peaks, USA (Fig.2.10). In the present author's opinion, Knopf in fact managed to obscure totally what lamprophyres are, and what distinguishes them from other igneous rocks; many of his so-called "lamprophyres" are nothing but common andesites, basalts and porphyries. He also derived quite erroneous deductions from a confusion between two quite distinct suites of lamprophyres of different ages, which happen to coexist in this one area (Section 1.7). Unfortunately, Knopf's alternative, "simple descriptive terminology" (a recipe for confusion) was all too often adopted: from Fenner (1938), who described ordinary Antarctic basalts and basanites as "fourchites" (Appendix B), to Currie (1976). It has lurked behind much subsequent literature on the Spanish Peaks itself (e.g. Johnson 1968), to the extent that it is all but impossible to understand second-hand what these rocks are. More insidiously, Knopf's views infected generations of petrologists with the notion that lamprophyres are difficult to define, by not being cleanly separated from common igneous rocks. An unhappy result was that all kinds of other obscure rock-types also became lumped with lamprophyres, merely because they also were difficult to classify: for example, Moorhouse (1959) merged lamprophyres with *beerbachites* (a type of hornfels)! Knopf's influential work has only finally been expiated through reappraisal of the Spanish Peaks by Jahn *et al.*(1979).

When petrology began to recover from World War II, this prior history proved enough to discourage recovery in the study of lamprophyres. These rocks had become a 'problem', and indeed, the very words "lamprophyre problem" appeared in many titles of this period (e.g. H.G.Smith 1946; Bederke 1947; Eskola 1954; Watznauer 1964). Consequently, the number of publications dealing with lamprophyres had barely recovered by 1960 from its immediate pre-War peak (Fig.1.1a). Papers from the late 1940s and 1950s are mostly period pieces, concerned largely with reassessment of earlier work, with local problems, or with obsolete debates such as the granite controversy. They present few new data, and some of the issues so hotly debated in them seem quaint nowadays, even in translation. In textbooks of the period, lamprophyres are treated as little more than obscure petrological curiosities, being relegated to the footnotes or considered unworthy of attention at all.

1.1.3 The late 1960s to the present: lamprophyres come of age

The first serious attempts to resurrect interest were made during the late 1960s and early 1970s by Dr.D.Velde (née Métais) in France, Dr.D.Nemec in Czechoslovakia and Prof.W.

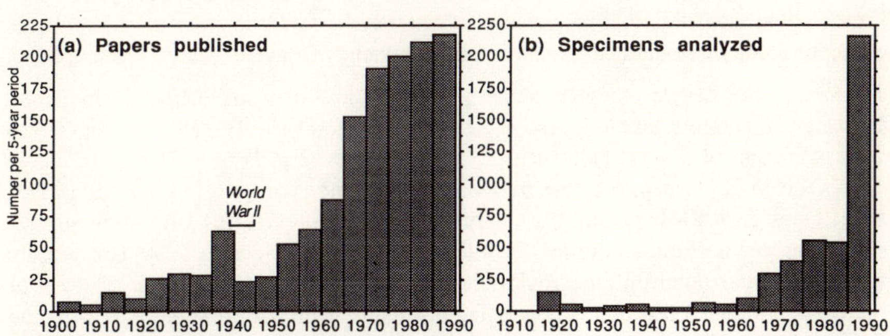

Fig.1.1 *Volume of lamprophyre research during the 20th century.*
(a) Number of publications concerned substantially with lamprophyres over 5-year periods.
(b) Lamprophyre specimens analyzed over the same periods.
Data in (a) derived from non-selective version of the Bibliography in this book, and in (b) from LAMPDA (Appendix D). No distinction is made in (a) between books, papers and abstracts, but papers mentioning lamprophyres only *en passant* are excluded. In (b), total data available for each specimen counted might actually reach several *hundred* individual data-values (e.g. one whole-rock analysis for up to 50 major and trace elements and isotopes, plus dozens of mineral analyses for perhaps 10-15 oxides). The minima in the World War II years are a predictable phenomenon of the literature in general.

Wimmenauer in Germany. Even though lamprophyre specialists were still very few, the prolific output of these workers in particular caused publications to multiply $2\frac{1}{2}$ times over the 10-year period (Fig.1.1a), even though the amount of analytical data did not grow to anything like the same extent for the time being (Fig.1.1b).

The surge of interest had become far more widespread by the late 1970s, and then became firmly established as a long-term phenomenon, after the discovery of the world's richest diamond deposit in the Argyle pipe of NW Australia (Fig.2.8). For a full century, prospectors had relied on the maxim that diamonds only occur in kimberlite (Appendix B), first named after Kimberley (S.Africa) in 1887. Purely for this reason, early descriptions of Argyle referred to it as a "kimberlite" or "kimberlitic rock" (Atkinson *et al.* 1984a). Soon, however, it was realized that Argyle is not a kimberlite but a lamproite (Appendix B), and with that came the implication that diamonds could occur in related rock-types — such as lamprophyres. More or less concurrently, the old prospectors' notion that lamprophyres might be related in some way to gold deposits was briefly resurrected by Boyle (1979, p.250) and reappraised with full force by McNeil & Kerrich (1986).

The result of these developments has been remarkable. The volume of analytical data published on lamprophyres (*excluding* kimberlites) since 1985 has exceeded the total volume accumulated up to that time (Fig.1.1b). The first-ever symposium devoted to lamprophyres was convened in 1985 by the late Prof.Sharon Bachinski at the annual GAC/MAC meeting at Fredericton, New Brunswick, and a special issue of *Canadian Mineralogist* (**26**, no.1, 1988) has since been published to commemorate Sharon's untimely death. Papers dealing exclusively or substantially with lamprophyres (even excluding kimberlites) have proliferated remarkably: whereas the 1st International Kimberlite conference (IKC) in 1973 resulted in only one or two, the 2nd, 3rd and 4th IKCs generated 4, 5 and 26 papers respectively. About 45 are currently published each year (Fig.1.1a), with dozens more mentioning or documenting lamprophyres more superficially. Collectively, these papers now yield new data for some 450 lamprophyre specimens annually (Fig.1.1b), and of the order of 1,000 whole-rock and mineral analyses, which of course means many thousands of new data-values for individual variables, each year.

There is little indication that the present bubble may burst, for interest in lamprophyres now extends along *three* quite different and complementary fronts.

(1) *Diamonds*. The Argyle pipe now alone generates >30% of world diamond production by weight, so there is ample incentive for further exploration. By 1987, this had led to reclassification of several "kimberlites" as lamproites (e.g. Prairie Creek, Arkansas, USA) and to confirmation of the presence of diamond not only in other lamproites (e.g. Luangwa Valley, Zambia), but also in two varieties of lamprophyres (damkjernites and monchiquites) in Western Australia (Section 9.1). The present position is that lamprophyric rocks are the only confirmed *magmatic* sources of diamond, and much research is now being directed towards determining how wide the field of diamondiferous rocks extends, how the extension of the field affects traditional diamond exploration by indicator minerals, and how it can be determined whether a given lamprophyre body is potentially diamondiferous at the earliest stages of its discovery. The diamond exploration industry has had its ups and downs, but seems currently (end-1989) to be in a recovery phase.

(2) *Gold*. The high gold price sustained a 'boom' in research and exploration in the 1980s. The recognition of a recurrent space-time association between mesothermal gold and

lamprophyres has already led to the hypothesis that lamprophyres might transport gold up from the deep mantle in the same way that they bring diamonds. Lamprophyres have consequently made their entry into the biennial Gold conferences (with 5 papers at Gold '88, in Melbourne, Australia), in the same way that they had already entered the Kimberlite Conferences.

(3) *Granitoids*. This most recent front of interest is primarily academic, though it too has economic overtones. The origin of post-orogenic granitoids has of course been debated for decades, but Suzuki & Shiraki (1980) were the first to suggest of late that lamprophyres might be parental to certain granitoids. By 1985, exhaustive isotopic work on British Caledonian granitoids had clearly implicated parent magmas which perfectly fit the compositions of widespread associated lamprophyre dyke-swarms. However, a throwback attitude to the 1960s among some granite specialists, who continued to regard lamprophyres as "grotty mafic dykes" not worth a second look, delayed the forging of the granite–lamprophyre connection itself, and it fell to other groups of workers in 3 countries to do this (Section 8.6). The lamprophyric affinities of widespread mafic enclaves in some granitoids have also recently begun to be realised. Now that lamprophyres have at last begun to be studied in the same petrological and isotopic detail as their associated granitoids, their potentially parental status in some areas is becoming much clearer.

The present author first encountered lamprophyres during his PhD research at Monchique (the *monchiquite* type locality), although his principal study concerned the host nepheline syenite intrusion. In 1977, however, he was given the task of surveying the vast Permo-Carboniferous lamprophyre dyke-swarms in the Scottish Highlands and Islands (Fig.2.7). Although his initial interest was thus imposed rather than chosen, the change in lamprophyres' fortunes had taken place by the time this regional survey was completed, and has now led inexorably to the publication of this book.

1.2 Classification and nomenclature: the Lamprophyre Clan

It is pointless here to recite the endless stream of contradictory classifications of lamprophyres over the past century. Suffice it to say that confusion probably attained its nadir with the Arrow Peak dyke, Montana (USA), variously described as "orthoclase-camptonite" (Rosenbusch 1897), "minette" (Pirsson 1905), "leucite-monchiquite" (Beger 1923), "diopside-lamprophyre" (Knopf 1936) and even "mafic phonolite" (Buie 1941). Mercifully, the IUGS Subcommission on Igneous Rock Systematics has now recognized three groups of lamprophyres (calc-alkaline, alkaline and "melilitic"), each comprising several rock-types (Streckeisen 1979). It has subsequently approved the grouping of lamprophyres, lamproites and kimberlites together as *lamprophyric rocks* (Le Maitre 1989). Overall, this means that we have a *clan* of lamprophyric rocks, to be divided into 5 *branches* (Fig.1.2). IUGS-approved definitions of individual rock-types in Fig.1.2 are outlined in Table 1.1, and amplified in Appendix B.

This book adheres to the latest IUGS recommendations (Le Maitre 1989), which include several minor changes to those in Streckeisen (1979):

(1) Streckeisen's (1979) attempt to fit lamprophyres into the QAPF double-triangle has been abandoned, and Table 1.1 is therefore constructed without reference to QAPF. In truth, QAPF is inapplicable to lamprophyres because: (i) melilite and carbonates can cause

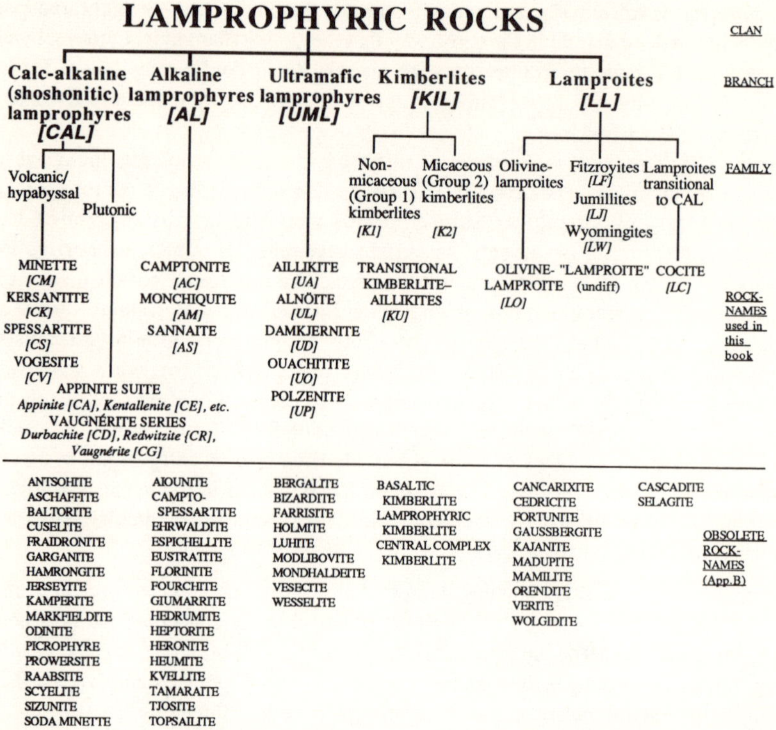

Fig.1.2 *Hierarchical classification of lamprophyric rocks adopted in this work.* After Streckeisen (1979), Le Maitre (1989). Hierarchical levels *clan, branch, family* defined by Rock (1981). Individual rock-names explained in Appendix B. Codes *[in square brackets]* as used throughout this book.

confusion, being light-coloured yet 'mafic'; (ii) melilite has often also been considered as a feldspathoid (Streckeisen 1976, p. 6); (iii) as the colour index of UML is around the critical value of 90%, small modal variations change their QAPF field in an exaggerated fashion, or cause the rock to enter fields which the scheme does not cover; (iv) monchiquites and many lamproites carry essential glass; (v) the feldspar ratio in lamprophyres is commonly indeterminate.

(2) Streckeisen's *melilitic lamprophyres* have been redefined as *ultramafic lamprophyres*, because melilite-rich and melilite-free lamprophyres commonly coexist (Rock 1986).

(3) *Ouachite* has been redefined as an ultramafic, not alkaline, lamprophyre.

(4) *Fourchite* is no longer approved, presumably because 21% of "monchiquites" (including the type-rocks) are olivine-free, and because there is no corresponding term for the 43% of described "camptonites" lacking olivine (Gallagher 1963; Rock 1977).

Three incidental adjustments are also made to Le Maitre's (1989) recommendations:

(a) Merely for brevity's sake, *lamprophyric rocks* is generally shortened to *lamprophyres*.

(b) Following Dawson (1987, 1989) and Skinner (1989), kimberlites are considered to represent a branch (i.e. several rock-types), and *not* a single rock-type (Section 1.3.2).

(c) Three rock-names (*aillikite*, *damkjernite*, *ouachitite*), not apparently approved by the

IUGS, are retained here. (In fact, there is doubt as to which names *are* IUGS-approved: for example Le Maitre's glossary (p.41) states that aillikite is among the approved names in his table B.3, but it is in fact missing).

Careful use of this IUGS terminology should overcome much of the lamprophyre mystique, provided that: (i) similarities and differences between the 5 branches (Chapters 4–5) are properly balanced; (ii) the term *lamprophyre* itself is reserved primarily for field use, and the more precise rock names in Fig.1.2 employed in more detailed descriptions. Both changes are necessary because *lamprophyre* alone covers an inordinate range of compositions from ultramafic to mesotype, ultrabasic to intermediate, Na- to K-rich, melilite- to quartz-bearing, and peraluminous to perpotassic. Among rocks requiring wholesale renaming are "central complex kimberlites", which become *aillikites* (Rock 1986), together with "peralkaline minettes" (Velde 1967; Hall 1982), "leucite-lamprophyres" and "mica-peridotites" (e.g. Sarkar *et al.* 1980), which become *lamproites.*

1.3 In defence of the Lamprophyre Clan

Although the clan concept has effectively been sanctioned by the IUGS, and Chapters 2–9 will hopefully illustrate to the reader the hard factual basis behind it, the following general arguments are nevertheless proferred in its favour at this stage, for the benefit of sceptics:

(1) Fig.1.2 allows similarities and differences to be balanced. 'Splitters' and 'lumpers' can be accommodated. The hierarchical structure allows rocks to be classified accurately, at every stage of their description, *according to how much information is available.* Thus

Table 1.1 Classification of lamprophyric rocks by light-coloured minerals present in the mode[1]
(expanded from Streckeisen 1979, Le Maitre 1989)

Light-coloured minerals	Plagioclase > Af	Af > plagioclase	No feldspars	+Melilite,no Fp
No Na-foids[3] or leucite;	{KERSANTITE	MINETTE		
↓ *minor quartz may occur*	{SPESSARTITE*	VOGESITE*	VERITES[4,8]	—
↓ *Leucite occurs, no Na-foids[3]*	—	JUMILLITES[4,6]	ORENDITES[4,7]	—
↓ *Na-foids occur[3], no leucite*	CAMPTONITE*	SANNAITE*	MONCHIQUITE*[4]	POLZENITE
↓ *Carbonate[5], Na-foids[3] occur*	—	DAMKJERNITE[10]	OUACHITITE	ALNÖITE
↓ *Carbonate[5] essential, no foids*	—	—	{AILLIKITE, {KIMBERLITE[9]	—

→→→→→→→→→→→→→→→→→→→→→→→→→→→→→→→→→→[2]

[1] As dark-coloured minerals, *all* lamprophyres carry essential phlogopite-biotite, amphibole (calcic, sodic-calcic or alkali, e.g. kaersutite, richterite, arfvedsonite), or both, together commonly with olivine and clinopyroxene. Phlogopite-biotite normally dominates amphibole except in rocks marked *, where the reverse holds true.

[2] → arrows indicate direction of increasing undersaturation with SiO_2

[3] Na-foids = analcime, cancrinite, haüyne, nepheline, nosean and/or sodalite.

[4] Rocks commonly containing glass, which may occlude foids/feldspars, but is not considered in classification.

[5] Primary magmatic carbonate (calcite, dolomite or other Ca-Mg-Fe-Ba carbonates), where distinguishable.

[6] Includes cancarixite, cocite, jumillite.

[7] Includes cedricite, fitzroyite, kajanite, mamilite, orendite, wolgidite, wyomingite.

[8] Includes fortunite, madupite, verite.

[9] These two rock-types carry the same olivine-phlogopite-calcite-ilmenite±diopside mineral assemblage; they can only be distinguished on textural features (kimberlites have more rounded olivine macrocrysts, higher Si/Al and Mg/Ca whole-rock ratios, generaly more Mg-rich ilmenites, lack amphibole, etc.; see Chapters 4-5).

[10] Accessory alkali feldspar only, no plagioclase.

lamprophyre is a broad field term implying knowledge only of mode of occurrence and macroscopic petrology, whereas *kimberlite* is a precise petrological term implying detailed knowledge of both rock and mineral chemistry (R.H.Mitchell 1986, 1989). At present, geologists involved in diamond exploration are obliged to call a rock "kimberlite" before they have enough information to do so confidently; this has led to incessant arguments as to what is and what is not kimberlite, to endless reclassification of many diamond pipes, and to considerable resulting confusion in the literature.

(2) Lamprophyres are currently unified as the only confirmed magmatic source of diamond (Section 9.1), and hence have the deepest origins of igneous rocks. This unity can only be embodied in the present type of clan concept.

(3) As detailed in Chapters 4–5, the petrology of lamprophyres is a consistent picture of unity in diversity: of small (though important) differences between the branches, within a continuous overall gradation between one lamprophyre and another. Differences between the 5 branches are in every significant respect far less than those between the lamprophyre clan as a whole and common igneous rocks. Boundaries between the branches themselves are not sharp but gradational (Fig.1.4a), and populated by numerous transitional occurrences (Fig.1.4b). Two groups in particular are sufficiently common to warrant special symbols on Fig.1.2: namely, 'LC' and 'KU'. Again, this picture can only be accurately represented within a hierarchical clan concept.

(4) If kimberlites/lamproites were separated from lamprophyres, rare alkaline rocks would receive 5 classification schemes (+ those for carbonatites, melilitic rocks), as against only two (plutonic and volcanic) for 99% of igneous rocks: a *reductio ad absurdum*!

1.3.1 Further arguments for including lamproites within the lamprophyre clan

Comment is necessary because Bergman (1987) — alone among recent published commentaries, and contrary to the IUGS recommendations — considers lamproites *not* to be lamprophyres. However, Niggli (1923) used the prefix *lampro-* in his original definition, precisely to embody lamproite/lamprophyre affinities, and most lamproite definitions have retained the word *lamprophyric*, for example (author's italics):

Tröger (1935): "group name for K- and Mg-rich extrusive rocks = syn. *lamprophyric* extrusive rock"
Wade & Prider (1940): "ultrapotassic leucite-bearing rocks, commonly of *lamprophyric* aspect"
Sørensen (1974): "A group name for K- and Mg-rich extrusive rocks = *lamprophyric* extrusives..."
R.S.Mitchell (1985): "...a *lamprophyric* extrusive rock..."
Jaques *et al.*(1984a,1986): "...a potash and magnesia-rich *lamprophyric* rock..."
MacKenzie *et al.*(1982, p.133): "...extrusive equivalent of potassic *lamprophyre*..."
Hughes (1982, p.322): "...volcanic *lamprophyre* with mica phenocrysts..."
Middlemost (1987): "....*lamprophyre-like* rocks..."

Quite apart from this historical consensus, arguing that lamproites are not lamprophyres is like arguing that granites are not granitoids — an etymological as well as petrological cul-de-sac. In all major bibliographical databases (e.g. *Bibliography and Index of Geology*), lamproites are classified under lamprophyres for this very good etymological reason. Furthermore, there is complete mineralogical and geochemical gradation between lamproites and other lamprophyres, particularly minettes (Fig.1.4; Chapters 4–5). For this reason, many recent papers (e.g. Middlemost *et al.* 1988) have indepdendently affirmed the IUGS view that lamproites must logically be grouped with lamprophyres (Le Maitre 1989, p.11).

1.3.2 A brief note on lamproite nomenclature

Lamproite nomenclature is burdened with a plethora of locality-based names (see Appendix

B), most of which fail to recognize similarities between varieties (e.g. *jumillite* and *wyomingite*). Unfortunately, neither the IUGS (by its own admission) nor recent lamproite reviews have arrived at a more satisfactory alternative. R.H.Mitchell (1985) recommends distinguishing *phlogopite-lamproites,* with resorbed phenocrystic phlogopite (cedricites, orendites, etc.) from *madupitic lamproites,* with poikilitic groundmass phlogopite (madupites, jumillites, etc.), whereas Bergman (1987, p.108) states that this tends to "over-simplify the mineralogy of a given rock". Meanwhile, the IUGS considers *madupite* to be obsolete. Furthermore, by no means all lamproite descriptions specify whether the phlogopite is phenocrystic or groundmass. Bergman (1987) suggests following Scott Smith & Skinner (1984b), in using the modal abundance of principal primary minerals, so as to yield phlogopite–sanidine-lamproites, olivine–leucite-lamproites, etc., but this suffers from the same drawback of inadequate modal data in most lamproite descriptions.

Since the fine details of lamproite classification are largely immaterial in this book, Fig.1.2 divides lamproites for convenience into three families based on locality, and a fourth (olivine-lamproites) based on a natural bimodal distribution of MgO content (Fig.8.2); the latter division is analogous to the distinction between basalts and komatiites.

1.3.3 Further arguments for including kimberlites within the lamprophyre clan

Although the IUGS view, grouping kimberlites with lamprophyres, is more novel than grouping them with lamproites, it does also have a strong historical precedent. This case has been argued at length by Rock (1989a), and will only therefore be summarised here.

(1) Wagner (1914) divided kimberlites into *basaltic* and *lamprophyric* types. Many subsequent texts describe them as lamprophyric. Hughes (1982) — and presumably also the IUGS — took this as one reason for grouping kimberlites and lamprophyres.
(2) Kimberlites and aillikites (Appendix B) are very similar in thin section, and have long been confused (i.e. true kimberlites versus "central complex kimberlites"). They can only be distinguished on fine mineralogical details such as *trend* of spinel or mica compositions. Classifying them into totally separate rock clans is thus taking 'splitting' to absurd extremes, and making classification impossible for the non-specialist.
(3) Kimberlites share all the characteristics of lamprophyres defined in Section 1.4, and their mineralogy and geochemistry overlap substantially with other lamprophyres (Chapters 4–5). Even rare mica-free (Group I) kimberlites are sufficiently rich in volatile components to be retained within the clan.
(4) R.H.Mitchell (1979,1986,1989) has argued that true kimberlites are nowhere associated with contemporaneous lamprophyres. This is now known to be false, as there are in fact several well-documented examples (Fig.1.4c; Table C6).
(5) R.H.Mitchell (op.cit.) has also argued that kimberlite–lamprophyre transitions do not occur. This too is untrue (Fig.1.4b,c), and leads to self-contradiction: R.H.Mitchell & Meyer (1989) called one group of rocks "micaceous kimberlites" and yet stated (p.96): "it would seem unreasonable to continue referring to them as kimberlites".

1.4 Summary of criteria for identifying lamprophyres as a clan

A hierarchy of criteria, in roughly decreasing importance, is outlined below and detailed in Chapters 2–5; criteria for each branch are summarised in (Rock 1977,1984,1986,1987a).

1.4.1 Mineral assemblage (for further details, see Chapters 3–4)

Lamprophyres carry *essential* amphibole and/or biotite-phlogopite, and may also carry abnormal amounts of minerals rich in F, Cl, SO_3, CO_2 and H_2O such as carbonates, halides, sulphates and zeolites. High Ba, Fe^3, Al and/or Ti in K-feldspars, amphiboles, phlogopites and pyroxenes (as appropriate) are characteristic. Mg-rich mafic minerals (e.g. diopside, forsterite) commonly coexist with Na-K-feldspars and even quartz. Cognate, *primary* phases found in some igneous rocks but absent from lamprophyres include cummingtonite–gedrite, tremolite-actinolite (*sensu* IMA 1978), fayalite, kalsilite, kaliophilite, muscovite, orthopyroxene, pigeonite and wollastonite.

1.4.2 Texture (see Chapter 3 for further details, plus definitions in Appendix B)

Lamprophyres are mesotype to hypermelanocratic (C.I. > 33%), and porphyritic (commonly panidiomorphic), with no feldspar or quartz phenocrysts. Feldspathoid or melilite phenocrysts, and groundmass olivines, are almost always absent. Phenocryst olivines may be euhedral, polycrystalline, rounded, or surrounded by reaction rims (Fig.3.2k–m). Felsic globular structures, and bent, pseudohexagonal or castellated biotites (commonly with deep-brown rims and pale cores) are characteristic (Fig.3.2b–d). Most fine-grained mafic dykes with abundant macrocrysts of biotite or amphibole are lamprophyres (though many only have groundmass amphibole or biotite).

1.4.3 Mode of occurrence (see Chapter 3 for further details)

Most lamprophyres form subvolcanic dykes, sills, plugs, stocks, vents, or, exceptionally, margins to larger intrusions. Associated intrusive and explosion breccias are common. Individual intrusions more than a few sq.km in outcrop area remain unconfirmed, but regional pipe-clusters or dyke-swarms may respectively contain hundreds to many thousands of bodies. Lamprophyre lavas, tuffs and pyroclastics are uncommon, but do exist (Section 3.1.7). Lamprophyre dykes are characterized by unusual intrusive form, such as offsetting, segmentation, and banded or zoned internal structures (Section 3.1.1).

1.4.4 Whole-rock composition (see Chapter 5 for further details)

Chemical screens are given for each branch by Rock (1977,1984,1986,1987a); no overall screen is feasible. Lamprophyres typically carry H_2O, CO_2, F, Cl, LILE, P, Rb, Ba, LREE and Th at levels 2–3 orders of magnitude higher than MORB, but MORB-like levels of Y, Ti, HREE and Sc (Fig.5.6).

1.4.5 Tectono-magmatic association (see Section 2.4 for further details)

Rocks satisfying Sections 1.4.1–1.4.4 are restricted to a few, well-defined igneous associations (Table 2.2). "Lamprophyres" outside these are almost always apochryphal.

1.4.6 Miscellaneous features

Alteration is characteristically selective: if mafic (phenocryst) minerals are fresh, felsic (groundmass) minerals are commonly altered, or occasionally vice-versa. Lamprophyres exfoliate characteristically, weathering to a red, mica-rich soil. Abundant and varied inclusions may occur (Chapter 6). A few lamprophyres emit a pitched note under the hammer, and some outcrops may support anomalous plant communities (e.g. Sieber 1956).

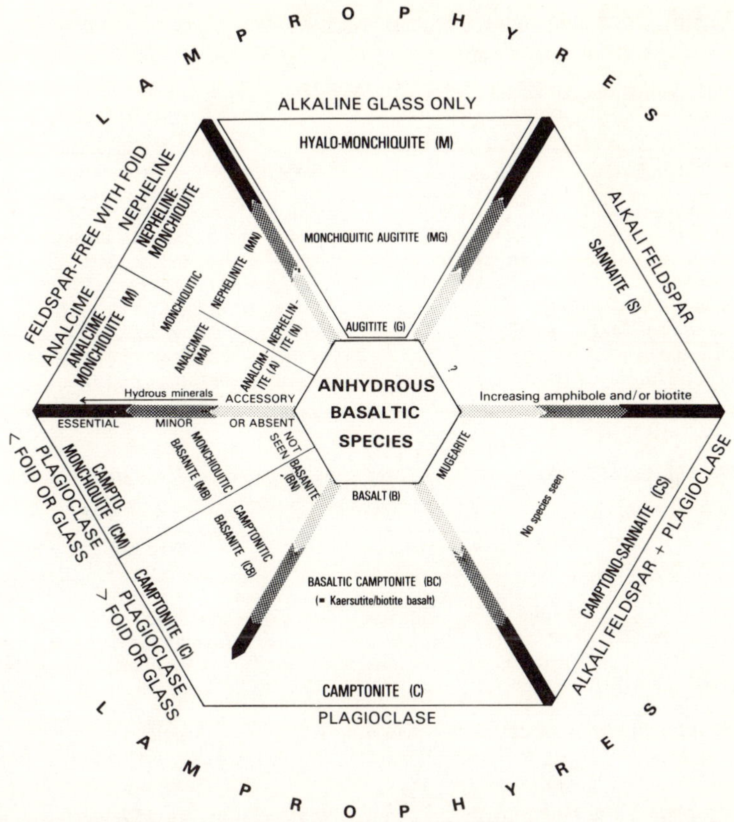

Fig.1.3 *Classification of alkaline lamprophyres (AL) and their volatile-poor and volatile-free relatives.* Rocks are classified according to the ratios of [amphibole + biotite] : [pyroxene + olivine] and of the various felsic phases. After Rock (1983b).

1.5 Identification and nomenclature of closely related rock-types

1.5.1 Volatile-poor lamprophyre relatives (Fig.1.3; Table 1.3)

A persistent disservice to the term lamprophyre has been to apply it to *anhydrous* rocks, so blurring lamprophyres' very real differences from ordinary basaltic rocks: the so-called "monchiquites" and "monchiquitic basalts" of the Midland Valley of Scotland are a classic example, being merely analcimites or analcime-basanites, with neither lamprophyric texture nor essential amphibole/biotite (Cameron & Stephenson 1985). Baxter (1987) more correctly restricted "monchiquite" in Scotland to rocks with primary amphibole/biotite. Richards (1988) has fallen into the reverse trap with lamprophyres at Porgera, PNG (Table C1), calling them "alkali basalts" when they in fact show abundant, panidiomorphic amphibole and biotite, together with globular structures, and yet carry very little olivine.

Nevertheless, some lamprophyres (notably AL and UML) do grade into rocks of similar bulk composition but with lesser amounts of amphibole, biotite and carbonate; this is commonly accompanied by loss of globular structures (Section 3.2.4) and by appearance of

Table 1.2 Volatile-poor and volatile-free petrological equivalents of lamprophyres

(a) *Volatile-free equivalents of various rock-types from the 5 branches*

Branch	Lamprophyre	≡	Volatile-free equivalent(s)
CAL	*Spessartite/kersantite* {Am,Pl,Af,Cpx(Ol)}	≡	*Shoshonite /absarokite*{Pl,Af,Cpx(Ol)}*
AL (see	*Camptonite* {Am,Bi Pl,Cpx,(Ol,Ne,Ac)}	≡	(Olivine)-*basalt,basanite,tephrite* {Pf,Cpx,(Ol,Ne)}
also	*Monchiquite* {Am,Bi Cpx,Ne,Ac(Ol)}	≡	(Olivine)-*nephelinite/analcimite* {Cpx,Ne,Ac(Ol)}
Fig.1.3)	*Sannaite* {Am,Bi,Af,Pl,Cpx,Ne,Ac(Ol)}	≡	?*Mugearite* {Af,Pl,Cpx,Ne,Ac(Ol)}
UML	*Alnöite* {Bi,Cpx,Me(Ol)}	≡	(Olivine)-*melilitite* {Cpx,Me(Ol)}
	Polzenite {Bi,Cpx,Ne,Me(Ol)}	≡	(Olivine)-*melilite-nephelinite* {Cpx,Ne,Me(Ol)}
KIL	*Kimberlite* {Ol,Bi,Cb(Cpx)}		No obvious equivalent
LL	*Jumillite,wyomingite* {Bi,Am,Cpx,Lc(Ol)}	≡	(Olivine)-*leucitite* {Cpx,Lc(Ol)}

(b) *Nomenclature for gradations between ultramafic lamprophyres and volatile-free equivalents*

	Melilite only	Melilite+nepheline	Nepheline dominant	Carbonate dominant
UML (essential Am+Bi)	*Alnöite*	*Polzenite*	*Ouachitite*	*Aillikite*
Volatile-poor (minor Am+Bi)	*Lamprophyric melilitite*	*Lamprophyric melilite-nephelinite*	*Lamprophyric nephelinite*	*Lamprophyric carbonatite*
Volatile-free (accessory Am+Bi)	*Melilitite*	*Melilite-nephelinite*	*Nephelinite*	*Carbonatite*

* See Appendix A for explanation of mineral abbreviations

phenocryst plagioclase or groundmass olivine. Nomenclature covering such gradations is given in Fig.1.3 and Table 1.2b; it could be extended to rare volatile-poor CAL or LL. Table 1.2a identifies nearest 'dry' equivalents of each lamprophyre, where these exist.

1.5.2 Ultramafic (cumulate) and felsic (differentiated) lamprophyre relatives (Table 1.3)

Widespread mafic-enriched rock types in lamprophyric suites (Rock 1979, 1983b) can be covered simply by IUGS-approved prefixes such as *picritic monchiquite* (as in Jaques *et al.* 1989b) or *mela-spessartite.* In a few examples, sufficiently unwieldy terms result that special names may be warranted: for example, Rock (1983b) used the name *yamaskite* (Appendix B) as a shorthand for some Scottish pyroxene-rich mela-monchiquites. The prefix *mela-* appears tautological with UML since the rocks are already ultramafic (i.e. mafic index > 90%). It is nevertheless useful for examples with *colour* index > 90% (as in Upton & Emeleus 1987), remembering that the abundant melilite and carbonates in these rocks are counted in the mafic index but *not* in the colour index.

Unequivocal felsic relatives of LL and UML are not yet recognized. A few KIL are associated with carbonate-rich rocks which can reasonably be termed *calcite-kimberlites* or *kimberlite-carbonatites* (R.H.Mitchell 1986). Intermediate to felsic rock types are much more commonly associated with CAL (Section 8.5); they have been termed *malchites, porphyrites, porphyries* (the names preferred here), and also *odinites* and *semi-lamprophyres* (Appendix B). A felsic relative of the Navajo minettes (USA) has been

Table 1.3 Differentiated and cumulate equivalents of various lamprophyres[1]

Lamprophyre branch	Cumulate (ultramafic) equivalents	Mesotype equivalents	Felsic (leucocratic) equivalents
CAL	*Hornblendite, pyroxenite*[1] *hornblende-peridotite, etc.*	*Malchite/porphyrite* (andesitic/dacitic)[2]	*Porphyry* (rhyolitic)
AL	*Picrite, yamaskite* (picrite)	*Maenite* (trachyandesitic)	*Bostonite* (trachytic)
UML	Not applicable (already ultramafic)	None known	{Carbonatite}
KIL	Not applicable (already ultramafic)	None known	{Calcite-kimberlite}
LL	*Mica-peridotite*	None known	None known

[1] For definitions of individual rock-types, see Appendix B [2] These terms *sensu* LeBas *et al.*(1986)

termed both *felsic minette* and "sanidine trachyte" (Bachinski & Scott 1980). Mugearitic, trachytic or phonolitic rocks also commonly associated with AL have been termed *bostonites, maenites* (the terms preferred here), and also *hedrumites* and *heronites.* Although CAL→porphyrite→porphyry and AL→maenite→bostonite rock series partly mirror andesite→dacite→rhyolite and alkali basalt→mugearite→trachyte series respectively, textural and compositional peculiarities may justify the separate terminology.

1.5.3 Plutonic and volcanic lamprophyre equivalents (see Chapter 7 for further details)

Lamprophyres are not simply textural variants of common plutonic or volcanic rocks, partly because of their high volatile and LILE contents (Streckeisen 1979). AL are the most like common igneous rocks, being volatile-rich versions of alkali basalts, basanites and nephelinites, and of amphibole-bearing gabbros (Section 7.2). A minority of CAL have relatives among shoshonitic rocks (Appendix B), but differ from common calc-alkaline diorites and andesites (Sections 1.6.2, 7.2.3). KIL, LL and UML have no straightforward volcanic or plutonic equivalents: UML reach lower SiO_2 and higher CaO than any other silicate igneous rocks (grading into carbonatites), while LL attain higher K_2O (up to 13%), and KIL higher MgO than other rocks of similar SiO_2 (Chapter 5).

1.6 Distinctions between some petrographically similar rock-types

Parts of Table 1.1 and Figs.1.2–1.4 likely to cause most difficulty are now considered.

1.6.1 Coexisting lamprophyres from different branches

Intimate associations between 2 or more branches within one igneous suite are increasingly widely recognized (Table C6), underlying the continuum within the lamprophyre clan. Associations between *adjacent* branches on Fig.1.4b are the most common (Fig.1.4c), but in fact, all possible combinations of 2 coexisting lamprophyre branches, and several even of 3 or more branches, have now been amply documented (see for example the CAL/UML and AL/UML/LL associations of the Navajo and Ouachitas provinces (USA), respectively, in Table C6). Different branches are perhaps most intimately juxtaposed in carbonatite complexes, where UML are commonly associated with AL and, much more rarely, with CAL. The coexisting AL and UML commonly form a compositional continuum, but can be

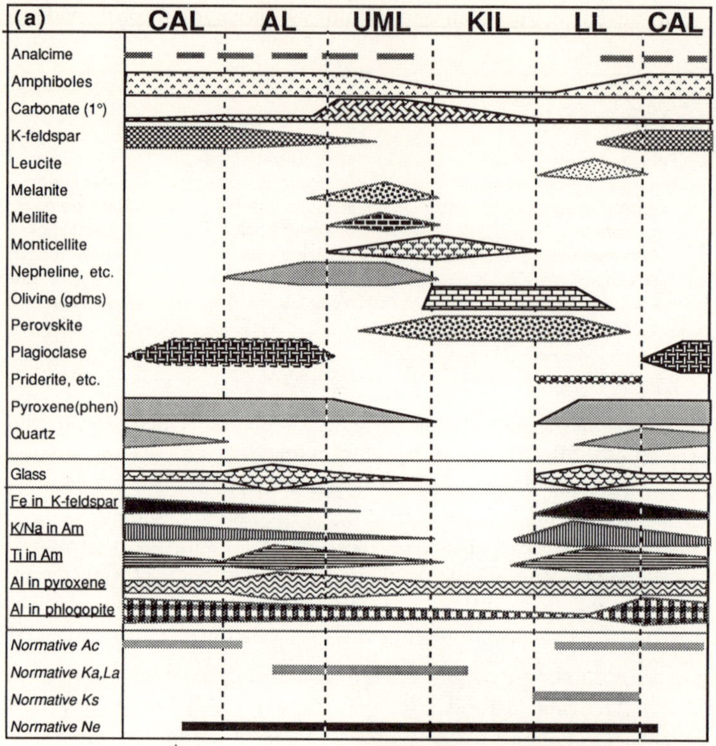

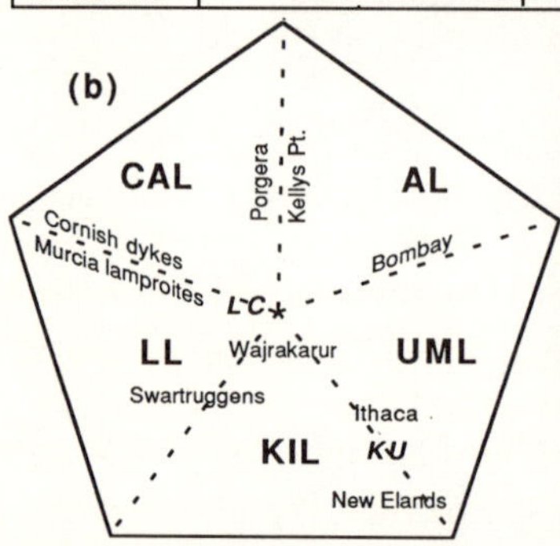

(c)	*Inter-branch transitions*	*Inter-branch associations* (App.C, Table C6)
CAL /AL	Porgera, PNG (Table C1) Kellys Point, Australia (Table C6)	Spanish Peaks, USA
AL/ UML	Bombay, India; Aillik Bay, Canada (Table C6)	20 cases in Table C6 (eg Kaiserstuhl,Germany)
UML /KIL	All KU; New Elands, S.Africa; Ithaca, USA (Table C3,C6, Fig.5.5)	Kuruman, S.Africa
KIL /LL	Swartruggens, S.Africa (Table C4)	Bobi, Ivory Coast
LL/ CAL	All LC; Murcia, Spain (Table C4); Cornwall, UK (Table C1)	8 cases in Table C6 (eg Gondwanas, India)

Fig.1.4 *The compositional spectrum between the 5 lamprophyre branches.*

(a) The proportions and ranges of important distinguishing modal/normative mineral phases, together with significant changes of mineralogical composition, are indicated diagrammatically. Olivine, pyroxene and biotite–phlogopite extend across the entire spectrum. The vertical dividing lines particularly between CAL and LL are essentially arbitrary, but follow Rock (1984) and Bergman (1987). CAL repeated in part at right to indicate that diagram is in fact pentagonal, as in **(b)**.

(b) Examples of petrographically transitional lamprophyres on the criteria of **(a)**, indicating their approximate positions with respect to the 'boundaries'. For example, the Cornish (UK) minettes can be logically classified as CAL but show some features of LL, the Murcia (Spain) lamproites the converse. LC (cocites) such as Coc-Pia, Vietnam (Table C4), and all other rocks crossing the boundaries, can only arbitrarily be assigned to one side or the other.

(c) Further illustration of the lamprophyre continuum. Left-hand column indicates where details of examples on **(b)** may be found. Right-hand column gives examples where two adjacent branches on **(b)** are intimately juxtaposed in the same suite, with or without petrographical transition.

Table 1.4 Some differences between calc-alkaline lamprophyres, diorites and andesites

Feature	Andesite	Diorite	CAL (spessartite/vogesite)
Olivine	Rare	Rare	Pseudomorphs may occur
Orthopyroxene	Common	Rare	Absent
Clinopyroxene	Augite	Augite	Ti-Al-salite/diopside
Amphibole	Hornblende may occur, but normally resorbed/oxidized	Hornblende common, normally anhedral	Pargasite/tschermakite, normally panidiormorphic
Mica	Rare resorbed anhedral biotite	Anhedral biotite common	Castellated phlogopite common
Plagioclase phen	Common	Rare	Absent
1° baryte, carbonate	Not recorded	Not recorded	Common accessory
Globular structures	Not recorded	Not recorded	Common
Crustal xenoliths	Rare	Rare	Common

arbitrarily divided at the appearance of melilite, perovskite, picroilmenite and tetraferriphlogopite in the UML, which are compositionally closer to carbonatites.

1.6.2 Calc-alkaline lamprophyres versus shoshonites and common calc-alkaline rocks (see Section 7.2.3 for further discussion; also Table 1.4)

A minority of CAL (especially spessartites and vogesites) share the same (plagioclase-hornblende–feldspar ±clinopyroxene±biotite) mineralogy as andesites, diorites, absarokites and shoshonites. However, the following criteria, in brief, distinguish the lamprophyres:

— castellated, globular and panidiomorphic textures;
— lack of orthopyroxene but greater abundance of olivine;
— amphibole, clinopyroxene and mica which are far more Mg-rich;
— presence of alkali-rich pyriboles, Zn-rich spinels, primary carbonates and sulphates.
— higher *mg*, K_2O (Fig.7.4), Ba, Rb, Sr, V, Cr, Ni and LREE/HREE.

1.6.3 Ultramafic lamprophyres versus melilitites and related rocks (see also Section 7.3)

No self-respecting petrologist (including any who follow IUGS recommendations) would call a rock with essential amphibole or biotite an alkali "basalt" or "gabbro"; the correct name is alkaline lamprophyre (Streckeisen 1979). Similarly, therefore, a pyroxene-nepheline ±melilite ±olivine assemblage with essential amphibole or biotite should not be called melilitite, melilite-nephelinite or nephelinite, but ultramafic lamprophyre (Table 1.3b). Unfortunately, some petrologists have illogically tolerated more biotite in "melilitites" and "nephelinites" than in "basalts" (e.g. McIver & Ferguson 1979). Even if this were allowed (which it is not in this book), the following secondary distinctions in practice separate UML from 'normal' melilitites and similar:

(1) UML commonly carry amphiboles and primary carbonates; melilitites do not.
(2) UML carry essential (>10%) phlogopite, usually in two generations; melilitites generally carry only minor groundmass phlogopite.
(3) UML only vary rarely carry groundmass olivine or phenocryst melilite; melilitites commonly carry two generations of both minerals.
(4) UML almost never form lavas; melilitites commonly do so.

1.6.4 Ultramafic lamprophyres versus leucite-bearing ultramafic rocks

Some K-rich rock-types such as potash-ankaratrite, venanzite (Appendix B) and the 'kamafugitic rocks' (katungite, mafurite, ugandite) of Uganda (Holmes 1950) carry a similar paragenesis to UML (olivine-pyroxene-phlogopite ±nepheline ±melilite). They are distinguished primarily by the additional presence of leucite and/or kalsilite, by higher K_2O, and by a lack of associated sodic alkaline rocks (carbonatites, ijolites, etc). Nevetheless, it is possible to regard kamafugitic rocks as in some respects linking UML with LL, and their omission from this book is therefore more for reasons of space than of petrology.

1.6.5 Alnöites versus aillikites and carbonatites

As primary and secondary carbonates cannot always be distinguished (Section 4.12), Table 1.1 does not discriminate 'carbonated alnöite', wherein melilite has been replaced by secondary carbonates, from 'aillikite'. Fortunately, most UML are either melilite-*rich* or melilite-*free*; and features such as cebollite, lath-shaped pseudomorphs, or secondary garnet, usually provide adequate evidence of former melilite. *Aillikite* should be used where such evidence is lacking. UML and carbonatites grade into one another, both in global terms and within single intrusions (Kapustin 1981), carbonatitic globular structures being common in UML (Section 3.2.4). A limit of 15% SiO_2, approximating the IUGS-recommended limit of 50% carbonate minerals in carbonatites (Streckeisen 1979; Le Maitre 1989), is therefore used to separate the two in this book.

1.6.6 Ultramafic lamprophyres versus Group II kimberlites

Published names for ultramafic olivine-phlogopite-calcite±spinel rocks are confused. Ferguson & Sheraton (1979) described the Jugiong pipes (Table C3) as "kimberlitic rocks" showing "petrographic affinities with olivine-analcitites and monchiquites" and yet "chemical affinities with the olivine nephelinite suite". Dawson's (1971,1980) term "central complex kimberlite" (Appendix B) has been rather indiscriminately promulgated — despite Dawson's own clear *caveats* — for what are here termed UML. This reflects the still widely held view that kimberlitic and carbonatitic–lamprophyric magmatism are compositionally continuous and genetically related. Kimberlites, alnöites and carbonatites have even been reported together within single dykes (e.g. Le Bas 1977).

By contrast, R.H.Mitchell (1979,1986) has delineated major differences between "central complex kimberlites" (UML) and archetypal (e.g. S.African) KIL, and denied any such kimberlite–carbonatite relationship. The following are the most important features which distinguish UML from 'true' KIL (for further details see table 4 of Rock 1986):

- — UML have lower whole-rock SiO_2/Al_2O_3, MgO/CaO, Cr and Ni, but higher Zn, P_2O_5 and CO_2 (Fig.5.5a,5.7), reflecting higher modal carbonates, apatite, but lower olivine;
- — UML have shallower REE profiles (lower La/Yb) at given enrichments (Fig.5.10);
- — alkali feldspars, feldspathoids, Ti-Zr-melanite, melilite, Mg-Fe-rich primary carbonates, and Na-Fe-Ti-Al-rich amphiboles are common in UML but absent from KIL (Fig.1.4);
- — olivine macrocrysts tend to be euhedral in UML (Fig.3.2k), rounded in KIL;
- — UML pyroxenes have consistently higher Al_2O_3 and TiO_2 (Fig.4.3);
- — some UML phlogopites reach much higher TiO_2 (Fig.4.5);
- — some UML spinels follow different trends (Fig.4.9) and are richer in Zn and Mn;
- — some UML ilmenites are richer in Mn (Fig.4.10);
- — some UML perovskites are richer in Na and Fe (Section 4.15).

Building on recommendations by R.H.Mitchell (1979,1986,1989) and Clement *et al.* (1984), the term "central complex kimberlite" is abandoned in this book, and such rocks are termed *aillikites* (UML) instead. Nevertheless, a balance must be maintained, since UML (and olivine-lamproites) remain the closest allies to KIL among igneous rocks; close relationships and gradations between the two *do* exist (Fig.1.4; Chapter 2).

1.7 Genetic connotations for lamprophyre names: are they reasonable?

Rosenbusch (1897) asserted that minettes are associated primarily with granites, spessartites with diorites, and camptonites with nepheline syenites, thus attaching genetic connotations to these lamprophyre names. These connotations were soon used to infer the presence of hidden syenite bodies near camptonite dyke-swarms (e.g. Kemp & Marsters 1893), and are still widely recognised (e.g. Wimmenauer 1973a).

Later, Knopf (1936) challenged these genetic connotations on 3 grounds:

(1) the lamprophyre branches are not sharply demarcated, but overlap considerably;
(2) mixed lamprophyre swarms invalidate the connotations;
(3) minettes can be found associated with alkaline rocks, and camptonites with granites.

Knopf based his case on the Spanish Peaks (USA), where a dyke-swarm including both camptonites and minettes radiates from a granitic stock (Table C6). He contended that this would be a "realm of petrographic anarchy" if Rosenbusch's connotations were valid. Unfortunately, Knopf failed to recognize two clearly independent magmatic phases in the area — one calc-alkaline (granitoids), and the other alkaline (analcime gabbros, shonkinites and teschenites). Many of the camptonites do *not* in fact radiate from the granitoid stock (Johnson 1961) and isotope studies (Jahn 1973; Jahn *et al.* 1979) evince *two* independent magmas — one a crustal granite, the other a mantle lamprophyre. Far from refuting genetic connotations, therefore, the Spanish Peaks illustrates that where *both* granites *and* alkaline rocks occur together, they may be accompanied by *both* CAL *and* AL.

Once many rocks misidentified in the literature have been reclassified, Rosenbusch's original assertion can be reaffirmed in modified form: broadly, CAL connote calc-alkaline granitoids or shoshonites, AL connote mildly alkaline rocks, and UML connote strongly alkaline rocks (including carbonatites). However, there is complete intergradation between the branches, and mixed lamprophyre associations do commonly occur because the connoted igneous rocks are themselves mixed in the same suites.

1.8 A new definition of the term 'lamprophyre'

The following definition is based on R.H.Mitchell's (1986, 1989) latest definition of kimberlite. Phrases *in italics* are reproduced exactly, or merely by replacement of *kimberlite* with *lamprophyre.* The dominance of these phrases in the revised definition further underlines the case for including kimberlites within the lamprophyre clan. The final 2 sentences are added in deference to the kimberlite definition of Clement *et al.* (1984); it includes geological as well as purely petrographical criteria, which are regarded as critical.

> *Lamprophyres are a clan of H_2O and/or CO_2-rich,* alkaline rocks ranging from sodic to *potassic* and from *ultrabasic* to intermediate. *Commonly, they exhibit a distinctive inequigranular texture resulting from the presence of macrocrysts (and in some instances megacrysts) set in a fine-grained matrix. The megacryst/macrocryst assemblage consists of rounded* to euhedral *crystals of magnesian ilmenite, Cr-poor titanian pyrope,* forsteritic *olivine, Cr-poor clinopyroxene, phlogopite, enstatite, Ti-poor chromite,* kaersutitic amphibole and alkali feldspar. *The matrix minerals may include: second*

generation euhedral primary olivine; calcic, sodic-calcic or sodic clinopyroxenes and amphiboles; *phlogopite*-biotite (commonly rich in Ti and/or Ba)*;* feldspars; feldspathoids; melilite; Mg-Mn-bearing ilmenite; *perovskite, spinel, monticellite, apatite, and primary late-stage* Ca-Mg-Fe carbonate and chlorite/*serpentine*. Amphibole and/or phlogopite-biotite are essential primary phases. Irregular to spherical, felsic globular structures, filled with combinations of carbonates, chlorite, feldspars, feldspathoids and zeolites, are widespread. *The replacement of early-formed minerals by deuteric chlorite-serpentine, carbonates, epidote and zeolites is common.* Lamprophyres typically form en echelon dykes, sills, pipes and vents which, though individually small, may aggregate into extensive swarms or clusters. Brecciation of country-rocks or of lamprophyre intrusions themselves is common.

2 When and Where? — Global Distribution, Igneous Associations

2.1 A preliminary *caveat*: metamorphism and lamprophyres

Because they are partly characterized by peculiar textural features and by high contents of mobile elements (Chapters 1,3,5), lamprophyres are among the *least* likely of igneous rocks to be recognizable after even low-grade metamorphism. The older and/or more foliated they are, the more easily they may be confused with other metabasites: for example, foliated Caledonian microdiorites in northern Scotland were long called "foliated lamprophyres" and "lamproschists" in the mistaken belief that they were metamorphosed equivalents of younger Caledonian lamprophyres (D.Smith 1979). In the converse sense, "amphibolites" long known to be useful marker bands in the Limpopo Belt of Zimbabwe turned out on closer examination to be metalamprophyres (Watkeys & Armstrong 1985). This intrinsic confusion was exacerbated by the former tendency of some geologists to call any peculiar or altered igneous rock a "lamprophyre" (Chapter 1).

Until recently, no-one had detailed the effects of metamorphism on lamprophyres: no swarms had been traced regionally up metamorphic gradients; no single bodies displaying relict igneous features had been examined. Furthermore, though lamprophyres are metabasites in the broad sense, it *cannot* be assumed that they will follow the well-documented metamorphic behaviour of, say, metabasalts. For example, a typical kersantite (CAL), with 30–50% phlogopite, feldspar and minor quartz, is mineralogically much more akin to a *pelite* than to a basalt, and examples are now accumulating in which lamprophyres in fact behave much more like pelites, deforming incompetently rather than boudinaging like typical metabasalts. The end result is that some Precambrian "lamprophyres", and many

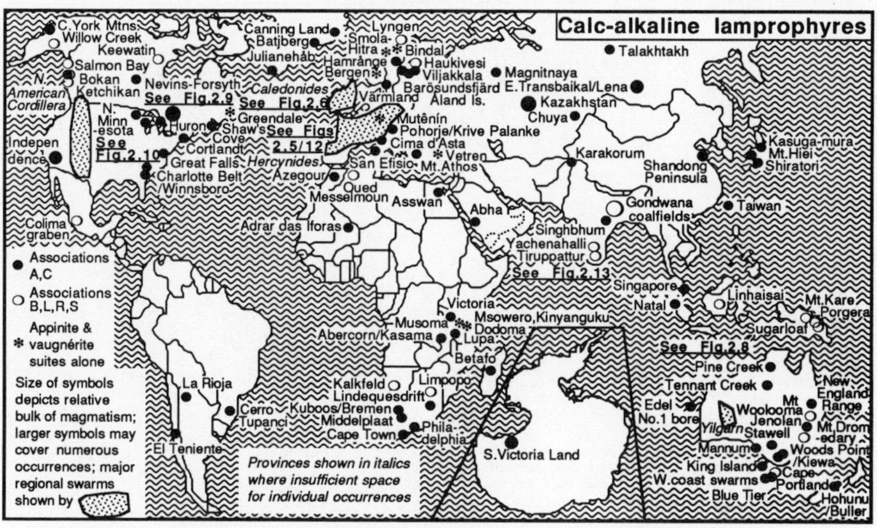

Fig.2.1 *Global distribution of confirmed calc-alkaline (shoshonitic) lamprophyres (CAL).* For details of individual occurrences, see Appendix C, Table C1.

"meta-lamprophyres" (Steiner 1984) in the literature are doubtful (Table C8), while conversely, many genuine metalamprophyres have almost certainly been overlooked.

Nevertheless, now that lamprophyres' petrographical and geochemical characteristics have been better defined, schists and even gneisses representing metalamprophyres are being confidently identified up to mid-amphibolite facies, and as far back as the late Archaean (≈2.7 Ga). Metamorphosed dyke-swarms and pipes in the Norseman-Wiluna greenstone belt of Western Australia have been documented in particular detail (Perring *et al.* 1989a,b; Rock *et al.*1988d). Whilst no granulite or blueschist facies lamprophyres have yet been identified, textural and/or geochemical diagnostics can definitely be well enough preserved even in these ancient greenstone belts to enable metalamprophyres to be confidently distinguished from other types of metabasites.

2.2 A compilation of worldwide lamprophyre occurrences

Appendix C summarises and Figs.2.1–2.13 locate occurrences documented in the Bibliography references. To distinguish 'fact' from inference and conjecture as far as possible, 4 groups of rocks are separately compiled, as explained in the introduction to Appendix C: (1) *confirmed lamprophyres;* (2) *unconfirmed lamprophyres*; (3) *apochryphal lamprophyres;* (4) *newly suggested (additional) lamprophyres.* There are *very few* examples of group 4, and the relative sizes of these groups is 1 » 2 » 3 » 4. In short,

this book countenances and compiles *fewer* examples of lamprophyres than are actually described in the literature as such.

Far from inventing some new, 'catch-all' definition of lamprophyres, this book effects a vigorous *pruning* of "lamprophyres" already described, leading to a *smaller* number of remaining, confirmed occurrences.

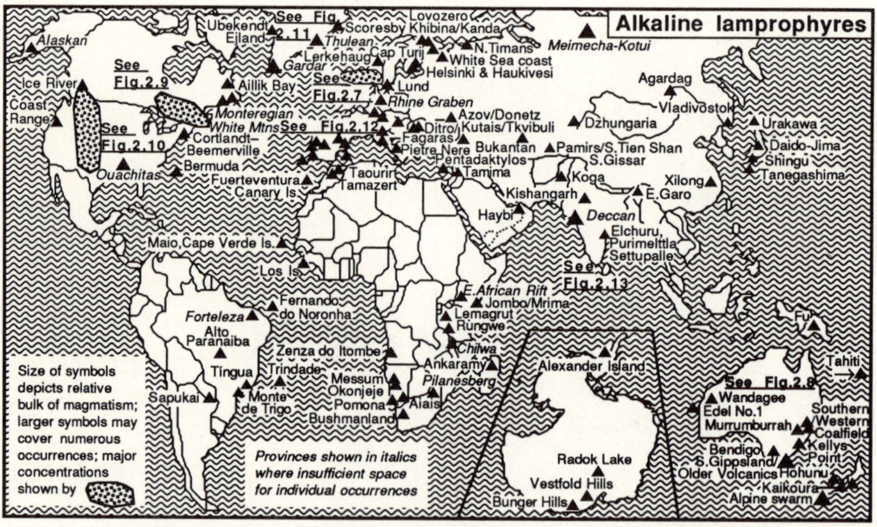

Fig.2.2 *Global distribution of confirmed alkaline lamprophyres (AL).*
For details of individual occurrences, see Appendix C, Table C2.

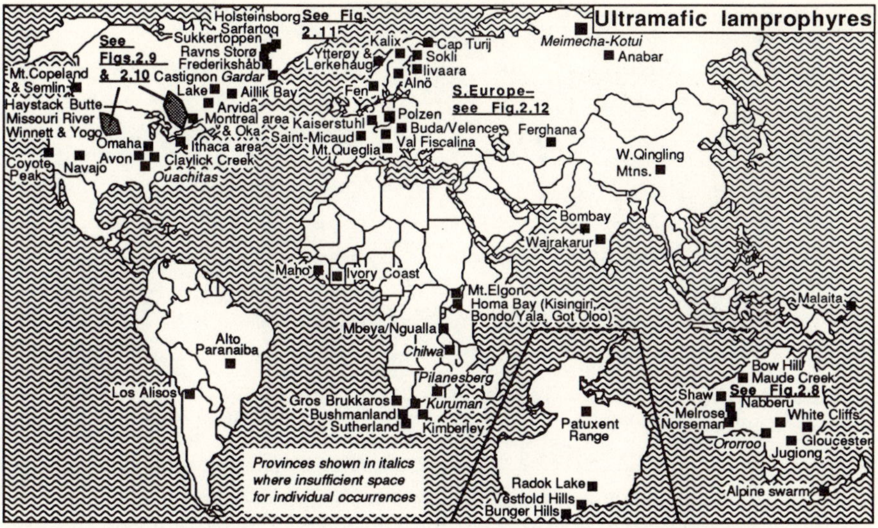

Fig.2.3 *Global distribution of confirmed ultramafic lamprophyres (UML).* For details of individual occurrences, see Appendix C, Table C3.

2.3 Global distribution and abundance of lamprophyres

This Section tries briefly to summarise some of the wealth of information in Appendix C.

2.3.1 Overall distribution through geological time (Table 2.1; Fig.2.14)

Age data summarised in Fig.2.14 suggest that lamprophyres are more-or-less continuously

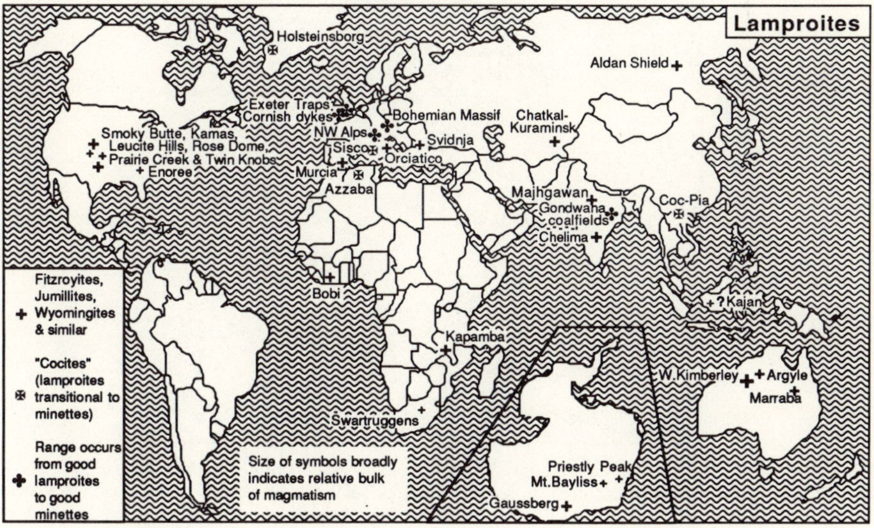

Fig.2.4 *Global distribution of confirmed lamproites (LL).* For details, see Appendix C, Table C4.

Fig.2.5 *Distribution of Late Hercynian (Upper Carboniferous–Lower Permian) calc-alkaline lamprophyres in Western Europe and eastern North America* (at different scales). For details of individual occurrences, see Appendix C, Table C1.

distributed from late Archaean (CAL) or Proterozoic (AL, UML) to Recent times. The marked secular decrease with age in all 4 parts of Fig.2.14 most probably reflects the difficulty of recognizing old, metamorphosed lamprophyres (Section 2.1), and suggests that the detailed distribution of lamprophyres through geological time may never be

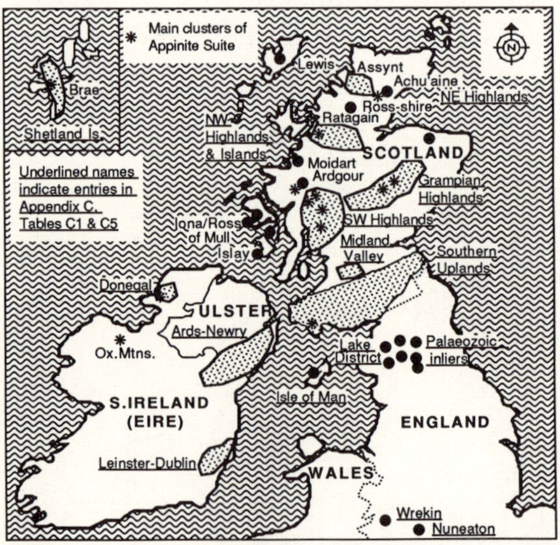

Fig.2.6 *Distribution of Late Caledonian (Siluro–Devonian) calc-alkaline lamprophyres in the British Isles.* For details of individual occurrences, see Appendix C, Table C1. Simplified from British Geological Survey 1"/1:50,000 maps and author's unpublished data. The reliability of the diagram in some areas (e.g. SW Highlands, Islay, Grampian Highlands) is less than in others because of the poor quality of outcrop and the antiquity of available maps. For larger-scale distribution in northern Britain, see Rock *et al.*(1988a).

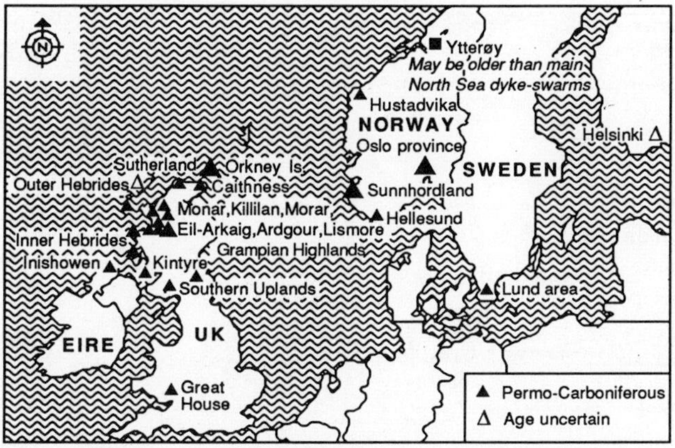

Fig. 2.7 *Distribution of Lower Carboniferous–Triassic alkaline lamprophyres in the North Sea 'super-province'*. For individual details, see Appendix C, Table C2. For larger-scale distribution, see Rock (1983b, 1988b).

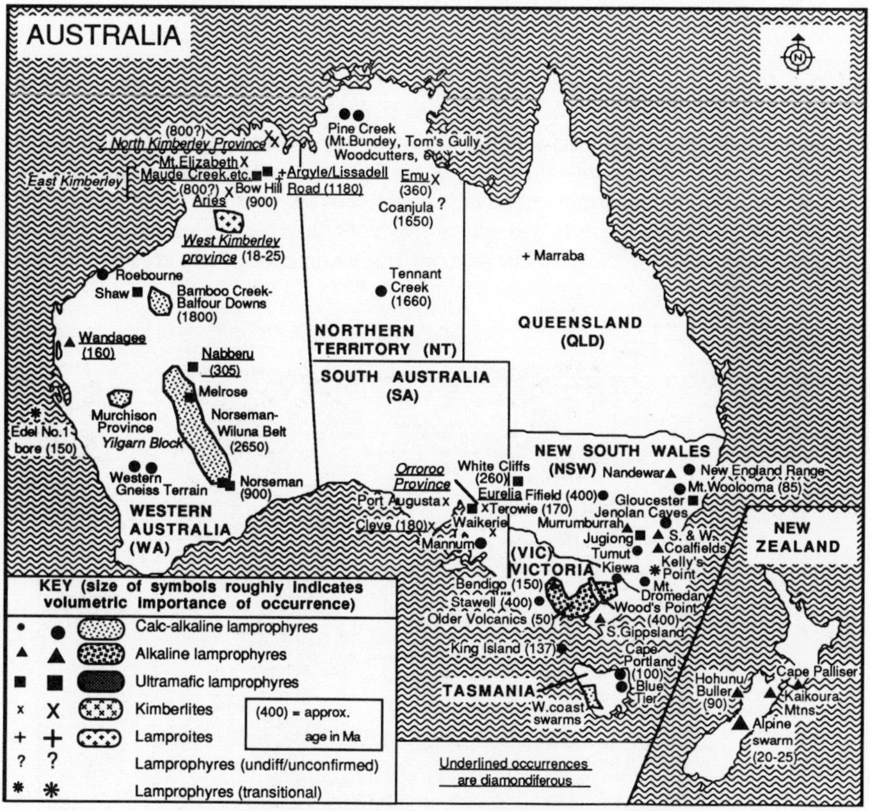

Fig. 2.8 *Distribution of all types of lamprophyres in Australasia.* For details of individual occurrences, see appropriate tables in Appendix C.

Fig.2.9 *Distribution of all types of lamprophyres in Canada and southernmost Alaska.* For key see Fig.2.8; for details of individual occurrences, see Appendix C.

completely defined. Not least because of bias in literature coverage, relative heights in Fig.2.14 should be treated ordinally (i.e. rankings may be meaningful, but ratios almost certainly are not). For example, the length of the Caledonian/Hercynian bar in Fig.2.14a at least in part mirrors the disproportionate amount of work over 100 years carried out in western Europe (Section 1.1.1), and may or may not reflect the true volumetric abundance of lamprophyres in these orogens. Furthermore, it is extremely difficult to estimate magma-

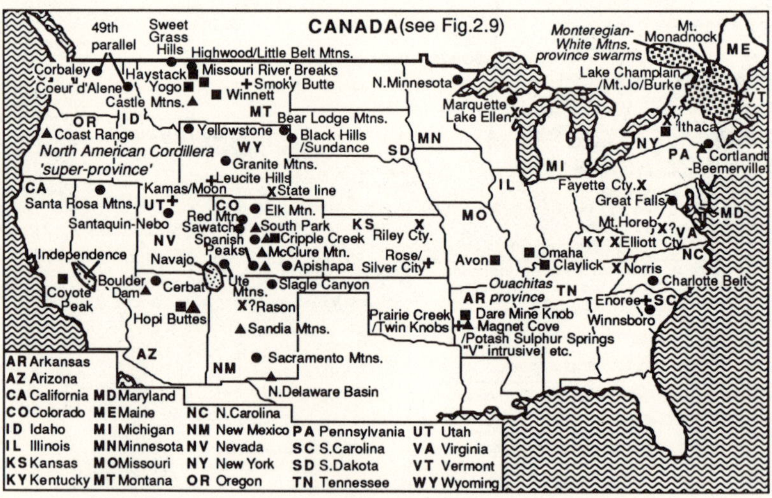

Fig.2.10 *Distribution of all types of lamprophyres in the coterminous USA.* For key see Fig.2.8; for details of individual occurrences, see Appendix C.

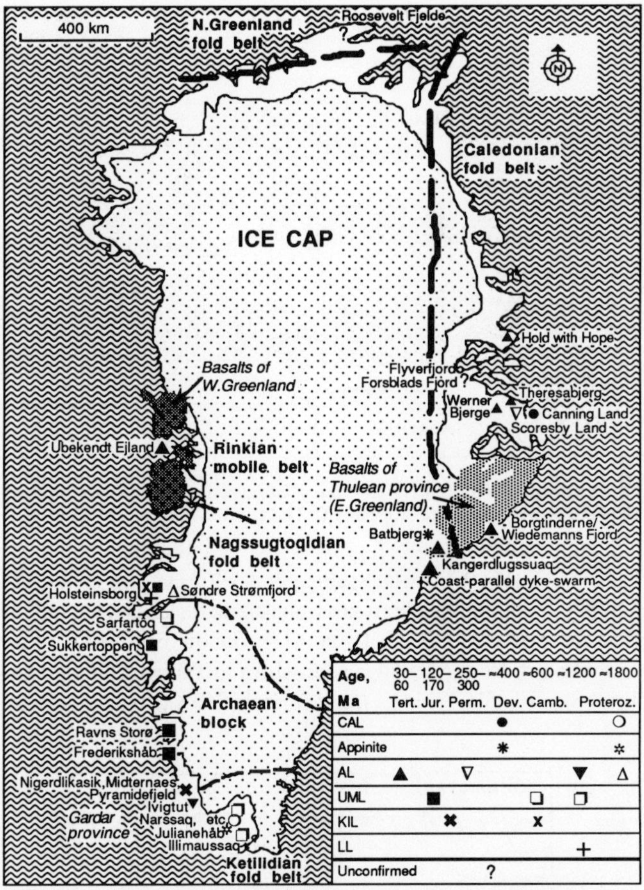

Fig.2.11 *Distribution of all types of lamprophyres in Greenland.*
For key see Fig.2.8; for details of individual of occurrences, see Appendix C.

volumes with rocks occurring mainly as dykes, and hence it is impossible to compare one occurrence with another, in terms of the total size of the magmatic event.

Extremes are fortunately unaffected by these *caveats*. The oldest known lamprophyres

Table 2.1 Age-ranges and global abundance estimates for the 5 lamprophyre branches

	Age-range	Oldest	Youngest	Lavas	Dykes	Pipes	Magma-volume
CAL	Recent–2.7Ga	Yilgarn*	Colima*	A few km^3	$\approx 10^6$	$\approx 10^2$	Many batholiths
AL	Recent–2.0Ga	Azov?	Boulder Dam#	Very small	$\approx 10^5$	$\approx 10^2$	Several batholiths
UML	Pleistocene–1.9Ga	Castignon†	Homa Bay†	None?	$\approx 10^2$	$\approx 10^2$	A few stocks
KIL	Tertiary–1.6Ga	Kuruman§	Colorado	None?	$\approx 10^2$	$\approx 10^3$	A few plutons
LL	Recent–1.3Ga	Ivory Coast§	Leucite Hills‡	A few km3	≈102	≈103	A few stocks

Explanation: $\approx 10^2$ means of the order of a few hundred; dykes includes sills, sheets, veins
* See Table C1 # Table C2 † Table C3 ‡ Table C4 § Table C6

Fig.2.12 *Distribution of all types of Mesozoic–Cainozoic lamprophyres in southern Europe.* For key see Fig.2.8; for details, see Appendix C. There are no known kimberlites in this region.

are late Archaean CAL (Fig.2.14) which, along with ≈2.7 Ga carbonatites from Greenland (Larsen *et al.* 1983), represent the oldest known alkaline rocks, and the oldest rocks of deep mantle origin (contrast the oldest KIL at ≈1.6 Ga on Fig.2.14). Recent (≤ 1 Ma) lamprophyres are known from Antarctica (Gaussberg in Table C4), Papua New Guinea and Mexico (Colima, Sugarloaf in Table C1), and from the USA (Boulder Dam in Table C2).

Fig.2.13 *Distribution of all types of lamprophyres in the Indian subcontinent.* For key see Fig.2.8; for details of individual occurrences, see Appendix C.

LL, like KIL, share with carbonatites an apparent concentration at certain times (Fig.2.14d), although the *periodicity* observed by many authors with carbonatites is not yet evident. Voluminous KIL were emplaced in the Upper Jurassic–Cretaceous and, more strikingly, *all known* Group II KIL (Appendix B) are 110–200 Ma old (Skinner 1989). By contrast, the two major LL peaks are in the Tertiary (< 30 Ma) and Proterozoic (≈1200 Ma).

2.3.2 *Overall volumetric abundance* (Table 2.1)

The occurrences in Appendix C vary from single isolated dykes or pipes, to regional swarms with thousands of dykes covering many thousands of square kilometres, volumetrically equivalent to substantial plutons (Rock 1983b). J.Watson (1984) estimated that lamprophyres and associated felsic minor intrusions make up 15% of Scottish Lake Caledonian igneous rocks (i.e. several batholith equivalents), and crustal extensions reach 50% locally (Barnes *et al.* 1986; Rock *et al.*1986a,b;1987,1988a). As well as being very widely distributed, therefore, lamprophyres are far from being rare in a *volumetric* sense. Although quantitative comparisons are not yet possible, they must collectively have an aggregate global bulk considerably greater than carbonatites, melilite- or leucite-bearing rocks, and probably approaching that of nepheline syenites or alkali granites (the most abundant types of alkaline rocks: Rock 1987b). Within the lamprophyres themselves, CAL and AL each probably constitute at least 40–45% of the total bulk, with the much less abundant UML, LL and KIL together making up the remaining 10–15% (Table 2.1).

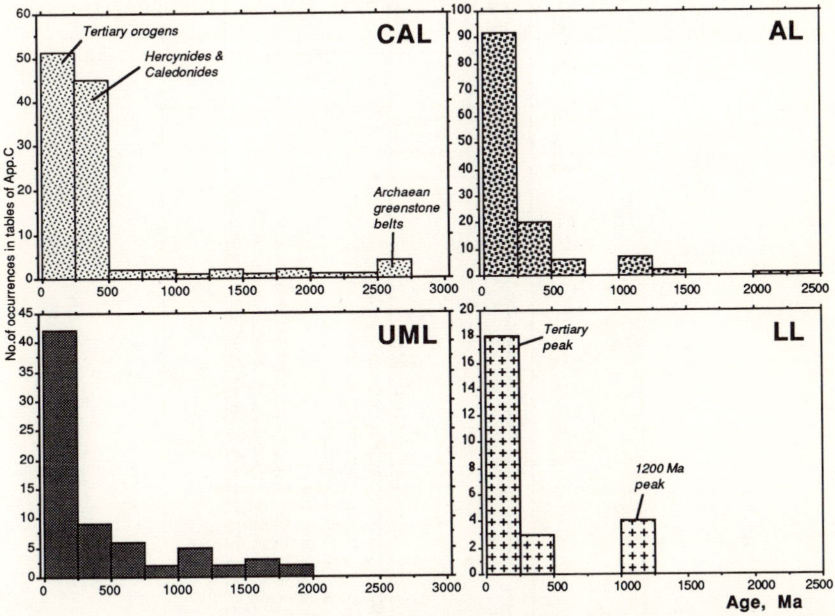

Fig.2.14 *Secular distribution of the 5 lamprophyre branches through geological time.* Individual occurrences in each table of Appendix C contribute one 'count', no allowance for differing magma-volumes being feasible. Occurrences are included as long as the uncertainty on their age is not significant at the scale of the diagram; for example, an occurrence known to be 'Tertiary' can be confidently plotted in the leftmost class interval, whereas one known only to be 'Palaeozoic' has to be omitted altogether.

Table 2.2 Classification and coding of lamprophyre tectono-magmatic associations

LETTER	NUMBER	1	2	3	4	5	6	7
	Associated igneous rocks	***Convergent margins*** *(Andean-type continental orogens)*	***?Convergent margins*** *(Greenstone belts, mobile belts)*	***Convergent margins*** *(Island -arcs)*	***Passive margins*** *(Transform faults)*	***Divergent margins*** *(Rifts, triple junctions)*	***Intra-plate*** *(Oceanic islands, hot-spots)*	***Intra-plate*** *(Old stable or recently stabil -ized cratons)*
A	Calc-alkaline (granitoid) intrusive/extrusive suites	CAL	CAL	CAL	CAL	—	—	—
B	Shoshonitic (monzonite-shonkinite) intrusive/extrusive suites	CAL	?	CAL	CAL	—	—	—
C	Intrusive breccia pipe-complexes	CAL	?	?	CAL	—	—	—
L	Other lamprophyres only	AL	CAL,AL	CAL,AL	CAL,AL	AL,UML	AL,UML	All branches
M	Mildly alkaline (gabbro-syenite) intrusive/extrusive suites	—	—	—	—	AL	AL	
R	None (isolated, local or regional lamprophyric magmatism)	LL	—	—	AL	LL	UML	All branches
S	Strongly alkaline (carbonatite-ijolite) extrusive/intrusive suites	—	—	—	—	UML	AL,UML	LL,UML

2.3.3 Geographical distribution

Figs.2.1–2.13 locate over 500 *confirmed* lamprophyre occurrences, covering all 6 continents, over 60 countries, and numerous oceanic islands. Among the world's largest nations, lamprophyres are recorded in most of the 15 republics of the USSR, 11 of the 12 Canadian provinces and territories, at least 24 of the 50 US states, and in all 8 states and territories of Australia. The apparent paucity of lamprophyres in parts of the developing world (Africa, the Middle East and S.America) is almost certainly a misleading result of low or patchy information density (less numerous and detailed geological maps, etc.), further exacerbated by generally poor outcrop in these mostly tropical or arid countries. Minor intrusions like lamprophyres are unlikely to be mentioned at all on the reconnaissance (e.g. 1:250,000) geological maps which are all that are available for many developing nations; indeed, many maps of *developed* countries at these scales quite understandably overlook even substantial lamprophyric dyke-swarms (Rock *et al.* 1988d). That large numbers of lamprophyres are being found in the only two areas of the globe where almost 100% outcrop is available — the ice-free coasts of Greenland and Antarctica — implies that the *global* density of lamprophyres is underestimated even by that in the developed countries on Figs.2.1–2.4. Clearly, therefore, lamprophyres are very widely distributed indeed.

The distribution in the USSR and China is perhaps the least reliable feature of Figs.2.1–2.4, partly because of the inaccessibility of much of the literature, and partly because locality descriptions in available papers are frequently difficult to relate to western maps. Spelling variants, translation and transliteration problems, different places with the same name, and a lack of sketch-maps or latitude and longitude data, are among problems faced in compiling data for these nations.

2.4 Tectono-magmatic associations

Included in the tables of Appendix C is a short column "**As**", which codifies a wide range of associations that lamprophyres exhibit (Table 2.2). In the two-letter codes used (e.g. A1, B7), the first letter indicates the coeval igneous rocks (A, B, C, L, M, R or S), the second the tectonic setting (1–7). For both aspects, the association is implied to be spatial, temporal and genetic: that is, the lamprophyres are comagmatic with the specified igneous rocks and were emplaced at a time when the particular tectonic regime was (or is still) in force. Because this scheme necessarily mixes what is generally an observable 'fact' (the associated igneous rocks), with what — except in the youngest lamprophyres— is at best an inference (the tectonic setting), there are many examples in the tables where only the initial letter can be given objectively. Even here, however, information is locally insufficient to separate igneous rocks which are coeval/comagmatic with the lamprophyres from those which are merely spatially coincidental, and question marks indicate this kind of doubt.

About 30 of the 49 possible associations (A1→S7) are observed in Nature (Table 2.2). As a clan, lamprophyres are emplaced in constructive, destructive, intra-plate and passive regimes, but CAL are most typical of convergent, AL/UML of divergent or passive, and LL/KIL of intra-plate settings. We now describe the major associations (by letter) in more detail, but assessment of their genetic significance is left to Chapter 8.

2.4.1 Association A: with calc-alkaline granitoid plutons

This is by far the most common association for CAL, but does not occur with other lamprophyre branches (Table 2.2). Any of the CAL rock-types in Fig.1.2 may be present, and the lamprophyres almost always form the mafic members of voluminous dyke-suites, which themselves define a compositional continuum through malchites and porphyrites to porphyries, acidic aplites and pegmatites (Section 8.5; Appendix B).

Although, in some cases, lamprophyric dyke-swarms appear immediately to postdate associated granitoid plutons, in many others they *bracket* plutons in both time and space, as shown by well-constrained field evidence and often, nowadays, by direct dating. For example, dykes may appear as xenoliths in, but may also cut, the granitoids; or they may cut some phases of a polyphase pluton but not others — see Rock *et al.* (1988a) for accumulated evidence on this point from the British Caledonides. The dyke-swarms may radiate from the granitoid pluton, or intersect it in a linear belt.

Many orogenic belts contain a prodigious variety of granitoids — from foliated pre-kinematic *in situ* melts, to post-tectonic forcefully emplaced bodies. This is not the place to delve into the numerous (and often conflicting) classifications of such rocks. Suffice it to say that the varieties most commonly involved in Association A appear to be late- to post-orogenic, unfoliated, forcefully emplaced bodies (including ring-complexes) which display a wide spectrum of basic–acidic granitoid compositions (i.e. from diorites or ultramafic rocks to granites *sensu stricto*). Lamprophyric dyke-swarms also show some preference for the *I*-type ('igneous') granitoids of Chappell & White (1974), although there are well-developed swarms around some *S*-type (e.g. two-mica) granites, notably the Cornubian Batholith in the Hercynides of SW England (Table C6).

2.4.2 *Association B: with shoshonitic suites*

Here, lamprophyres are more immediately associated with silica-saturated or mildly under-saturated, somewhat potassic rock-types such as monzonites, shoshonites, shonkinites and syenites (Appendix B). Granitoids are generally absent. The shoshonitic rocks most commonly occur as extrusives and/or as high-level intrusives.

2.4.3 *Association C: with appinite–breccia pipe-complexes*

Many granitoid plutons in the Caledonides are surrounded by clusters of appinite–breccia–hornblende lamprophyre pipes or vents, which tend to overlap the *earliest* plutonic phase in time (Section 7.2.1). Contemporaneous and/or later lamprophyric dyke-swarms may also radiate from or cut the pluton. Similar associations, described under different names, exist in other countries and orogens (Section 7.2.2). As with Association A, the lamprophyres are mafic members of continuous rock-series through porphyrites to acidic dykes, and Association C is probably just a special case of Association B, mainly confined to hornblendic CAL.

2.4.4 *Inter-relationships between Associations A, B and C*

These associations are not only related but gradational. For example, the Hercynian Exeter Trap minettes are spatially and temporally related not only to shoshonitic lavas (Knill 1969; Cosgrove 1972) but also to the Cornubian Batholith (Table C6). Again, several Scottish Caledonian granites (e.g. Barnesmore, Ballachulish) occur with both pre-granite (Association C) and post-granite (Association A) lamprophyre dykes. Furthermore, lamprophyre swarms are often best developed around rather K-rich or quartz-poor (i.e. shoshonitic) plutons: for example, Åland Isles, Finland (which includes potassic monzonites), and Ratagain, NW Scotland (which includes syenites) in Table C1. In the 49th parallel occurrence (Table C1), Association A lamprophyres related to the Trail and other granitoid batholiths, and Association B lamprophyres related to the K-rich Coryell intrusion, form such an intimate mix that "it is impossible to say which batholithic type has produced a given dyke" (Daly 1912, pp.317,376). Association B is particularly closely related to Association A in provinces where granitoid plutons and shoshonitic volcanic suites coexist (e.g. occurrences of Colorado, Montana and Wyoming, USA, in Table C1).

2.4.5 Association L: lamprophyres with each other alone

A good example is the Indian Gondwanas (Table C6), where 2 or perhaps 3 lamprophyre branches are well represented, but there are no other contemporaneous plutonic or volcanic rocks except for minor vein-rocks derived directly from the lamprophyres themselves.

2.4.6 Association M (mildly alkaline): with alkaline syenite-gabbro plutons

This is the most characteristic association for AL, but is not found with any other lamprophyre branch (Table 2.2). There are two sub-associations: (1) *mafic-felsic* (**M**), where the pluton contains a wide compositional range (alkali gabbros, monzonites, syenites, teschenites, theralites ± feldspathoidal varieties thereof); and (2) *felsic-dominated* (**M***), where mafic rocks are rare or absent and the pluton is dominated by syenitoids.

2.4.7 Association R: regional lamprophyric magmatism unrelated to other igneous activity

This is the association for most KIL and LL, and many AL, but few CAL or UML. It varies considerably in size, from *regional* associations (dyke-swarms with hundreds or thousands of dykes, sill-complexes, fields of pipes and diatremes), to *local* associations, where there are only a handful of isolated minor intrusions. However in both cases, the lamprophyres are essentially unconnected with contemporaneous volcanism or plutonism.

2.4.8 Association S (strongly alkaline): with carbonatite–ijolite–nephelinite complexes

As already mentioned (Section 1.6.1), this contains the widest lamprophyre spectrum: it is most characteristic of UML, but plagioclase-free AL (i.e. monchiquites, sannaites) also commonly occur, and there are at least two reasonably well authenticated examples of CAL, although most rocks described as "minettes" or "vogesites" in this setting are misnamed AL or UML (Rock 1977, table IV). The tectonic environment is essentially identical to that of Association M, and indeed, the two associations are commonly represented within the same igneous province (e.g. Monteregian Hills, Canada in Tables C2, C3).

2.4.9 Some spurious or equivocal associations

Many further associations reported in the literature are apochryphal — either because the so-called "lamprophyres" are themselves spurious, or because the spatial association has since proved to be a fortuitous juxtaposition of rocks with quite different ages. A "diabase-lamprophyre" association in the Hercynides (Wimmenauer 1973a,b) is negated by the differing ages of the diabases and lamprophyres (D'Amico & Guidicini 1961; Leutwein *et al.* 1972). A "potassic lamprophyre"– kimberlite link made by Scott (1979) at Holsteinsborg (Greenland) is now overruled because the kimberlites are ≈580 Ma old and the "lamprophyres" (lamproites) ≈1230 Ma (Tables C3, C4). A supposed association in

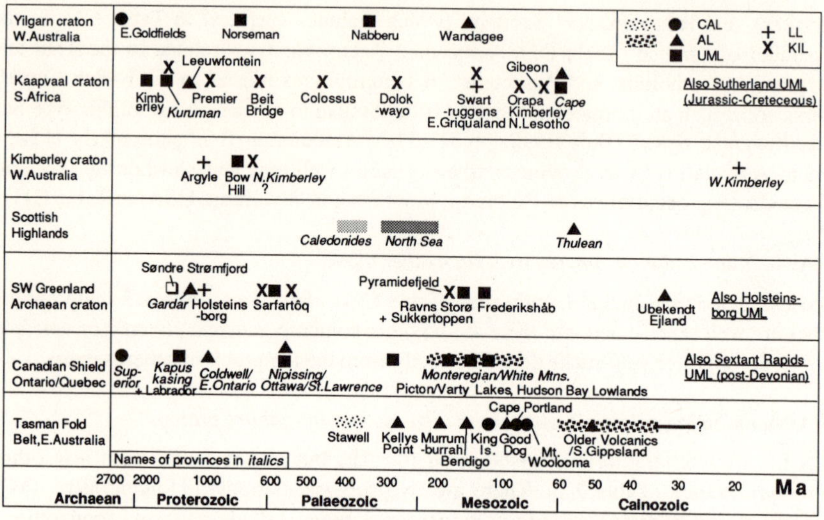

Fig.2.15 *Repeated lamprophyric magmatism in selected areas.* In some cases (e.g. Scottish Highlands), lamprophyre dykes of different ages may be intimately juxtaposed, even within the same fracture; in others (e.g. Kimberley craton, W.Australia), the different magmatic episodes occupy different sub-areas within the same region. Note logarithmic X-axis representing geological time. Data from Appendix C, Dawson (1980, 1989), Smith *et al.* (1985b), R.H.Mitchell (1986).

Canada between "kimberlites" and spessartites (Dawson 1980, p. 27) is a spatial juxtaposition of UML and CAL of quite different ages (Lesuer in Table C1, Bachelor Lake in Table C3). "Kimberlites" formerly reported with the Navajo minettes (Watson 1967a) are now described as "serpentinized microbreccias" closely related to the minettes themselves (Table C6; R.H.Mitchell 1986). "Lamprophyre" veins in eclogites from the Roberts Victor kimberlite (S.Africa) are equally suspect (Table C8). A few other associations appear to be genuine but very unusual: for example, the Lesuer (Canada) spessartites are apparently coeval with a 5 x 2 km 'eucrite'–gabbro stock (Table C1).

2.5 Repeated lamprophyre injection in one area over geological time

Fortuitous juxtapositions of lamprophyres with different ages are common, simply because many parts of the globe show repeated injection of lamprophyric magmas over very long periods of geological time (Fig.2.15). This phenomenon is not confined to the ancient cratons (Canadian Shield, Kaapvaal, SW Greenland, Yilgarn Block, etc.), but is also found in orogenic belts (Alps, Caledonides, Hercynides, Tasman Fold Belt in Fig.2.15). Melts may even intrude exactly the same fractures at several different times, for example:

— *Kimberley, S.Africa:* Proterozoic UML dykes are cut by Cretaceous KIL pipes at De Beers and Wesselton (Clement *et al.* 1979);

— *Scottish Highlands:* Devonian CAL are cut by Permo-Carboniferous AL, while a few 'pseudo-composite' dykes show parallel injections of *both* types (Rock 1983b);

— *Guinea:* Cretaceous KIL are cut by younger "alkaline lamprophyres" (Rombouts 1987).

This further underlines the preliminary nature of the information compiled in Appendix C, for there are many other areas where age-dating and field evidence (e.g. cross-cutting relationships) are simply inadequate to distinguish or even to recognize these multiple phases of intrusion. Some of the less well-documented cases in Table C6 may in particular be fortuitous juxtapositions, rather than truly 'mixed' associations.

3 Field Geology and Petrography — Macroscopy and Microscopy

3.1 Field geology of lamprophyres: diagnostic forms of occurrence

A general exposition of minor intrusion characteristics is superfluous here, since many of the descriptions of basaltic dyke-swarms in Halls & Fahrig (1987), Maaløe (1987) and textbooks, plus comprehensive models of vents in R.H.Mitchell (1986), Bergman (1987) and Milashev (1988), already apply to lamprophyres. Detailed descriptions of lamprophyre bodies are available for example in Geldon (1972), Wallace (1975), Mykura (1976), Rock *et al.*(1986a) and Milashev (1988). This section therefore emphasises aspects peculiar to lamprophyre intrusions, which help to distinguish them from basaltic intrusions. Such aspects need to be considered at scales ranging from single bodies to entire dyke-swarms.

3.1.1 Small-scale characteristics of lamprophyre intrusions (Fig.3.1)

Nearly all lamprophyres form dykes, pipes, sheets, sills or vents (Appendix B). Commonly, these occur together, representing different facies of one magmatic event (cf. R.H.Mitchell 1986, fig.3.1). Dykes and sills are typically 0.1–5 m wide, rarely exceeding 10 m, while pipes and diatremes are generally tens, hundreds or, at most, a few thousand metres across. Nevertheless, in some well exposed terranes (e.g. Antarctica, Greenland), lamprophyre dykes can be traced for tens of kilometres, either as individual bodies or *en echelon.* In some mining areas, they are known to extend to depths of several kilometres (e.g. Drysdale 1914; Marsden 1988). The largest outcrop defined by a single, dominantly lamprophyric body is 11x1 km, at Lindequesdrift, S.Africa (Table 3.1, Table C1), although the aggregate outcrop areas of some regional dyke-swarms and pipe-clusters are orders of magnitude larger — even excluding the intervening country-rocks (Section 2.3.3). As far as scanty data allow generalizations, the ratio of magma-volumes contained in dykes+sills versus pipe+vents probably decreases in the order CAL > AL > LL > UML ≈ KIL: that is, the typical CAL is a dyke or sill, whereas the typical UML or KIL is a pipe or vent.

As with most dyke-rocks, lamprophyres can either weather out among their host-rocks, forming deep clefts, or remain upstanding as walls. There is little obvious explanation (climatic or petrological) for which type of behaviour occurs, since adjacent dykes in the same swarm commonly show opposite behaviour, and examples of upstanding dykes or pipes can be found in desert, subarctic and even tropical terranes alike. Spectacular examples are found in the arid terranes of the south-western USA, where dykes form walls reaching tens of metres high and extending for several kilometres, which may radiate from pipes forming 'monuments' hundreds of metres high (e.g. Ship Rock, New Mexico).

In many terranes where lamprophyric and basaltic dykes coexist, the lamprophyres are more-or-less readily distinguishable in the field, even where unequivocal cross-cutting relationships are lacking. Some or all of the following differences separate lamprophyres from basalts in all climates and terranes, although several are difficult to describe in words and are best appreciated via the classic field geologist's art of 'getting one's eye in':

—*Rock texture:* this is richly porphyritic in lamprophyres, with a fine-grained grey-black groundmass, as against mostly aphyric with fine/medium-grained greenish matrix in basalts; feldspar phenocrysts in basalts are diagnostic, as are panidiomorphic, mafic macrocrysts (*especially* large books of black mica) in lamprophyres.

—*Joint patterns:* these are commonly more irregular in lamprophyres.

—*Width:* lamprophyres are narrower overall but more variable within individual bodies (Fig.3.1d), and their length/width ratios are mostly higher than basalts.

—*Sinuous margins:* lamprophyre dykes pinch and swell, taper, or show serpentinous, uneven shape with irregular branching, segmentation into discrete 'blobs', and variations along one margin commonly not reflected in the other (Fig.3.1d). Basalts typically have straight or smoothly curving borders, and branch in a more regular fashion.

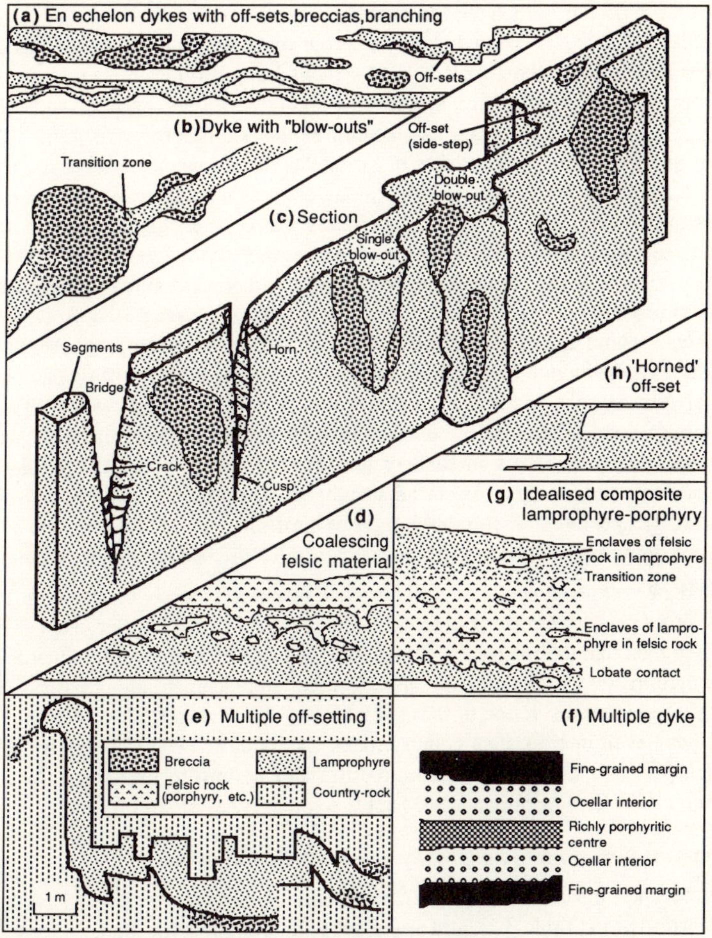

Fig.3.1. *Sketches of typical field aspects of lamprophyre intrusions.*

(a) Modelled after accurate maps of Ship Rock minette dyke, USA (Delaney & Pollard 1981).
(b) As (a), but also accommodating features in Scottish Permo-Carboniferous monchiquitic vents (Rock 1981).
(c) Substantially modified from sketches of Ship Rock minette dyke (Delaney & Pollard 1981).
(d) Modelled on monchiquite dykes from Canada (Philpotts 1972).
(e) Simplified sketch of actual "Loch Ness Monster" kersantite dyke from SW Scotland (Rock et al.1986a).
(f) Modelled after camptonite dykes from Greenland and Portugal (Upton 1965; Rock 1979).
(g) Modelled after descriptions in Table 3.1.
(h) Modelled after monchiquite dykes from Canada (Currie & Ferguson 1970).

—*Chilled margins:* these are much rarer in lamprophyres than in basalts.
—*Carbonation, bleaching, mineralization of dykes:* these, by contrast, are more frequent in lamprophyres.
—*Contact alteration of country-rocks:* this may be more apparent with lamprophyres, but hornfels development may be *less* obvious, for dykes of similar size.
—*Banding:* several macroscopically distinct bands are commonly found in lamprophyres (e.g. Fig.3.1f). Internal contacts may be sharp or gradational, the latter commonly reflecting the gradual inward concentration of a phenocryst or macrocryst phase, or even a xenolith type (e.g. Greenough *et al.*1988), from margin to centre. Banding of either type is rarer in basalts, which tend to be homogeneous.
—*Non-faulted off-sets (side-steps):* both lamprophyre and basalt dykes may side-step on various scales between host-rock layers, *without being faulted*; however, this is much more common in lamprophyres and more irregular. Unique 'horned' ends (Fig.3.1h) may occur, which totally preclude any tectonic explanation for the off-set.
—*Budding and brecciation:* extensive but irregular development of breccias, notably in 'buds' on the main body, or as isolated pods or in sharp angles, is far more common with lamprophyres (Fig.3.1a,b,c,e). In many cases, the buds appear to represent incipient explosion of the lamprophyre melt, where more complete explosion would have given rise to a fully developed vent. Vents which may represent a more stage of this process commonly occur along strike from some such dykes (Fig.3.1b; Rock 1983b).
—*Weathering:* onion-scale weathering is finer in texture with lamprophyres, and leads ultimately to a distinctive rich, red-brown soil, commonly replete with glistening biotite or hornblende crystals; in some areas this soil is distinctive enough to identify a former lamprophyre even where no trace of solid rock remains. Less intensely weathered lamprophyres may display a mottled or knobbly (so-called 'pustular') surface. Where basalt and lamprophyre dykes occur adjacently, the lamprophyre is commonly the more weathered, irrespective of their relative ages of intrusion.
—*Pitch:* some lamprophyres, like phonolites, may emit a pitched note under the hammer.

These features may generate overall intrusive forms so patently bizarre (e.g. Fig.3.1e), that they can be considered diagnostic of lamprophyres. Currie & Ferguson (1970) and others have argued that their common explanation lies in the uniquely high volatile contents of lamprophyre melts, which facilitate rapid, forceful emplacement along narrow fractures. Alternating release and resolution of these volatiles by *P-T* changes allows lamprophyres to 'drill' their way even into resistant country-rocks, paying little heed to pre-existing fracture patterns. In short, all the features of kimberlite and lamproite intrusions described by R.H.Mitchell (1986) and Bergman (1987) apply to other types of lamprophyres.

3.1.2 Composite intrusions of lamprophyres with felsic rock-types

As well as *multiple* intrusions (Fig.3.1f), *composite* bodies (Appendix B) have now been recognized worldwide (Table 3.1), and are of substantial interest for the light they throw on petrogenetic relations between lamprophyric and felsic melts (Section 8.5). The most common varieties are composite CAL–porphyry bodies; these are so remarkably similar in occurrences of widely differing ages worldwide that descriptions of one can almost be taken piecemeal for another, allowing an idealised form to be easily defined (Fig.3.1g). Almost invariable features of these bodies include the following:

—The more acidic rock (porphyry) forms the much thicker centre of the composite body.

Table 3.1 Examples of composite and differentiated lamprophyre intrusions

Country	Province	Occurrence	No.	Width	Margins	Interior of body	Centre	Basic/acidic relationships	References
Antarctica	— (Fig.2.1)	S.Victoria Land (Dry Valley)	1	?	CK	Transition zone	Qz-porph	CK xen in Qz-porph	Campbell Smith (1924)
Australia (WA)	Yilgarn (Fig.2.8)	Norseman-Wiluna (Kambalda Dome)	>2	2–3 m	CC	Variable; CC carries xen of andes-porph and vice-versa	andes-porph	Mutual xenoliths, lobate contacts indicate magmas contemporaneous	Perring et al.(1989a,b); Perring & Rock (in prep)
Czechos-lovakia	Hercynides (Fig.2.5)	Bohemian Mass (Cercany)	1	≈3 m	CM (25 cm)	Transition syenite porphyry zone (gradational chemistry)	grdi-porph	Interpreted as differentiation in situ	Nemec (1973b,1988b)
Germany (DDR)	Hercynides (Fig.2.5)	Lausitz	1	?	grt-peg	?	Foliated CS/CV	Peg forms "edges of the lower dykes", lamp forms "inner parts"	Hentschel (1934); Wimmenaeur (1973b)
Germany (FDR)	Hercynides (Fig.2.5)	Schwarzwald (Fensterstollen)	1	≈20 m	CM (1 m)	Porphyrite transition zone	grt-porph	Continuous and symmetrical change across dyke in most major dykes	Wimmenauer (1972, 1973a,b)
Germany (GDR)	Hercynides (Fig.2.5)	Erz Mtns. (Brand-Erbisdorf)	1	8–10 m	CM (0.5-2 m)	Transition zone (10-20 cm)	Qz-porph	CM & grt xen in porph; mica-schist xen in CM; magmas	Kramer (1966)
Portugal	Iberian (Fig.2.12)	Monchique	A few	≈1	AC	Tongues of tinguaite and foyaite aplite	AC	Related indirectly by differentiation	Rock (1983a)
S.Africa	— (Fig.2.1)	Lindequesdrift, Vredefort Dome	1	≈1 km	CS with peg schl.	Transition zone over < 1 m	Syenodio, sye, peg	Felsic rocks increase towards centre; interpreted as differentiation	Bisschoff (1972)
UK (England)	Caledonides (Fig.2.6)	Palaeozoic inlier (Cross Fell)	2	≈3.7 m	CM, CK	Transition zone	Qz-porph, Fp-porph	Interpreted as hybridisation	Harker (1892); Arthurton & Wadge (1981)
UK (Scotland)	Caledonides (Fig.2.6)	Inner Hebrides (Iona)	5	2-5 m	CS,CC	Transition zone or sharp contacts	Qz-porph	Mutual xenocrysts, lobate contacts indicate magmas contemporaneous	Rock & Hunter (1987)
UK (Scotland)	Caledonides (Fig.2.6)	Southern Uplands (Newmains)	1	2–5 m	CS	Variable due to melting, hybridization and mixing	syenite, pegmatite	CS melted/hybridized country-rock gwk while differentiating to syenite	Kennedy & Read (1936); Macdonald et al.(1986)
USA (MT)	N.American Cordillera	Little Belt Mtns. (Fig.2.10)	3	8-10 m	CK (60 cm)	Transition zone (0-60 cm)	micro-grt (5.5-8 m)	CK has xen of Qz & Sa from grt; one of grt has Cpx xen from CK	Witkind (1970,1973)
USA (SC)	Hercynides (Fig.2.5/10)	Winnsboro'	2-3	0.6-2.2 m	peg (2-8 cm)	Complex net-veined zone with CS pillows in granitic matrix	CS	Result of (1) liquid immiscibility, or (2) comingling of separate melts	Vogel & Wilbrand(1978) McSween et al.(1979)
USSR	— (Fig.2.1)	E.Transbaikal (Shiviya)	Many	?	CS	Transition zone of Gd-porph	grt-porph	Interpreted as hybridisation	Solov'yeva (1973)
USSR (Uzbek)	— (Fig.2.1)	Almalyk-Bezirkes	3	?	CS	Hybrid transition zone	grdi-porph	CS xenoliths in grdi-porph	Matschanow (1964); Wimmenaeur (1973b)

—Although the porphyry can usually be taken to postdate the lamprophyre (from transgressive contacts), there is always strong evidence that the lamprophyre was still largely molten at the time the porphyry invaded: in other words, two *melts* coexisted in these composite bodies, due to multiple injection. This evidence include lobate contacts (typical of liquid-liquid interaction), the presence of xenoliths or xenocrysts of *both* melts in one another, and commonly, a transition zone along at least one of their mutual contacts which shows all the conventional signs of hybridization between two melts.

3.1.3 Differentiated lamprophyre intrusions

There are two common forms here: (1) *gradational (zoned):* bodies in which a concentration of macrocryst phases takes place towards the centre; these do not generally cover a very large compositional range; (2) *discontinuous:* bodies carrying subordinate felsic rock (aplite, porphyry, syenite, etc.) as globules, patches, schlieren, or veins; lithologies intermediate between the dominant host lamprophyre and vein are mostly lacking. Composite intrusions (Section 3.1.2), can be regarded as a variant of type (2), in which felsic rock volumetrically dominates, and is more regularly emplaced into the lamprophyre.

Mal'kov (1967,1970) has provided among the best documentation for gradational types. He finds vertical dykes to be symmetrically zoned, with an increase in both the size and proportion of phenocrysts towards the centres, which tend to be fissile. In inclined dykes, the zonation is asymmetric, with the zone of maximum phenocryst concentration displaced towards the footwall. The resulting chemical zonation tends to follow a second-order (parabolic) relationship with distance from the margin. He rejects ionic diffusion and migration mechanisms, and interprets these findings instead in terms of *flow differentiation*, which would generate a parabolic curve of laminar flow velocity. Gravitational effects also come into play in inclined dykes.

Discontinuous types are perhaps best exemplified by Monteregian sills (Philpotts 1972), in which felsic globules form larger and larger masses (including pipes) towards the top (Fig.3.1d). These have been interpreted in terms of liquid immiscibility (Section 8.5.3).

Layered intrusions, though well known in large basaltic/ultramafic dykes (e.g.Great Dyke, Zimbabwe), have only once been described in lamprophyres. The Dead River pluton (Geldon 1972) has a discontinuous margin of pyroxenite around a spessartite which shows both a flor foliation, and a crude, cryptic layering marked by upward increase in grain-size, K-feldspar and Fe/Mg in the mafic phases. That this sole layered body is one of the largest single lamprophyre intrusions (0.6 x 1 km) suggests that the smaller size of other bodies may render layer-forming processes inoperative (e.g. convection currents).

3.1.4 Contact metamorphism and metasomatism around lamprophyre intrusions

It seems that intrusions may affect their country-rocks either very little or rather intensely. Descriptions of *minimal* effects, confined to minor growth of biotite or carbonate, with or without baking or bleaching, are legion (e.g. D.Williams 1923). Again, known examples of lamprophyres intruding coals generate cokes which indicate temperatures below 600°C (Pareek 1966; Podwysocki & Dutcher 1971). By contrast, a few lamprophyres induce pyroxene-hornfels facies contact metamorphism, coupled with widespread anatexis, back-veining, intense potash-metasomatism, hybridization, contamination with anatectic melt, and related processes (e.g. Kennedy & Read 1936; Walter 1950; Macdonald *et al.*1986).

There is again little logical explanation as to why one dyke causes intense contact effects, but an adjacent one virtually none: the former may perhaps have acted as a long-lived feeder to lavas, so allowing the country-rocks to heat up gradually. However, few lavas overlying lamprophyre feeder dykes have as yet been unequivocally identified (Section 3.1.7).

Alkali metasomatism, strongly akin to the fenitization around carbonatites (Heinrich 1966) but on a smaller scale, is increasingly reported around UML dykes (e.g. Ferguson et al.1975; Fielding & Jaques 1989; Hamilton & Rock 1990). It is marked by desilication of the country-rocks, and crystallization of aegirine and alkali amphiboles within them.

3.1.5 Larger-scale groupings of lamprophyre intrusions

Individual bodies normally group into dyke-swarms, sill-complexes and pipe-clusters. Dyke-swarms can be very intense as well as widespread: crustal extensions commonly reach 5%, and locally as much as 50%. One series of swarms in SW Scotland (Leake & Cooper 1983) is so dense that it is mapped at 1:50,000 scale (British Geological Survey sheet 5W) as a solid 'blob' of lamprophyres and porphyrites covering some 30 km^2. Whilst magma-volumes for such swarms are extremely difficult to estimate, the aggregate surface area of $X \times 10^3$ dykes, averaging Y m wide and traceable for Z km, is XYZ km^2; hence some larger CAL and AL swarms almost certainly attain the equivalent surface area of a batholith (>100 km^2), which requires 5,000 dykes averaging 4 m wide and traceable for 5 km. In turn, the *total* magma-volume of CAL in orogens such as the Caledonides and Hercynides is probably equivalent to several of the coexisting granitoid plutons.

Among the most extensive pipe-clusters are those of the West Kimberley, Australia (Table C4). Three fields (Calwynyardah, Ellendale and Noonkambah) together contain over 100 pipes, whose surface areas range from a few hectares to a few sq.km (Walgidee Hills).

3.1.6 Inferred intrusion mechanisms

A complete discourse on lamprophyre intrusion mechanisms would occupy an entire book. The preceding sections have already demonstrated that multiple intrusion, flow differentiation, gravitational separation and liquid immiscibility may all be involved in producing the diversity of internal structures observed in lamprophyre dykes. Fortunately, some of the most painstaking and exemplary published studies on dyke intrusion mechanisms have been on lamprophyric dyke-swarms in North America (Johnson 1961; Delaney & Pollard 1981; D.Smith 1987; Greenough *et al.*1988); readers may thus refer to these for more comprehensive geological and mathematical details. Among the more interesting and perhaps surprising conclusions from these studies are the following:

— xenolith orientations can provide a good guide to melt emplacement mechanisms;
— regional joint patterns may play little or no part in determining the orientation of dykes;
— latest magma injections may be dominantly *lateral* and not, as might be expected, vertical;
— magma discharge is dominantly from cusps and buds rather than from linear segments;
— dykes can be emplaced from several kilometres depth in a matter of hours;
— high length/width ratios of dykes can be explained by high volatile contents.

The relationship between dykes and plugs does, however, appear to be more complex than the simple 'blow-out' formerly envisaged, and further work is needed on this aspect.

3.1.7 Volcanic lamprophyres: lavas, pyroclastics and tuffs

No 'lamprophyre eruption' has ever been observed, but the Recent LL at Gaussberg,

Antarctica (Table C4) can unequivocally be taken as volcanic rocks, and a fair proportion of older LL are also confidently assumed to be volcanic. By contrast, despite intensive searches, there is only one *possible* example of a kimberlite lava, and only 2 probable tuffs (R.H.Mitchell 1986, p.31–2), contrasting with hundreds of known kimberlite pipes. Other types of lamprophyres are characteristically plutonic to *sub*volcanic, but the occurrence of vesicular breccias and hydrovolcanic features indicates that they commonly approached or even breached the palaeosurface. Overall, the proportion of lavas among the 5 branches appears to decrease in the order LL » CAL > AL » UML > KIL, but it is worth listing *all* plausible cases here, in order of decreasing age (and perhaps commensurate credibility).

(1) *Sugarloaf volcanics, Papua New Guinea* (Table C1): This extensive Pleistocene lava –maar field appears to include plausible CAL associated with shoshonitic basalt lavas.

(2) *Mt.Amiata, Italy* (Table C1): Minettes here occur as inclusions in rhyodacite lavas from a Pleistocene volcano; they therefore constitute undoubted volcanic ejecta, but not lavas.

(3) *Mexican volcanic belt* (Table C1): Recent lavas and pyroclastics reported from here as "minettes" were not initially convincing to Rock (1984), to Bachinski & Simpson (1986), or to Dr.D.Velde (pers.comm.), largely because some of the "minettes" had $Na_2O > K_2O$ (an impossible condition for rocks defined as being dominated by biotite and sanidine). However, erroneous transposition of published Na_2O and K_2O values was responsible for this anomaly (Luhr *et al.*1989; Luhr & Kyser 1989, p.218); it is now clear that the "minettes" do have the characteristics of CAL. Simply because of the dearth of alteration ubiquitous in older lamprophyres, these rocks nevertheless have certain mineralogical peculiarities, such as the presence of ?primary analcime (Section 4.10.1); it could thus be argued either that they are unrepresentative of typical, older lamprophyres, or that they are the freshest lamprophyres known, and thus the *only* true guides to the nature of lamprophyre melts. This is critical, for it makes all the difference between deciding whether lamprophyres are actually erupted as volatile-*rich* melts, or acquire their hydrous mineralogy by interaction with an aqueous phase during cooling.

(4) *Hopi Buttes, New Mexico, USA* (Table C2): Pliocene feeder vents and dykes to maars consist of monchiquites and "limburgites". Although some of the "monchiquites" are really anhydrous basanites, others carry sufficient amphibole or biotite to justify this name. They were erupted into a shallow lacustrine basin, then "maar craters were formed at the surface of the vents" which were filled with water-lain tuff, claystone and marl deposits, finally "culminating in the quiet upwelling of lava" (Suda *et al.*1982, p.1219). Monchiquite lavas do therefore appear to exist in this occurrence.

(5) *Daido-Jima, Japan* (Table C2): Camptonite forms a coeval vein in Oligocene basalt lava.

(6) *Gorringe Bank, Atlantic Ocean* (Table C2): Earliest Tertiary dredge samples from here are assumed to be lavas, but the evidence is scanty.

(7) *Gross Brukkaros, Namibia* (Table C3): This Cretaceous "carbonatitic volcano in an incipient stage" resulted from phreatic eruptions of alkaline ultrabasic magma (Janse 1969). As conceded by R.H.Mitchell (1986, p.36), the presence of melilite and of associated fenitization preclude the name "kimberlite" (cf. Ferguson *et al.*1975), and the rocks fit the present definition of UML. Encircling the central vent are stratified epiclastic breccias, dominated by country-rock fragments. Although actual UML appear to be confined to radial dykes, sills and satellite intrusions, this is the only known example in which UML-related magmatism has apparently breached its palaeosurface.

(8) *Tamama, NE Syria* (Table C2): This is the oldest known (Jurassic) occurrence of

volcanic AL. Monchiquite lavas lie in a 100 m-thick volcano-sedimentary succession:

>20 m	Radiolarites
≈30 m	Phonolite lavas
≈10 m	Radiolarites
≈20 m	Volcanic breccias
≈15 m	Monchiquite lavas (including pillow lavas)
≈15 m	Basanite lavas
>10 m	Middle Jurassic limestone

Parrot (1974) was well aware that this occurrence was "assez inhabituel", but his careful documentation of both chemistry (typically AL) and mineralogy (phenocrysts and macrocrysts of Am, Bi, Cpx in an analcimic-glass matrix with oligoclase, albite or

Table 3.2 Modal composition of lamprophyres (phenocrysts and groundmass not separated)

(a) *Range of typical primary modal compositions in the 5 lamprophyre branches*

	Calc-alkaline	Alkaline	Ultramafic	Kimberlites	Lamproites
Amphibole, %	}20–40	2–50	<35	<5	<25
Biotite/phlogopite	}	2–30	5–70	<40	<25
Clinopyroxene	<25	20–50	<50	<5	<50
Feldspars	33–67	25–50‡	<5*	Absent*	<50
Feldspathoids	<5*	<15	<35	Absent*	<20
Garnet (melanite)	Absent*	<5	<25	Absent*	Absent*
Glass	<5	<30	<20	Absent*	<50
Melilite	Absent*	Absent*	<35	Absent*	Absent*
Monticellite	Absent*	Absent*	<25	<25	Absent*
Olivine	<10	<10†	<50	25–75	<60
Quartz	<10*	Absent*	Absent*	Absent*	<10

* by definition ‡ zero by definition in monchiquites, and <25 in cumulate varieties
† may reach much higher amounts in picritic varieties (e.g. Wandagee, Australia (WA) in Table C2)

(b) *Relative abundance of various mafic mineral parageneses among ≈350 AL in the literature*

No. of thin sections with	*Camptonites* Olivine-free	Olivine-bearing	*Monchiquites* Olivine-free	Olivine-bearing
Amphibole only	14	14	7	1
Amphibole + clinopyroxene	40	56	22	42
Amphibole + clinopyroxene + biotite	13	18	19	27
Amphibole + biotite	10	6	1	3
Clinopyroxene + biotite	1	2	5	48
Totals	78	92	54	133
Overall totals	170		187	
% with olivine	57		71	
% with amphibole	98		65	
% with biotite	31		55	
% with clinopyroxene	75		94	
% with globular structures	39		36	
% with groundmass analcime	17		most§	

§ Provided secondary products are included and identifications without substantiation are accepted.

sanidine microlites in some rocks) leaves little doubt that this succession does indeed include AL lavas.

(9) *Devon lavas, UK* (Table C6): These Permian rocks have been erratically named, but some are unequivocal minettes. Unfortunately, poor exposure and intense alteration make it difficult to determine which outcrops represent lavas and which are intrusions,

Table 3.3 Some good-quality published illustrations of lamprophyre textures and rock-types

Rock-type	Features depicted in photomicrograph/sketch	No.*	Reference
Aillikite	Globules, rounded/euhedral olivines, resorbed phlogopite	2	Von Eckermann (1966, p.12ff)
Aillikite/alnöite	Numerous views of textures and minerals, inc. type-rocks	2	Von Eckermann (1948, plate 42ff)
Alnöite	Euhedral zoned clinopyroxene phenocrysts; matrix melilite	1	MacKenzie et al.(1982, p.134)
Alnöite	General view of texture	3	Williams et al.(1982, p.237)
Alnöite	General view of olivine macrocrysts	2	Mitchell (1986, p.26)
Camptonite	Panidiomorphic kaersutite	3	Nockolds et al.(1978, p.206)
Camptonite	General view of texture	3	Williams et al.(1982, p.237)
Camptonite	Zoned kaersutite	3	Chatterjee (1974, p.412)
Camptonite	General views of many different varieties	2	Rock (1979,1983a,b)
Camptonite	General view of texture	3	Moorhouse(1959, p.327)
Damtjernite	General view showing various phenocrysts	2	Mitchell(1986, p.25)
Kentallenite	General view: euhedral olivine, pyroxene, irregular biotite	1	MacKenzie et al.(1982, p.105)
Kersantite, etc.	Battlemented/zoned biotite	3	Moorhouse (1959, p.327)
Kersantite	Various views of biotite textures	1	Rock et al.(1987,1988d)
Kimberlite	Rounded olivine macrocrysts	1	MacKenzie et al.(1982, p.81)
Kimberlites	General views of various important textural variants	2	Mitchell (1986, p.12-14)
Kimberlites	General views of various important textural variants	2	Boyd & Meyer (1979a, e.g. p.130ff)
Kimberlites	Contrasts between Group I and Group II kimberlites	2	Skinner (1989)
Lamproite	Irregular biotite + leucite phenocrysts	1	MacKenzie et al.(1982, p.136)
Lamproite	Pseudohexagonal and battlemented phlogopite phenocrysts	1	MacKenzie et al.(1982, p.137)
Lamproite	General view of texture	1	MacKenzie et al.(1982, p.138)
Lamproite	General view of texture	3	Williams et al.(1982, p.206)
Lamproite	General view of various textures	2	Scott (1981)
Lamproite	General view of various textures	2	Scott Smith & Skinner (1984a)
Lamproites	General views of thin sections and hand specimens	2	Bergman (1987, p.115-116)
Lamproschist	Various views of foliated and schistose rocks	1	Rock et al.(1988d)
Metalamprophyre	Various views of progressive stages of metamorphism	2	Perring et al.(1989b)
Minette	Pseudohexagonal, zoned, battlemented biotite	1	MacKenzie & Guilford (1980, p.56)
Minette	Quartz-feldspar segregation	1	MacKenzie et al.(1982, p.133)
Minette	Superb battlemented biotite,octagonal diopside	3	Hatch et al.(1972, p.417)
Minette, etc.	Battlemented biotite	3	Williams et al.(1982, p.230)
Minette, etc.	Battlemented biotite, octagonal diopside	3	Nockolds et al.(1978, p.205)
Minette	Pseudohexagonal, zoned biotite, euhedral diopside	2	Viterbo & Zanettin (1959)
Monchiquite	Pyroxene comb texture; euhedral zoned Cpx phenocrysts	1	MacKenzie et al.(1982, p.25,44)
Monchiquite	Panidiomorphic kaersutite	3	Nockolds et al. (1978, p.208)
Monchiquite	General view of texture	3	Williams et al.(1982, p.233)
Monchiquite	General views of various textures	2	Rock (1979,1983a,b)
Olivine lamproite	General view of texture	2	Mitchell (1986,p.27)
Sannaite	General views of various textures	2	Rock (1979,1983a,b)
Spessartite	Panidiomorphic hornblende; globular structures	1	Rock et al.(1987,1988d)
Spessartite	Panidiomorphic hornblende; progressive alteration	2	Perring et al.(1989)
Vogesite	Panidiomorphic hornblende	3	Nockolds et al. (1978, p.206)
Vogesite	Panidiomorphic hornblende	3	Hatch et al.(1972, p.420)

* 1 = colour photomicrograph; 2 = black-and-white photomicrograph; 3 = *camera lucida* monochrome drawing

but it remains reasonably probable that both feeder minettes and extrusive equivalents are present. They therefore represent the oldest *probable* volcanic lamprophyres known.

Other occurrences such as Hold with Hope, Greenland (Table C2), and Saint-Micaud, France (Table C3) are only described as "probable" lavas. Further probable lavas are concealed elsewhere by obtuse nomenclature. For example, the only difference between "mafic trachyte" lavas and intimately associated minette dykes and sills in Keewatin, Canada (Table C1) appears to be a coarser-grained groundmass and lack of flow foliation in the lamprophyres (Blake 1980, p.29). The lavas are primitive (*mg* ≈70–80%), phlogopite-bearing rocks with ≈30% mafic minerals, in no way resembling trachytes *sensu* Streckeisen (1979) or Le Bas *et al.*(1986). The use of such an oxymoron as "mafic trachyte" appears merely to reflect an anachronistic aversion to describing lavas as "lamprophyres". Similar arguments apply to the "mafic phonolites" of the Highwood Mtns., USA (Table C1).

3.2 Petrographical (thin section) characteristics of lamprophyres

3.2.1 Modal composition and paragenetic sequence (Fig.1.4; Table 3.2)

Table 3.2 gives modal ranges commonly observed, and Fig.1.4 shows the extents to which different minerals occur in the 5 branches of Fig.1.2. Some individual modal analyses are included in Table 8.1 and in Rock (1986). The paragenetic sequence for KIL (R.H.Mitchell 1986, p.57) applies equally to other lamprophyres, provided that feldspars, feldspathoids and melanite (as appropriate) are added at the low-T end of the sequence.

3.2.2 Lamprophyric (panidiomorphic) and related textures (Fig.3.2)

Since only a limited display of lamprophyre types and textures can be included in Fig.3.2, Table 3.3 lists some other useful illustrations. In the classic panidiomorphic texture (Fig.3.2b,d,e,n), many or most of the biotite and hornblende grains are pseudo-hexagonal, and the clinopyroxenes pseudo-octagonal (i.e. completely euhedral). Panidiomorphic texture is generally best developed in minettes, kersantites and lamproites, least developed in camptonites. Cumulophyric or glomeroporphyritic aggregates are common (Fig.3.2n). Joplin's (1966) suggestion that these textures trace rapid crystal growth facilitated by high volatile contents, probably cannot be bettered. It carries the curious implication that lamprophyre phenocrysts may not necessarily represent liquidus phases, but merely those grains which grew fastest in the volatile-rich melt.

Two minerals — mica and olivine — show additional textural characteristics. Biotite-phlogopites in *all* lamprophyres commonly show one or more of three distinctive features:

(1) *Shape:* pseudohexagonal *or* castellated form (Fig.3.2b,d) typifies crystals viewed down or across [001] respectively; also irregular textures (Fig.3.2a); none of these textures is known in other igneous rocks, nor is there any known explanation for them.
(2) *Fabric:* bent or kinked micas are common (Fig.3.2c) even in totally unfoliated hosts.
(3) *Zoning:* dark brown to black, Fe-Ti-rich (biotitic) rims commonly surround much paler, Mg-rich (phlgopitic) cores (Fig.3.2b,c,d,i).

Olivine textures are more varied among the 5 branches, but distinctive. KIL olivines include rounded macrocrysts and euhedral–anhedral microphenocrysts (R.H.Mitchell 1986), whereas other lamprophyres normally show one generation of phenocrysts only (Fig.3.2k). A few LL carry multiple growth aggregate and hopper olivines. Reaction rims may occur.

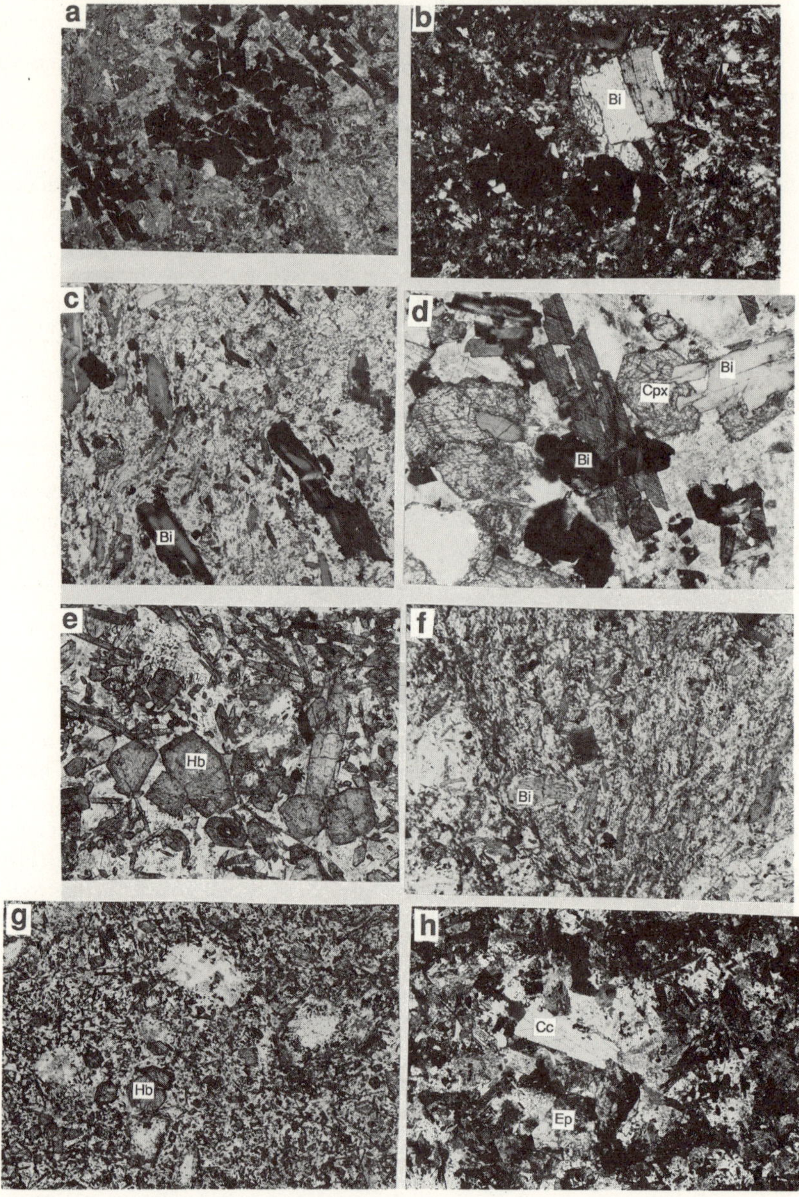

Fig.3.2 *Photomicrographs of selected lamprophyre rock-types and characteristic textures.* Bracketed numbers refer to registered specimens in the rock collections of the British ('S', 'F') and West Australian ('GSWA') Geological Surveys, the University of WA ('UWA') and the CSIRO.

Calc-alkaline lamprophyres

(a) *Minette*, Glen Affric, NW Scotland (S65134). Unusual irregular 'symplectic' micas.

(b) *Minette*, as (a) (S65176). Pseudo-hexagonal micas with dark rims, pale cores and etching.

(c) *Kersantite*, Mt.Morgans, W.Australia (GSWA 5684). Kinked and bent, strongly zoned micas.

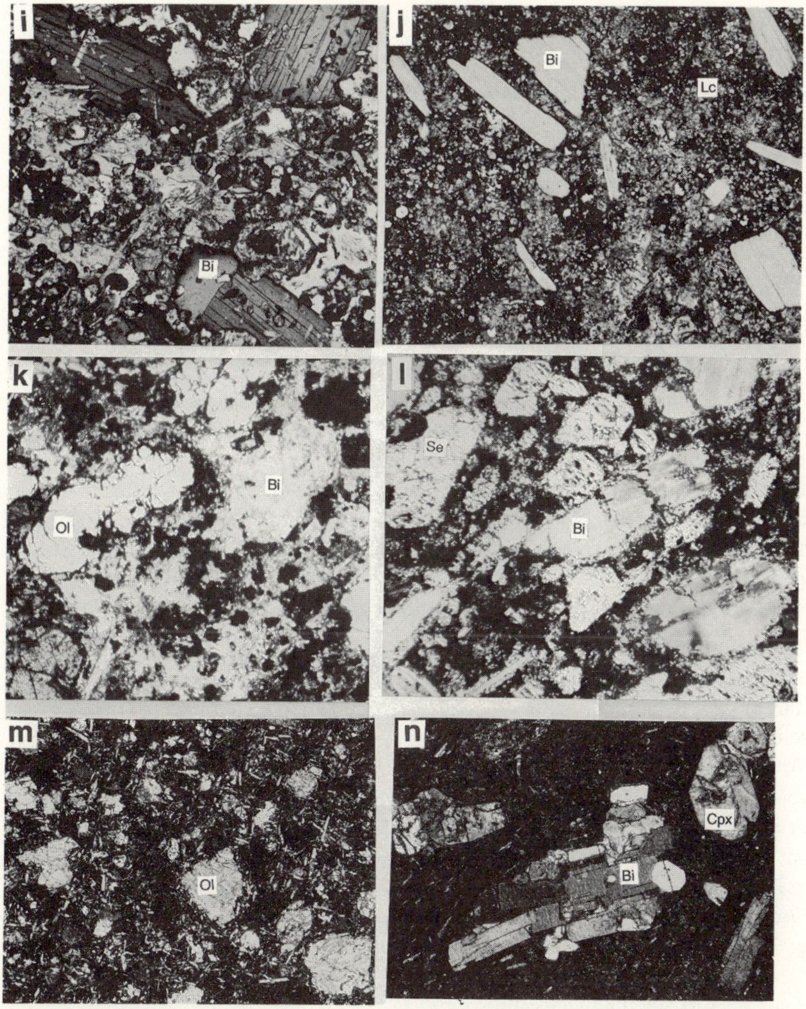

(d) *Minette*, Murrin, W.Australia (CSIRO 8012). Strongly zoned, pseudo-heaxgonal and castellated micas, pseudo-octagonal clinopyroxenes.
(e) *Spessartite*, Port William, SW Scotland (S70670). Panidiomorphic hornblende.
(f) *Sheared lamprophyre* (lamproschist), Leonora, W.Australia (UWA 103115).
(g) *Spessartite*, Port William, SW Scotland (S70509).
Type 2 (feldspathic) globular structures and panidiomorphic hornblendes.
(h) *Vogesite*, as (g) (S72338). Type 3 (carbonate-epidote) globular structures. (Crossed polars).

Lamproites
(i) *Lamproite*, W.Kimberley, W.Australia (GSWA 71366). Compare mica textures with (b), (d).
(j) *Lamproite*, as (i) (GSWA 71821). Compare mica textures with (b), (d).

Ultramafic lamprophyres
(k) *Aillikite*, type locality. Texture similar to kimberlite, except for more euhedral olivine.
(l) *Alnöite*, type locality (F2158). -do-.

Alkaline lamprophyres
(m) *Picritic monchiquite*, NW Scotland (S69862). Elongated microphenocrysts are biotite.
(n) *Hyalomonchiquite*, NW Scotland (S65107). Cumulophyric, panidiomorphic texture

3.2.3 Autometasomatic alteration

Late-stage or subsolidus alteration of primary minerals is developed to an extreme degree in lamprophyres, due to their high volatile contents. Minerals normally considered secondary (e.g. carbonates, chlorite, epidote, serpentine and zeolites), are commonly almost as abundant as magmatic phases, even in 'fresh' (i.e. unweathered) rocks (Fig.3.2h,l).

3.2.4 Globular structures

Sub-rounded, leucocratic structures — variously termed globules, ocelli, ovoids, schlieren, etc., but here grouped as *globular structures* (Appendix B) — appear in perhaps 40% of thorough lamprophyre descriptions (see summaries in Popov 1972; Zimmerle 1977; Rock 1977,1984,1986,1987a). In most cases they are also macroscopically apparent. Although similar structures do occur in basaltic rocks (Philpotts 1976), they appear to be much less widespread. In Scotland, for example, the presence of globular structures is "of the utmost service in the field for distinguishing camptonites from other (basaltic) types" (Bailey *et al.*1924). All globular structures are distinguished from their host by lower colour index, and some, usefully termed *ocelli* (Appendix B), are additionally defined by their subspherical shape and tangential biotite or hornblende (Fig.3.2g). Others, however, show no such delimitation, and are here termed *segregations*, since they grade into their host (Rock 1979; Fig.3.2h).

There are at least 3 distinct mineralogical types of globular structures in lamprophyres:

(1) *Carbonate-analcime*. These can be dominated by either mineral. Apatite or minor alkali feldspar may also occur (e.g. Carstens 1962). This type occurs commonly in UML and AL but has not been recorded in CAL, KIL or LL. Zoned structures are common, with analcime forming a central pool surrounded by carbonate.

(2) *Feldspathic* (Fig.3.2g). These are of syenitic composition, and may also carry minor quartz or feldspathoids (analcime, nepheline), amphibole, biotite or carbonates. They occur commonly in AL and more rarely in CAL, but hardly at all in UML, and they are absent from KIL and LL. Felspathoidal examples are probably sufficient (though not necessary) to identify a rock as an AL. Texturally they tend to be homogeneous.

(3) *Carbonate-chlorite/serpentine-epidote–quartz ±feldspar±zeolites* (Fig.3.2h). These occur in all types of lamprophyres. They carry a range of probably secondary phases, and are mineralogically analogous to amygdales in basalts. Internal textures vary from homogeneous to radiating-spherulitic.

3.2.5 Autoliths, lapilli, pellets and related structures

These add to globular structures in imparting gross inhomogeneity to the fabric and texture of many lamprophyres. Although they do not appear to be common in CAL, AL or LL, they are probably not much less widespread in UML than they are in KIL, and many of the features ascribed to them by R.H.Mitchell (1986, p.47ff) apply also to other lamprophyres.

4 Cognate Mineralogy

Mineral chemistry may provide the most effective way of distinguishing lamprophyres from each other and from common igneous rocks. The aim of this chapter is therefore to identify both unusual mineral species and compositional fields of common minerals which are characteristic either of lamprophyres collectively, or of particular branches. Only a broad overview can be attempted in the available space; those interested in fine mineralogical detail can refer to specific reviews such as Bergman (1987), and R.H.Mitchell (1985,1986).

This Chapter is as usual primarily descriptive, and the genetic implications of the reported compositions are not fully discussed until Chapter 8. However, because minerals completely unrelated to a host lamprophyre are obviously both misleading and irrelevant in this context, an immediate problem lies in recognizing the *cognate* phases in the first place, and it is paradoxically necessary to begin with some genetic deductions.

All the available evidence to be presented below suggests in fact that a *continuum* exists between truly cognate and truly foreign (accidental) minerals in lamprophyres. This is arbitrarily divided here into 6 types, in order of increasing 'foreign-ness' to the host:

(A) *Cognate (low-P) phenocrysts/autoliths*: these crystallized from the host lamprophyre melt, and are in equilibrium with the enclosing low-*P* groundmass (i.e. fully cognate).

(B) *Largely cognate (high-P) inclusions*: these crystallized from the host melt at greater depths, and are no longer in equilibrium with the host groundmass (largely cognate).

(C) *Semi-cognate (differentiate) inclusions:* these crystallized not exactly from the host lamprophyre melt itself, but from an equivalent formed as it evolved in some magma-chamber; they were then picked up by subsequent batches of more primitive melt ascending from below. They are out of equilibrium with the groundmass, like Type **B**, but may themselves have crystallized under either high-*P* or low-*P*, depending on where the magma-chamber was situated within the mantle or crust.

(D) *Partly cognate (related magma) inclusions:* these crystallized from a genetically related magma (e.g. carbonatite, phonolite, etc.) which, though within the same cogenetic suite as the lamprophyre, has also evolved in its own right.

(E) *Mantle-type mafic-ultramafic inclusions:* these could represent either source for the lamprophyre melt, or restite after extraction of the melt from original mantle source, or mantle material accidentally torn away during the melt's ascent (i.e. mainly foreign).

(F) *True (crustal) xenocrysts/xenoliths:* these have been torn from the lamprophyre's conduit as it rose, and have no genetic connection with the host (i.e. wholly foreign).

Present terminology is rather inadequate for this continuum, since the terms 'phenocrysts/autoliths' and 'xenocrysts/xenoliths' only strictly cover Types **A** and **F**. Table 6.1 lists the nomenclature adopted here, while Appendix B defines individual terms.

The present chapter concentrates as far as possible on Type **A** inclusions and groundmass minerals. Types **B** to **F**, covered in Chapter 6, are taken to be identified by resorption or reaction rims, by compositional, geothermometric or geobarometric disparity with groundmass phases, or by other obvious incompatibilities (e.g. orthopyroxene macrocrysts in a foid-bearing rock). However, it turns out that different parts of one grain, as well as separate grains, can sometimes belong to different Types, so that the essential arbitrariness (and occasionally the impossibility) of this division must always be borne in mind. Fortunately, wherever the distinction between these Types is most difficult (as in the case of olivine: Moore 1988), the chemical differences are commensurately slight, so that the chemical summaries and arguments presented below are robust to the adopted divisions.

The quantities of data now accumulating on the mineral chemistry of lamprophyres are staggering. Rock (1977) could locate a mere 25 analyses on which to base comparisons between alkaline lamprophyres, but LAMPDA (Appendix D) now contains some 5,000 (mainly microprobe) analyses, the vast majority published since 1984. The only lamprophyres for which data are still sparse are vogesites, kersantites and, to a lesser extent, spessartites. Clearly, a full description of such a volume of data would take a complete book, so this chapter attempts a brief synopsis by providing summaries and discussions from three, complementary, points of view:

— *Compositional fields* of each mineral are identified, along with inter-element correlations (Figs.4.1–4.11), to illustrate substitutions and to provide discriminatory diagrams.
— *Extreme compositions* of each mineral are compiled (Table 4.1), to show how 'exotic' the mineralogy of lamprophyres is, and to indicate the extent of solid solution.
— *Coexisting minerals* in representatives of each major rock-type (Fig.1.2) appear in Table 4.2, to illustrate both the more peculiar and the more 'normal' features of lamprophyres.

4.1 Mineralogical features diagnostic of the whole lamprophyre clan

Lamprophyres carry one or more of the following minerals which inextricably link them with alkaline rocks: alkali pyriboles, Ba-Ti-phlogopites, Ba-K-feldspars, feldspathoids, melanitic garnet, melilite, monticellite, perovskite and silicates of alkalis with transitional

Table 4.1 Some assumed cognate minerals of extreme composition in lamprophyres

	(4.1a) *Olivines*				**(4.1b)** *Ilmenites*			
	Max.FeO	Max.MgO	Max.CaO	Max.NiO	Max.Al_2O_3	Max.Mg/Cr	Max.MnO	Max.Nb_2O_5
Rock‡	AC	LW	AM	KU	CM	KIL?	KIL	LL
Mode†	PH	??	??	PH	??	??	GD	GD
Occ.#	Vestfold Hills	Leucite Hills	N. Timans	Ithaca	Exeter Traps	Vajra-karur	Skerring	Argyle
SiO_2	38.16	•	39.58	41.55	1.3	bdl	0.04	0.16
Al_2O_3	•	•	1.36	•	**2.9**	0.24	bdl	bdl
Fe_2O_3	•	•	1.70	•	5.4	•	•	1.95
FeO	**40.94***	6.5*	13.22	10.74*	35.1	21.7*	26.0*	41.46
MgO	19.34	**51.9**	40.16	48.87	6.2	**18.7**	0.05	1.11
CaO	bdl	0.02	**1.01**	bdl	0.1	0.34	0.04	0.02
Na_2O	•	•	0.25	•	bdl	•	•	•
K_2O	•	•	0.17	•	0.1	•	•	•
TiO_2	0.15	•	•	•	49.7	54.7	53.0	51.2
MnO	0.2	0.08	0.32	•	bdl	0.54	**11.16**	3.67
Cr_2O_3	•	•	0.01	•	bdl	**3.7**	0.08	0.04
NiO	bdl	0.32	0.30	**0.81**	•	•	bdl	•
Nb_2O_5	•	•	•	•	•	•	•	**0.90**
Total	98.78	58.82	98.97	101.97	100.8	99.92	90.37	100.51

(4.1c) *Pyroxenes*

	Max.Al_2O_3	Max.FeO*	Max.MgO	Max.Na_2O	Max.TiO_2	Max.V_2O_3	Max.Cr_2O_3
Rock	AC	AC	AC	AC	AC	LF	AN
Mode				GD			
Occ.	Mt. Megantic	Callander Bay	Vestfold Hills	Coast-parallel	Monch-ique	W. Kimberley	Bombay
SiO_2	39.57	•	57.5	52.69	45.33	52.30	43.82
Al_2O_3	**16.55**	•	0.37	1.29	3.96	0.23	3.26
Fe_2O_3	•	•	•	•	•	22.33	0.62
FeO	10.70*	**31.7***	9.80*	25.96*	7.27*	3.75	12.55
MgO	6.12	0.3	**32.32**	•	12.06	1.94	13.52
CaO	23.01	2.8	0.74	0.98	23.21	2.60	21.23
Na_2O	0.87	•	bdl	**13.87**	0.61	11.81	0.19
K_2O	•	•	bdl	•	0.01	0.16	•
TiO_2	2.24	•	0.19	4.13	**7.38**	3.80	1.78
MnO	0.22	•	bdl	0.07	0.01	0.08	bdl
V_2O_3	•	•	•	•	•	**1.05**	•
Cr_2O_3	bdl	•	bdl	•	•	bdl	**1.35**
Total	99.28	34.8	100.92	98.99	99.84	100.05	98.32

metals (e.g. priderite). Below are listed minerals and compositions which do not seem to occur in other *igneous* rocks, and can thus be used to recognize lamprophyres collectively:

(1) *High-Al pyroxenes* (Section 4.6): calcic pyroxenes with Al_2O_3 > 10% (fassaites of Deer *et al.* 1978) occur only in lamprophyres, metalimestones, lime-contaminated

(4.1d) *Amphiboles*

	Max.Na_2O	Max.K/Sr	Max.TiO_2	Max.MnO	Max.F	Max.Cr_2O_3	Max.BaO
Rock	UA	LW	AC	AC	CB	AS	UA
Mode	??	??	GD	GD	PH	PH	??
Occ.	Aillik Bay	Leucite Hills	Vestfold Hills	Taourirt	Southern Uplands	Agardag	Narssaq
SiO_2	52.38	54.6	40.21	38.13	52.48	38.75	40.94
Al_2O_3	0.11	0.45	9.87	12.56	0.98	10.90	11.27
FeO*	28.86	2.6	15.41	28.78	3.14	16.77	10.91
MgO	2.10	19.3	12.77	2.03	21.65	9.73	13.06
CaO	1.09	6.6	10.54	10.16	6.99	11.52	11.52
Na_2O	**13.23**	3.2	2.05	2.16	3.84	2.53	2.73
K_2O	0.09	**6.5**	0.75	2.62	4.93	1.65	1.02
TiO_2	0.09	4.4	**8.61**	0.12	3.28	3.17	5.37
MnO	0.01	0.06	0.19	**1.52**	0.06	•	0.15
F	•	•	•	•	**4.12**	•	•
Cr_2O_3	•	0.01	bdl	•	0.11	**1.13**	•
NiO	•	•	•	•	0.05	•	•
SrO	•	0.74	•	•	0.02	•	•
ZrO_2	•	•	•	•	0.04	•	•
BaO	•	.59	•	•	0.05	•	**0.66**
Total	97.96	99.05	100.46	98.08	100.01	96.15	97.63

(4.1e) *Nephelines*

	Max.SiO_2	Max.K_2O
Rock	AC	UL
Mode	GD	GD
Occ.	Ubekendt Ejland	Navajo
SiO_2	**49.84**	41.91
Al_2O_3	31.45	32.55
FeO*	•	2.19
MgO	•	0.31
CaO	0.53	0.16
Na_2O	14.98	14.71
K_2O	3.12	**8.01**
TiO_2	•	0.04
MnO	•	•
F	•	•
Cr_2O_3	•	•
NiO	•	•
SrO	•	•
ZrO_2	•	•
BaO	•	•
Total	99.92	99.88

(4.1f) *Phlogopite/biotites*

	Min.Al_2O_3	Max.FeO*	Max.MgO	Max.TiO_2	Max.MnO	Max.F	Max.Cl	Max.Cr_2O_3	Max.BaO
Rock	LC	LL	LJ	LL	AS	LC	CM	LC	UA
Mode	GD	GD	??	??	PH	PH	PH	??	??
Occ.	Holstein-sborg	Kapamba	Murcia	Smoky Butte	Agardag	Holstein-sborg	Exeter Traps	Coc-Pia	Narssaq
SiO_2	45.26	36.51	37.24	38.64	37.76	41.82	38.5	39.98	30.48
Al_2O_3	**0.13**	4.88	13.26	8.86	19.32	6.38	15.8	13.71	17.97
FeO*	17.82	**33.12**	3.31	6.46	24.83	8.06	10.7	4.71	4.34
MgO	19.39	4.98	**30.52**	17.03	3.06	22.72	15.5	22.36	22.25
CaO	•	0.09	•	0.04	0.15	0.03	0.09	0.02	•
Na_2O	0.09	0.52	•	0.15	0.81	0.69	0.5	0.23	0.12
K_2O	9.17	9.18	9.51	8.55	9.27	10.11	8.6	9.62	5.98
TiO_2	0.65	6.46	2.57	**12.59**	0.88	4.45	3.81	1.09	3.55
MnO	0.06	0.28	•	0.02	**2.93**	0.05	0.04	0.01	0.08
F	•	0.38	•	•	•	**5.74**	2.09	•	1.97
Cl	•	•	•	•	•	•	**0.29**	•	bdl
Cr_2O_3	0.02	bdl	•	0.05	bdl	bdl	0.04	**2.09**	•
NiO	•	0.02	•	0.10	•	0.11	bdl	0.09	•
BaO	•	0.39	•	2.37	•	0.44	0.17	0.26	**12.20**
Total	92.59	96.65	96.41	94.86	99.01	98.18	95.18	94.17	98.11

basalts or, occasionally, mantle xenoliths. The highest Al_2O_3 of 16.6% in a lamprophyre (Table 4.1c) is among the highest recorded from *any* natural pyroxene.

(2) *High-Ti pyroxenes, amphiboles and phlogopites* (Sections 4.6–4.8): pyroxenes or amphiboles with $TiO_2 > 5\%$, or phlogopites with $TiO_2 > 8\%$; the *maximum* TiO_2 contents of lamprophyre pyroxenes (7.4%), amphiboles (8.6%) and phlogopites (12.6%) are again among the highest known for these minerals (Table 4.1c,d,f; see also Figs.4.2–4.4). If high pyroxene Al_2O_3 is combined with high TiO_2 (criteria 1 and 2), then the compositions become even more uniquely diagnostically of lamprophyres.

(3) *High-Ba amphiboles, phlogopites and K-feldspars* (Sections 4.7–4.9): K-feldspars with BaO > 2%, and amphiboles with BaO > 0.5%, are diagnostic. Phlogopites with BaO > 5% occur in carbonatites (e.g. Gaspar & Wyllie 1982), a few nephelinitic rocks (e.g. Mitchell & Platt 1984), and very rarely in marbles (Bol *et al.*1989), but are far more common in UML, underlining the UML–carbonatite connection (Section 2.4.8).

(4) *High-F amphiboles and phlogopites* (Sections 4.7–4.8): maximum F in lamprophyre amphiboles reaches the very high value of 4.1%, in phlogopites 5.7% (Table 4.1d,f).

(5) *Micas rich in* Fe^{3+} (Section 4.8): micas rich in tetraferriphlogopite [$KMg_3Si_3Fe^3O_{10}(OH)_2$], though also occurring in some other peralkaline, Si-poor rocks (e.g. carbonatites), are particularly common in UML, KIL and LL; however, micas rich in ferriannite [$KFe^2{}_3Si_3Fe^3O_{10}(OH)_2$] appear to be confined to lamprophyres among igneous rocks.

(6) *Mg-Fe-Sr-Ba-carbonates* (Section 4.12): these also occur in carbonatites but are again confined to lamprophyres among *silicate* igneous rocks.

(7) *High-Zn spinels* (Section 4.13): spinels with ZnO > 2% or MnO > 2% are diagnostic.

(8) *Mg-Mn-ilmenites* (Section 4.14): those with MgO > 5% or MnO > 2% are diagnostic.

(4.1g) *Feldspars*

	Max.Na_2O	Max.CaO	Max.Fe_2O_3	Max.K_2O	Max.SrO	Max.BaO
Rock	AM	CK	CM	LW	AM	UD
Mode	GD	GD	GD	GD	GD	GD
Occ.	Corvara in Badia	Hautes-Pyrénées	Jharia	W. Kimberley	Ubekendt Ejland	Nabberu
SiO_2	49.38	47.39	62.21	64.91	60.0	50.69
Al_2O_3	30.69	32.92	14.01	18.11	20.9	22.29
Fe_2O_3	•	•	**5.91***	0.32*	0.1*	•
FeO	•	•	•	•	•	0.07
MgO	0.02	0.03	0.09	•	•	0.02
CaO	0.17	**17.38**	0.01	0.03	0.71	<0.01
Na_2O	**12.16**	2.11	0.75	0.06	4.29	0.18
K_2O	0.06	0.01	15.49	**17.03**	7.33	10.14
TiO_2	•	•	0.24	•	•	<0.01
MnO	•	•	•	•	•	<0.01
V_2O_3	•	•	•	•	•	•
Cr_2O_3	•	•	•	•	•	•
SrO	•	•	•	•	**1.46**	•
ZrO_2	•	•	•	•	•	•
BaO	•	•	•	•	4.93	**17.40**
Total	92.6	100.08	98.71	100.46	99.72	100.79

(4.1h) *Garnets*

	Max.Cr_2O_3	Max.ZrO_2	Max.MgO	Max.TiO_2
Rock	UA	UA	UL	AM
Mode	GD	GD	GD	GD
Occ.	Ile Bizard	McKellar Harbour	Los Alisos	Tugtilik
SiO_2	35.8	20.97	36.7	28.58
Al_2O_3	10.2	3.32	0.45	•
Fe_2O_3	15.1*	11.48	•	•
FeO	•	1.83	21.5*	17.22*
MgO	0.7	2.52	**11.6**	•
CaO	34.5	30.26	24.4	32.97
Na_2O	bdl	•	•	•
K_2O	bdl	•	•	•
TiO_2	1.9	9.83	0.12	**19.85**
MnO	0.1	.1	0.27	0.22
V_2O_3	•	•	•	0.09
Cr_2O_3	**2.6**	bdl	bdl	•
SrO	•	•	•	•
ZrO_2	•	**19.53**	•	•
BaO	•	•	•	•
Total	100.9	99.84	95.04	98.93

(9) *Sr-Ba-rich apatites* (Section 4.17): apatites and fluor-apatites with exceptionally high BaO and SrO are increasingly reported.

(10) *Sulphates*: Ca-Sr-Ba sulphates tend to be abundant in lamprophyres compared with other igneous rocks, and may even be relatively easily detectable in thin section.

(4.1i) *Spinels*

	Max.Al_2O_3	Max.MgO	Max.TiO_2	Max.MnO	Max.V_2O_3	Max.Cr_2O_3	Max.ZnO
Rock	LJ	AM	LJ	UA	AC	LJ	CM
Mode	??	MA	GD	??	??	IO	??
Occ.	Murcia	Ubekendt Ejland	Murcia	McKellar Harbour	Mt. Megantic	Murcia	NW Alps
SiO_2	•	0.18	•	•	4.84	•	•
Al_2O_3	**65.0**	62.3	1.32	7.97	0.98	3.04	2.98
Fe_2O_3	•	•	8.45	•	•	•	•
FeO	16.5*	13.4*	12.97	52.13*	56.54*	20.33*	23.29*
MgO	18	**22.3**	8.00	7.10	0.18	10.41	0.15
CaO	•	0.07	•	•	0.86	•	•
Na_2O	•	•	•	•	0.02	•	•
K_2O	•	•	•	•	0.04	•	•
TiO_2	0.31	0.6	**66.12**	15.23	27.24	1.23	0.90
MnO	0.14	0.04	0.15	**9.69**	0.10	0.69	3.48
V_2O_3	•	•	•	•	**1.48**	•	•
Cr_2O_3	0.13	0.4	0.33	4.01	bdl	**65.34**	56.95
NiO	•	•	•	•	0.08	•	•
ZnO	•	•	•	•	•	•	**8.74**
Total	100.08	99.29	97.34	96.13	92.36	101.04	96.49

(4.1j) *Analcimes*

	Max.CaO	Max.K_2O
Rock	AC	AC
Mode	GD	GD
Occ.	Cape Verde	Mt. Megantic
SiO_2	52.35	51.53
Al_2O_3	22.65	23.52
Fe_2O_3	•	•
FeO	0.12	1.47
MgO	•	0.78
CaO	**1.61**	0.35
Na_2O	10.86	13.00
K_2O	2.65	**3.15**
TiO_2	•	0.49
MnO	•	•
V_2O_3	•	•
Cr_2O_3	•	•
NiO	•	•
ZnO	•	•
Total	90.24	94.09

(4.1k) *Carbonates*

	Max.FeO	Max.MgO	Max.CaO	Max.SrO	Max.BaO
Rock	AC	AC	AS	AS	LL
Mode	GD	GD	??	??	OS
Occ.	Vestfold Hills	Mt. Megantic	Aillik Bay	Aillik Bay	Argyle
SiO_2	0.33	0.04	•	•	•
Al_2O_3	0.14	0.01	•	•	•
FeO*	**32.61**	3.77	0.07	0.36	bdl
MgO	21.82	**45.84**	0.03	0.02	bdl
CaO	0.10	1.66	**58.99**	55.1	50.5
Na_2O	0.38	•	•	0.04	•
K_2O	bdl	0.02	•	•	•
TiO_2	•	•	•	•	•
MnO	0.46	0.06	0.03	0.08	bdl
Cr_2O_3	•	bdl	•	•	•
SrO	•	•	1.78	**2.36**	1.29
Nb_2O_5	•	•	•	•	•
BaO	•	•	0.04	0.14	**3.93**
Total	•	•	•	•	•

(4.1l) *Perovskites*

	Max.MgO	Max.MnO	Max.Nb_2O_5
Rock	KIL?	UA	AM
Mode	GD	??	GD
Occ.	Vajra-karur	McKellar Harbour	Wandagee
SiO_2	0.75	0.59	bdl
Al_2O_3	0.32	•	0.08
FeO*	1.23	8.03	•
MgO	**4.50**	•	bdl
CaO	36.92	28.24	32.96
Na_2O	0.32	•	1.81
K_2O	0.20	•	•
TiO_2	56.47	52.73	53.52
MnO	0.10	**1.77**	0.06
Cr_2O_3	•	•	0.04
SrO	•	•	0.49
Nb_2O_5	•	2.66	**4.19**
BaO	•	•	•
Total	100.81	94.02	93.15

bdl below detection limit
* Total Fe as Fe_2O_3 or FeO
‡ See Fig.1.2 for explanation of codes
\# See appropriate table of Appendix C for further details of occurrence
† GD = groundmass; IO = inclusion in cognate olivine; MA = macrocryst (believed cognate); PH = phenocryst; OS = ocellus; OV = overgrowth; ?? = unspecified in source reference

These peculiarities are partly but not wholly linked. For example, high Ti contents contrast strikingly with near-basaltic whole-rock Ti contents (Fig.5.6a). This implies that most Ti in lamprophyres is carried in silicates, and may explain the unusual Mg-Mn-Zn-rich, relatively Ti-poor oxides as a corollary. However, high Ti in phlogopites is believed to reflect *physical* (high-temperature, high-fO_2, low-pressure) conditions (Arima & Edgar 1981), whereas high Ti in pyroxenes and amphiboles is believed to reflect *chemical* (Na-rich, Si-undersaturated) conditions, which apply only to AL and UML. This may explain why AL have high Ti in all mafic silicates, whereas CAL and LL have high-Ti phlogopites but lower-Ti pyroxenes and amphiboles (see below).

Another peculiarity of many lamprophyres is the coexistence of one or more minerals normally found in primitive, *mafic-ultramafic* rocks such as peridotites (e.g. Mg-olivine, pyroxene, phlogopite), with one or more minerals normally typical of highly evolved, *felsic* rocks such as peralkaline granites (e.g. orthoclase, albite, Na-rich pyriboles). Table 4.2 illustrates this in minette (rocks a1,a2), spessartite (a4), monchiquite (b1), sannaite (b3,b4) and damkjernite (c2). To the author's knowledge, no igneous rocks other than lamprophyres show *apparent* parageneses which encompass such strange combinations as forsterite–albite, diopside–orthoclase or phlogopite–aegirine.

4.2 Minerals and parageneses diagnostic of each lamprophyre branch

Below are listed features distinguishing each lamprophyre branch *individually* from other branches and from common rocks, but are not diagnostic of lamprophyres collectively.

Table 4.2 Coexisting mineral compositions in some individual lamprophyre rock-specimens§

(4.2a) *Calc-alkaline lamprophyres*

(4.2a1) Minette (Karamu River, Indonesia: Table C1)

Mineral‡	Cpx	Bi	Or	Sp(1)	Sp(2)
Mode‡			Gd	Gd	Gd
SiO_2	52.6	39.1	64.2	0.11	0.1
Al_2O_3	1.51	13.7	19.3	0.11	0.88
Fe_2O_3	•	•	0.48	6.37	50.2
FeO	4.11	8.04	•	54.3	34.6
MgO	17.3	20.7	0.01	1.59	1.39
CaO	21.5	0.03	0.89	0.5	0.4
Na_2O	0.39	0.59	4.17	•	•
K_2O	0.02	9.1	10.4	•	•
TiO_2	0.5	3.56	•	30.5	8.28
MnO	0.09	0.08	0.06	0.68	1.04
F	•	1.51	•	•	•
Cl	•	0.02	•	•	•
Cr_2O_3	0.43	0.15	•	0.04	0.02
NiO	•	•	•	0.06	0.07
ZnO	•	•	•	•	•
SrO	•	•	0.11	•	•
BaO	•	bdl	bdl	•	•
Total	98.45	95.94	99.62	94.26	96.98
mg	88.2	82.1			
Ab,%			37.9		

(4.2a2) Minette (NW Alps, Italy: Table C1)

Mineral‡	Cpx	Am	Bi	Or	Sp
Mode‡	??	??	??	Gd	??
SiO_2	54.57	53.6	42.04	63.74	•
Al_2O_3	0.27	0.47	12.08	17.91	3.09
Fe_2O_3	•	•	•	1.16	•
FeO	3.58	17.2	5.29	•	21.58
MgO	18.57	11.1	24.9	•	9.09
CaO	22.7	0.61	•	•	•
Na_2O	•	7.83	•	.54	•
K_2O	•	2.18	9.74	15.74	•
TiO_2	0.25	4.09	2.08	•	0.94
MnO	0.23	0.43	•	•	1.79
F	•	•	•	•	•
Cl	•	•	•	•	•
Cr_2O_3	•	•	•	•	63.96
NiO	•	•	•	•	•
ZnO	•	•	•	•	0.06
SrO	•	•	•	•	•
BaO	•	•	0.32	0.85	•
Total	100.17	97.51	96.45	99.94	100.51
mg	90.2	84.5	89.4		
Ab,%				5.0	

(4.2a3) Spessartite (Kasuga-mura, Japan: Table C1)

Mineral	Cpx	Am	Sp
Mode	Ph	Gd	??
SiO_2	52.6	43.6	•
Al_2O_3	2.47	8.34	8.98
Fe_2O_3	•	•	4.33†
FeO	8.33*	20.9*	15.3†
MgO	15.1	9.35	11.9
CaO	20.8	10.9	•
Na_2O	0.16	1.89	•
K_2O	•	0.70	•
TiO_2	0.61	1.56	0.18
MnO	0.14	0.38	0.19
Cr_2O_3	•	•	59.4
Total	100.21	97.62	100.28
mg	76.4	44.4	
Ab,%			

(4.2a4) Spessartite (Limpopo,Zimbabwe:Table C1)

Mineral	Cpx	Am	Or	Pl
Mode	??	??	Gd	Gd
SiO_2	51.45	41.69	63.04	59.49
Al_2O_3	2.03	10.49	19.45	26.34
Fe_2O_3	•	•	•	•
FeO	10.2*	16.55*	0.05*	0.17*
MgO	13.1	11.32	•	•
CaO	21.63	11.33	0.09	6.80
Na_2O	0.94	2.38	1.89	7.48
K_2O	•	2.01	13.75	0.25
TiO_2	0.35	2.54	0.01	0.02
MnO	0.55	0.35	•	•
Cr_2O_3	•	•	•	•
Total	100.25	98.66	98.28	100.53
mg	69.6	54.9		
Ab,%			17.3	49.9

(1) CAL and AL: Neither are known to carry any uniquely diagnostic minerals, but [quartz + plagioclase] and [nepheline + plagioclase] parageneses respectively distinguish them from other lamprophyres, since none of the other branches carries plagioclase (Fig.1.4).

(4.2b) *Alkaline lamprophyres*

(4.2b1) Monchiquite (Wandagee, Australia: Table C2)

Mineral Mode	Ol Ph	Cpx Ph	Am Gd	Bi ??	Af ??	Ac Gd	Sp(1) Gd	Sp(2) Gd	Sp(3) Gd
SiO_2	40.49	44.23	38.77	32.94	63.48	51.68	0.10	0.03	0.31
Al_2O_3	•	8.37	14.15	15.74	20.24	25.6	6.19	5.34	3.93
Fe_2O_3	•	•	•	•	•	•	•	47.47†	22.86†
FeO	10.6*	8.22*	11.53*	13.53*	0.87*	0.07*	21.17*	25.55†	44.81†
MgO	48.65	11.75	12.29	15.14	0.17	bdl	11.69	9.91	1.68
CaO	0.22	23.35	11.86	bdl	0.84	0.18	0.07	0.16	0.27
Na_2O	•	.57	2.15	0.55	5.14	12.76	•	•	•
K_2O	•	bdl	1.18	8.3	8.87	0.04	•	•	•
TiO_2	•	3.28	5.25	5.45	0.24	bdl	1.05	10.43	17.73
MnO	0.2	0.18	0.15	0.24	•	bdl	0.24	0.59	0.83
Cr_2O_3	0.03	bdl	bdl	bdl	•	•	59.28	0.06	6.75
NiO	0.17	bdl	bdl	bdl	•	•	0.13	0.04	0.06
Total	100.36	99.95	97.33	91.89	99.85	90.33	100.01	99.72	99.62
mg	89.1	71.8	65.5	66.6					
Ab, %					46.8				

(4.2b2) Camptonite (Mt.Megantic, Canada: Table C2)

Mineral Mode	Cpx Gd	Am Ph	Pl Gd	Sp Gd	Cb Gd
SiO_2	47.36	38.10	57.99	0.17	0.02
Al_2O_3	8.65	14.56	25.66	6.90	0.03
FeO*	5.80	12.21	0.30	70.09	7.22
MgO	13.29	11.39	•	2.47	43.62
CaO	21.07	11.70	7.52	0.16	0.94
Na_2O	0.83	2.62	6.65	•	0.07
K_2O	0.01	1.31	1.07	0.01	•
TiO_2	2.38	3.28	0.07	12.63	0.01
MnO	0.10	0.32	•	1.14	0.06
F	•	0.44	•	•	•
Cl	•	0.06	•	•	•
V_2O_3	•	•	•	0.75	bdl
Cr_2O_3	0.18	•	0.01	0.02	0.01
NiO	•	•	•	0.02	0.10
SrO	•	•	•	•	•
Nb_2O_5	•	•	•	•	•
Total	99.67	95.79	99.27	94.34	•
mg	80.3	62.4			
Ab,%			90.4		

(4.2b3) Sannaite (Aillik Bay, Canada: Table C6)

Mineral Mode	Cpx ??	Bi ??	Kf Gd	Sp ??	Cc Gd
SiO_2	45.86	36.52	63.88	bdl	•
Al_2O_3	4.30	15.09	18.63	0.87	•
FeO*	5.87	12.34	0.14	76.41	0.36
MgO	15.38	17.50	•	3.56	0.02
CaO	24.05	0.01	0.02	•	55.10
Na_2O	0.45	0.36	0.32	•	0.04
K_2O	0.02	8.82	15.48	•	•
TiO_2	2.34	2.47	•	11.93	•
MnO	0.14	0.19	•	0.77	0.08
F	•	•	•	•	•
Cl	•	•	•	•	•
V_2O_3	•	•	•	•	•
Cr_2O_3	•	•	•	bdl	•
NiO	•	•	•	•	•
SrO	•	•	•	•	2.36
Nb_2O_5	•	•	•	•	0.14
Total	98.41	93.3	98.47	95.54	•
mg	82.4	71.7			
Ab,%			3.1		

(4.2b4) Sannaite (Monchique, Portugal: Table C2)

Mineral	Ol	Cpx	Am	Bi	Ab	Ao	Ne
Mode	Ph	Ph	Ph	Gd	Gd	Gd	Gd
SiO_2	43.76	47.72	37.96	34.33	65.78	66.07	46.16
Al_2O_3	•	4.89	14.70	15.74	20.62	18.92	32.24
FeO*	14.78	6.33	10.49	19.09	0.01	0.25	0.71
MgO	39.74	12.99	12.54	9.25	•	•	•
CaO	0.36	23.60	12.54	bdl	0.76	0.15	0.80
Na_2O	•	•	2.30	0.59	10.81	5.17	16.17
K_2O	•	•	1.65	8.16	0.11	8.57	4.65
TiO_2	•	3.04	7.36	7.92	0.01	0.01	•
MnO	0.25	•	bdl	•	•	•	•
Total	98.64	98.57	99.54	95.08	98.1	99.14	100.73
mg	82.7	78.5	68.1	46.3			
Ab,%					99.3	47.8	

(2) UML: Minerals almost confined to UML (among igneous rocks) include Ti-Zr-garnet (kimseyite), high-Ba-Ti/low-Si phlogopite, ferri-diopside ($Fe_2O_3 > 2\%$), and Na-K-Fe sulphide minerals such as bartonite and rasvumite (Czamanske *et al.*1978–1981). Melilite and Mg-Mn-spinels distinguish UML from other lamprophyres (Rock 1986).

(3) KIL: Olivines are rounded and occur as macrocrysts, microphenocrysts, xenocrysts, and in the groundmass, whereas in other lamprophyres they tend to be euhedral (or resorbed) and confined to the phenocrysts (Fig.1.4). Phenocryst pyroxenes are virtually absent, and amphiboles of any description are far rarer than in other lamprophyres (Fig.1.4).

(4.2c) *Ultramafic lamprophyres*

(4.2c1) Aillikite (Los Alisos, Argentina: Table C3)

Mineral	Gt(1)	Gt(2)	Ol	Cpx	Bi	Pv	Cb	Sp(1)	Sp(2)
SiO_2	26.3	36.7	41.8	52.9	38.6	0.41	0.39	1.45	1.1
Al_2O_3	0.39	0.45	0.02	1.11	13.4	0.27	bdl	0.04	20.1
Fe_2O_3	•	•	•	•	•	•	•	55.17†	9.6†
FeO	15.5*	21.5*	12.6*	4.45*	7.27*	0.96*	0.08*	32.56†	20.1†
MgO	2.59	11.6	45.3	15.4	21.6	0.26	0.33	0.57	14.3
CaO	31.7	24.4	0.36	25	•	37.8	54.6	0.27	0.09
Na_2O	•	•	•	0.58	0.28	0.19	•	0.55	0.02
K_2O	•	•	•	•	9.77	0.06	0.07	•	•
TiO_2	18.5	0.12	0.02	1.11	3.53	58.5	0.03	5.73	6.93
MnO	0.1	0.27	0.25	0.14	0.09	0.06	•	1.57	0.49
Cr_2O_3	0.05	bdl	bdl	•	0.06	0.03	•	0.16	26.8
NiO	0.52	bdl	0.21	•	•	•	•	•	•
Total	95.65	95.04	100.56	100.69	94.6	98.54	•	98.07	99.53
mg	22.9	49.0	86.5	86.1	84.1				

(4.2c2) Damkjernite (Yogo Gulch, USA: Table C3)

Mineral Mode	Cpx(1)	Cpx(2) Ph	Am	Bi Ph	Or Gd	Sp Gd	Il
SiO_2	51.83	47.6	51.2	38.51	65.2	•	•
Al_2O_3	0.38	5.39	1.5	12.32	16.9	2.71	0.02
FeO*	24.58	5.81	20.61	16.95	0.71	79.1	41.34
MgO	3.37	13.9	9.3	13.24	0.05	3.36	0.03
CaO	5.03	25.1	0.93	•	0.07	0.1	•
Na_2O	10.15	•	9.02	0.37	0.28	•	•
K_2O	•	0.01	1.25	8.46	15.00	•	•
TiO_2	1.22	1.39	3.4	3.92	0.04	6.76	49.96
MnO	0.35	0.04	0.47	0.21	0.01	1.04	9.17
Cr_2O_3	•	0.04	•	0.03	•	1.1	0.01
Nb_2O_5	•	•	•	1.13	•	•	•
Total	96.91	99.28	97.68	95.14	98.26	94.17	100.53
mg	19.6	81.0	44.6	58.2			
Ab,%					2.8		

(4) LL: These carry a unique suite of K-Ba-Fe-titanates, including jeppeite, priderite, shcherbachovite, and transitional metal silicates such as jeppeite and wadeite, which reflect their perpotassic condition; minettes transitional to lamproites carry similarly exotic phases such as baotite (Section 4.17). Potassium titanian richterite (*sensu* IMA 1978) distinguishes LL from all other igneous rocks, including lamprophyres, while leucite distinguishes them from all except the most K-rich volcanic rocks.

These features can be made into a series of quick decision rules for classifying a rock more precisely, *once it has been recognized as a lamprophyre.* For example (cf. Fig.1.4):

— if the rock carries leucite, potassium richterite, or priderite, etc., it can only be an LL;

(4.2c3) Alnöite (Malaita, Solomon Is.: Table C3)

	Ol	Cpx	Bi	Me	Sp(2)	Sp(2)
SiO_2	40.0	54.4	39.6	42.5	•	•
Al_2O_3	bdl	0.7	14.5	6.1	36.3	0.8
Fe_2O_3*	•	•	•	•	19.9	76.1
FeO*	15.3	3.8	8.1	3.1	•	•
MgO	45.5	16.4	21.7	8.8	16.7	5.2
CaO	0.4	23.7	•	34.5	•	•
Na_2O	•	0.6	1.0	4.0	•	•
K_2O	•	•	8.8	0.2	•	•
TiO_2	•	0.7	1.9	0.1	0.6	11.0
MnO	0.1	0.1	0.1	0.1	0.3	1.0
Cr_2O_3	bdl	bdl	0.2	bdl	26.5	1.3
NiO	0.2	bdl	0.2	bdl	•	•
Total	101.5	100.4	96.1	99.4	100.3	95.4
mg	84.1	88.5	82.7			

(4.2d) *Lamproites*

(4.2d1) Cocite (Pendennis, England: Table C4) **(4.2d2) Lamproite** (Murcia, Spain: Table C4)

Min. Mode	Am	Bi	Or	Cb	Cb	Ol Ph	Cpx Ph	Am Gd	Bi	Or Gd	Sp Inc in Ol
SiO_2	51.60	38.14	63.13	•	•	40.30	53.60	53.02	39.57	64.40	•
Al_2O_3	0.30	11.39	17.27	•	•	•	0.48	1.05	9.78	16.60	0.94
Fe_2O_3	•	•	2.3*	•	•	•	•	•	1.56†	3.06*	•
FeO	18.29*	15.59*	•	26.72*	5.12*	10.70*	2.63*	5.92*	5.95†	•	49.42*
MgO	10.12	15.11	•	25.72	18.33	48.00	17.20	18.19	19.33	0.15	5.39
CaO	0.92	0.10	•	0.58	29.6	0.16	23.70	5.53	•	0.07	•
Na_2O	7.44	0.53	0.66	•	•	•	0.28	5.39	0.30	1.81	•
K_2O	3.53	10.14	15.34	•	•	•	•	3.63	10.30	14.1	•
TiO_2	3.17	4.57	•	•	•	•	0.48	6.31	8.55	0.36	2.76
MnO	0.55	0.22	•	1.09	0.18	0.23	0.04	0.10	0.10	•	0.95
F	•	•	•	•	•	•	•	1.13	2.09	•	•
Cl	•	•	•	•	•	•	•	•	0.05	•	•
Cr_2O_3	•	0.05	•	•	•	•	0.94	•	•	•	39.09
NiO	•	•	•	•	•	•	•	•	•	•	0.23
Nb_2O_5	•	0.82	1.63	•	•	•	•	•	•	•	•
Total	95.92	96.66	100.33	•	•	99.39	99.35	99.79	96.69	100.55	98.78
mg	49.7	63.3				88.9	92.1	84.6	85.3		
Ab,%			6.1							16.3	

— if the rock carries melilite or Ti-(Zr)-garnet, it can only be an UML (alnöite or polzenite);
— if the rock carries nepheline or sodalite minerals, it can only be an AL or UML;
— if the rock carries perovskite, it can only be an LL or UML (or, rarely, transitional AM);
— if the rock carries plagioclase, it can only be a camptonite, spessartite or kersantite;
— if the rock carries quartz, it can only be a CAL or LL.

4.3 Olivines (Table 4.1a; Fig.4.1)

Fo contents in 359 analyses from LAMPDA, combined with data in R.H.Mitchell (1986), indicate that KIL and LL olivines have similar *mg*, while AL, CAL and UML have equally similar, but slightly less magnesian, compositions. The overall range is $Fo_{45.8\text{-}93.5}$, although olivines with Fo < 70% are very rare indeed. NiO and CaO contents are generally 0.2–0.4%, but both may reach 0.8–1% in AL and UML olivines (Table 4.1a; Fig.4.1). Though olivines in many lamprophyres exist only as relicts, those in CAL are the most commonly pseudomorphed — by a combination of carbonates, chlorite, fibrous actinolite, montmorillonite, silica and talc commonly known as pilite (Velde 1968b). Spectacular zoned resorption rims are visible in some AL (e.g. Rock 1979, 1983a), in which the olivine is enveloped successively by magnetite, then titanaugite, kaersutite and titanbiotite.

4.4 Monticellites

Monticellite is relatively common in KIL (R.H.Mitchell 1986), and may be more common in UML than reported hitherto (Fig.1.4). Its occurrence as rims round olivine, or as a

(4.2d3) Olivine-lamproite (Prairie Creek, USA: Table C4)

Mineral Mode	Ol Ph	Cpx Gd	Am(1) Gd	Am(2) Ph	Bi Gd	Sp(1) Gd	Sp(2) Gd	Sp(3) Gd	Pv Gd
SiO_2	40.7	52.8	39.7	52.94	42.1	0.19	0.37	•	0.84
Al_2O_3	0.03	0.33	14.1	0.75	6.98	14.4	0.06	0.37	0.09
Fe_2O_3	•	•	•	•	•	11.1†	37.4†	22.56†	•
FeO	8.19*	2.87*	26.2*	3.05*	7.39*	15.5†	36.6†	24.95†	1.18*
MgO	50.1	16.7	3.3	21.86	21.8	12	3.14	4.93	•
CaO	0.05	24.3	9.84	6.55	0.03	0.06	0.06	0.14	37.4
Na_2O	•	0.52	2.37	4.01	0.74	•	•	0.04	0.53
K_2O	•	•	0.56	5.00	9.86	•	•	0.09	0.34
TiO_2	•	1.78	0.29	3.34	6.03	0.79	11.4	4.46	55.8
MnO	0.18	0.07	0.48	0.07	0.1	0.45	1.23	1.65	0.06
F	•	•	•	3.95	•	•	•	•	•
Cr_2O_3	0.03	0.25	•	bdl	0.08	44.4	11.3	37.68	0.43
NiO	0.27	0.19	0.15	0.03	•	0.37	0.05	0.08	•
Total	99.55	99.81	96.99	99.98	95.11	99.26	101.61	96.95	96.67
mg	91.6	91.2	18.3	92.8	84.0				

* Total Fe as Fe_2O_3 or FeO † Total Fe partitioned via site occupancy considerations bdl = below detection limit
‡ See Appendix A for explanation of abbreviations §For data-sources see relevant references in Appendix C

Explanation

The above are selected from ≈5,000 mineral analyses in LAMPDA (Appendix D) according to the following criteria:

— the analyzed minerals must coexist in *one rock specimen* from the particular body;
— there must be *at least four* minerals analyzed in full from the rock, including all its essential phases;
— where a particular mineral is of essentially uniform composition, or demonstrates continuous zoning, a single analysis is chosen from all those available to represent the *middle* of the range;
— where a particular mineral appears to show two or more quite separate compositions in the one rock specimen, without intergradation, then representatives of *all* these are included;

There were no sets of mineral analyses from vogesites or kersantites obeying these criteria (the maximum number of separate minerals analyses in *any* specimens of these rock-types is 3). Where there were several sets of analyses fitting the bill for a particular rock-type, the set or sets with the largest numbers of analyzed mineral species were chosen.

'replacement' matrix mineral with augite (e.g. Bowen 1922), together with consideration of systems such as Ak-Ne-Di, suggests it is often a product of reactions such as:

Melilite + olivine = monticellite + diopside *or* Olivine + melilite = monticellite + biotite

However, it may be primary in some pyroxene-free alnöites (Yoder 1979).

4.5 Garnets (Table 4.1h; Fig.4.2)

Undoubtedly cognate, primary garnets are essentially confined to UML (Fig.1.4), and are most abundant in certain "ouachitites" (e.g. Von Eckermann 1958). Minor amounts of matrix garnet have been reported from one sample of the Prairie Creek olivine-lamproite (Gogineni *et al.*1978), from a single Greenland monchiquite (Rucklidge *et al.*1980), and from the "kimberlites" of Homa Bay, Kenya (Ito 1986), but even these may either be xenocrystic, as admitted by the original authors and confirmed below, or may express host-rocks which are transitional between other lamprophyres and UML (Section 5.1.3).

Primary garnets occur as groundmass grains, mats or poikilitic pools, commonly

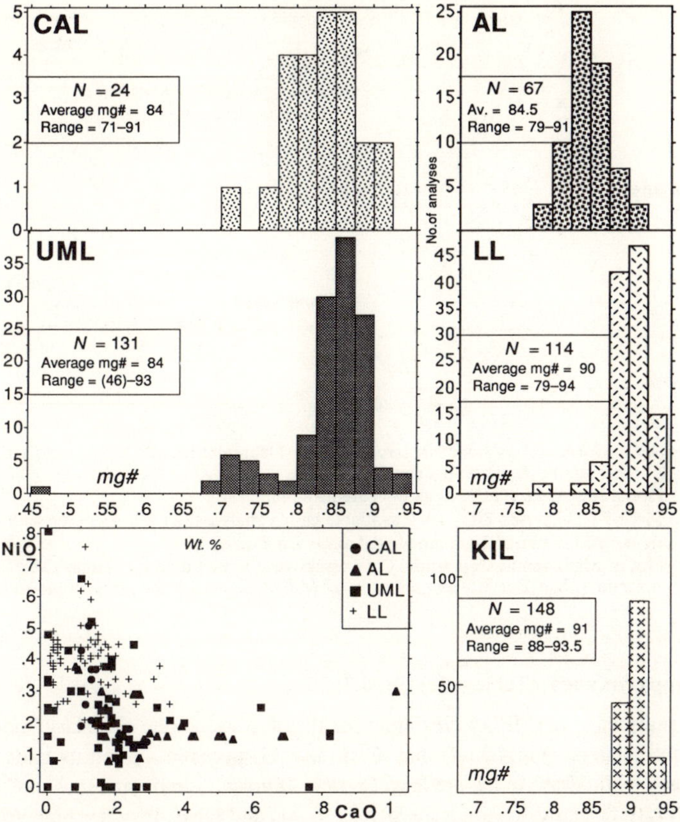

Fig.4.1 *Olivine compositions in lamprophyres. mg* = mole Mg/[Mg+Fe^{2+}]. Data from LAMPDA (Appendix D). For comparison, cognate kimberlite olivines show a range Fo_{85-93} (R.H.Mitchell 1986).

described as skeletal, mossy, etc. The 25 analyses available in LAMPDA are all Ca-rich, Mg-Mn-Cr-poor garnets, but cover three main compositions, illustrated in Tables 4.1–4.2 and Fig.4.2: (1) *andradite* with rather variable proportions of Al-Mg-Fe (Fig.4.2a); (2) *melanite* with up to 20% TiO_2; (3) *kimseyite* with up to 19.5% ZrO_2 (Fig.4.2b). A single Cr-pyrope (MgO = 20%, Cr_2O_3 = 1.9%), and two almandines (one with MnO = 3.3%), all from Prairie Creek and Homa Bay, are so different from the others that a xenocrystic origin seems probable. Three other relatively magnesian garnets (MgO > 4%) are of equally uncertain (i.e. again possibly xenocrystic) status. The remaining analyses have $Cr_2O_3 \leq$ 0.4%, except for one melanite in tuffisite breccia from Ile Bizard (Raeside & Helmstaedt 1982) with 2.6% Cr_2O_3. None of these garnets can be classified under the Dawson & Stephens (1975) scheme, devised to cover pyrope garnets in kimberlites.

In some rocks, zoning may be complex, and a wide compositional range be present in one specimen, while in others, distinct compositions may occur in a suite but *not* coexist, suggesting different origins (e.g. primary versus replacement: Meyer & Villar 1984).

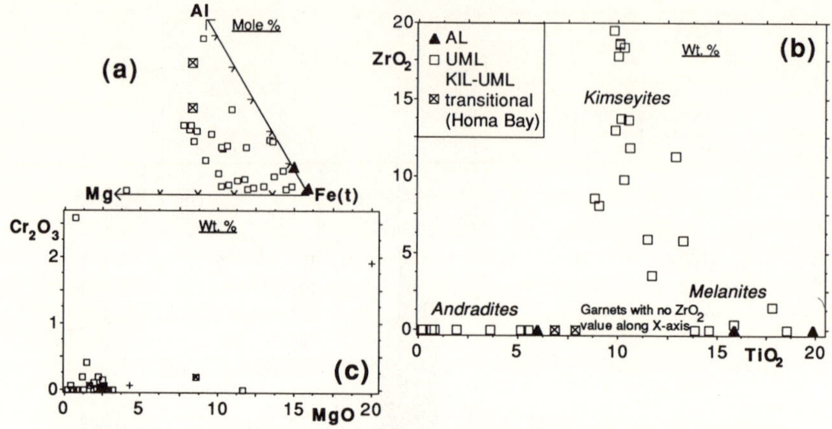

Fig.4.2 *Garnet compositions in lamprophyres.* Data from LAMPDA (Appendix D).
Note: other lamprophyres do not carry garnets.
(a) Plot of Mg-Al-Fe(t) in Ti-Zr-poor andradites (mole %).
(b) Plot of TiO_2 versus ZrO_2 in all available garnet analyses (wt.%). Analyses with no figure published for ZrO_2 are plotted along the *X*-axis.
(c) Plot of MgO versus Cr_2O_3; the CaO versus Cr_2O_3 plot used by Gurney (1984) is not appropriate as nearly all garnets in LAMPDA lie outside the range of the latter.

4.6 Clinopyroxenes (Table 4.1c; Fig.4.3)

The 1,095 analyses in LAMPDA cluster across the diopside-augite boundary in the standard pyroxene quadrilateral (Fig.4.3c–f), but CAL and AL pyroxenes trend towards less calcic compositions, while those in LL are least Fe-rich. 'Typical' clinopyroxenes in CAL and LL are thus low Al-Ti diopsidic augites, whereas in AL and UML they are high Al-Ti-salites, and reasonable discrimination can be affected (Fig.4.3a).

Ca-Na pyroxenes (sodian augite and aegirine-augite, *sensu* IMA 1988) have been reported in all lamprophyres except KIL (Fig.4.3g), but are particularly noteworthy in AL as 'green cores' to phenocrysts (see below). Na pyroxene also occurs, as groundmass grains or phenocryst rims, and reaches over 90% acmite molecule (13.9% Na_2O; Table 4.1c, Fig.4.3g). Pyroxenes with up to 1% V_2O_3 and 1.35% Cr_2O_3 have also been recorded (Table 4.1c), although the latter are much less magnesian than the 'Granny Smith' chromian diopsides found as macrocrysts in kimberlites (Schulze 1987).

Although high Fe_2O_3 is of course present in the more sodic pyroxenes, *ferrian diopsides* (*sensu* IMA 1988), have also been reported from AL (e.g. Lupanova 1934) and especially UML (Upton & Thomas 1973); the most ferrian has 9.7% Fe_2O_3 at only 0.06% Na_2O. This high Fe_2O_3 is clearly genuine: it has been checked chemically, agrees with stoichiometric calculations, and is accompanied by unusual yellowish pleochroism and inky-blue interference colours. It appears to reflect crystallisation from high temperature, oxidizing, Si-poor magmas (Upton & Thomas 1973).

Pyroxene zoning varies from slight to major, and may be very complex, notably in AL and UML (e.g. P.W.Scott 1980; Bédard 1988; Dessai *et al.*1990). Different grains even within a single thin section may variously display normal (*mg*$\downarrow$, Na$\uparrow$ from core$\rightarrow$rim), reverse (*mg*$\uparrow$, Na$\downarrow$ core$\rightarrow$rim), inverse (*mg*$\downarrow$, Na$\downarrow$ core$\rightarrow$rim), oscillatory, sector or plain irregular zoning (especially with respect to Al and Ti variation). Parallel strips or irregular patches of different compositions may also be present. These complexities mean that 6–10 analyses may be required to characterise each single grain. In addition, green (sodic augite) cores to AL pyroxenes are common, surrounded by a more magnesian titanaugite and rimmed by a still more sodic pyroxene; they are considered in Section 6.2. Amphibole overgrowths may also occur (e.g. Suzuki & Shiraki 1980).

Every *orthopyroxene* reported in CAL, AL or UML has proved apochryphal: the host "lamprophyre" is illusory (e.g. Grosser 1966, Oji & Takeshita 1970 in Table C8), or reaction rims imply xenocrystic origin (e.g. Fisera 1974; Rogers & Longshore 1960). Orthopyroxenes are most commonly reported in fortunite LL, but even here there are good grounds for considering them xenocrystic (Venturelli *et al.* 1984a; R.H.Mitchell 1985).

4.7 **Amphiboles** (Table 4.1d; Fig.4.4)

The 723 analyses in LAMPDA cover a wide range of compositions, but reasonably effective discrimination is possible (Figs.4.4a,c). The following generalizations are warranted [all amphibole names below are *sensu* IMA (1978) and Rock & Leake (1984)]:

(1) CAL: CAL amphiboles are described in most papers merely as 'brown' or 'green hornblende'. Of 137 analyses, over 90% are diverse calcic amphiboles, dominated by hastingsitic (32%), tschermakitic (14%), magnesio-hornblende (13%) and pargasitic (9%) compositions; most of these are also 'potassian' and/or 'titanian'. A further 16% are known or presumed secondary amphiboles, namely tremolite-actinolites or actinolitic hornblendes. Phenocrysts tend to show slightly higher Al, Mg/Fe and K, lower Si and estimated Fe^{3+} than groundmass phases. Sodic-calcic (katophorite/anophorite) and alkali amphiboles (arfvedsonite/riebeckite) form about 9% of available analyses; primary alkali amphiboles are largely confined to minettes (Tröger 1932; Waldmann 1935; Velde 1971a; Nemec 1973a,b; Wierzcholowski 1979), but may occur as a secondary phase in all CAL rock-types (Seymour 1900; Horne & Hinxman 1914).

(2) AL: AL amphiboles are the most restricted in IMA *name*: almost all of 333 analyses are calcic, some 65% are variants of kaersutite, 15% of pargasite and 13% of hastingsite. Many of these kaersutites are 'subsilicic,' and one enters the field of sadanagaite, the most Si-poor amphibole known, with a mere ≈5.48 Si per formula unit.

(3) UML: Amphiboles are rarer in UML, and occur only as macrocrysts in melilite-bearing UML, or in breccias (Alnö, Oka). The 68 analyses are dominated by kaersutites (47%), and magnesio-hastingsites (28%), but the remainder include diverse calcic, alkali and sodic-calcic amphiboles (eckermannites and arfvedsonites 10%; anophorites and richterites 4%). In these respects, therefore, UML amphiboles occupy an intermediate position between those of CAL and AL, although amphiboles combining high BaO with low K_2O appear to be confined to UML (Fig.4.4c).

(4) KIL: Cognate primary amphiboles are not even mentioned in R.H.Mitchell's (1986) book, and can be considered as absent from KIL, although secondary tremolite–actinolites do occur.

(5) LL: The 100 available analyses are entirely sodic-calcic (richterite 44%, katophorite 20%, winchite 1%) and alkali amphiboles (arfvedsonite 22%, riebeckite 10%, eckermannite 4%). Contents of TiO_2 > 4% at SiO_2 > 50%, F > 1.5% and K_2O > 3.5%, are diagnostic (Fig.4.4a,b,c).

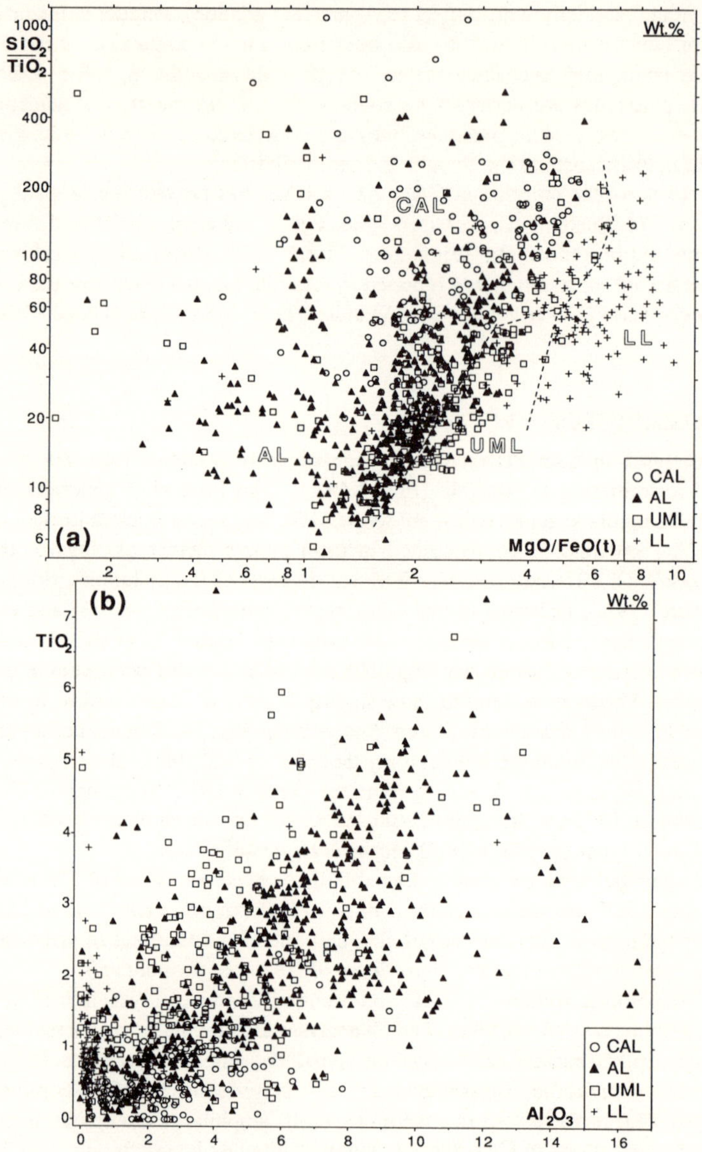

Fig.4.3 *Pyroxene compositions in lamprophyres.* Data from LAMPDA (Appendix D):

(a) Diagram allowing CAL,AL,UML & LL pyroxenes to be discriminated; note logarithmic scales; wt.% oxides used in ratios.

(b) Plot showing high Al_2O_3, TiO_2 contents achieved by many lamprophyre pyroxenes (wt.%).

(c)–(f) Plots of Na-poor pyroxenes (< 2 wt.% Na_2O) in the standard pyroxene quadrilateral (mole %). Inset shows position of larger plots near the diopside point. Fields as defined by IMA (1988).

(g) Plot of Na-rich pyroxenes (> 2 wt.% Na_2O) according to Mg-Ca-Na (mole %), showing the complete trend from aegirine-poor to aegirine-rich pyroxenes in lamprophyres.

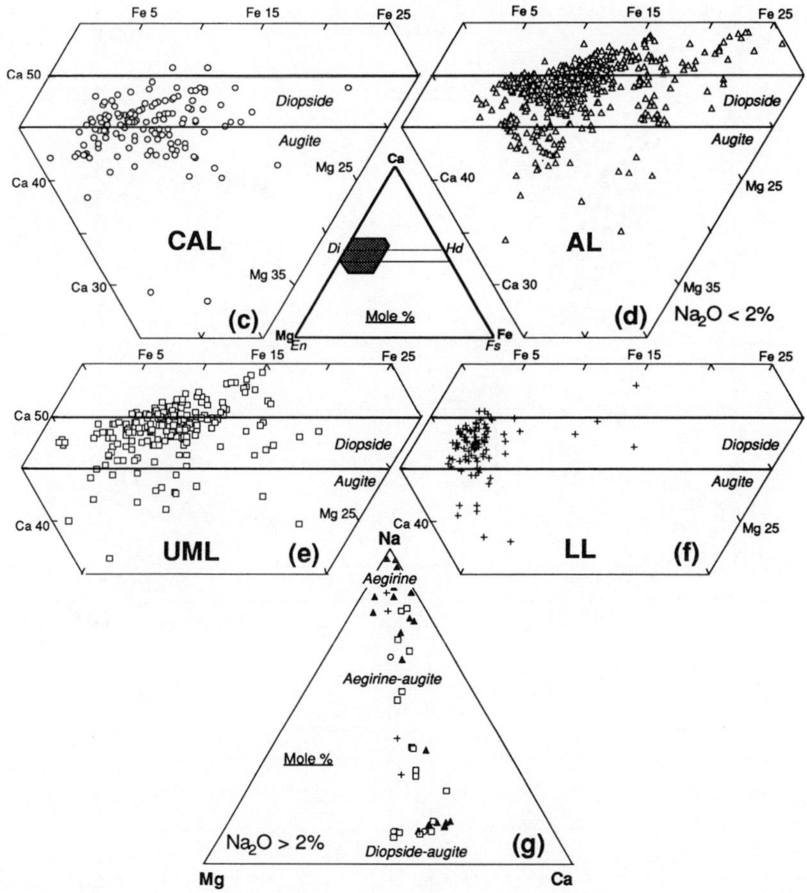

Fig 4.3 *cont'd*

Considering alternatively by *amphibole* type, further generalizations are warranted:

— *Alkali amphiboles* are common in LL, present in CAL and UML, but virtually absent from AL; *magnesio*-riebeckite/arfvedsonites are diagnostic of lamprophyres among igneous rocks, being otherwise confined to metasomatic rocks (fenites) and metamorphosed evaporites.
— *Calcic amphiboles* occur in all lamprophyres except LL.
— *Fe-Mg-Mn (cumingtonite)* and *tremolite-actinolite* occur only as secondary phases.
— *Kaersutite* (especially when 'subsilicic') is diagnostic of AL and UML.
— *Katophorite* is confined to LL among lamprophyres.
— *Sodic-calcic amphiboles* dominate LL, occur in CAL, but are rare in AL and UML.
— *Titanian potassium richterites* are diagnostic of LL.
— *Tschermakitic* amphiboles are virtually confined to CAL.
— Ti contents decrease in the general order AL ≈ LL > UML > CAL, but a large number of lamprophyre amphiboles are 'titanian' (i.e. more than ≈2.5 % TiO_2: Fig.4.4a).

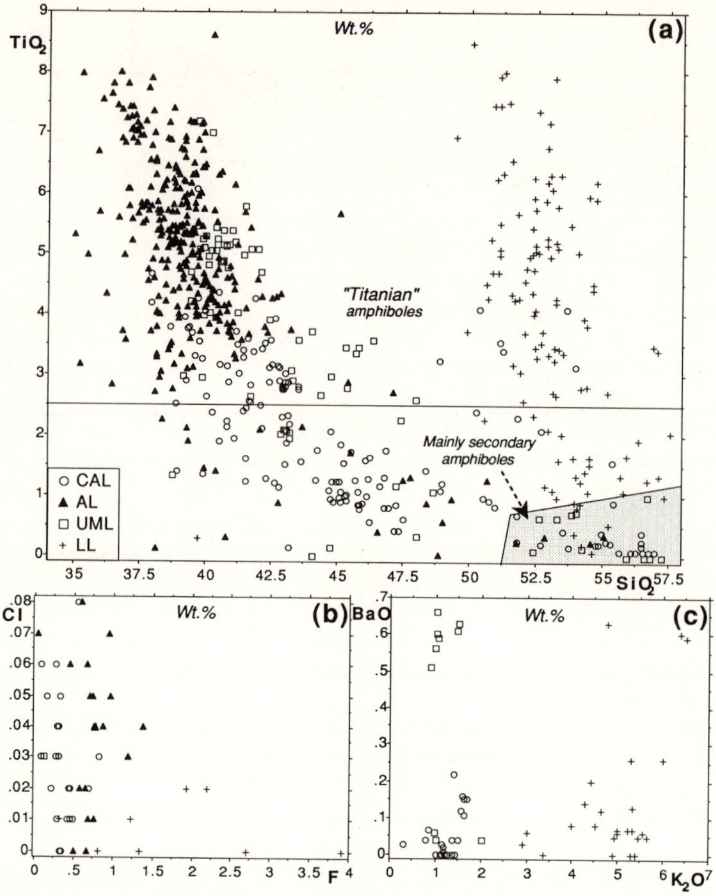

Fig.4.4 *Amphibole compositions in lamprophyres*. Data from LAMPDA (Appendix D).
(a) SiO_2–TiO_2 plot (wt.%) allowing CAL, AL, UML and LL amphiboles to be discriminated.
(b) Cl versus F covariation (wt.%).
(c) K_2O versus BaO covariation (wt.%).

As well as high TiO_2 and (in LL) K_2O, features of lamprophyre amphiboles are locally high MnO (1.5%), F (4.1), Cr_2O_3 (1.1) and BaO (0.7: Table 4.1d). However, there is little covariation between F and Cl (Fig.4.4b), or between BaO and K_2O (Fig.4.4c).

4.8 **Biotite-phlogopites** (Table 4.1f; Fig.4.5)

As with pyroxenes, a great diversity among 987 analyses in LAMPDA extends to single grains in one rock (e.g. Nixon *et al.* 1980,1986). Sufficiently detailed synopses of lamprophyre micas are available (Bachinski & Simpson 1984; R.H.Mitchell 1985,1986) to obviate full discussion here, but the following are useful generalizations:

— Phlogopites with $Al_2O_3 \approx$ 12–16% and $TiO_2 \approx$ 3–5% are common to *all* lamprophyre varieties (including KIL), and most commonly form grain cores (Fig.4.5a).

— Evolution of CAL and AL phlogopites from these compositions tends to be towards increasing TiO_2 at stable Al_2O_3, whereas LL phlogopites show decreasing Al_2O_3; R.H.Mitchell (1986) defined 3 evolutionary trends for KIL phlogopites, of which one (decreasing Al_2O_3 *and* TiO_2) is also followed by UML phlogopites (Fig.4.5a).
— UML and LL phlogopites often carry extensive tetraferriphlogopite molecule [$KMg_3Si_3Fe^3O_{10}(OH)_2$], as seen in points plotting below the phlogopite-annite join in Fig.4.5g,i; indeed, LL micas reach down to a mere 0.1% Al_2O_3 content (Table 4.1f). Like those in carbonatites, such micas commonly display reversed pleochroism, which is explained by the entry of Fe^{3+} on to Al^{VI} sites (cf. Fay & Hogarth 1969).
— Micas rich in ferriannite [$KFe^2{}_3Si_3Fe^3O_{10}(OH)_2$], which also reach low Al_2O_3 contents but at $FeO_t > 30\%$, are known from CAL, AL (Mokhtari *et al.*1985) and LL (Scott Smith *et al.*1989).
— High Ba and high-Ti micas can have Ba > K and/or Ti > Al (Table 4.1e) and appear to contain major proportions of unusual end-members such as $BaMg_3Si_2Al_2O_{10}(OH)_2$ (kinoshitalite), $KMg_3TiSi_2AlO_{10}(OH)_2$, and even an anhydrous end-member, $BaMg_2TiSi_2Al_2O_{12}$ (Bol *et al.*1989).
— Discrimination between the lamprophyre branches is rather difficult, except for the most extreme (low Al/Ti) tetraferriphlogopites, which are diagnostic of LL (Fig.4.5a,b), and the most Ba-rich, Si-poor micas, which are diagnostic of UML (Fig.4.5e).
— Micas with CaO up to 3% have been reported from CAL, AL and UML, but most examples appear to be impure, and there are no clearly authenticated Ca-bearing micas.
— Characteristic phenocryst zoning is from pale yellow cores to reddish-brown pleochroic rims, which have higher Fe, Ti, Mn, Ba and REE, lower Mg, Ni, Cr and F, and correspond to groundmass micas.
— Lamprophyre phlogopites may reach high contents of F, MnO, NiO and Cr_2O_3 (Table 4.1f; Fig.4.5c), but Cl/F ratios remain low, significant Cl contents having only been reported from CAL and (rarely) LL micas (Fig.4.5d).
— High NiO and Cr_2O_3 (Fig.4.5c) are commonly coupled with the high *mg*.

4.9 Feldspars (Table 4.1g; Fig.4.6)

Feldspars are confined to the groundmass of lamprophyres, where they may display morphologies ranging from laths to spherulitic, plumose and dendritic aggregates. Fig.4.6c–f shows that 349 available analyses in LAMPDA cover almost the entire range possible in subvolcanic rocks (except for KIL, which never carry feldspars). However, many unusual features are apparent when this range is examined in detail.

4.9.1 Plagioclases

Plagioclases more calcic than albite are present only in camptonites, kersantites and spessartites, but albite itself occurs also in sannaites and (as an accessory phase) in damkjernites. The overall range of plagioclases recorded spans $An_{0.5\text{-}90}$ (Table 4.1g; Fig.4.6c-f), and normal zoning even in individual grains can exceed 50% An content (Shannon 1920; Biljon 1940; Nemec 1975a; Suzuki & Shiraki 1980). Although 'typical' plagioclases are thus hard to define, it is at least clear (from compositional statistics and the presence of relict calcic plagioclase and secondary epidote minerals) that almost pure albite in numerous CAL, AL and UML (Hieke 1945; Phillips 1956; Nemec 1966) resulted from

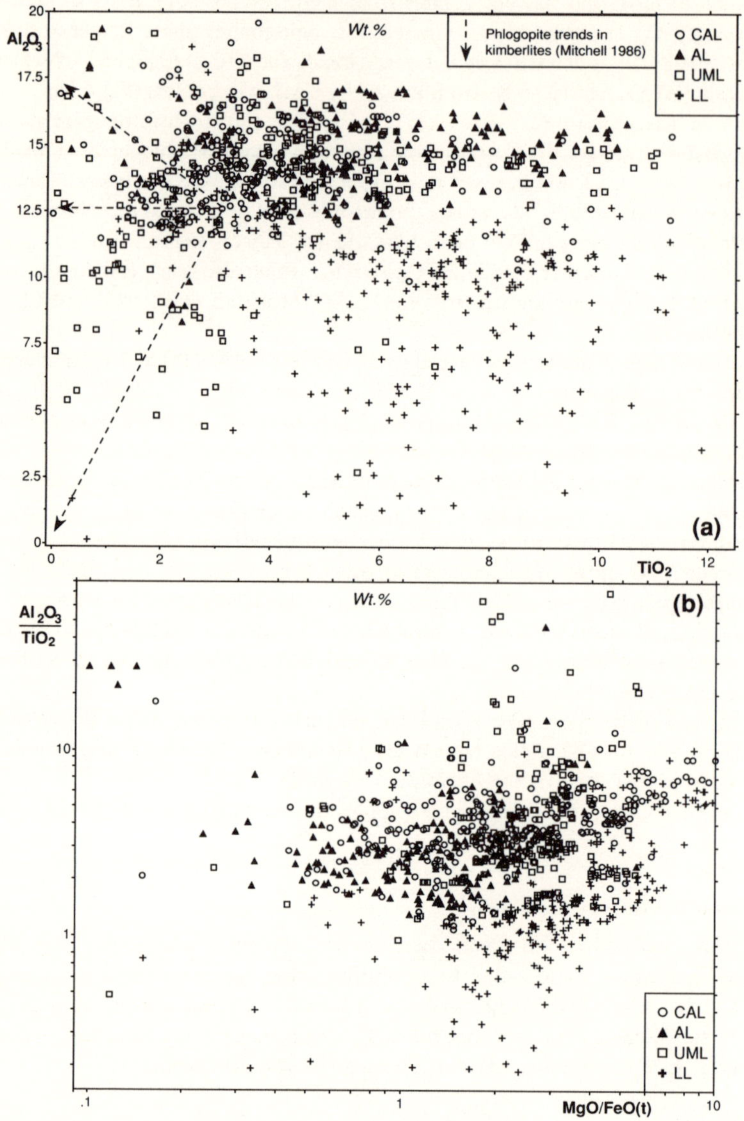

Fig.4.5 *Mica compositions in lamprophyres.* Plots (c), (d) and (e) have fewer point, since fewer analyses include values for the elements plotted. Data from LAMPDA (Appendix D).

(a) Diagram after R.H.Mitchell (1986, p.215) purporting to separate the evolutionary trends of KIL, LL and CAL phlogopites; note the overlap between UML and KIL micas (the latter shown by arrows).

(b) Diagram allowing better discrimination than (a); note logarithmic scales; wt.% oxides in ratios.

(c) NiO–Cr_2O_3 covariation (wt.%).

(d) Cl–F covariation (wt.%).

(e) BaO–SiO_2 covariation (wt.%); note logarithmic scale for BaO.

(f)–(i) Plots of CAL, AL, UML and LL micas in the Al-Mg-Fe triangle (mole %).

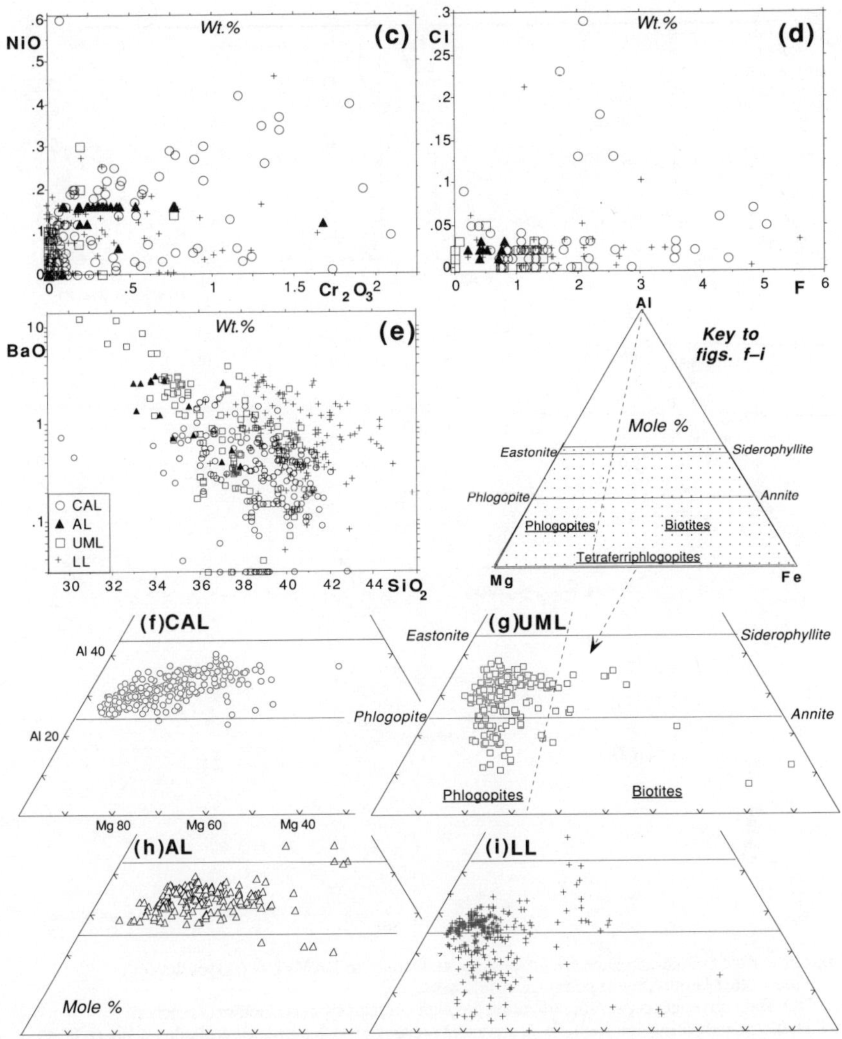

Fig. 4.5 *cont'd*

albitization of an originally more calcic plagioclase. This is particularly true of CAL in Archaean greenstone belts, which have been subjected to mesothermal alteration in gold-mineralizing systems (e.g. Perring *et al.* 1989b). Nevertheless, a few lamprophyres do carry clear, apparently primary, albite (Kennedy & Read 1936; Woodward 1970), and these pose problems for the Streckeisen (1979) nomenclature (Table 1.1): they could be called either vogesite or spessartite (if Hb > Bi), minette or kersantite (if Hb < Bi). A further problem is that the terms camptonite, kersantite and spessartite all cover rocks whose plagioclases can range from An_{100} to An_5, and so tend to mask evidence of differentiation in a way which is precluded by the nomenclature of less exotic igneous rocks.

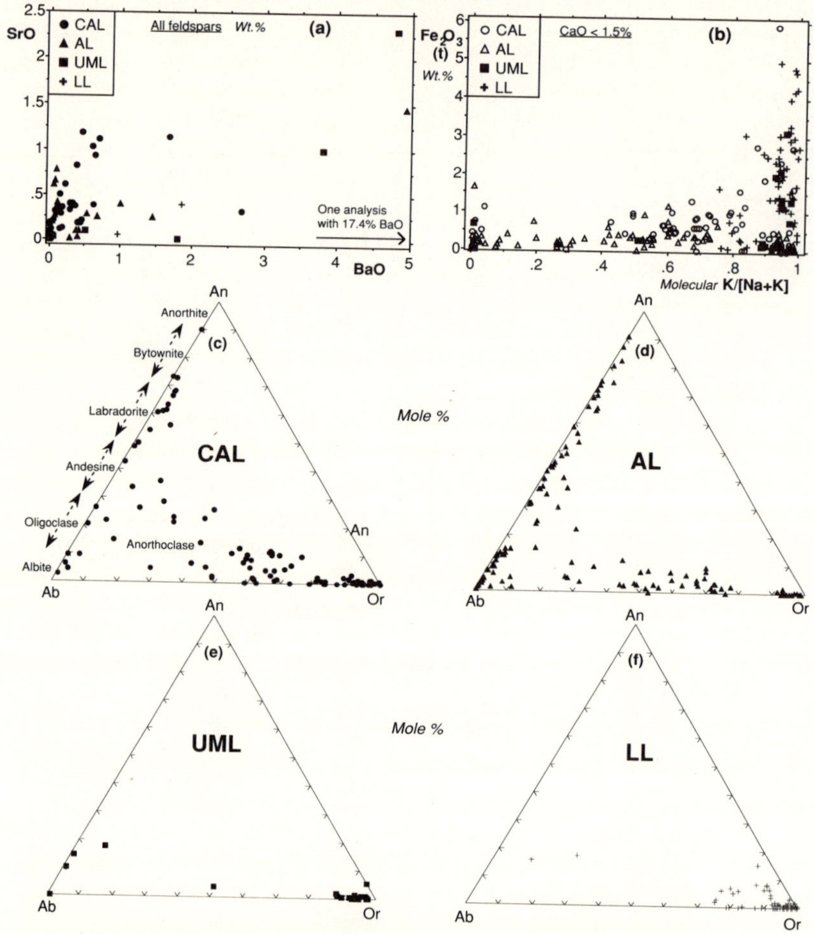

Fig.4.6 *Feldspar compositions in lamprophyres.* Data from LAMPDA (Appendix D). *Note:* other lamprophyres do not carry feldspars.
(a) SrO–BaO covariation (wt.%), indicating the high contents of these elements reached.
(b) Diagram indicating the high Fe_2O_3 contents achieved in lamprophyre alkali feldspars (CaO < 1.5%), which also allows reasonable discrimination between CAL, AL, UML and LL feldspars.
(c)–(f) Plots of CAL, AL, UML and LL feldspars in Ab-An-Or space (mole %).

4.9.2 Na-K-Ba feldspars

These are absent from KIL, UML (except damkjernites) and some LL (e.g. fitzroyites, olivine-lamproites). The range in other lamprophyres extends from essentially pure albite to pure orthoclase (Table 4.1g; Fig.4.6b–f), and to hyalophane with ≈50% celsian. Fe_2O_3 contents are highest in LL (Fig.4.6b), whereas SrO and BaO (based on limited data) reach maxima in UML followed by CAL (Fig.4.6a). Comments by branches follow, in part based on an invaluable posthumous review by Bachinski & Simpson (1986):

(1) CAL: Minettes and vogesites mostly carry a single K-feldspar, but, structurally, this can

be *microcline* (Waldmann 1935; Marco 1958; Velde 1971a), *orthoclase* (numerous reports), *sanidine* (many reports, including Ehrenberg 1979; Jones & Smith 1983), occasionally *microperthite* (Wierzcholowski 1979), or, in globular structures (Section 3.4), *anorthoclase* or *adularia.* Bulk compositions mostly range from Or_{40} to Or_{98}, with up to 1.2% SrO, 2.7% BaO (Fig.4.6a) and $KFeSi_3O_8$ (Fig.4.6b). Rarely, minettes may also carry minor, nearly pure, albite (e.g. Reynolds 1931). Although there are virtually no published data for feldspars in kersantites or spessartites, optical data indicate that 3 feldspars may coexist in one rock: Na-feldspar, K-feldspar, and intermediate plagioclase. Furthermore, what appear optically in some lamprophyres to be single-phase K-feldspars can in fact contain minute lamellae of Na-feldspar.

(2) AL: Many AL also carry 2 or even 3 alkali feldspars as separate, essentially unzoned, grains (Ab+Or, Ab+Ao as in Table 4.2, rock b4; or Ab+Or+Ao: Delor & Rock 1990), and in some camptonites plagioclase occurs as well. There are few data on structural states, but Cooper (1979) reports sanidine.

(3) UML: Damkjernites may carry accessory, nearly pure orthoclase (Table 4.2, rock c2), albite–sodic oligoclase (Griffin & Taylor 1975), or hyalophane (Table 4.1g). One intermediate Na-K feldspar is reported (Fig.4.6e) from the "anomalous lamprophyres" of Holsteinsborg (Scott 1979,1981). There are no data on structural states.

(4) LL: These carry only *one* feldspar and, apart from 2 anorthoclases (Fig.4.6f), it is universally Na-Ca-poor, but Fe-rich (Fig.4.6b). In fact, atomic Fe commonly exceeds Ca, and even exceeds Na in the most K-rich examples. In contrast to CAL, sanidine is almost always the structural modification — except in one apparently unique Antarctic occurrence, where it is intermediate-maximum microcline (Sheraton & England 1980).

4.9.3 A further note on feldspar assemblages and structural states

Overall, lamprophyre feldspars are peculiar or even bizarre in the following ways:

(1) They have exceptionally high contents of SrO, BaO and Fe_2O_3 (Fig.4.6a,b).

(2) 'Subsolvus' 2–3 feldspar assemblages commonly occur in what are undoubtedly volcanic or subvolcanic rocks. Exsolution phenomena (perthites, etc.), myrmekite and similar post-consolidation replacement textures have *not* been widely reported, so that these several feldspars all appear to have crystallized from the lamprophyre melt.

(3) Sanidine is the reported structural state even in subsolvus assemblages, yet *microcline* occurs in the chilled margins of some minette dykes (Velde 1971a).

(4) Between-grain zoning in Na-K-Ba feldspars often vastly exceeds within-grain zoning.

4.10 **Feldspathoids** (Fig.4.7)

The modal foid varies considerably between different branches (Fig.1.4):

— CAL carry occasional minor analcime only;
— AL and UML commonly carry analcime ± nepheline ± sodalite minerals, but no leucite;
— KIL carry no feldspathoids at all;
— LL may carry leucite ± minor analcime, but no nepheline.

4.10.1 Analcimes (Table 4.1j)

Analcime has been reported optically in all branches except KIL, and ranges from a few modal per cent in minettes (Velde 1971a; Bachinski & Scott 1979; Roden & Smith 1979) to a major phase in some monchiquites (Evans 1901). It occurs almost exclusively in the matrix, rarely as microphenocrysts. However, of 26 actual analyses in LAMPDA originally reported as analcime, a mere 4–5 approach the ideal formula $NaAlSi_2O_6.H_2O$: several others appear to be affected by impurities (e.g. FeO up to 2.8%), a further large group have been spoiled by the well-known propensity of analcime to decompose under the probe beam, and yet more have such high Al/Si, Al/[Na+K+Ca], Ca/Na and K/Na molecular ratios (up to 2.5, 1.6, 0.21 and 0.33 respectively), that they are almost certainly various zeolites (e.g. natrolite, phillipsite, ?gmelinite), misidentified as analcime. In analyses which approach the correct stoichiometry, maximum reported CaO and K_2O contents reach 1.61% and 3.15% respectively (Table 4.1j), although the latter analysis almost certainly has significant biotite impurities and the real maximum may be lower (≈ 2.5% K_2O?)

Experimental and textural studies of other mafic analcimic igneous rocks, such as analcimites (e.g. Velde 1975; Roux & Hamilton 1976; Rock 1977, 1979; Wilkinson 1977), have suggested that the analcime is unlikely to have crystallized directly from precursor melts. However, Roden & Smith (1979) have suggested that it may locally form high pressure microphenocrysts in minettes, and Luhr & Kyser (1989) have confirmed from a detailed study that "primary analcime is most likely to occur in lamprophyres" among *any* igneous rocks. More specifically, they have singled out microphenocrysts in the Quaternary Colima minettes, Mexico (Table C1) as representing "the strongest case for primary igneous analcime". These have high Al_2O_3 but low $\delta^{18}O$ (+9.2‰) and, unlike analcimes in other igneous rocks, do not lie on the isotopic line for meteoric water.

4.10.2 Nephelines, cancrinites and sodalite group minerals (Table 4.1e; Fig.4.7)

These are reported only from AL and UML, where modal amounts exceptionally reach 35% (Table 3.2). Analyses are available only for AL and UML, where they are richer in Na and Si than the Morozewicz composition (Fig.4.7), implying ≈1,000°C formation temperatures.

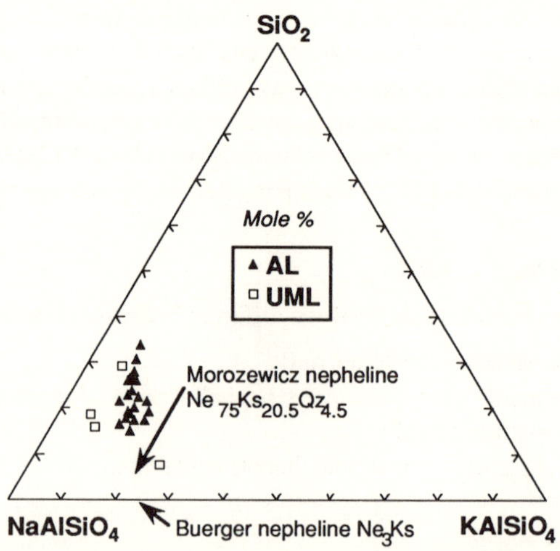

Fig.4.7 *Nepheline compositions in lamprophyres.* Plotted in Petrogeny's Residua System. Data from LAMPDA (Appendix D). *Note:* other lamprophyres do not carry nepheline.

4.10.3 *Leucites*

Leucites occur only in glass-poor LL (by definition). Analcime and even feldspathic ocelli have been repeatedly misidentified as leucite in other lamprophyres (cf. Tidmarsh 1932 with Velde 1971c; Williams 1936 with Roden & Smith 1979; Rock 1984; R.H.Mitchell 1986). The absence of modal leucite in minettes and many UML, despite high average K/(Na + K) and normative *lc* (Tables 5.1,5.3), is presumably due to occlusion of potential leucite in abundant phlogopite (phlogopite ≡ leucite + olivine + H_2O). *Pseudoleucite* has been confirmed as macrocrysts and in the matrix of some lamproites (e.g. B.H.Scott 1981).

4.11 Melilites

By definition, melilite is confined to UML (alnöites, polzenites), where modal amounts may be up to 40%. Despite a long history of reports, *melilite does not occur in KIL* (R.H.Mitchell 1986), nor does it seem to occur in amphibole-bearing UML (D.Velde, pers.comm. 1986). Its presence does, however, allow alnöites and polzenites to be readily distinguished from other lamprophyres, and indeed from most other igneous rocks. Only 8 analyses are published (Nixon *et al.*1980), which plot near the $\frac{1}{3}$Na-melilite–$\frac{2}{3}$akermanite composition typical of most igneous melilites (El Goresy & Yoder 1974). Melilite phenocrysts are not seen, because whole-rock [$Na_2O + K_2O$] contents fall below the critical limit of 7.25% (Yoder & Velde 1976), and *mg* numbers are high (Fig.5.5a).

4.12 Carbonates (Fig.4.8)

Some lamprophyres carry far more carbonate than other silicate igneous rocks, reflecting high whole-rock CO_2 (Table 5.1). Probe data and staining have begun to reveal previously unsuspected compositional variety, including not only the usually reported 'calcite', but also breunnerite, dolomite, ferrodolomite, and strontian or barian calcite with up to 2.4% SrO and 3.9% BaO (Table 4.1k).

Carbonates may occur: (a) in the groundmass; (b) in globular structures (Section 3.2); (c) as pseudomorphs after olivine, melilite and many other minerals; (d) as intergrowths

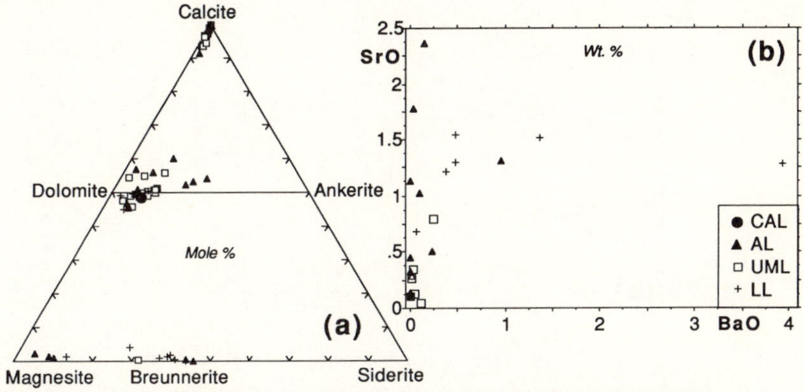

Fig.4.8 *Carbonate compositions in lamprophyres.* Data from LAMPDA (Appendix D).
(a) Plot in the simple Ca-Mg-Fe system (mole %) indicating the 3 or 4 main compositional fields.
(b) SrO–BaO covariation (wt.%), indicating the high contents of these oxides reached locally.
Note that the carbonate in kimberlites is reported by R.H.Mitchell (1986) to be calcite.

with talc, garnet, etc.; and (e) as late veins. Compositions can vary according to mode of occurrence, so that several carbonate phases may occur within a single rock (e.g. Table 4.2, rock d1). At least four possible *origins* must also be considered for these carbonates: (1) primary crystallization; (2) carbonation of primary silicates; (3) dolomitization of primary calcite; (4) dolomitization of secondary calcite (dolomitization having already been recognized in carbonatites, e.g. Le Bas 1984). Cathodoluminescence and further isotopic studies are urgently needed to clarify which origin(s) are likely to be correct, but a few general comments may be made from available data.

In general, it is the older literature which regards lamprophyre carbonates as wholly secondary (e.g. Jérémine 1927), and although many carbonate pseudomorphs and veins *are* indeed clearly secondary, recent work has also provided good evidence for the primary status of much other lamprophyre carbonate. Although textural evidence, such as euhedral

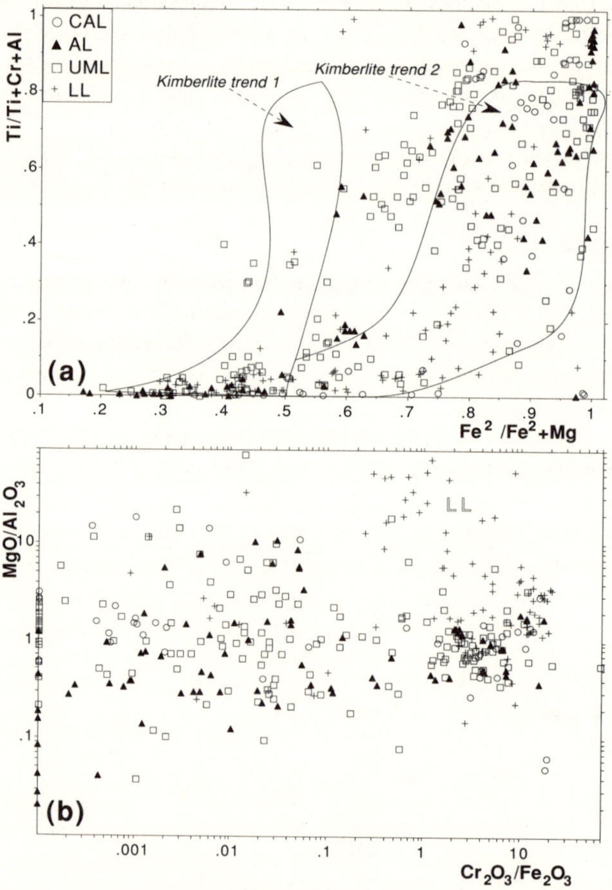

Fig.4.9 *Spinel compositions in lamprophyres.* Data from LAMPDA (Appendix D). Fe reallocated in all analyses by adjustment to 24 cations, 32 oxygens using the general method of Droop (1987).
(a) Molecular ratios used by R.H.Mitchell (1986, p.238) to distinguish kimberlite spinel trends 1 & 2.
(b) Discrimination of LL from other lamprophyre spinels using simple wt.% oxide ratios; analyses with Cr_2O_3 missing or below detection limits shown along *Y* axis.

shape or absence of replacement features (Permingeat 1954; Velde 1971a; Bachinski & Scott 1979) can be debatable, stable isotope studies (e.g. Frenzel 1971; Kirkley *et al.*1989) are now revealing 'magmatic' $\delta^{18}O$ and $\delta^{13}C$ values: in Fig.5.13, the CAL carbonate data are mainly from gold camps, where mesothermal carbonation has affected the rocks, but the range for both CAL and AL obviously extends into the fields normally considered primary. More generally, Bailey (1984) ingeniously demonstrated the primary origin of KIL calcites in a 'calcite mixing line' on a kimberlite whole-rock CaO-CO_2 plot; this implies that CO_2 is combined in kimberlites as $CaCO_3$, *not* as added CO_2. Even though such a treatment is largely precluded for other lamprophyres by: (i) the entry of CaO into melilites, plagioclases, pyroxenes, etc.; and (ii) the entry of Mg, Fe and Sr into carbonate minerals, *aillikite* whole-rock compositions still yield CaO-CO_2 correlations which are high enough to suggest that much of their CO_2 also is present as primary $CaCO_3$.

4.13 Spinels (Table 4.1i; Fig.4.9)

A detailed comparative study of lamprophyre spinels is in progress, and this section reports only preliminary observations. The 405 available analyses of spinels in LAMPDA are extremely variable, covering variously magnesian, titanian or zincian magnetites, chromites, ulvöspinels, etc. Compositional ranges overlap thoroughly between the branches, but phenocryst cores and rims, groundmass spinels, and primary and secondary spinels may all differ widely in one rock (e.g. Table 4.2, rocks a1,b1,c1,c3,d3). The following tentative generalizations are nevertheless warranted:

— Many CAL and AL spinels, together with some groundmass spinels in UML, appear to be 'common' igneous types (i.e. magnetite–ulvöspinel series).
— Some Cr-rich CAL spinels are similar to those in boninites (Suzuki & Shiraki 1980).
— Zn-rich spinels with up to 8.7% ZnO (Table 4.1h) occur in CAL, AL and UML, and are similar only to spinels associated with Fe-Ni sulphide ores.
— Spinels are relatively rare in LL, being incompatible with priderite, etc.; where they do occur, they may extend to higher Mg and Cr contents that those in other lamprophyres, overlapping with peridotite and kimberlite spinels only (Table 4.1i; Fig.4.9b).
— Although most spinels lie within R.H.Mitchell's (1986) 'trend 2', which is reportedly found in kimberlites, lamproites and related rocks, a few (mostly from UML) do lie within his 'trend 1' (Fig.4.9a), which is supposed to be unique to kimberlites.
— Subtle groupings may be indicated by Fig.4.9 in, for example, AL spinels, where there is a distinct gap in the middle of the compositional range.
— Spinels rich in *both* Mg *and* Mn are known only from UM (as in Section 4.14).

4.14 Ilmenites (Table 4.1b; Fig.4.10)

A detailed comparative study of lamprophyre ilmenites is also in progress, and this section again reports only preliminary observations. It is clear from 102 analyses available in LAMPDA, and over 1200 for KIL ilmenites, that the 'picroilmenites' formerly regarded as diagnostic of KIL are in fact widespread in UML also, averaging $\approx$ 7% MgO; it is only the most Mg-rich picroilmenites (> 13% MgO) which are confined to KIL (Fig.4.10). In most respects, UML ilmenites overlap substantially with those of KIL and show further affinity with carbonatite ilmenites. However grains approaching pure $FeTiO_3$, and relatively Mn-rich ilmenites alike, are rare in KIL relative to other lamprophyres (Fig.4.10), even though

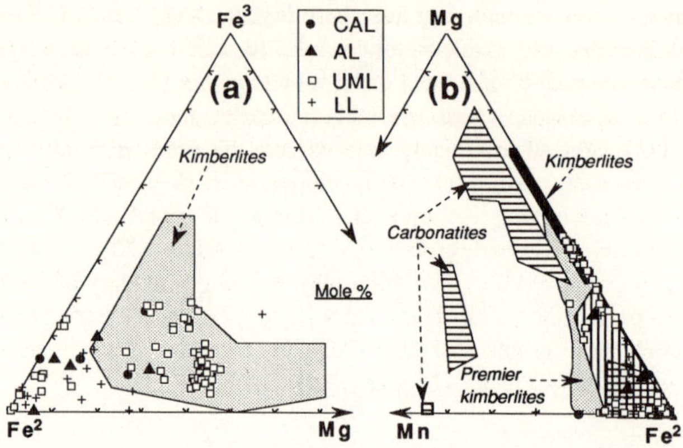

Fig.4.10 *Ilmenite compositions in lamprophyres.* Data from LAMPDA (Appendix D). Fe reallocated by adjusting to 2 cations, 3 oxygens after Droop (1987). Fields from R.H.Mitchell (1986).

the most MnO-rich lamprophyre spinel comes from a KIL (Table 4.1b). Typical contents of minor elements are as follows (for maxima see Table 4.1b): $Al_2O_3 \approx 0.35\%$, $CaO \approx 0.1\%$, $V_2O_3 < 0.1\%$, $Cr_2O_3 \approx 0.2\%$, $NiO \approx 0.05\%$, $Nb_2O_5 \approx 0.4\%$.

4.15 Perovskites (Table 4.1l; Fig.4.11)

Perovskite is essentially a mineral of UML, KIL and rocks transitional between these. It can be abundant in some madupitic LL (R.H.Mitchell 1985), but very few confirmatory data are available, and its distribution in LL as a whole is incompletely documented. It is absent from CAL and AL, although monchiquites transitional to UML may carry a few grains (Fig.1.4). Among 31 analyses of UML and LL perovskites in LAMPDA, MgO, MnO or Nb_2O_5 contents exceed those so far recorded in KIL perovskites (R.H.Mitchell 1986 records a maximum of 2% Nb_2O_5 in KIL). Perhaps even greater differences with KIL perovskites, however, lie in mostly elevated Fe and Na contents (Fig.4.11), which ally

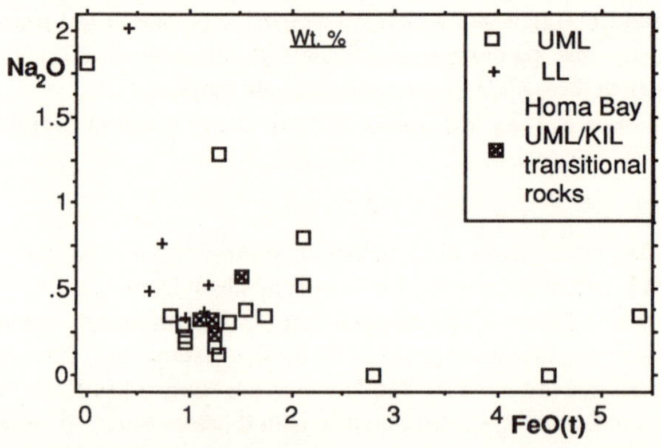

Fig.4.11 *Perovskite compositions in lamprophyres.* Data from LAMPDA (Appendix D). *Note:* other lamprophyres do not carry perovskites.

UML and LL perovskites more closely with carbonatites. However, carbonatite perovskites appear to be enriched in *both* Na and Fe, whereas UML perovskites are enriched in one *or* the other, and in this respect seem intermediate between those of KIL and carbonatites. Data on the REE content of lamprophyre (other than KIL) perovskites are unavailable.

4.16 Other oxides and sulphides

Anatase, armalcolite, baddeleyite, bravoite, chalcopyrite, cubanite, haematite, linneite, millerite, pentlandite, pseudobrookite, pyrrhotite, rutile, sphalerite, and vallerite have all been identified (Bautsch 1963; Rohde & Ullrich 1969; Bautsch & Rohde 1970; Frenzel 1971; Kresten *et al.* 1981; Rohde 1972; Velde 1975; Webb & Kerrich 1988).

Among the most unusual sulphides are alkali-rich phases conspicuous in the Coyote Peak UML, USA (Table C3): they include *bartonite* ($K_3Fe_{10}S_{14}$), *coyoteite* ($NaFe_3S_5.2H_2O$), *djerfisherite* ($K_6Na(Fe,Cu,Ni)_{24}S_{26}Cl$), *erdite* ($NaFeS_2.2H_2O$), and *rasvumite* ($KFe_2S_3$). Some of these minerals have also been found very locally in LL (Novgorodova *et al.* 1987), in ophiolites, and in mantle xenoliths from KIL (Distler *et al.* 1987), but are known otherwise only from chondrites. No reports of *cognate* grains have yet been located in any other igneous rocks, and "kimberlites" reported to contain these sulphides have proved to be misclassified UML.

4.17 Miscellaneous primary minerals

Allanite (Dal Piaz *et al.* 1979), *anhydrite* (e.g. H.G.Smith 1929), *barite* (Kaitaro 1953; Joplin 1966; Nemec 1971a, 1977b), *celestine* (Pelczar 1973), *fluorite* and *zircon* (Embey-Isztin 1972) have all been reported. Barite is a commonly overlooked accessory, reflecting whole-rock Ba measurable in % rather than ppm (Chapter 5). *Serpentine* and *zeolites* are common in the matrix and globular structures, and may be at least partly primary.

Apatite may reach essential amounts (whole-rock P_2O_5 up to 6%: e.g. Mueller *et al.* 1988), and characteristically forms phenocrysts. Lamproite apatites and fluor- apatites with up to 12.5% BaO and 5.8% SrO have been reported (Edgar 1989). Even higher SrO values in zoned lamproite apatites have been found recently (Edgar & author's unpubl.data).

Rare K-Ba-Fe-titanates, including *priderite* $(K,Ba)(Ti,Fe)_8O_{16}$ and *jeppeite* $(K,Ba)_2(Ti,Fe)_6O_{13}$, together with transitional metal silicates such as *shcherbakovite* $(Na,K)(Ba,K)Ti_2Si_4O_{14}$ and *wadeite*, $K_4Zr_2Si_6O_{18}$, appear to be virtually confined to lamproites. However, other transitional metal silicates such as *baotite* $Ba_4Ti_7NbSi_4O_{28}Cl$, have been reported in some "hyperpotassic minettes" (Nemec 1988a), which appear to be transitional to lamproites. The chemistry of all these rare minerals has been sufficiently well reviewed elsewhere (R.H.Mitchell 1985; Bergman 1987) to avoid repetition here. Some early described "rutiles" may also be misidentified priderites (cf. Prider 1939).

4.18 Secondary minerals

Autometasomatism (Section 3.2.3) leads for example to pilitic (talcose) alteration of olivine (Velde 1968b); chlorite-carbonate-tremolite alteration of pyroxenes, amphiboles and micas; albitization or saussuritization of feldspars; and analcimization of leucite. Carbonate, chlorite and epidote in particular are not only abundant in lamprophyres, but may form well-crystallized grains or aggregates, particularly in globular structures (Section 3.2.4). Textural distinction between primary and secondary phases may thus become blurred.

More unusual assumed secondary minerals include *calcium-gümbelite* (a phyllosilicate) pseudomorphing plagioclase (Frenzel 1971), *pectolite* (Davies 1952), *prehnite* plus *grossular* after mafic minerals and melilite (Saha *et al.* 1973), and *scapolite*. *Sphene* (e.g. Davies 1952) is quite common, and could be primary or secondary.

4.19 Glass

Matrix glass is absent from KIL, and has been reported from only a few CAL in small quantities (e.g. Shannon 1920; Reynolds 1931), but it occasionally reaches 20 modal % in UML (Nixon *et al.* 1980), is much more common in LL, and constitutes an essential phase by definition in monchiquites (AL). This distribution may, however, be purely an artefact of respective age distributions (Fig.2.14), for glass in such volatile-rich rocks as lamprophyres will be peculiarly susceptible to devitrification, and is unlikely therefore to survive in rocks more than a few million years old. The high proportion of glassy lamproites in particular may merely reflect the high proportion of *young* lamproites. It is significant in this respect that the youngest known (Quaternary) minettes carry matrix glass (Luhr & Kyser 1989, p.217), although glass is virtually unknown from older minettes.

No data on CAL or UML glass compositions have been located. Glasses in LL have proved difficult to analyse, and surprisingly variable in Na/K ratio. However, most appear to be evolved versions of their host-rocks — enriched in Si, Al, LILE and Fe, depleted in Mg and Ca, and with ≤90% normative *ab, or* ± *lc* (Carmichael 1967; Velde 1975; Roden & Smith 1979; Prider 1982; Jones & Smith 1983; Venturelli *et al.*1984a; Mitchell *et al.*1987). High FeO_t (3–8%) may explain the Fe-rich K-feldspars which crystallize in less glassy examples (Section 4.9.2).

The nature of the base in monchiquites has generated a confusing literature, which is discussed at length in Rock (1977, 1979, 1983b) and therefore only summarised here. Of existing monchiquite descriptions, some 45% explicitly specify the base as analcime seemingly because of its quasi-isotropic optics alone, but implicitly they follow Pirsson's (1896) influential assertion that the base in the type monchiquites is analcime. In fact, the type rocks from Monchique have an altered *glassy* base equivalent to a plagioclase-nepheline mixture (Hunter and Rosenbusch 1890; Tröger 1939; Rock 1979). Reports of plagioclase microlites in the base of other monchiquites (e.g. Flett 1900; Murthy 1948; Woodland 1962; Bloomfield and Garson 1965; Forsyth and Chisholm 1978), and of refractive indices well above that of analcime (Craig *et al.* 1911; Stillwell 1912; Flett 1935), are confirmed by recent microprobe studies which indicate *glassy* bases of *phonolitic* composition (e.g. Cooper 1979). Given experimental evidence alluded to earlier (Section 4.10.1), which makes direct crystallization of analcime from most lamprophyre melts unlikely, all this suggests that most monchiquites may initially crystallize *glassy* bases, from which analcime can undoubtedly be one of the devitrification products. Probe analyses of such analcime are in fact available in LAMPDA from 4 monchiquites.

5 Whole-rock Geochemistry

As in Chapter 4, this chapter concentrates on *descriptive* geochemistry; *interpretation* is left to Chapter 8. Table 5.1 presents averages for the various branches and rock-types (Fig.1.2), based on data for over 5,000 specimens contained in LAMPDA (Appendix D). Because coverage varies so much (e.g. some specimens are analyzed only for major elements, some only for isotopes), numbers of values for individual variables vary greatly within Table 5.1, and are therefore clearly annotated. For KIL, averages calculated from LAMPDA are basically in agreement with the consensus of published averages (e.g. Dawson 1980; R.H.Mitchell 1986; Bergman 1987), so the latter are reproduced in Table 5.1. For CAL, AL, UML and LL, the new averages differ little from arithmetic means already presented in Rock (1977,1984,1986,1987a), but are felt preferable for two reasons:

— they are in most cases based on vastly larger numbers of data-values;

— an overall figure has been chosen from a *range* of location estimates (arithmetic and geometric means, median, etc.), which should provide better estimates for the mostly non-Gaussian distributions than can arithmetic means alone (cf. Rock 1987a, 1988c).

5.1 Major and minor oxides (Table 5.1; Figs.5.1–5.5)

Judging from Table 5.1a, UML are the most Si-poor and Ca-rich of all silicate igneous rocks, and compositionally closest to carbonatites, while some LL are among the most K-rich of igneous rocks, and certainly the most K-rich among *mafic* rocks. The ultrabasic, basanitic-nephelinitic composition of AL and the shoshonitic composition of CAL are also evident. Fig.5.1 discriminates crudely between the 5 branches: CAL fall mainly in the basaltic, andesitic and trachyandesitic fields, AL in the basanite and foidite fields, UML and KIL in the foidite field, but LL in the phonolite, phonotephrite and trachyte fields. Apart from a few CAL (mainly kersantites and spessartites), all lamprophyres lie on the alkaline side of the alkaline-tholeiitic divide. Very few lamprophyres are 'sodic' *sensu* Le Bas *et al.*(1986), as the definition is somewhat severe, but the proportions emerging as 'potassic' are LL ≈ KIL > UML > CAL > AL; in other words, LL and KIL are extremely potassic, UML strongly and CAL mildly potassic, but AL sodi-potassic.

5.1.1 Discrimination between the 5 lamprophyre branches based on major elements

To obtain the best possible discrimination, we must turn to multigroup linear discriminant analysis (MDA: Le Maitre 1982; Rock 1988d), which finds simple linear functions of the oxides in multivariate space such that differences between the branches are maximized. Métais & Chayes (1964b) have already illustrated the efficacy of MDA with lamprophyres, and Table 5.2 and Fig.5.2 confirm its effectiveness with the very much larger data-sets in LAMPDA. From 3,770 major element analyses in LAMPDA, 3,508 were retrieved which have a value for all of the 9 oxides SiO_2,Al_2O_3, $Fe_2O_3{}^*$, MgO, CaO, Na_2O, K_2O, TiO_2, and P_2O_5. (To include MnO, H_2O, CO_2 or S would have meant decreasing the number of analyses substantially, because of missing data). The resulting 3,508 x 9 matrix was then subject to MDA using the respected statistical package Systat (Wilkinson 1976).

Table 5.1 New average whole-rock compositions of lamprophyre branches and rock-types

(a) *Whole branches*

	Calc-alkaline lamprophyres	Alkaline lamprophyres	Ultramafic lamprophyres	Kimberlites§	Lamproites
Major elements					
N #	*1590*	*854*	*456*	*550*	*846*
SiO_2,%	51.0	42.5	32.3	38.4	51.1
Al_2O_3	14.0	13.7	6.7	4.7	7.6
Fe_2O_3*	8.2	12.0	13.6	12.4	7.8
MgO	7.0	7.1	15.0	28.7	11.4
CaO	7.0	10.3	14.0	11.3	4.8
Na_2O	2.7	3.0	1.0	0.5	0.65
K_2O	3.1	2.0	1.9	1.4	7.3
H_2O	2.4	3.1 *(785)*	3.5 *(392)*	6.6	3.2
TiO_2	1.1	2.9	3.1	2.6	4.1
P_2O_5	0.6	0.74 *(793)*	1.0	0.9	1.3
MnO	0.13	0.20 *(782)*	0.22	0.18	0.10
CO_2	2.0 *(582)*	2.0 *(496)*	6.5 *(289)*	5.6	0.5 *(632)*
S	0.12 *(176)*	0.20 *(58)*	0.20 *(73)*	-	0.04 *(279)*
	99.4	99.7	99.0		99.9
mg◊	75.	67.	77.	90.	83.
Ox.ratio¤	35.*(1000)*	40.*(767)*	45.*(371)*	55.*(39)*	30.*(680)*
Trace elements (in order of atomic number; ppm except where indicated, to at most 3 sig.figs.)					
Li‡	44.*(121)*	24.*(32)*	20.*(2)*	29	12.*(524)*
Be‡	3.5*(61)*	1.*(3)*	3.*(2)*	1.6	6.*(6)*
F‡	1100.*(79)*	1800.*(19)*	3150.*(9)*	2774	3300.*(450)*
Sc	20.*(321)*	21.*(132)*	20.*(118)*	15	17.*(501)*
Cl ‡	105.*(64)*	330.*(12)*	350.*(8)*	202	10.*(462)*
V	170.*(591)*	285.*(291)*	250.*(161)*	120	150.*(603)*
Cr	370.*(772)*	97.*(338)*	480.*(208)*	1100	510.*(635)*
Co	36.*(406)*	38.*(99)*	75.*(53)*	77	40.*(444)*
Ni	150.*(820)*	65.*(339)*	430.*(209)*	1050	435.*(605)*
Cu	43.*(587)*	50.*(159)*	65.*(185)*	93¶	52.*(629)*
Zn	88.*(571)*	98.*(250)*	112.*(203)*	69	76.*(614)*
Ga	18.*(249)*	19.*(69)*	15.*(50)*	5.7	12.*(477)*
As‡	3.*(151)*	2.*(2)*	2.*(12)*	.	2.*(297)*
Rb	70.*(868)*	50.*(387)*	65.*(225)*	65¶	325.*(672)*
Sr	715.*(833)*	990.*(375)*	950.*(230)*	740	1250.*(661)*
Y	23.*(687)*	31.*(285)*	26.*(172)*	22	20.*(631)*
Zr	190.*(812)*	313.*(270)*	311.*(206)*	250	1160.*(637)*
Nb	13.*(655)*	101.*(234)*	120.*(212)*	110¶	135.*(583)*
Mo‡	2.*(98)*	8.5*(4)*	2.*(8)*	1.7	0.5*(334)*
Ru‡ (ppb)	1.2*(4)*	.	.	6.*(1)*	3.*(2)*
Rh‡ (ppb)	0.3*(4)*	.	.	7.*(7)*	.
Pd‡ (ppb)	4.2*(26)*	.	.	15.*(21)*	7.*(3)*
Sn‡	2.*(66)*	.	.	5.4	9.*(421)*
Sb‡	0.6*(84)*	.	0.6*(14)*	.	0.4*(8)*
Cs‡	3.*(155)*	2.*(33)*	3.*(35)*	2.3	7.*(41)*
Ba	1050.*(836)*	930.*(270)*	1100.*(221)*	1000	7700.*(645)*

(Table 5.1a contd).

La	53.(452)	66.(186)	125.(75)	150	310.(614)
Ce	110.(513)	125.(175)	230.(82)	200	470.(616)
Pr	11.(36)	14.(42)	29.(13)	22	35.(25)
Nd	56.(292)	54.(120)	95.(51)	85	157.(237)
Sm	10.5(230)	10.8(115)	18.(61)	13	23.(101)
Eu	3.1(199)	3.1(113)	5.0(57)	3.0	5.0(81)
Gd	11.(3)	8.2(60)	13.(24)	8.0	12.7(28)
Tb	1.1(178)	1.2(69)	1.547)	1.0	1.6(48)
Dy	3.7(48)	5.4(42)	7.9(25)	.	6.0(27).
Ho	0.9(61)	0.90(45)	1.3(19)	0.55	1.0(24)
Er	1.6(36)	2.7(41)	3.2(12)	1.45	2.3(27)
Tm	0.24(39)	0.38(16)	0.4(3)	0.23	.
Yb	1.8(199)	1.8(107)	1.9(45)	1.2¶	1.6(82)
Lu	0.26(182)	0.29(110)	0.27(51)	0.16	0.22(57)
Hf	5.2(157)	6.9(72)	6.5(31)	7	31.(60)
Ta	0.9(131)	5.0(47)	9.5(20)	9	6.3(59)
W‡	2.7(47)	2.5(5)	5.(6)	.	.
Os‡ (ppb)	0.27(4)	.	.	6.(13)	.
Ir‡ (ppb)	0.6(4)	.	.	5.(21)	5.(4)
Pt‡ (ppb)	3.0(26)	.	.	150.(12)	4.(3)
Au‡ (ppb)	6.(155)	22.(10)	12.(19)	3.(71)	21.(5)
Pb	13.(537)	7.(58)	7.(65)	15.3	50.(530)
Th	9.(607)	9.(113)	10.(134)	16¶	32.(601)
U	3.(435)	2.2(68)	5.(71)	3.1	2.(526)

Percentages of sodic ($Na_2O - 4 > K_2O$) *and potassic* ($K_2O > Na_2O$) *analyses* (Le Bas *et al.* 1986)

Sodic†	2%	3%	<1%	0%	0%
Potassic‡	53%	23%	78%	96%	97%

Isotopic averages

$^{87}Sr/^{86}Sr_{initial}$	0.7068(72)	0.7046(54)	0.7046(32)	0.704 & 0.708	0.709(154)
ε_{Nd}	−1.2(56)	+3.0(20)	+3.8(11)	+1.6(13)	−13(84)
$^{206}Pb/^{204}Pb$	18.5(29)	19.6(8)	18.9(6)	18.95(12)	17.5(39)
$^{207}Pb/^{204}Pb$	15.68	15.66	15.55	15.62	15.63
$^{208}Pb/^{204}Pb$	38.8	39.3	38.6	38.84	38.1
$\delta^{18}O$ (‰)$_{WR}$	+9	+7.5	—	Variable	—
$\delta^{18}O$ (‰)$_{CB}$	+11	—	+17.5	Variable	—
$\delta^{13}C$ (‰)$_{CB}$	−4	—	−4	Variable	

Molecular % $Fe^3/[Fe^3+ Fe^2]$

Molecular % $Mg/[Mg+ Fe^2]$, with molecular % $Fe^3/[Fe^3+ Fe^2]$ set to average lamprohyre (0.45)

Number of records (i.e. analysed rocks), ignoring missing data for individual elements; where numbers of data-values applicable to individual elements are substantially fewer, they are given with individual entries thus: *(45)*.

Averages based on a range of location estimates (arithmetic, geometric and harmonic means, median, MODE, etc; Rock 1987c). In most cases, either the arithmetic mean approximates the median, so that a value between them is chosen; or the median lies between the geometric and arithmetic means, so that a value close to the median is chosen.

Quality of data doubtful to varying degrees: inter-laboratory differences appear to be significant to very substantial and numbers of values are relatively small.

¶ Averages for kimberlites represent consensus between tabulations of R.H.Mitchell (1986) and Bergman (1987); values marked ¶ differ slightly (but not greatly) from averages estimated using data in LAMPDA (Appendix D).

no data or data inadequate (inconsistent or too few) to establish reliable average.

	(b) *Alkaline lamprophyres*			**(c)** *Ultramafic lamprophyres*				
	Camptonites	Monchiquites	Sannaites	Aillikites	Alnöites	Damkjernites	Ouachitites	Polzenites
N#	*456*	*239*	*37*	*232*	*69*	*45*	*36*	*21*
SiO_2,%	43.3	41.2	42.5	29.5	32.0	35.0	36.5	33.7
Al_2O_3	14.2	13.3	13.9	5.4	8.3	7.2	9.1	8.7
Fe_2O_3*	11.7	12.0	11.8	14.8	12.3	13.6	13.2	12.3
MgO	6.4	8.2	7.0	15.9	14.4	12.0	11.4	15.4
CaO	9.9	11.0	9.9	13.5	16.1	13.3	13.4	15.9
Na_2O	3.0	3.0	4.0	0.45	1.5	1.6	1.7	2.5
K_2O	2.1	2.0	2.3	1.6	2.2	2.2	2.1	1.9
H_2O	3.1 *(422)*	3.3 *(208)*	2.9	3.8	3.0	3.6	3.2	3.8
TiO_2	2.9	2.7	2.7	3.4	2.5	3.7	2.8	2.4
P_2O_5	0.70	0.84 *(210)*	0.75	0.82	1.5	1.1	1.1	1.2
MnO	0.19*(402)*	0.20 *(219)*	0.20	0.25	0.20	0.20	0.19	0.25
CO_2	2.0 *(259)*	1.5 *(159)*	1.0 *(18)*	10.5*(137)*	5.0 *(53)*	6.0 *(19)*	4.5	1.5 *(17)*
S	0.20 *(17)*	0.06 *(9)*	[0.22 *(1)*]	0.20*(40)*	0.15 *(15)*	.	0.22 *(9)*	0.40 *(3)*
	99.7	99.3	99.2	100.1	99.2	99.5	99.4	100.0
Li‡	27.*(29)*	16.*(3)*	.	.	.	.	.	.
Be‡	1.3*(3)*	.	.	.	.	.	.	.
F‡	1280.*(11)*	2150.*(6)*	.	5600.*(3)*	2475.*(3)*	.	.	.
Sc	20.*(56)*	21.*(41)*	28.*(5)*	20.*(96)*	22.*(4)*	23.*(5)*	17.*(6)*	19.*(2)*
Cl‡	1150.*(4)*	650.*(2)*	.	60.*(3)*	325.*(3)*	.	.	.
V	320.*(178)*	280.*(58)*	245.*(14)*	270.*(98)*	280.*(3)*	265.*(26)*	235.*(10)*	.
Cr	80.*(186)*	240.*(64)*	175.*(11)*	450.*(132)*	575.*(10)*	600.*(22)*	263.*(10)*	.
Co	37.*(59)*	35.*(35)*	60.*(5)*	90.*(23)*	65.*(6)*	77.*(5)*	.	40.*(2)*
Ni	50.*(191)*	170.*(57)*	110.*(12)*	450.*(132)*	400.*(10)*	280.*(25)*	300.*(10)*	.
Cu	65.*(35)*	68.*(31)*	25.*(12)*	70.*(120)*	52.*(6)*	70.*(26)*	56.*(10)*	.
Zn	90.*(118)*	82.*(35)*	105.*(14)*	105.*(132)*	120.*(8)*	120.*(26)*	146.*(10)*	195.*(2)*
Ga	22.*(18)*	15.*(15)*	16.*(10)*	10.*(23)*	.	15.*(17)*	21.*(6)*	.
As‡	2.*(2)*	.	.	3.*(7)*	.	.	.	.
Rb	56.*(215)*	46.*(64)*	61.*(17)*	60.*(139)*	70.*(13)*	75.*(26)*	55.*(11)*	60.*(3)*
Sr	1050.*(212)*	1100.*(55)*	1630.*(17)*	950.*(147)*	1400.*(11)*	1300.*(26)*	1300.*(11)*	.
Y	31.*(184)*	29.*(43)*	45.*(16)*	25.*(99)*	40.*(6)*	46.*(26)*	33.*(10)*	.
Zr	300.*(202)*	270.*(61)*	400.*(17)*	315.*(127)*	245.*(12)*	465.*(26)*	350.*(11)*	.
Nb	97.*(147)*	93.*(25)*	135.*(17)*	100.*(139)*	125.*(12)*	130.*(22)*	117.*(11)*	.
Mo‡	.	8.5*(4)*	.	1.*(3)*	.	.	.	.
Sn‡	.	.	.	.	.	.	.	.
Sb‡	.	.	.	0.4*(8)*	.	.	.	0.5*(2)*
Cs‡	4.*(11)*	1.5*(17)*	13.*(5)*	2.*(23)*	2.*(2)*	22.*(5)*	.	1.5*(2)*
Ba	850.*(106)*	1025.*(58)*	1250.*(15)*	1180.*(132)*	1700.*(13)*	1100.*(26)*	950.*(11)*	1390.*(3)*
La	60.*(96)*	86.*(52)*	.	140.*(38)*	115.*(16)*	86.*(9)*	106.*(9)*	.
Ce	110.*(93)*	160.*(45)*	.	270.*(45)*	225.*(16)*	160.*(9)*	220.*(9)*	.
Pr	10.*(16)*	18.*(22)*	.	47.*(5)*	24.*(6)*	.	13.*(2)*	.
Nd	49.*(70)*	63.*(42)*	.	104.*(26)*	88.*(16)*	.	66.*(6)*	.
Sm	10.6*(68)*	12.*(41)*	.	20.*(33)*	16.*(16)*	.	18.*(9)*	.
Eu	3.1*(67)*	3.2*(40)*	.	5.1*(31)*	4.2*(15)*	.	6.1*(8)*	.
Gd	8.3*(38)*	8.0*(20)*	.	12.*(10)*	10.*(8)*	.	24.*(4)*	.
Tb	1.2*(35)*	1.1*(29)*	.	1.5*(24)*	1.5*(14)*	.	2.1*(8)*	.
Dy	5.4*(22)*	5.3*(18)*	.	10.*(12)*	6.8*(10)*	.	.	.
Ho	0.93*(18)*	0.82*(24)*	.	1.1*(8)*	1.0*(6)*	.	3.1*(5)*	.
Er	2.6*(20)*	2.6*(19)*	.	4.5*(5)*	2.8*(7)*	.	.	.
Tm	0.41*(6)*	0.3*(8)*	.	.	0.4*(3)*	.	.	.
Yb	1.9*(59)*	1.8*(41)*	.	1.6*(20)*	1.9*(14)*	.	1.9*(8)*	.
Lu	0.28*(64)*	0.29*(40)*	.	0.26*(29)*	0.28*(11)*	.	0.29*(8)*	.
Hf	7.1*(47)*	5.8*(23)*	8.*(2)*	5.4*(23)*	6.5*(3)*	.	.	14.*(2)*
Ta	4.8*(28)*	5.6*(19)*	.	10.*(12)*	6.4*(3)*	.	.	14.*(2)*
W‡	2.4*(5)*	.	.	.	3.5*(2)*	.	.	.
Pb	8.5*(14)*	7.*(8)*	12.*(5)*	7.*(43)*	.	11.*(5)*	8.*(9)*	.
Th	8.*(41)*	9.*(26)*	16.*(14)*	8.*(88)*	13.*(7)*	11.*(21)*	14.*(9)*	23.*(3)*
U	2.*(15)*	1.5*(12)*	17.*(10)*	4.*(33)*	4.*(6)*	16.*(5)*	3.*(8)*	5.*(3)*

	(d) *Lamproites*		(e) *Calc-alkaline lamprophyres*					
	Olivine-lamproites	Cocites	Minettes	Vogesites	Kersantites	Spessartites	Appinites	Kentallenites
N #	*160*	*120*	*319*	*34*	*185*	*327*	*143*	*60*
SiO_2,%	42.2	49.8	51.5	50.8	51.2	52.2	48.3	49.1
Al_2O_3	3.8	9.0	12.8	14.9	14.8	15.1	13.8	10.7
Fe_2O_3*	9.0	6.6	7.3	8.6	8.4	8.4	9.2	9.6
MgO	25.0	9.3	7.1	5.9	6.6	6.8	8.0	14.8
CaO	4.9	5.4	6.7	6.8	6.2	6.9	9.8	8.1
Na_2O	0.38	1.6	2.0	3.0	2.8	3.4	2.8	2.4
K_2O	3.7	6.8	5.6	3.8	3.4	2.1	1.8	2.4
H_2O	6.0	2.5	2.1	2.4	2.6	2.4	2.2	1.7
TiO_2	3.2	2.0	1.3	1.2	1.2	1.1	1.1	0.64
P_2O_5	1.3	1.0	0.95	0.70	0.60	0.38	0.45	0.30
MnO	0.13	0.09	0.12	0.14	0.11	0.14	0.16	0.15
CO_2	0.5 *(142)*	5.0 *(74)*	2.0 *(144)*	1.3 *(17)*	1.2 *(102)*	0.8 *(95)*	2.4 *(66)*	1.0 *(6)*
S	0.04 *(113)*	[0.02 *(1)*]	0.07 *(40)*	[0.26 *(2)*]	0.17 *(8)*	0.10 *(20)*	0.12 *(39)*	.
	100.1	99.1	99.5	99.8	99.3	99.8	100.1	100.9
Li	11.*(120)*	40.*(28)*	34.*(39)*	.	85.*(19)*	36.*(32)*	40.*(11)*	.
Be‡	.	.	5.*(21)*	.	5.*(10)*	2.(11)	1.2*(4)*	.
F‡	4300.*(117)*	.	2400.*(16)*	.	1400.*(13)*	770.*(22)*	410.*(4)*	.
Sc	19.*(127)*		16.*(21)*	18.*(98)*	.	24.*(18)*	18.*(69)*	.
Cl‡	110.*(117)*	.	300.*(15)*	.	125.*(2)*	60.*(10)*	50.*(4)*	.
V	75.*(125)*	235.*(76)*	165.*(83)*	.	173.*(31)*	170.*(128)*	250.*(66)*	.
Cr	1060.*(139)*	475.*(70)*	360.*(129)*	.	440.*(67)*	329.*(157)*	290.*(78)*	.
Co	70.*(119)*	26.*(25)*	37.*(86)*	.	31.*(52)*	34.*(63)*	44.*(26)*	.
Ni	960.*(131)*	290.*(68)*	200.*(113)*	.	150.*(67)*	90.*(176)*	85.*(72)*	490.*(59)*
Cu	51.*(125)*	36.*(76)*	50.*(79)*	.	40.*(34)*	37.*(121)*	54.*(65)*	73.*(4)*
Zn	70.*(125)*	83.*(76)*	120.*(73)*	.	109.*(16)*	88.*(119)*	90.*(72)*	85.*(4)*
Ga	4.*(115)*	.	20.*(39)*	.	21.*(24)*	14.*(20)*	17.*(39)*	11.*(3)*
As	2.*(26)*	.	4.*(25)*	.	1.*(3)*	3.*(36)*	7.*(8)*	.
Rb	380.*(146)*	210.*(68)*	193.*(159)*	.	140.*(51)*	50.*(180)*	45.*(81)*	61.*(60)*
Sr	1240.*(140)*	1440.*(69)*	950.*(149)*	.	665.*(35)*	650.*(172)*	730.*(88)*	710.*(60)*
Y	16.*(126)*	25.*(74)*	25.*(130)*	.	26.*(16)*	19.*(149)*	20.*(76)*	.
Zr	880.*(134)*	660.*(76)*	300.*(144)*	.	270.*(38)*	148.*(162)*	150.*(86)*	84.*(60)*
Nb	174.*(133)*	34.*(28)*	19.*(141)*	.	19.*(18)*	8.*(144)*	10.*(45)*	.
Mo‡	0.7*(107)*	.	6.*(12)*	.	.	3.*(10)*	2.*(7)*	.
Sn‡	7.*(108)*	.	7.*(19)*	.	11.*(8)*	2.*(20)*	2.*(6)*	.
Sb‡	0.4*(7)*	.		.				.
Cs	6.5*(14)*	.	2.*(35)*	.	6.*(12)*	4.*(27)*	.	.
Ba	6600.*(142)*	3000.*(65)*	1800.*(159)*	.	1650.*(51)*	900.*(165)*	640.*(75)*	735.*(60)*
La	340.*(123)*	300.*(56)*	86.*(138)*	.	71.*(18)*	35.*(100)*	42.*(37)*	.
Ce	600.*(123)*	400.*(58)*	176.*(144)*	.	145.*(20)*	80.*(121)*	83.*(36)*	.
Pr	.	38.*(7)*	24.*(7)*	.	.	6.8*(8)*	25.*(11)*	.
Nd	95.*(23)*	158.*(14)*	100.*(96)*	.	67.*(14)*	30.*(14)*	32.*(37)*	.
Sm	14.*(11)*	28.*(12)*	18.*(80)*	.	9.*(9)*	7.5*(21)*	8.*(36)*	.
Eu	2.9*(6)*	5.6*(10)*	4.4*(77)*	.	2.5*(9)*	2.2*(18)*	3.3*(11)*	.
Gd	.	16.*(6)*	12.*(15)*	.	.	4.5*(8)*	9.5*(11)*	.
Tb	.	2.2*(3)*	1.5*(66)*	.	0.9*(9)*	0.48*(9)*	4.0*(11)*	.
Dy	.	7.7*(8)*	5.2*(22)*	.	.	2.5*(8)*	2.9*(4)*	.
Ho	.	1.3*(6)*	1.4*(30)*	.	.	0.45*(8)*	0.8*(11)*	.
Er	.	3.0*(6)*	2.4*(7)*	.	.	1.2*(8)*	2.0*(11)*	.
Tm	.	.	0.44*(9)*	.	.	0.19*(8)*	0.18*(4)*	.
Yb	1.0*(7)*	2.1*(10)*	1.9*(77)*	.	2.2*(9)*	1.7*(18)*	1.8*(11)*	.
Lu	0.18*(6)*	0.32*(8)*	0.27*(60)*	.	0.39*(9)*	0.34*(18)*	0.18*(11)*	.
Hf	17.*(15)*	17.*(3)*	11.*(51)*	.	7.2*(9)*	1.7*(8)*	2.1*(4)*	.
Ta	8.*(15)*	2.0*(3)*	1.3*(54)*	.	0.84*(9)*	.	.	.
W‡	.	.	2.*(18)*	.	.	.	.	.
Pb	45.*(121)*	.	24.*(58)*	.	24.*(40)*	10.*(129)*	10.*(75)*	.
Th	47.*(137)*	108.*(18)*	26.*(105)*	.	19.*(24)*	6.*(131)*	7.*(69)*	.
U	2.*(129)*	27.*(3)*	6.*(74)*	.	5.*(18)*	2.*(66)*	3.*(70)*	.

MDA yields 4 discriminant functions (DF), simple linear functions of the variables:

$$DF_1 = w_{11}SiO_2 + w_{12}Al_2O_3 + w_{13}Fe_2O_3 + w_{14}MgO + w_{15}CaO + w_{16}Na_2O + w_{17}K_2O + w_{18}TiO_2 + w_{19}P_2O_5$$

and similarly for F_{2i} through F_{4i}, where the w are simply weights, as given in Table 5.2a.

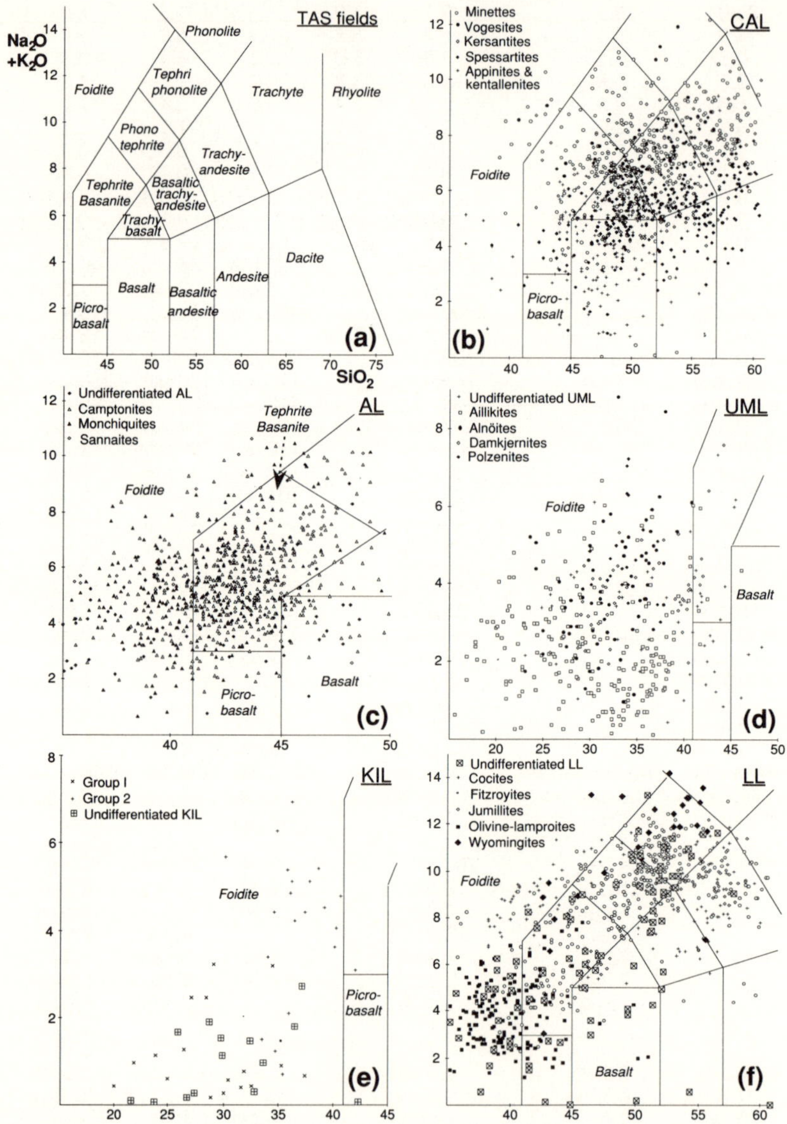

Fig.5.1 *Classification of 3,770 major element analyses of lamprophyres in the TAS diagram.* Diagram recommended by the IUGS (Le Bas *et al.*1986). Based on raw data without recalculation to 100% free of H_2O and CO_2, since otherwise anomalously high SiO_2 results for UML and KIL. Because many lamprophyres violate the criteria of < 0.5% CO_2 and < 2% H_2O suggested for use of the diagram, comparison with the IUGS-defined fields in **(a)** should be made with great caution. In **(b)**, undifferentiated CAL are omitted to avoid clutter. Data from LAMPDA (Appendix D).

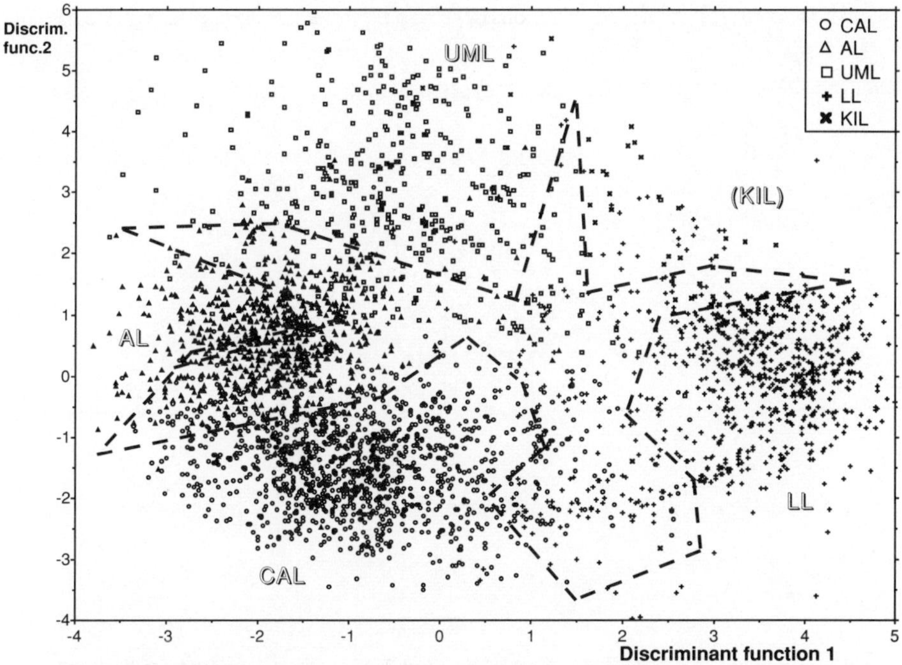

Fig.5.2 *Best-possible discrimination between the 5 lamprophyre branches using major oxides.* Obtained via multigroup discriminant analysis (MDA). For more detailed results see text and Table 5.2. The star-like distribution shows how the 5 branches are related in multidimensional chemical space.

For each lamprophyre analyzed, a value is calculated for DF_1 to DF_4, and hence the average value (centroid) for each of the branches. From this, it is possible to view the overlap in multidimensional space between the groups, and to calculate the distance between each individual analysis and each of these 5 centroids. This leads to a probability that a given analysis belongs to CAL, AL, UML, KIL or LL, and the greatest probability (nearest centroid) is then chosen as the 'calculated' group membership.

To obtain maximum consistency, MDA was actually performed on the analyses in LAMPDA in a multi-stage procedure, detailed in Appendix D. Table 5.2a compares the final predicted groups (calculated from MDA) with the observed groups (assigned mineralogically), and shows for example that 90% of the 1,327 CAL analyses are correctly assigned by MDA: in other words, given values for these 9 oxides *alone* (without any petrographical or other information), it is possible to classify 9 out of 10 analyses of CAL correctly. The classification efficiencies for AL are slightly higher still, and for UML and KIL only slightly lower (values along diagonals in Table 5.2a). Remembering that 'observed' branches rely largely on authors' descriptions in the literature (Appendix D), this remarkable result implies that 4 of the 5 lamprophyre branches are chemically distinct.

Various other tests for the efficacy of the MDA were calculated but, because they are dependent on statistical assumptions such as multivariate Gaussian distributions and equality of dispersion matrices, are more questionable with the present data. Nevertheless, test statistics such as Wilks' Λ, the Pillai and Hotelling-Lawley traces (analogous to

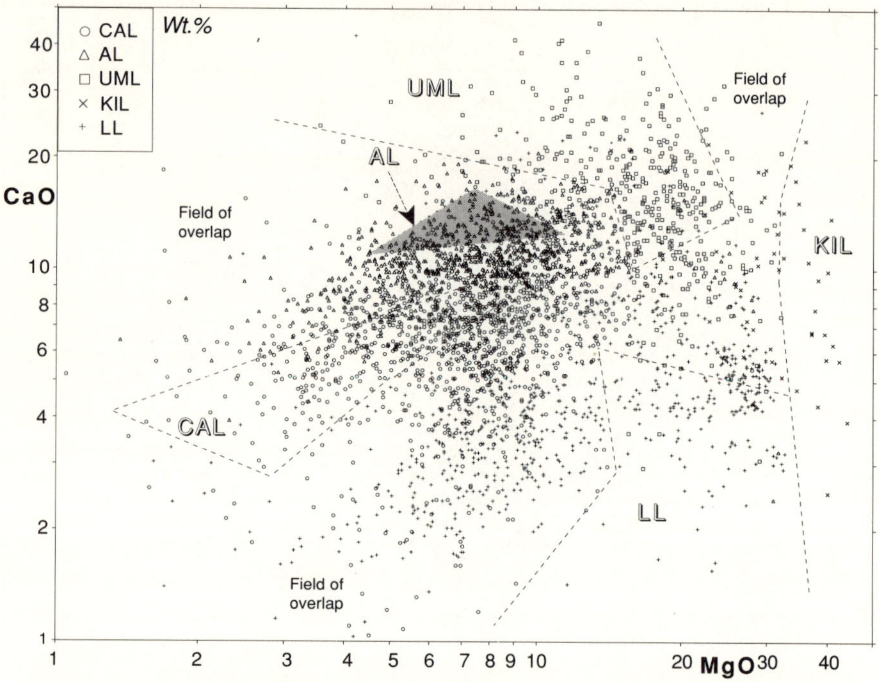

Fig.5.3 *Simple discrimination between the 5 lamprophyre branches using major oxides.* Analyses recalculated to 100% free of H_2O, CO_2 and MnO prior to plotting. Note logarithmic scales. If an analysis lies within one of the dotted fields, then it can be assigned to that lamprophyre branch with ≈90% confidence. The central areas exhibit too much overlap to be used effectively in classification. No other simple bivariate plots have been found to provide better discrimination.

univariate *t*-tests), confirm that the overall separation is highly meaningful, while univariate *F* tests indicate that the each of the 9 oxides contributes significantly to the discrimination (i.e. no individual oxide could be removed without harming the overall efficiency). Again, χ^2 tests indicate that each of the 4 discriminant functions is statistically meaningful.

As these 4 discriminant functions carry successively less of the variance, a plot of DF_1–DF_2 (Fig.5.2) provides the most useful visual check of the separation, projecting from 9 dimensions into 2. The 5 branches form a radial arrangement in which there is a field diagnostic of each branch, and an area of overlap with the adjacent ones. Fig.5.2 provides a firm mathematical basis for Fig.8.7, and provides yet more incontravertible evidence for including both lamproites and kimberlites within the lamprophyre clan.

Although plots based on two oxides or oxide ratios cannot intrinsically provide such a high separation efficiency as MDA, they are certainly simpler to comprehend and to use on an everyday basis. Figs.5.3–5.4 therefore provide two such plots, based on the oxides contributing most to the above discriminant functions. For example, Fe and Si have the highest (but opposite sign) canonical loadings on DF_1 in Table 5.2a, and so are ratioed to provide one axis on Fig.5.4. The resulting plots still allow a given analysis to be classified with reasonable confidence, although the fields of overlap (and consequently the numbers of 'unknowns' which cannot be unambiguously assigned) are larger than on Fig.5.2. In general, Fig.5.4 is to be preferred to Fig.5.3, since, by using ratios, it overcomes some of the inherent statistical problems in closed percentage data (Rock 1988d), and can be used not only for complete analyses totalling ≈100%, but also with *reliable* partial analyses (e.g. those without LOI or CO_2 determinations), which necessarily fall short of 100%.

5.1.2 Overlap between branches: the effects of accumulation, fractionation and alteration

The groups of misclassified analyses in Table 5.2a throw further light on the nature and extent of overlap in Figs.5.2–5.4. Perhaps the most interesting group is LL, correctly assigned in only 73% of cases. The reason for this is the almost total overlap between olivine-lamproites and Group II kimberlites (Fig.1.2), which results in many of the former being misclassified as the latter. For example, the Prairie Creek pipe (Table C4) is an olivine-lamproite on the present definitions, yet shows mineralogical features transitional towards KIL, as confirmed by MDA, which classifies 18 of 28 available analyses from the pipe as KIL. Dawson's (1987, 1989) suggestion of a continuum might even be extended into a suggestion that olivine-lamproites and Group II kimberlites are really the same thing!

Of the 60 LL misclassified as CAL in Table 5.2a, 42 are from the Murcia province, Spain (Table C4), and comprise about 30% of the 142 analyses available from that region.

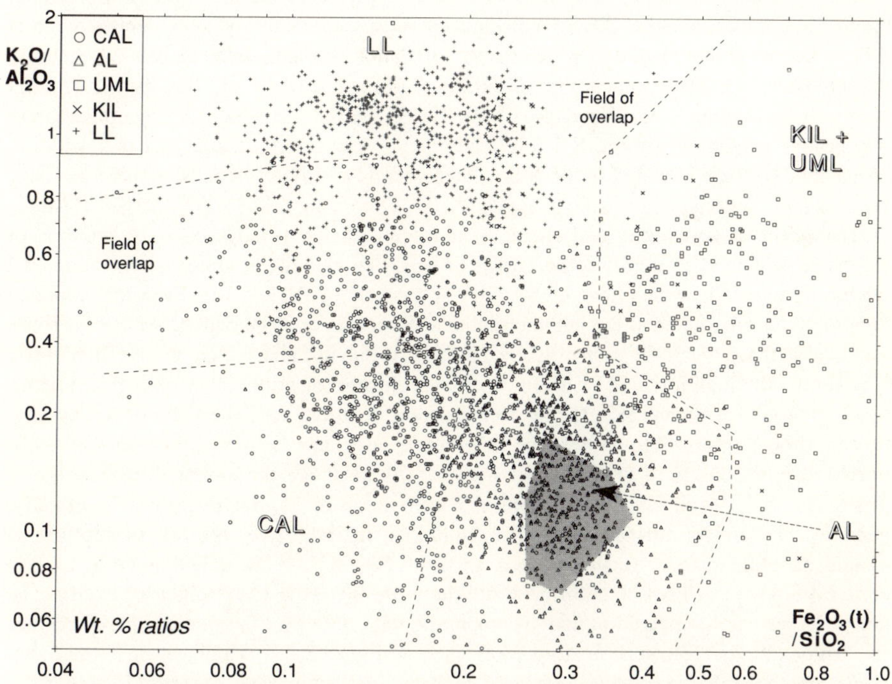

Fig.5.4 *Simple discrimination between the 5 lamprophyre branches using major oxide ratios (wt.%).* Note logarithmic scales. If an analysis lies within one of the fields outlined by dotted lines, then it can be assigned to that lamprophyre branch with ≈ 90% confidence. The central areas exhibit too much overlap to be used effectively in classification. No other simple ratio-ratio plots have been found to provide more effective discrimination. This plot is more convenient than Fig.5.3 for use with incomplete or interlaboratory data, as it does not require recalculation prior to plotting.

They illustrate that some of what have conventionally been termed "lamproites" are in fact no different in major element chemistry from ordinary minettes; this is confirmed by the general absence of diagnostic lamproite mineral phases (priderite, wadeite, etc.) from these Spanish rocks. Conversely, the 37 CAL misclassified as LL come mostly from certain parts of the Hercynides (notably Brittany, Jersey, and SW England). Since true lamproites also occur in SW England at Holmead and Pendennis (Table C6), the Hercynides clearly illustrates a transition in the opposite direction, with a dominantly CAL province exhibiting a few extreme members of LL affinities. Given the number of other provinces in Table C6 in which unequivocal CAL and LL coexist (NW Alps, etc. — Fig.1.4c), and given the existence of acknowledged transitional lamproites (cocites) in many other areas, it is clear that a complete spectrum exists from lamproites, via minettes, into plagioclase-bearing CAL. If the boundary between CAL and LL were drawn not at the point of appearance of normative *ks* and exotic minerals, as in Fig.1.4a, but at a lower K/Al ratio (around 0.8), this would reclassify many Spanish "lamproites" as minettes, and quite a few Hercynian "minettes" as lamproites, and would result in a much cleaner separation between the two branches in Table 5.2. However, even the degree of overlap on present definitions strongly confirms the speciousness of any attempt to hive lamproites off from the lamprophyre clan.

These then, are illustrations of *intrinsic* overlap between the 5 branches. However, this intrinsic overlap can be even further accentuated by crystal-liquid processes. For example, the 88 CAL misclassified as AL in Table 5.2a are mostly appinites and kentallenites (Section 7.1) — coarse-grained rocks in which the accumulation of mafic minerals has led naturally to more basic, AL-like compositions. Similarly, 8 of the the 10 AL misclassified as KIL are picritic monchiquites from Wandagee, Australia (Table C2), in which the accumulation of olivine has generated a more KIL-like chemistry. Conversely, if analyses of bostonites (Appendix B) were included in the MDA, these would nearly all be misclassified as CAL, because the fractionation of mafic phases has led to a more andesitic (CAL-like) chemistry.

The effect of weathering and alteration is more difficult to judge. Clearly, addition of carbonate will tend to mimic more UML- or KIL-like chemistry, while lateritization will tend to mimic CAL, which have highest Al_2O_3 contents in Table 5.1a. Given that so many lamprophyres are altered either autometasomatically or after consolidation, it is all the more remarkable that a discrimination efficiency as high as that in Table 5.2a is actually achieved.

Overall, although it would be useful to obtain 'cleaner' discriminatory plots than Fig.5.2 by eliminating specimens affected by accumulation, fractionation, alteration or weathering, this is almost impossible to achieve in practice. Because lamprophyres are so richly porphyritic, the kinds of criteria used with basalts to gauge the extent of crystal-liquid processes (i.e. porphyritic versus aphanitic or hyaline rocks) are inapplicable. Equally, because they are so intrinsically rich in volatiles, lamprophyres are not susceptible to simple, absolute tests of 'freshness' (e.g. $CO_2 < 0.5\%$, $H_2O < 2\%$ as laid down by Le Bas *et al.* (1986) for the TAS diagram). An aillikite with, say, 10% CO_2, might for example be nearer its primary composition than a camptonite with only 5%. In practice, therefore, it is felt better to provide discriminatory plots such as Figs.5.2–5.4 which, even if relatively crude, are applicable to the whole range of lamprophyre whole-rocks observed in practice.

5.1.3 Discrimination between aillikites and kimberlites based on major oxides

Because the compositional variability of lamprophyres is so high, MDA on all 5 branches at once is likely to be far less efficient than that on only 2 branches at a time. To complement

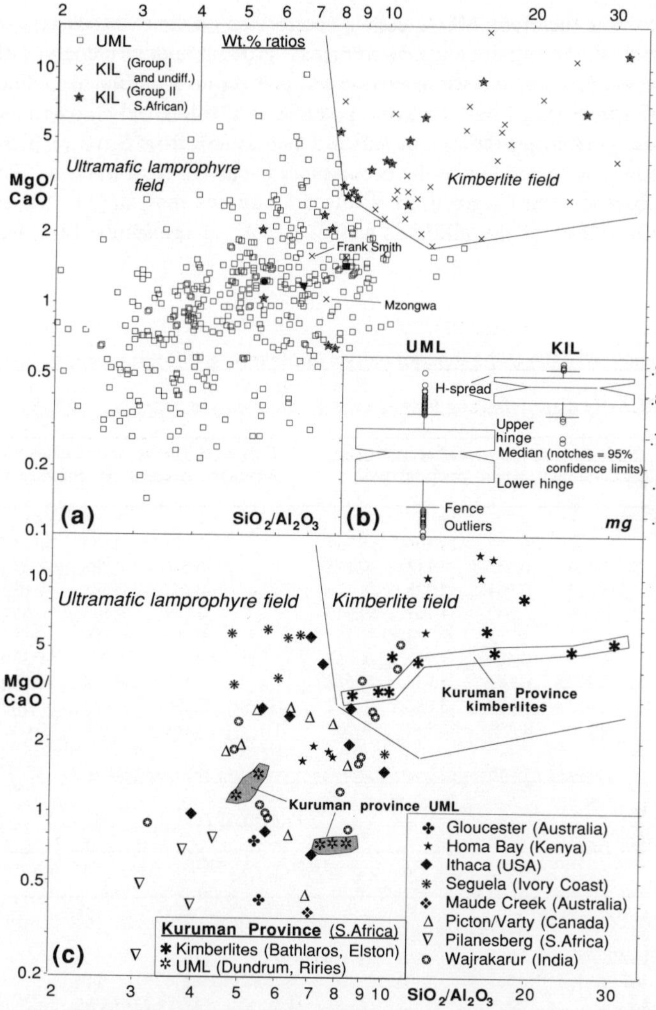

Fig.5.5 *Simple discrimination between UML and KIL using major oxide ratios (wt.%).* See Section 5.1.3 for further discussion. Data from LAMPDA (Appendix D).

(a) All available UML and KIL. Field boundary drawn to exclude all UML from KIL field, which leaves minor overlap with Group II KIL in UML field. Diagrams should *only* be used with analyses of uncontaminated, xenolith-free, ultramafic olivine-calcite-phlogopite-(spinel-perovskite-diopside) rocks which could be either UML or KIL on a *mineralogical* basis.

(b) Discrimination based on *mg* using notched box-plots. The central horizontal line represents the *median*, and the outer edges of the boxes the *hinges* (≈quartiles). Half the data occur above and half below the median; half the data above the median occur below the upper hinge and half above it (similarly for lower hinge). The H-spread (≈interquartile range), or width of the box, represents the spread of the distribution, and the symmetry its skew. Thus UML have a much more symmetric distribution with higher spread (variance) than KIL. Data-values beyond 1.5 H-spreads from the hinges are shown as individual outliers. The notches represent nonparametric 95% confidence intervals; here, the notches do not remotely overlap so the average *mg* of UML and KIL are statistically distinguishable. For further explanation of box-plots see Tukey (1977), Rock (1988d).

(c) Illustrating efficacy of (a) with data for several equivocal occurrences variously called "kimberlites", "kimberlitic rocks", "alnöites", etc. KIL data mainly from C.B.Smith *et al.*(1985a); Kuruman Province (S.Africa) data for coexisting UML and KIL from Shee *et al.* (1989).

the previous section, therefore, MDA was also performed on the two rock-types which can be most difficult to distinguish on *mineralogical* grounds alone: namely, aillikites and kimberlites, which can both consist of olivine and phlogopite ± carbonates, clinopyroxene, perovskite and spinel (Fig.1.4a). To avoid potential confusion, only *modern* analyses of *uncontaminated* kimberlites (criteria of R.H.Mitchell 1986) from S.Africa (C.B.Smith *et al.*1985a) and from accepted kimberlite occurrences in eastern N.America on Fig.2.10 (e.g. Alibert & Albarède 1988) were used. Table 5.2b shows that, of the 276 analyses in LAMPDA, 204 (92%) of the aillikites and 51 (94%) of the kimberlites are correctly

Table 5.2 Efficiency of MDA using 9 major oxides in separating different types of lamprophyres

(a) *Efficiency of MDA in discriminating between the 5 lamprophyre branches (3,508 analyses)*

	Dependent variable canonical coefficients (standardized by within-group std.devs)				Canonical loadings (correlations between conditional dependent variables and dependent canonical factors)			
Discriminant function	1	2	3	4	1	2	3	4
SiO_2	4.233	0.752	14.166	55.984	0.159	-0.884	-0.155	-0.253
Al_2O_3	1.506	0.382	7.113	30.271	-0.566	-0.588	-0.416	0.082
$Fe_2O_3(t)$	1.637	0.904	7.228	28.158	-0.281	0.685	-0.053	-0.156
MgO	4.586	1.380	14.767	59.948	0.269	0.409	0.557	0.610
CaO	2.826	1.215	10.784	42.257	-0.268	0.556	0.332	-0.239
Na_2O	1.035	0.480	3.231	14.783	-0.387	-0.220	-0.380	0.175
K_2O	2.172	0.649	5.838	23.588	0.457	-0.224	-0.303	-0.399
TiO_2	1.452	0.632	3.082	16.102	0.296	0.433	-0.743	-0.117
P_2O_5	0.770	0.039	2.140	8.269	0.181	0.189	0.008	-0.084

Tabulation of observed branches (rows) by predicted branches (columns)

PREDICTED

		CAL	AL	UML	KIL	LL	*Observed total*
	CAL	**1195(90%)**	88	5	2	37	1327
	AL	25	**767(92%)**	34	10	0	836
OBSERVED	UML	2	29	**368(82%)**	45	6	450
	KIL	1	0	6	**47(85%)**	147	55
	LL	60	5	11	147	**617(73%)**	840
	Predicted total	1283	889	424	251	661	3508

(b) *Efficiency of MDA in discriminating between aillikites and kimberlites only (276 analyses)*

PREDICTED

		Aillikites	Kimberlites	*Observed total*
OBSERVED	Aillikites	**204(92%)**	18	222
	Kimberlites	3	**51(94%)**	54
	Predicted total	207	69	276

Canonical loadings: Si –0.211; Al 0.293; Fe 0.516; Mg –0.702; Ca 0.256; Na 0.270; K –0.075; Ti 0.296; P –0.047

Table 5.2 *cont'd*

(c) *Efficiency of MDA in discriminating between the 4 CAL rock-types (862 analyses)*

		PREDICTED				
		Minettes	Vogesites	Kersantites	Spessartites	Observed total
OBSERVED	Minettes	**246(77%)**	37	28	8	319
	Vogesites	6	**12(38%)**	6	8	32
	Kersantites	25	33	**81(44%)**	44	183
	Spessartites	2	31	52	**243(74%)**	328
	Predicted total	279	113	167	303	862

Explanation: in the efficiency tabulations, figures (%) in **bold** refer to analyses correctly assigned by MDA using whole-rock chemistry alone. Thus 767 of the 836 analyses classified petrographically as AL in (a) are correctly assigned by MDA. Of the remaining 69 AL analyses, 25 are incorrectly assigned by MDA to CAL, 34 to UML and 10 to KIL.

classified. The 3 incorrectly classified kimberlites are from Mzongwana (2) and Boshof (1), while the 18 incorrectly classified aillikites are from Aillik Bay (5), Alnö (2), Bow Hill (4), Bulljah (1), Claylick Creek (1), Kalix (1), McKellar (1), Mt.Queglia (1), Narssaq (1) and Ouachitas (1). In 10 of these 18 cases, the statistical probabilities of classification as aillikites exceed 0.3 (i.e. these are marginal cases). Furthermore, with the exception of Bow Hill, other analyses for each of these occurrences classify dominantly and correctly as aillikites. In the case of Bow Hill, 4 of the 5 analyses in Fielding & Jaques (1989) are incorrectly classified as kimberlites, but the geology and mineralogy of these rocks, including Ti-garnet, aegirine-augite and richterite with associated fenitization, quite clearly identify them as UML. Overall, whole-rock geochemistry is capable *on its own* of separating over 90% of aillikites and kimberlites, but a petrological continuum does exist, such that mineralogical confirmation is required in borderline cases.

As in Section 5.1.2, a simplification of the multivariate UML/KIL discrimination functions into a simple oxide ratio plot is provided in Fig.5.5a. This plot in fact includes *all* analyses of UML from LAMPDA (alnöites, etc. as well as aillikites), to illustrate its efficacy. Analyses plotting within the KIL field can be confidently assigned to kimberlite, as there is no overlap with UML. With two exceptions (named on Fig.5.5a), all the KIL plotting in the UML field are Group II KIL, once again illustrating the clear transition between KIL and UML envisaged by Dawson (1987, 1989), and the absurdity of divorcing kimberlites from the lamprophyre clan. As yet further confirmation of the usefulness of this plot, all available analyses from the Kuruman Province (Shee *et al.* 1989), where KIL and UML of the same age *coexist* (Table C6), are also plotted: rocks called kimberlites plot well within the KIL field and those called lamprophyres within the UML field (Fig.5.5a).

5.2 CIPW normative composition (Table 5.3)

CIPW norms were calculated on 3,357 analyses in LAMPDA which are complete enough for this purpose. The computer program used (Rock 1989c) has been specially extended to cope with highly undersaturated rocks, by calculating *ka* and *ge* in addition to standard normative minerals. KIL were excluded, because they lie chemically way beyond the useful limit of CIPW norms. Results are discussed separately below for the other 4 branches:

Table 5.3 CIPW normative character of lamprophyre branches and rock-types

(5.3a) *Silica saturation parameters*

	N	*qz* > 0	*hy* > 0, *qz* = 0	*ne* > 0 *lc* = 0	*lc* > 0 *ne* = 0	*lc* > 0 *ne* > 0	*ka* > 0‡	*la* > 0‡	*ge* > 0‡	Impossible
Calc-alkaline lamprophyres (CAL)										
Appinites (CA)	78	33	47	17	0*	3	3‡	2‡	0*	0*
Kentallenites (CE)	60	0	32	68	0*	0	0*	0*	0*	0*
Kersantites (CK)	169	46	24	30	0*	<1	0*	0*	0*	0*
Minettes (CM)	286	34	23	36	0*	8	<1‡	0*	0*	0*
Spessartites (CS)	286	50	29	21	0*	<1	0*	0*	0*	0*
Vogesites (CV)	30	30	27	37	0*	7	0*	0*	0*	0*
All CAL#	**1446**	**41**	**27**	**28**	**0***	**3**	**<1**	**<1**	**0***	**0***
Differentiated relatives of CAL										
Malchites (CC)	53	87	2	11	0*	0*	0*	0*	0*	0*
Porphyrites/porphyries (CP)	86	97	0	3	0*	0*	0*	0*	0*	0*
Alkaline lamprophyres (AL)										
Doleritic (AD)	45	25(0)	0(25)	80(80)	0*	0	0*	0*	0*	0*
Camptonites (AC)	423	0*	7(3)	83(81)	0*	1(2)	<1(1)	0*	0*	0*
Monchiquites (AM)	225	0*	<1(<1)	68(68)	0*	30(30)	24(32)	11(17)	0*	0*
Sannaites (AS)	31	0*	0(0)	87(71)	0*	13(0)	55(55)	55(52)	0*	0*
All AL#	**753**	**<1**	**5(3)**	**78(76)**	**0***	**17(17)**	**14(20)**	**7(11)**	**0***	**0***
Differentiated relatives of AL										
Bostonites (AB)	40	45(30)	18(25)	38(45)	0*	0*	0*	0*	0*	0*
Ocelli (AO)	16	6(6)	6(6)	88(88)	0*	0*	0*	0*	0*	0*
Ultramafic lamprophyres (UML)										
Aillikites (UA)	191	0*	3†	56	<1	15	60	51	6	23
Alnöites (UL)	69	0*	0*	87	0	4	91	87	26	9
Damtjernites (UD)	12	0*	0*	83	0	17	92	83	8	0
Ouachitites (UO)	18	0*	0*	72	0	28	78	61	0	0
Polzenites (UP)	20	0*	0*	95	0	0	90	90	10	5
All UML#	**345**	**0***	**3†**	**63**	**«1**	**15**	**68**	**59**	**10**	**16**
Lamproites (LL)										
Fitzroyite family	369	76	19	<1	2	<1	4	2	0*	0*
Jumillite family	121	55	25	20	0	0	0	0*	0*	0*
Olivine-lamproites	115	1	84	3	11	1	0	0*	0*	0*
Wyomingite family	27	26	26	4	11	33	19	0*	0*	0*
All LL#	**813**	**48**	**33**	**6**	**8**	**4**	**6**	**2**	**0***	**0***

(1) CAL: are about $\frac{2}{3}$ *qz*-normative, $\frac{1}{3}$ *ne*-normative. The transition of minettes and vogesites to LL is indicated by the appearance of *ac* ± *ns* in around 10% of analyses, but spessartites and kersantites are, of course, *an*-normative because of their modal plagioclase. The appearance of *cm* in 8% of analyses is not necessarily a reflection of alteration, being quite common in subduction-related magmas (Gill 1982).

(2) AL: camptonites have almost exclusively basanitic norms (i.e. *ab+or+an+ne+di+ol*). By

(5.3b) *Alumina saturation and other parameters*

	N	*cm* > 0	*an* > 0	*ac* > 0 *ns* = 0	*ns* > 0 *ks* = 0	*ks* > 0 *ns* = 0	*ac* > 0 *ns,ks* >0	*wo* > 0	*hm* > 0	*ru* > 0
<u>*Calc-alkaline lamprophyres (CAL)*</u>										
Appinites (CA)	78	3	100*	0*	0*	0*	0*	0	0†	0*
Kentallenites (CE)	60	0	100*	0*	0*	0*	0*	0	0†	0*
Kersantites (CK)	169	18	100*	0*	0*	0*	0*	0	0†	0*
Minettes (CM)	286	6	91	7	2	0*	<1	3	0†	0*
Spessartites (CS)	286	5	100*	0*	0*	0*	0*	0	0†	0*
Vogesites (CV)	30	13	87	0	10	0*	3	3	0†	0*
All CAL#	**1446**	**8**	**96**	**1**	**<1**	**0***	**<1**	**1**	**0†**	**0***
Differentiated relatives of CAL										
Malchites (CC)	53	38	98	0*	0*	0*	2	2	0†	0*
Porphyrites & porphyries	86	78	99	0*	1	0*	0*	0	0†	0*
<u>*Alkaline lamprophyres (AL)*</u>										
Camptonites (AC)	423	<1†	100*	0*	0*	0*	0*	<1†	0†	0*
Doleritic (AD)	5	0†	100*	0*	0*	0*	0*	0†	0†	0*
Monchiquites (AM)	255	0†	100*	0*	0*	0*	0*	0†	0†	0*
Sannaites (AS)	31	<1†	94	6	0*	0*	0*	0†	0†	0*
All AL#	**753**	**<1†**	**99**	**<1**	**0***	**0***	**0***	**0†**	**0†**	**0***
Differentiated relatives of AL										
Bostonites (AB)	40	32.5	97.5	0*	0*	0*	0*	0†	0†	0*
Ocelli (AO)	16	12	100	0*	0*	0*	0*	19	0†	0*
<u>*Ultramafic lamprophyres (UML)*</u>										
Aillikites (UA)	191	6	75	1	0*	0*	<1	0*	0†	0*
Alnöites (UL)	69	0	83	7	1	0*	0*	0*	0†	0*
Damtjernites (UD)	12	0	100	0	0*	0*	0*	0*	0†	0*
Ouachitites (UO)	18	0	100	0	0*	0*	0*	0*	0†	0*
Polzenites (UP)	20	0	80	15	0*	0*	0*	0*	0†	0*
All UML#	**345**	**1**	**81**	**3**	**«1**	**0***	**«1**	**0***	**0†**	**0***
<u>*Lamproites (LL)*</u>										
Fitzroyite family	369	<1	8	8	<1	76	8	2	69	64
Jumillite family	121	2	53	17	15	<1	14	2	0†	0
Olivine-lamproites	115	0*	22	13	0	63	3	0	0†	0
Wyomingite family	27	0	15	15	15	0	56	0	0†	0
All LL#	**813**	**1**	**24**	**11**	**4**	**46**	**14**	**1**	**33**	**31**

Explanation: figures given are percentages of the total number of analyses *N* in the left-hand column; where two figures are given, the first (unbracketed) refers to norm calculations with the ratio $Fe^3/(Fe^3 + Fe^2)$ set to 0.3, and the second (in brackets) with the same ratio set to 0.1, *for all analyses*. All norms were calculated treating CO_2 as secondary (i.e. no *cc* was calculated). No figures are given for kimberlites because CIPW norms are quite meaningless for these rocks

* should be zero in all accurate analyses.

† should be zero in *fresh*, accurate analyses.

‡ *ka, la* or *ge* appear artificially in some highly carbonated samples because *cc* ($CaCO_3$) was not calculated.

\# includes all the specific rock-types in the rows above, plus 'undifferentiated' samples assigned to branches but not individual rock-types; metamorphosed and schistose varieties are excluded.

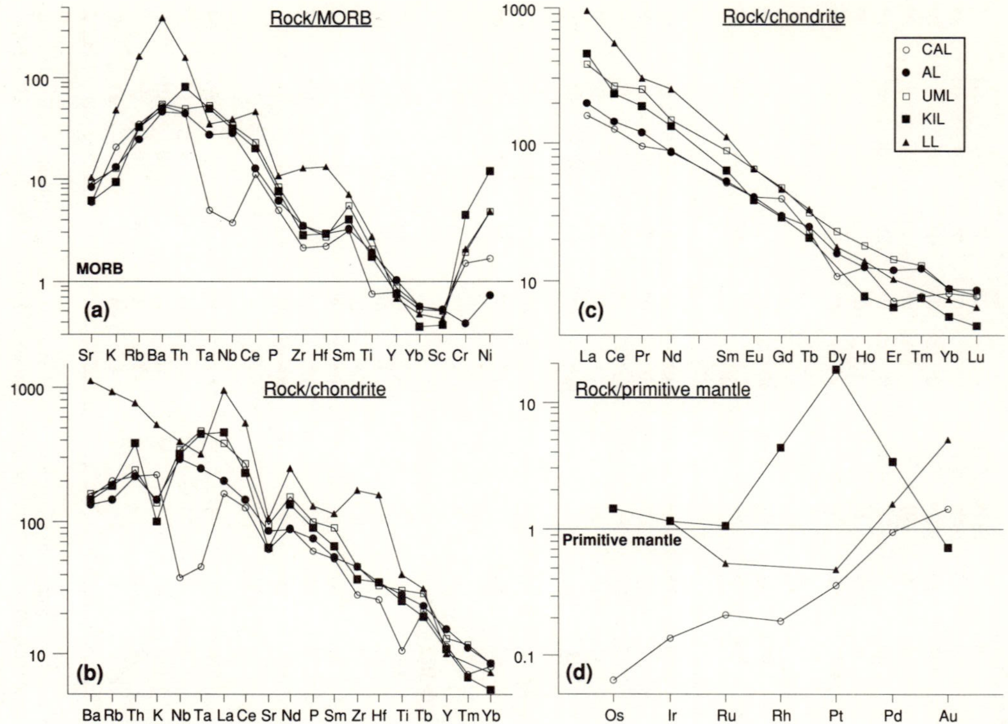

Fig. 5.6 *'Spidergrams' of average compositions for the 5 lamprophyre branches.* Data from Table 5.1.
(a) normalized to MORB, using element order and normalizing values of Pearce (1983);
(b) normalized to chondrites, using order and normalizing values of Thompson (1982);
(c) REE normalized to chondrites, using normalizing values of Nakamura (1974);
(d) PGE and Au normalized to primitive mantle, using order of decreasing melting point.

contrast, over half of the monchiquites, and rather more of the relatively few sannaites, carry *lc* ± *ka* ± *la*, indicating gradation into foidites and UML. Peralkaline norms (*ac*) are shown by only 6% of sannaites. All, by definition, are silica-undersaturated.

(3) UML: these are so SiO_2-poor and CO_2-rich that respectively 5% and 20% of alnöite and aillikite analyses have insufficient CaO to combine with the quoted CO_2; CO_2 must thus be assigned arbitrarily between CaO, MgO and FeO, to avoid negative results. If CO_2 is ignored altogether, 9% of the alnöites and 23% of the aillikites still have insufficient SiO_2 to form even the most Si-deficient normative assemblage possible (*ne* + *ka* + *ol* + *la* + *ge*). Of the remaining (calculable) analyses, 91% of the alnöites, but also 60% of the aillikites, yield *la* ± *ge*, confirming that 'normative melilite' is a necessary but not sufficient condition for crystallization of modal melilite (Velde & Yoder 1976, 1977). *Ac* occurs in 7% of the alnöites and 15% of the polzenites, but in almost none of the aillikites, although in contrast, 6% of the latter carry *cm*.

(4) LL: these uniquely combine *ac* ± *ks* with *ru*. They are about 80% silica-oversaturated, 20% undersaturated (i.e. more silicic than CAL). However, the Spanish lamproites carry *ks* much less commonly than the Australian or North American lamproites, and indeed over half of them are *an*-normative, all of which underlines the transition between the Spanish rocks and minettes already referred to in the previous section.

5.3 Trace elements (Table 5.1; Figs. 5.6–5.8)

Basic statistics are compared in Table 5.1, 'spidergrams' in Fig.5.6, and element distributions in Fig.5.7. The following are the most significant generalizations:

(1) Lamprophyres are linked together and with alkaline rocks by their high average contents of LILE (K, Rb, Sr, Ba, etc.), but near-basaltic levels of Sc, Ti, Y and HREE. Ultramafic elements (Cr, Ni) lie at levels typical of mafic-ultramafic rocks.
(2) Coupled negative Ta–Nb–Ti (TNT) anomalies are present only in CAL (Fig.5.6a,b).
(3) Sr and Ba commonly rise higher than in other silicate igneous rocks, the highest Sr value (7275 ppm) exceeding Wedepohl's (1978) maxima even for carbonatites.
(4) Rb, Zr and Th reach values approaching those in rare-metal pegmatites and agpaites.
(5) Although Y generally has levels characteristic of MORB, it also achieves a few high values (> 150 ppm) in AL and UML.
(6) The mean AL spidergram has an almost identical shape to that of typical alkali basalts, but at enrichments higher by a factor of 1.5–2.
(7) *Ranges* of most trace elements in the 5 branches overlap almost completely (Fig.5.7), although *median* contents particularly of Li, V, Cr, Co, Ni, Zr, Nb, La, Yb, Hf, Ta, Pb, Th and to a lesser extent Zn, Rb, Y, and U are statistically distinguishable.

This evidence of similar patterns and ranges, yet distinguishable averages, is once again thoroughly in accord with the lamprophyre 'clan' concept in Fig.1.2.

A lack of Cl-F covariation is illustrated in Fig.5.8. Although data are sparse other than for LL, the plot does nevertheless imply higher F contents in LL than in CAL.

Trace element discrimination (cf. Figs.5-2–5.4) is much more difficult to set up than for major elements. This is because different authors inevitably quote different ranges of trace elements in their analyses, so that the data-set in LAMPDA is highly inhomogeneous — that is, full of missing data (see different numbers of determined values for each trace element in Table 5.1). To subject this data-matrix to MDA, which has to have a measured value for

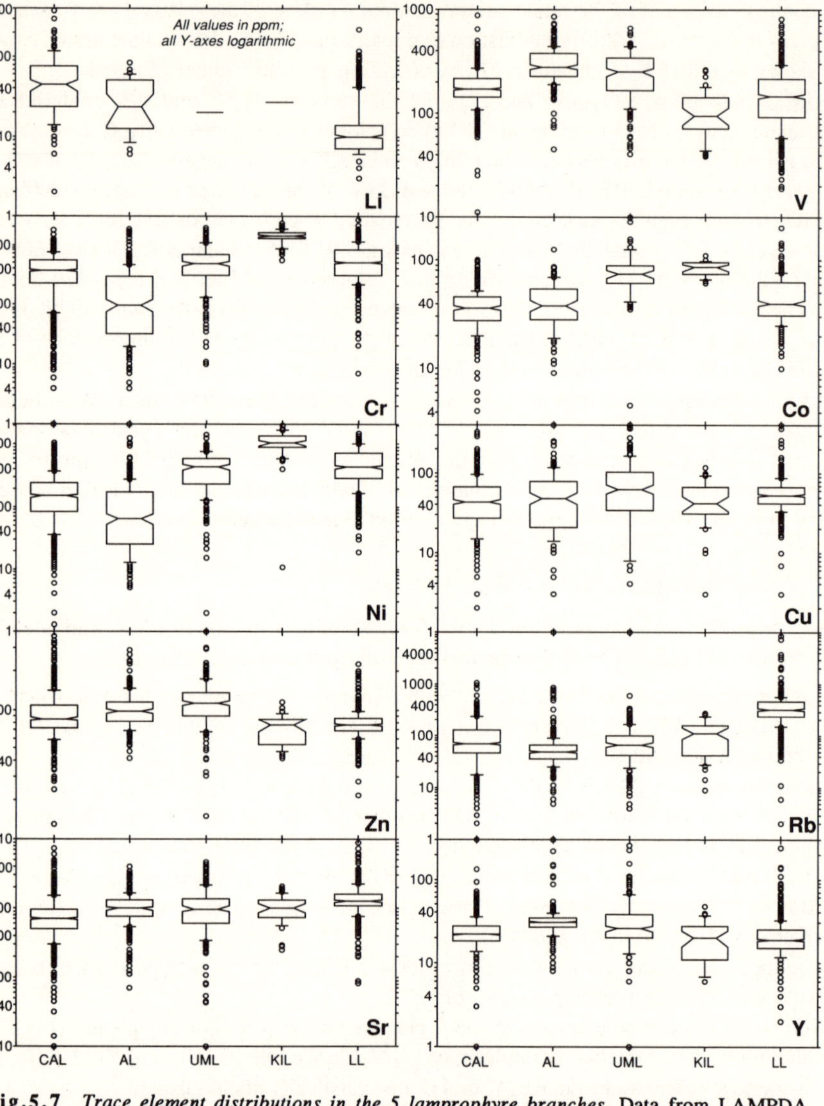

Fig.5.7 *Trace element distributions in the 5 lamprophyre branches.* Data from LAMPDA (Appendix D). All data in ppm. For explanation of box-plots see Fig.5.5b. Numbers of values given in Table 5.1. Elements ordered by atomic number. Be, Mo, Sn, Sb, Cs, W and PGEs omitted due to paucity of data, Sc and Ga due to statistical uniformity. Ce chosen as representative LREE and Yb as HREE because more data are available than for adjacent elements. For complete Au data see Rock *et al.*(1989). *Medians* for most elements are unequal for the 5 branches, but ranges are similar.

every element to be included in the discriminant functions, *either* (1) the range of trace elements has to be severely restricted to the few most commonly determined trace elements (Cr, Sr, Zr, Ba, etc.), *or* (2) the range of samples used has to be restricted to those with the most complete range of elements determined. Both restrictions have to be done subjectively,

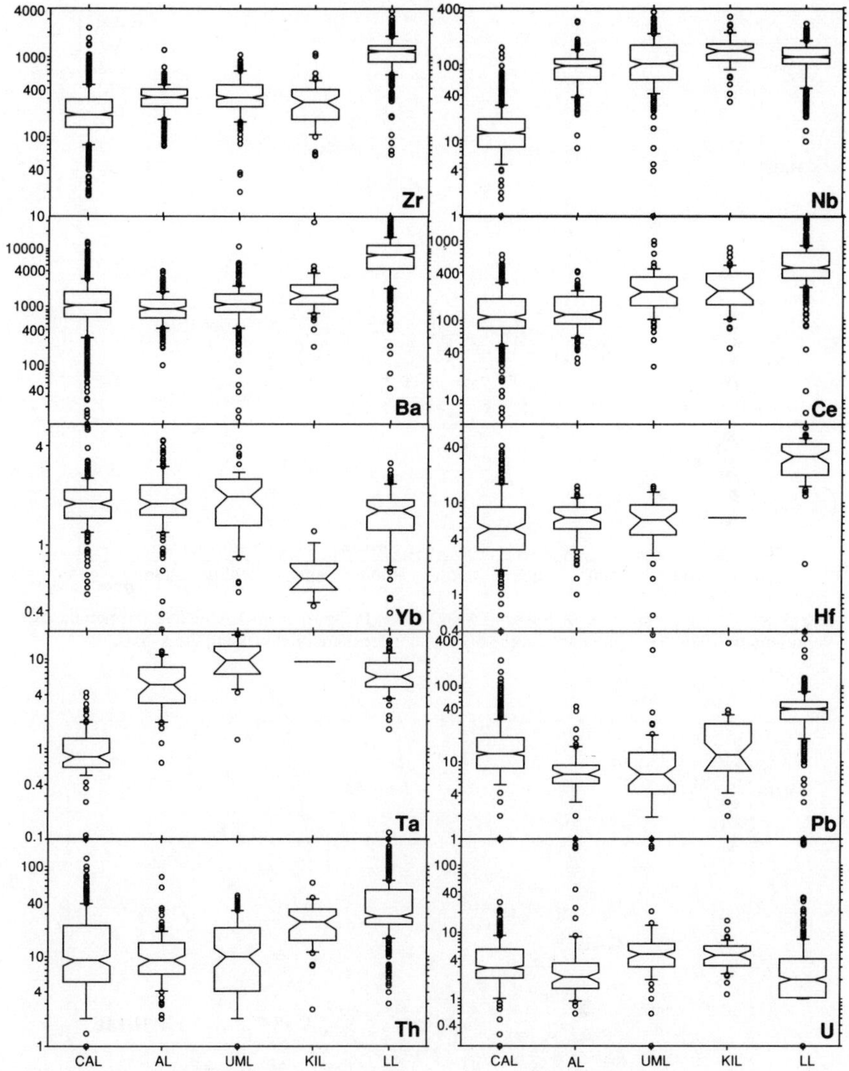

Fig. 5.7 *cont'd*

and lead to only a fraction of the total data being employed, with critical loss of information.

Discrimination via simple ratios is thus the best that can be expected (the tremendous overlap in ranges on Fig.5.7 obviously precludes discrimination based on simple element pairs as in Fig.5.3). Furthermore, it is more crucial to use ratios than with major oxides, since possible incomparabilities between different analytical methods are thereby minimized. Fig.5.9 illustrates the optimum separation achieved out of many ratios tried; it shows that many CAL can be distinguished by their low Nb/Pb, and at least some AL by the high V/Cr ratios, though the region of total overlap is still very large.

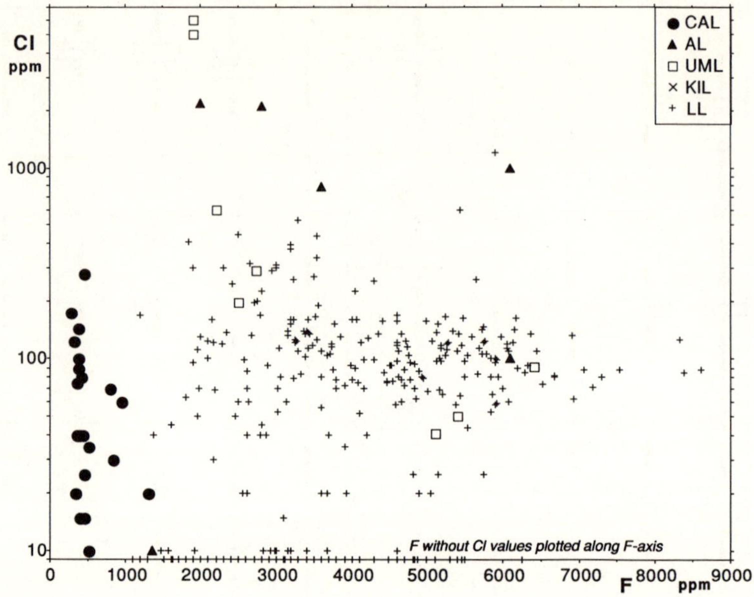

Fig.5.8 *F-Cl covariation in lamprophyres.* Data (in ppm) from LAMPDA (Appendix D). Representative data for F (without accompanying Cl values) are plotted along the *X*-axis.

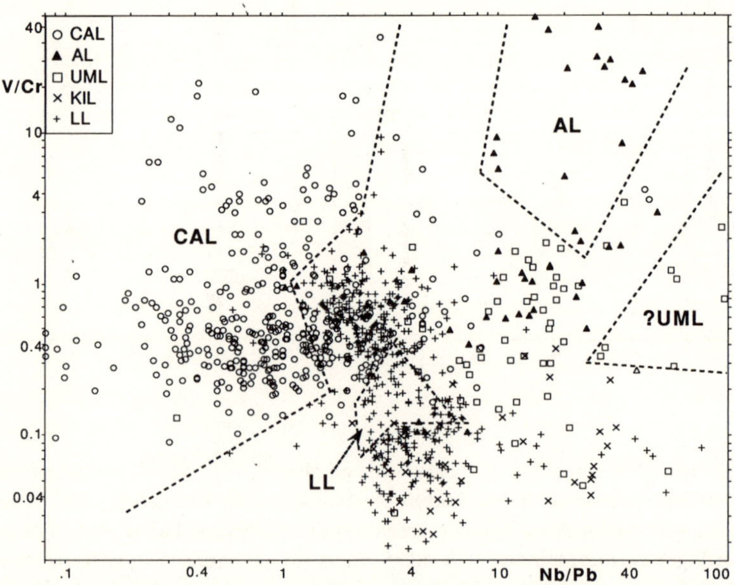

Fig.5.9 *Simple discrimination between the 5 lamprophyre branches using trace element ratios.* No other pair of ratios has yet been found to yield better discrimination. Analyses plotting within the dashed fields can be attributed to the particular branch with high confidence, but the central area shows total overlap between the branches. There is no diagnostic field for KIL.

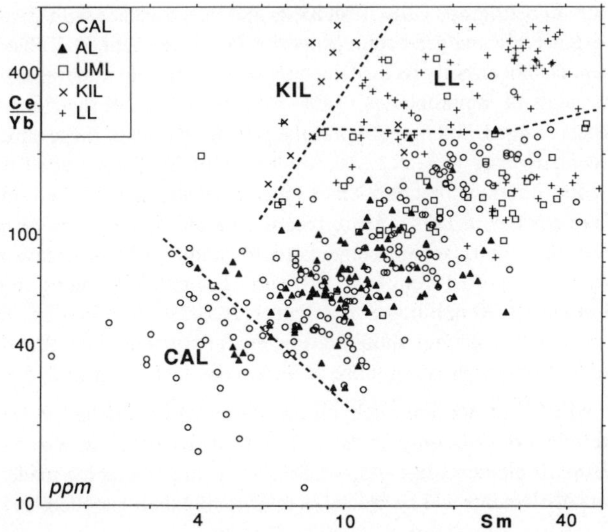

Fig.5.10 *Simple discrimination between the 5 lamprophyre branches using REE.* Ce/Yb represents the steepness of the pattern on Fig.5.6c, and Sm the total REE content. Analyses plotting within the dashed fields can be attributed to the particular branch with high confidence, but the central area shows total overlap between the branches. There is no diagnostic field for AL.

5.4 Rare-earth elements (REE) (Fig.5.6c, 5.10)

Although overall patterns on Fig.5.6c are very similar, minor differences are apparent in detail from Fig.5.10, which allow at the least the more extreme analyses to be distinguished by the lower Ce/Yb and Sm of CAL relative to KIL and LL. In the most general terms, plagioclase-bearing lamprophyres (spessartites and camptonites) tend to have substantially lower Sm and Ce/Yb than plagioclase-free types (Rock 1987), and total LREE increase with increasing K/Na and K-feldspar/plagioclase ratios (*cf.* Cullers & Graf 1984).

5.5 Precious metals (PGE, Au) (Fig.5.6d)

R.H.Mitchell (1986, p.290) believed that contents of these elements in KIL are still "very poorly characterized". There is certainly little doubt from recent interlaboratory comparisons that *precise, accurate* data for these elements at the required ppb levels are few and far between. As Rock & Groves (1988a) and Rock *et al.*(1988c,1989) have comprehensively listed and statistically examined all available data, only the briefest of summaries need be given here, but provisos over their accuracy should be borne in mind:

— The overall distribution of 243 Au determinations in lamprophyres is approximately log-normal, with a median of 5 and arithmetic mean of 43 ppb. Some 70% of these values exceed the range of 0.5–2 ppb Au now considered 'typical' of common igneous rocks, while 25% exceed 15 ppb and 10% exceed 70 ppb.

— Although high Au values do occur in CAL associated with gold deposits (Section 9.2), they may also occur in CAL away from goldfields, and, perhaps more significantly, in *all the other lamprophyre branches*, which are not associated with gold ores in any way.

— High Au occurs in petrographically fresh rocks and this, together with statistical analysis of the most extensively analyzed suite (Superior Province, Canada: Table C1), suggests that enrichment cannot entirely be explained by secondary gold mineralization processes.
— Average S contents of lamprophyres (Table 5.1) seem to be of the right order to allow Au-S complexes to form, and may support the primary nature of these enrichments.
— Although recent reviews (e.g. A.J.Macdonald 1986) have concluded that KIL have higher PGE concentrations than any other igneous rocks, this is based on questionable data (e.g. Kaminsky *et al.*1974). More recent data are still too few to make reliable generalizations, but significant enrichments (comparable to those in layered intrusions) have certainly been reported from both KIL and LL, including values up to 56 ppb Pd, with several in the 10–20 ppb range (e.g. Paul *et al.*1979). In principle, there seems no reason why some lamprophyres should not be PGE-enriched, given that they have Mg, Cr and Ni contents approximating those of ultramafic rocks (Figs.5.3, 5.5, 5.6).

To confirm whether or not lamprophyres are intrinsically enriched in precious metals, high-quality determinations not only of Au and PGE themselves, but also of so far barely determined chalcophile elements like As, Sb, Se and Te, are now being made. Covariations between these sets of elements will be crucial in interpreting their geochemistry.

5.6 Radiogenic isotopes (Figs.5.11–5.12)

As recently as 1985, only a handful of isotopic determinations were published for lamprophyres, but the database has now grown to cover data for fully 472 specimens. In the case of radiogenic isotopes, *coupled* Sr-Nd isotope determinations, corrected for both age and mantle evolution via the use of ε values, are more useful for interpretation than

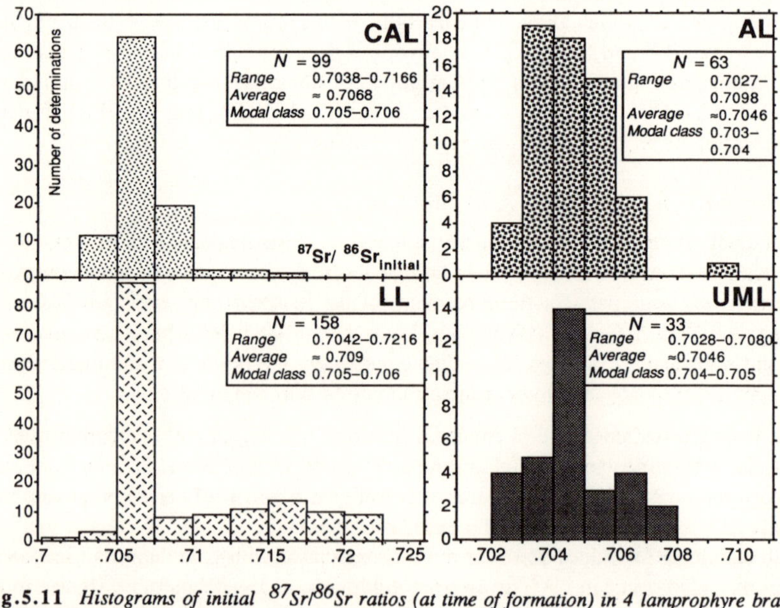

Fig.5.11 *Histograms of initial $^{87}Sr/^{86}Sr$ ratios (at time of formation) in 4 lamprophyre branches.* Numbers of included values indicated by *N*. Data from LAMPDA (Appendix D).

either ratio alone. Unfortunately, there are far fewer published isotope determinations for Nd (199) than for Sr (390), and some of the older papers which give Sr ratios only, do not provide sufficient information to calculate ε values. The present discussion is therefore in two parts: the first considers the 390 Sr determinations as initial $^{87}Sr/^{86}Sr$ ratios, and the second the 182 Sr-Nd determinations as coupled ε values.

For consistency, $^{87}Sr/^{86}Sr$ *initial* ratios have all been recalculated from published *measured* ratios, using known ages (Appendix C) and $\lambda_{Rb} = 1.42 \times 10^{-11}$. Although initial ratios change slightly with time, standard normalization procedures, and essentially insignificant interlaboratory differences for the benchmark NBC carbonate Sr standard, mean it is meaningful to group these calculated initial ratios and treat them statistically. The overall order of initial ratios can be taken as LL > CAL > UML ≈ AL, and comments on global distributions for the individual branches (Fig.5.11) follow:

(1) CAL: 56 determinations show a distribution clearly displaced towards higher values than uncontaminated basaltic or mantle rocks (Bulk Earth or CHUR is normally taken to have an initial ratio of 0.7045); however, the highest values (> 0.71) are in rocks which are either transitional to (and perhaps better classified as) LL (e.g. Dal Piaz *et al.* 1979).

(2) AL: 63 determinations are more typical of uncontaminated, mantle-derived rocks, with an average almost identical to Bulk Earth, although there is one value approaching 0.71.

(3) UML: 33 determinations show a similar distribution to AL, but no values above 0.708.

(4) LL: 158 determinations show the same modal class as CAL, but there is a much longer tail towards high ratios and a subsidiary mode at ≈ 0.7016; the arithmetic and geometric means are consequently rather higher than for CAL (≈ 0.709), although the median remains the same. Only two initial ratios below 0.704 were found, for two specimens from the Argyle lamproite pipe; these almost certainly reflect overcorrection for the Proterozoic (≈1100 Ma) age of the body due to the mismatch between whole-rock Rb/Sr and measured Sr ratios.

Discussion of coupled Sr-Nd ratios is hindered by the lack of standard interlaboratory procedures for Nd: neither measured nor initial $^{143}Nd/^{144}Nd$ ratios are comparable in the published literature, because some authors normalize to $^{146}Nd/^{144}Nd = 0.7219$ (which leads to CHUR ≈ 0.51264), whereas others normalize to $^{146}Nd/^{142}Nd = 0.63671$ (which leads to CHUR ≈ 0.51184). As a consequence, measured $^{143}Nd/^{144}Nd$ ratios for the Leucite Hills lamproites in Vollmer *et al.* (1984) are in the range 0.5118–0.5120, whereas those provided by Nelson & McCulloch (1989) are apparently lower, at 0.5110, even though there is virtually no age correction for these Recent lavas. A similar apparent inconsistency is evident between determinations for the Spanish lamproites by McCulloch *et al.* (1983) and Nixon *et al.* (1984): see Bergman (1987, p.154). All these values can in fact only be reconciled by recalculation to ε values, using the conventional tranformation. Because, in addition, many calculation and plotting errors are manifest in published papers, all data have been carefully checked and recalculated here, before replotting on Fig.5.12.

The following features are evident on Fig.5.12, but it should be remembered that the apparently distinct fields might well merge, given a larger database:

(1) CAL: 40 coupled determinations (mostly from the Caledonides and Hercynides and the Navajo, USA) almost all show a broad clustering on Fig.5.12a at somewhat +ve ε_{Sr} and ε_{Nd} straddling the zero CHUR value, to the right of the mantle array defined by MORB and Ocean Island Basalt. The Tertiary Navajo minettes (Table C6) are more tightly clustered than the older examples. A distinct, unique, field at near-chondritic ε_{Sr} but –ve

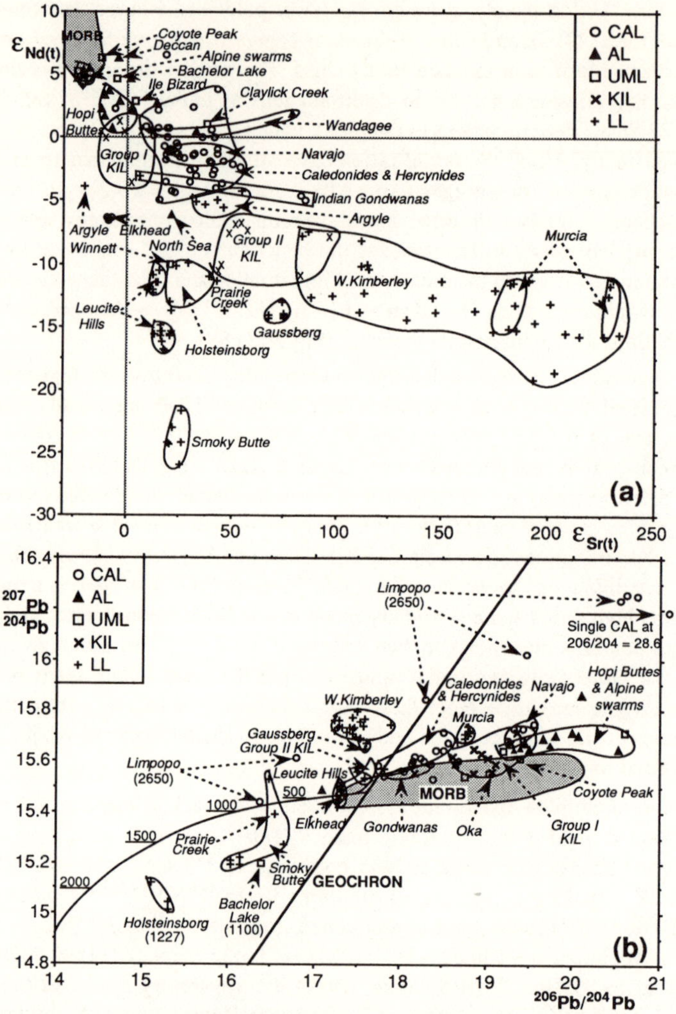

Fig.5.12 *Radiogenic isotope composition of lamprophyres.* Data from LAMPDA (Appendix D).
(a) ε_{Sr} versus ε_{Nd}. Individual occurrences annotated as in Appendix C.
(b) $^{207}Pb/^{204}Pb$ versus $^{206}Pb/^{204}Pb$ (*measured* ratios, without age corrections).

ε_{Nd} is formed by the Elkhead minettes, USA (Table C1). On Fig.5.12b, 37 determinations form three fields: the Navajo and Caledonian/Hercynian CAL both lie to the right of the geochron at higher (more radiogenic) $^{207}Pb/^{204}Pb$ than MORB or OIB, but the Navajo rocks have higher $^{206}Pb/^{204}Pb$. This cannot be an age effect (Fig.5.12b is not age-corrected, unlike Fig.5.12a), since the Navajo rocks are substantially younger. The third field is an errorchron from the Limpopo CAL, Africa (Table C1), which reflects their Archaean age.

(2) <u>AL:</u> 20 coupled Sr-Nd and 16 Pb values (mostly from the Hopi Buttes and Alpine swarms) define fields corresponding broadly with the mantle array (particularly OIB).

(3) UML: 9 coupled Sr-Nd and 6 Pb determinations lie entirely within the OIB field.

(4) KIL: determinations in C.B.Smith *et al.*(1985a) and Alibert & Albarède (1988), coupled with R.H.Mitchell's summary (1986), confirm that Group I fall around CHUR, whereas Group II have more +ve ε_{Sr}, more –ve ε_{Nd} and lower $^{206}Pb/^{204}Pb$.

(5) LL: 82 coupled Sr-Nd and 43 Pb determinations define several distinct fields. All lie well below the mantle array on Fig.5.12a at moderate to strongly +ve ε_{Sr} and –ve ε_{Nd}, with some suggestion of a hyperbolic trend including the W.Kimberley and Murcian rocks but a near-vertical trend for the remainder. On Fig.5.12b there is even less coherence, making generalizations unrealistic.

5.7 Stable isotopes (Fig.5.13)

Stable isotope data are fewer than for radiogenic isotopes. Only 5 δH values are known, from biotites and amphiboles in metamorphosed Archaean CAL from the Yilgarn block (Rock *et al.* 1989), which show a range from –59 to –79‰. Turning to $\delta^{13}C$ and $\delta^{18}O$, coupled determinations of *both* ratios are far fewer (26) than of $\delta^{18}O$ alone (62), so that, as in the previous section, it is necessary to discuss $\delta^{18}O$ separately before considering $\delta^{13}C$ and $\delta^{18}O$ together. On Fig.5.13a, CAL and AL whole-rocks both lie well within the range typical of magmatic rocks, namely +6 to +10‰, although CAL whole-rocks may be

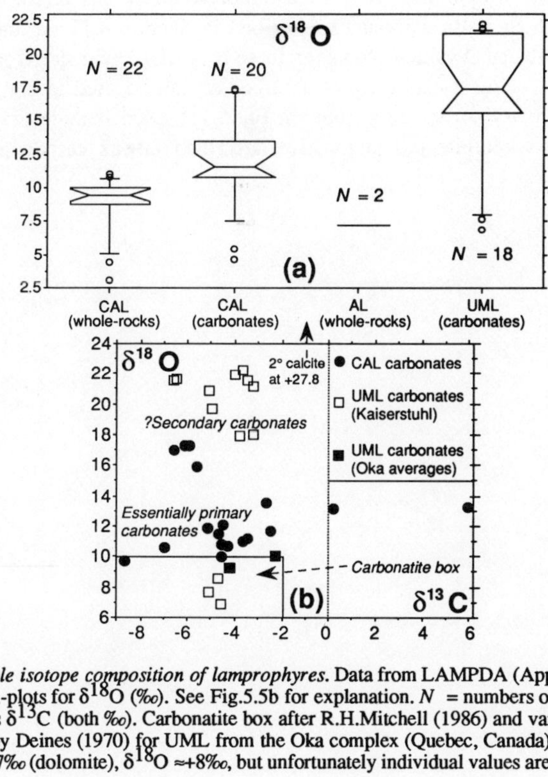

Fig.5.13 *Stable isotope composition of lamprophyres.* Data from LAMPDA (Appendix D).
(a) Notched box-plots for $\delta^{18}O$ (‰). See Fig.5.5b for explanation. N = numbers of values.
(b) $\delta^{18}O$ versus $\delta^{13}C$ (both ‰). Carbonatite box after R.H.Mitchell (1986) and various other authors. Ranges quoted by Deines (1970) for UML from the Oka complex (Quebec, Canada) are: $\delta^{13}C$ = –4.4‰ (calcite) and –2.7‰ (dolomite), $\delta^{18}O \approx$ +8‰, but unfortunately individual values are not quoted.

somewhat more enriched in ^{18}O. CAL carbonates are more enriched still, with many lying outside the normal range for carbonatites, and UML carbonates show a much wider range still, from magmatic values to those typical of sedimentary limestones.

The 26 coupled carbon–oxygen determinations (Fig.5.13b), all from carbonate minerals (ankerite, calcite and dolomite), show that this variation reflects several groupings. Of 12 values from UML carbonates on Fig.5.13b, 10 come from the Kaiserstuhl (Hubberten *et al.*1988); here, a group of 3 magmatic values lies within the 'carbonatite box', whereas another of 7 unequivocally *secondary* values overlaps low-temperature hydrothermal and sedimentary carbonates from the same complex. The other two UML carbonates (averages) on Fig.5.13b are from the Oka complex (Canada); Deines (1970) found these to be enriched in both ^{18}O and ^{13}C relative to those from associated carbonatites, but lighter than in rocks attributable to metasomatism (fenites, etc.) He also found evidence for contamination in the great variability of $\delta^{13}C$ in UML carbonates within the complex, particularly relative to dykes well outside it.

The lower $\delta^{18}O$ set of CAL values on Fig.5.13b are from the Yilgarn block, and clearly include a strong primary (magmatic) component, whereas the 3 values from the Hercynides appear to be predominantly secondary.

Stable isotopes thus reinforce the notion from the chemistry of carbonate minerals (Section 4.12), that a range of primary and secondary carbonate minerals may be present in lamprophyres, even within different specimens from the same occurrence. Although no coupled microprobe/isotopic data are as yet available to check this assertion, it is intuitively likely that the high Sr ± Ba carbonates described in Section 4.12 are the primary grains which also yield low $\delta^{18}O$ values. However, there is no clear evidence from the isotopes as yet that the *major* element chemistry of a carbonate can be used to gauge its origin: the primary and secondary carbonate groups on Fig.5.13b each include grains described as ankerite, calcite and dolomite, though most are described only as 'carbonate'.

6 Inclusion Suites — Macrocrysts, Xenocrysts, Xenoliths, etc.

The spectrum between truly cognate and wholly foreign inclusions was explained in the introduction to Chapter 4. Here we deal with inclusions which are largely foreign (Types **B**-**F**). Table 6.1 explains the fairly standard nomenclature used. Inclusions characteristically constitute tens of percent of the volume of individual lamprophyre dykes, and locally reach 70–80% (e.g. Reynolds 1936; Ayres & Higgins 1939; Elders 1957). They can be concentrated in the central portions of dykes and sills via flow differentiation (Section 3.1).

Nixon's (1987) bible of global mantle xenolith localities, and R.H.Mitchell's (1986, 1989) expansive treatment of inclusions in KIL, obviate a great deal of repetition here: in particular, no attempt is made to redescribe implications that these inclusions have for the nature of the mantle. In terms of the popular 'bus/passenger' dichotomy (Boyd & Meyer 1979a,b), this book is primarily concerned with the bus, and with the implications the passengers have for the bus's point of departure and itinerary, rather than with a detailed inventory of the passengers themselves. Nevertheless, many chapters in Nixon (1987) are based on older classifications of inclusion-bearing hosts: for example, many mafic–ultramafic diatremes in the eastern USA were formerly termed "kimberlite" *primarily* because of the presence of inclusions, and many of these have already been shown to be other types of lamprophyres (Appendix C). This chapter does therefore seek to complement Nixon's, Mitchell's and other related works in several ways, as follows:

(1) It considers material of *crustal* as well as mantle derivation (i.e. Type **F** of Chapter 4).
(2) It makes more accessible available information specifically on lamprophyres.
(3) It adds a significant number of inclusion localities not mentioned in any previous review.

Table 6.1 Summary of terminology for inclusions in lamprophyres
(see Appendix B for more precise definitions)

	Accidental	Cognate	Unknown (non-genetic term)
Crystals	*Xenocryst*	*Phenocryst*	*Macrocryst* (*megacryst* if >5 mm)
Rocks	*Xenolith*	*Autolith*	*Enclave*

Despite the vastly greater amount of data available from KIL inclusions, inclusions in other lamprophyres are important in several ways; for example:

— They cover the complete range of *eruption ages* of available mantle nodules: thus the varied inclusions in Scottish Permo-Carboniferous AL (Table 6.2) represent the oldest mantle nodules available in *basaltic* host-rocks, while those in the ≈1.6 Ga Marathon and Wawa UML (Canada) and Kuruman KIL (S.Africa) represent the oldest known nodules anywhere (Nixon & Davies 1987). At the other end of the range, mantle nodules also occur in very young (Pleistocene-Recent) LL such as the Leucite Hills (USA).

— In numerous areas, such as the Solomon Islands or the Pacific-Ocean side of Japan, lamprophyres provide the *only* mantle inclusion samples.

— Lamprophyres include inclusions unrivalled among KIL suites, such as "megacrystalline nodules" in the Malaita alnöite (Solomons) and Navajo (USA) minettes (Section 6.6.2).

Table 6.2 Inclusions in lamprophyres and lamproites

Country (Occurrence)	Fig	L*	Type D		T	y	p	e	E		Type F#	Est.T,°C	km	Me	gacryst		&	macro	cryst	(d	isc	ret	e n	odu	le)	s	ul	te	Comments/Interpretation
				ecl	dun	1†	2†	3†	4†	5†				Ol	Gt	Zc	Op	Cpx	Am	Bi	Mt	Sp	Il	Cm	Ru	Ap	Pl	Af	
Antarctica (Patuxent Range)	2.3	UL				√					ano	1090±20	50–90					√	√Rc	√						√			
Antarctica (Radok Lake)	???	AM				√		√																					
Australia (Argyle, WA)	2.8	LL				√	√					1156-1179	150		G1,3,9		√	√Cr-Di				√							Equilibrated within diamond field
Australia (Bendigo)	2.8	AM																										Ao	
Australia (Kelly's Pt.,NSW)	2.8	AL?							Sp-pxn						√			√	Kt/Pg			√							
Australia (Mt.Woolooma)	2.8	CM																√		√			√			√			
Australia (Norseman,WA)	2.8	UA												√						√			√						
Australia (S.Gippsland, VIC)	2.8	AM			√																							Ao	Ao = Or23Ab68An9
Australia (W.Kimberley,WA)	2.8	LF			√	√		√				1053–1117			G1,3,5,9		√	√Cr-Di				√			√				
Australia (Wandagee, WA)	2.8	AM			√									√	G1,9,11	√	√	√Cr-Di	√Pg			√	√						
Canada (Aillik Bay, LAB)	2.9	UA								√	amph																		glm=cognate high-P cum
Canada (Ayer's,White Mtns.)	2.9	AM									Ky-Gt-gm	≤ 1200°C	≤ 40 kb																Gm in 50% of 20 dykes examined
Canada (Coral Rapids,ONT)	2.9	UL	sye		√				√		dio,gbr,gns			√				√ '								√	√		megacrysts=disaggregated nodules
Canada (Ile Bizard, Montreal)	2.9	UA				√	√	√	√			880–980	60–100	√			√	√		√	√								web=cognate high-P cum
Canada (McKellar, NFLD)	2.9	UA			√																								
Canada (Meguma zone, NS)	2.9	CS									Various	≈600	14-18																Rich aluminous crustal xen suite
Canada (Wawa, ONT)	2.9	AM						√(Gt)				1007–1196	50–90	√															
Cape Verde Is.	2.2	AC																										Or	
China (Xilong)	2.2	AM			√	√	√	√							√			√					√						
Czech. (Brloh, Bohemia)	2.5	CM												√			√												Ol, Opx "phen" have kelyphitic rims
Finland (Sokli)	2.3	UA	cbt,fen,oliv																										
France (Corbières/Pyrenees)	2.12	AM				√				√																			
Gardar (Igdlutalik)	2.11	UA						√		√																			
Greenland (Coastal swarm)	2.11	AM						√	√									√	√										um rocks=cognate low-P cumulates
Greenland (Frederikshåb)	2.11	UA								√		850–1250	60–120	√					√	√									
Greenland (Holsteinsborg)	2.11	AL						√																					
Greenland (Holsteinsborg)	2.11	LL						√																					
Greenland (Kangerdlugssuaq)	2.11	AC	Kt-gbr																										
Greenland (Sarfartôq)	2.11	UA				√									√			√		√									
Greenland (Scoresby Land)	2.11	AL						√																					
Greenland (Ubekendt Ejland)	2.11	AM				√		√	√			820–1081	40–50																lhz=cognate high-P cum
Greenland (Wiedemanns)	2.11	AM				√		√(Sp)									√	√	√										
Hungary (Buda/Velence Hills	2.12	AM																√											High/low-P cognate Cpx
India (Bombay/W.coast)	2.13	AM							√																				xen porphyroclastic so not cognate
Ireland (Inishowen)	2.7	AM				√				√				√				√			√							Ao	
Italy (Altipiano di Tonezza)	2.12	AC																									Og	Sa	
Italy (Predazzo-Monzoni)	2.12	AC				√			√																			Sa	
Japan (Shingu)	2.2	AC									gbr,ano	950–1050						√	√										
Kenya (Homa Bay)	2.3	UA			√			√		hbd,glm																			
Killilan (Eilean Donain,Scot)	2.7	AM								√																			
Kintyre (Macrihanish, Scot.)	2.7	AM				√		√		√Bi-perid								√	√Kt	√	√	√				√		Ao	Ao = Or18Ab74Ab3
New Zealand (Alpine swarm)	2.8	AL								√								√	√									Ao	
New Zealand (Kaikoura)	2.8	AC	gbr-ano-pxn																										
Norway (Fen)	2.3	UD	cbt,sye			√				√Kt-Ph	gns	950–1070	30–50	√					√	√	√								lhz unusually fertile
Norway (Hellesund)	2.7	AL																										√	
Norway (Sunnhordland)	2.7	AC				√					gbr,Gt-gm	1070–1120	45-55					√				√							Reaction margins; 3 groups spinels
Portugal (Monchique)	2.12	AL																										Or	Disaggregated fragments of nesy

1=Sp-lhz; 2=Gt-lhz; 3=wehr/harz; 4=pxnt/websterite; 5=Am/Ph-bearing perid/pxn * see Tables C1–C4 for host-rock details # Immediate country-rocks excluded

Table 6.2 Inclusions in lamprophyres and lamproites *cont'd*

Country (Occurrence)	Fig	L*	Type D		T	y	p	e	E		Type F#	Est.T,°C	km	Me	gacryst		&	macro	cryst	(d	isc	ret	e n	odu	le)	s	ui	te	Comments/Interpretation
Solomon Is. (Malaita)	2.3	UL				√	√						≤ 135		√	√	√	int		√			int						"Unrivalled" macrocryst suite
Spain (Murcia)	2.12	LJ				√																							Depleted,low-T,shallow mantle
Sweden (Alnö)	2.3	UA			√			√		√	Gt-gm	760–1100			√			√	√Kt										Few reaction rims
Sweden (Kalix-Luleå)	2.3	UA				√																							
UK (Coire na Bà, Scotland)	2.7	AM				√																							
UK (Coombs Fell, Scotland)	2.7	AM				√																							
UK (Duncansby,Scotland)	2.7	AD				√		√	√	√Bi-perid									√Kt	√	√								
UK (Eilean Shona,Scotland)	2.7	AM				√			√(Gt)										√Kt		√								
UK (Great House, England)	2.7	AM								√																			
UK (Kilchattan, Scotland)	2.7	AM				√			√									int	√Kt	in	√					√			
UK (L.Roag, Scotland)	2.7	AM				√		√	√	√Bi-pxn					√			√	√(Kt)	√	√		√	√	√	√			
UK (Lauderdale, Scotland)	2.7	AM						√		√																			
UK (Morar, Scotland)	2.7	AD				√													√Kt										
UK (Orkney Is., Scotland)	2.7	AM				√												int	int										
UK (Riasg Buidhe, Scotland)	2.7	AM				√								√				√	√Kt	√									
UK (Rough Gill, Scotland)	2.7	AM				√																							
UK (Sanquhar, Scotland)	2.7	AM				√																							
UK (Stob a'Ghrianain, Scot.)	2.7	AM								√								√		√	√	√				√			
UK (Streap, Scotland)	2.7	AM				√		√	√	√Bi-pxn		1100–1200	40–70				√	√		√	√					√			
USA (Ashaway, White Mtns	2.10	AM				√		√				950–1200																	
USA (Boulder/Hoover dams)	2.10	AC																										Ao	Ao from lower crust/upper mantle
USA (Cambridge,White Mtn	2.10	AC			√	√		√		√									√Kt	√									
USA (Hartland,White Mtns)	2.10	AC			√	√		√	√																				
USA (Ithaca, NY)	2.10	UA				√	√		Gt-pxn		Gt-gm	759	75		√			√+int					int						
USA (Leucite Hills, WY)	2.10	LW						√	√		Px-gm			√				√		√			√			√			
USA (Missouri River)	2.10	UL			√Gt		√	√	√																				
USA (Navajo)	2.10	CM				√	√		√		Gt-gm	950–1250	100–120	√	√		√	√int		√			int						Unique megacrystalline nodules
USA (Twin Knobs, AR)	2.10	LL		√		√			√					√	√		√	√					√						
USSR (Azov/Donetz)	2.2	AC																√											
USSR (Bukantan)	2.2	AC							√		Gt-gm								√Kt										
USSR (Dzhungaria)	2.2	AC				√			√		gbr								√Kt										
USSR (Oktyabr'skiy,Azov)	2.2	AC		√																									
USSR (S.Gissar)	2.2	AM	Ol-dio			√			√	√	Gt-gm							√Cr-Di											xen assigned to crust/mantle boundar
Zambia (Kapamba)	2.4	LL													Cr-Po							√							

Three problems often lead to important contradictions between published accounts of inclusions in lamprophyres, which are dealt with in the present account as follows:

(1) *Origins* (cognate or foreign). Certain authors infer cognate, others foreign origin for what seem similar inclusions. This account describes *all* inclusions which substantially exceed the groundmass in grain-size, and show some clear evidence of foreign origin (corrosion, reaction rims, compositional difference with groundmass, etc.)
(2) *Petrological nomenclature*: Because one author's "mica-peridotite" is another's "glimmerite", separately compiled lists of inclusions from one intrusion rarely share one rock-name in common! Therefore, this account uses a primarily descriptive classification with broad petrological categories, unaffected by such difficulties. In particular, fine subdivisions of broad lithological groups adopted in Nixon (1987) — Group I and Group II spinel peridotites and pyroxenites, etc.— are not distinguished here.
(3) *Sampling*. Unless an entire pipe or dyke has been quarried and every available fragment counted, it is highly unlikely that sample suites will faithfully reflect the abundances of different inclusion types in the available outcrop, let alone those in the inaccessible parts of the intrusion. Given the additional sampling problem of a small intrusion sampling huge volumes of rock (the crustal and/or mantle), the relative *proportions* of different inclusions, even if accurately measurable, are virtually meaningless. Table 6.2 does not therefore attempt either to estimate or to draw significance from relative abundances, but merely lists those materials occurring as 'present'. Generally, lists of materials cited here are *aggregates*, and not highest common factors, of all available descriptive accounts: if one author reports, say, lherzolite, whereas another reports glimmerite, then *both* are assumed to be present even if neither account actually reports both.
(4) *Local versus regional information*. Whereas many localities in some provinces (e.g. Scottish Permo-Carboniferous), have been individually sampled in great detail, only regional information is available in others. The individual rows in Table 6.2 therefore represent occurrences ranging unavoidably from single dykes to regional dyke-swarms.

6.1 Type B (largely cognate): inclusions derived from host melt at high-*P*

These include: (1) unique zoned Ti-phlogopite 'eggs' ($\leq$3 cm) in the Keewatin (Canada) minettes (Table C1); (2) kaersutite–olivine–Cpx-plagioclase gabbroic enclaves from the fjord-parallel camptonites of Greenland, interpreted by Brooks & Platt (1975) as a cumulate sequence formed at depths of between 5 and 10 km; (3) a metre-square inclusion of layered pyroxenite-gabbro–anorthosite carrying titanaugite, minor biotite and brown amphibole in one New Zealand camptonite (Challis 1963); and (4) ultramafic (lherzolite, clinopyroxenite and wehrlite) enclaves from other lamprophyres (e.g. Wiedemanns, Greenland; Ile Bizard, Canada, in Table 6.2), which carry more Mg-rich minerals than the host but are still not sufficiently Mg-rich to represent mantle inclusions. *Banded* ultramafic inclusions in AL have mostly been interpreted as Type **B**, but *sheared* or *deformed* ones as Type **E**.

6.2 Type C (semi-cognate): inclusions derived from evolved melt

Among good examples are the green, sodic cores to phenocryst pyroxenes in many AL (Section 4.6); these are believed to have crystallized from felsic fractionates of lamprophyre, but were subsequently incorporated in later batches of lamprophyre magma and became overgrown with more Mg-rich, Type **A** or **B** pyroxenes (e.g. Bédard *et al.* 1988).

6.3 Type D (partly cognate): inclusions derived from related melts

The MARID inclusion suite (Appendix B) in KIL appears to represent one of the best examples. This may have crystallized at high-*P* from genetically related lamproite (Waters 1987). Glimmerites from the Aillik Bay (Canada) UML (Ap–Di–Am–Il–Ph) differ from true MARID in lacking rutile and carrying olivine, and could be Types **B**, **C** or **D**. Other examples of this Type are inclusions of carbonatite, nepheline syenite, etc. (Table 6.2).

6.4 Type E (related or foreign): mantle-type mafic–ultramafic inclusions

All 5 lamprophyre branches furnish good examples (Table 6.2). Mantle inclusions are distinguished from Type **B** inclusions of otherwise similar petrology by, for example, the presence of garnet or orthopyroxene, of reaction rims (e.g. Raeside & Helmstaedt 1982; Griffin & Kresten 1987), of porphyroclastic (sheared) textures (numerous references), or of chilled margins in the adjoining host (e.g. Griffin & Taylor 1975). Examples span much the same wide range of lithologies found in alkali basaltic rocks (i.e. dunites, spinel- and/or garnet-bearing peridotites and pyroxenites), but assemblages displaying metasomatism, in the development of phlogopite, kaersutite and other hydrous minerals, do seem to be particularly well developed. These include titanoclinohumite-bearing peridotites in the Navajo (USA) minettes (D.Smith 1979).

Among the most potentially controversial inclusions are *eclogites* from the Ukraine, USSR (Bayrakov 1964). These are 2–3 cm in diameter, and have sharp contacts with host camptonite. They are composed modally of ≈35% granoblastic clinopyroxene, 11–30% orange, pink and pale pink garnet with keliphytic rims (optically determined at 44% Po, 33% Gr, 23% Ai; 38–42% Po, 58–62% Ai; 18% Po, 13% Gr, 69% Ai respectively), 11–12% hypersthene, 12-17% plagioclase, 7–11% brown spinel, plus 5–12% secondary minerals indicating retrogression over a wide temperature range (carbonate, epidote, fuchsite, phlogopite, sericite and serpentine). Significantly, the host dyke, which also carries bright green clinopyroxene macrocrysts, occurs within the Ukranian crystalline shield — an area where diamonds of unknown source have been widely reported — and its inclusions are considered similar to eclogites from Siberian diamondiferous KIL. Eclogite inclusions are rare in rocks other than KIL, although they do occur in Chinese "limburgite" (a rock often confused with monchiquite), and in the "serpentinized microbreccias" intimately associated with Navajo minettes (Table C6). The latter rocks locally carry lawsonite, and are compositionally typical of high-*P*, low-*T* metamorphic regimes (Menzies *et al.* 1987a).

6.5 Type F: true xenocrysts and xenoliths of unrelated crustal rocks

Upper crustal inclusions are common in AL and UML, but reach maximum development in CAL, where they may be so abundant that the underlying stratigraphy and structure can be quite confidently inferred from them (e.g. Read *et al.* 1926; Flett 1929). Many lamprophyres, moreover, grade into (or are associated with) breccias carrying abundant country-rock material (e.g. Nixon *et al.* 1980; Rock 1983b). Among inclusions which can almost invariably be interpreted as crustal xenocrysts (since they cannot have crystallized from the host lamprophyre), are those of andalusite, kyanite, sillimanite and staurolite in Table 6.2; in some cases, these can actually be seen to have formed by disaggregation of pelitic xenoliths. Among the most curious of all crustal xenoliths, however, are quartzites forming up to 70% of certain lamprophyre dykes from Canada and Scotland (Drysdale

1915, p.31; Reynolds 1936; Rock *et al.*1986a), when quartzites are *totally absent* either at outcrop or inferred subcrop in the respective host terrains.

Fragments of *lower* crustal granulites have all now been reported from most types of lamprophyres. Perhaps best known are garnet granulites comprising ≈5% of total xenoliths in certain Navajo minettes. In Scotland, such xenoliths have proved crucial in elucidating the nature and boundaries between displaced terranes (Upton *et al.*1983). Plagioclase-rich (anorthositic and/or troctolitic Pl–Ol–Cpx) cumulate-textured xenoliths reported by Boyd (1986) differ from normally hydrous Type **B** mafic inclusions (Section 6.2); they may represent fragments of large tholeiitic layered intrusions within the lower crust.

6.6 The discrete nodule (macrocryst, megacryst) suite

This suite covers both single crystals and intergrowths, which may represent disaggregated inclusions of *any* of Types **A→F** above. Schulze (1987) divided the suite in basaltic rocks into two: "*Group A* (Al-augite, Al-bronzite, olivine, kaersutitic amphibole, pyropic garnet, spinel and plagioclase), some of which may have crystallized from their host magma, and *Group B* (anorthoclase, Ti-mica, Fe-Na-salite, apatite, magnetite, ilmenite, zircon, corundum, rutile and sphene), all of which are considered to be xenocrysts belonging originally to a more evolved magma from shallow depths." *Group A* is therefore analogous to Types **A** or **B** here, and *Group B* to Type **C**, except that Al-bronzite in Schulze's (1987) *Group A* cannot be cognate in lamprophyres and would have to be assigned to Type **E**.

6.6.1 Megacrysts of one mafic mineral

Most salitic clinopyroxene, hastingsitic amphibole and opaque megacrysts in lamprophyres have been interpreted as disaggregated Type **B** or **C** inclusions (e.g. Delaney *et al.* 1979; Raeside & Helmstaedt 1982), whereas Cr-pyrope, orthopyroxene, and Cr-diopsidic clinopyroxene megacrysts have been interpreted as Type **E** (e.g. Kresten 1976; Kresten & Persson 1975). Foreign origins become more likely, the greater the compositional disparity between megacrysts and the equivalent groundmass mineral, the more the evidence of reaction rims or resorption, and the less abundant the minerals is in the matrix (as with the hastingsite macrocrysts in the Oka and Alnö UML). For example, augite to subcalcic diopside megacrysts from the Malaita alnöite (Nixon & Boyd 1979; Nixon *et al.* 1980) show marginal zones spattered with glass and phlogopite, and are distinctly poorer in Ti, Ca, and richer in Na than groundmass pyroxenes, but also poorer in Mg, Cr than diopsides from associated Type **E** lherzolite inclusions; hence they most probably belong to Types **B** or **C**.

Isotope data on mafic macrocrysts in lamprophyres are virtually non-existent, but Barreiro & Cooper (1987) found present-day $^{87}Sr/^{86}Sr$ ratios of 0.7029–0.7084 and ε_{Nd} of +3.5–5.2 in amphibole and clinopyroxene macrocrysts, which fall within the range of initial ratios of the host Alpine lamprophyres (Fig.5.12a). Model Nd ages are Palaeozoic, but the minerals appear to have been in isotopic equilibrium at the time of emplacement of their hosts, ≈ 25 Ma ago. Hence these are probably examples of Types **A** or **B**.

6.6.2 "Megacrystalline rocks" and intergrowths of two or more mafic minerals

Among more noteworthy intergrowths reported from lamprophyres are the following:

(1) "An unusual very magnesian clinopyroxene with lamellae of pargasite" and "a unique clinopyroxene with oriented ilmenite and sulphide rods" (Boyd 1986).

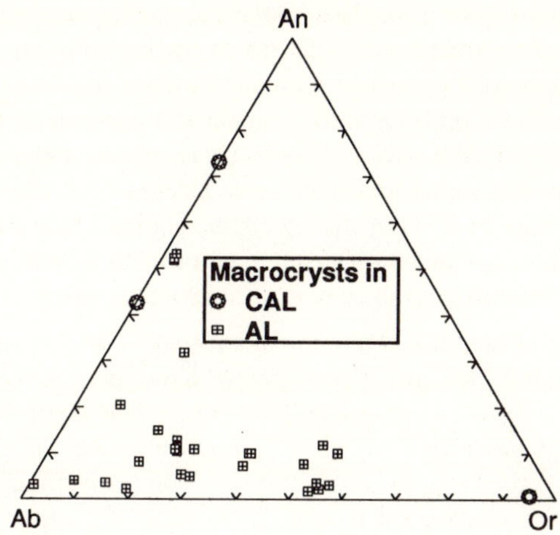

Fig.6.1 *Compositions of lamprophyre feldspar macrocrysts* (mole %). Data from LAMPDA (Appendix D).

(2) Ultra coarse-grained intergrowths of Cpx–Gt–Il–Opx–Ol–Ph from the Navajo minettes, which are probably related to the associated garnet-peridotites, and hence Type **E**.

6.6.3 Feldspar megacrysts and feldspar intergrowths

Rounded to almost perfectly euhedral, sometimes doubly terminated macrocrysts of alkali feldspar and (more rarely) plagioclase, some of gem quality and up to several centimetres in length, are fairly common in AL and an occasional feature of CAL (Table 6.2). They occur irrespective of whether the host-rock carries groundmass feldspar. Extremely widespread Scottish examples were for many years thought to be fossil fish teeth (Eyles 1924)! Feldspathic discrete nodules include corundum-bearing anorthoclasites (Upton *et al.* 1983), and sanidine–oligoclase aggregates (Lucchini *et al.* 1969).

Compositionally, 26 analyses from AL and 3 from CAL span a wide range, similar to that in their hosts and in alkali basalts in general (cf. Fig.6.1 with Fig.4.6d, fig.234 of Schulze 1987 and with Chapman & Powell 1976). Zoning in individual crystals tends to be not more than about 10% Or content. Contents of Fe_2O_3, SrO and BaO range up to 1.2, 1.5 and 4.9% respectively. Structurally, these megacrysts vary from high-sanidine and anorthoclase to maximally ordered microcline cryptoperthite, but individual crystals tend to be structurally homogeneous, the major reported inhomogeneity being thin anorthoclase rims around sanidines. Where coexisting plagioclase megacrysts occur, these tend to be high-oligoclases (e.g. Pieri *et al.* 1969).

Chapman (1975a) showed that a 9:1 mixture of anorthoclase and alkali basalt had liquidus anorthoclase under *dry* conditions and at <9 kbar, and therefore felt anorthoclase megacrysts might derive in part from arkose (Type **F**), in part by equilibrium crystallization (Types **A** or **B**). Val'ter & Yeremenko (1974) also inferred from consideration of solvus-solidus interactions that alkali feldspar megacrysts must reflect dry, high *P–T* crystallization conditions (Type **B**). Unfortunately, lamprophyres cannot crystallize under dry conditions, so that these experiments may be misleading. Furthermore, Chapman & Powell (1976) later argued that: "anorthoclase megacrysts ... did not crystallise from the host magma", and used density calculations to suggest that both anorthoclase and mafic megacrysts were "picked up from a high-level partly crystallised magma pool containing a more evolved alkali basaltic differentiate by a later surge of more basic liquid, which then carried both anorthoclase megacrysts and fragmented (mafic) cumulates to the surface" (i.e. Type **C**).

Two set of isotopic data have led to equally conflicting interpretations, although in both cases the megacrysts show low ratios which are in turn slightly lower than their host-rocks, closely reflecting data obtained from alkali basalts (Stuckless & Irving 1976):

(1) Lucchini *et al.*(1969) found one megacryst to have an $^{87}Sr/^{86}Sr$ ratio of 0.7051, in a host camptonite with 0.7059, and, after a detailed X-ray and chemical study, concluded that it represented Type **A**. Unfortunately, the error on their isotopic measurements was not given, so 0.7051 and 0.7059 may not in fact be statistically 'identical'.

(2) Foland *et al.* (1980) found ratios of 0.70298–0.70396 for anorthoclase megacrysts in one camptonite with whole-rock ratio 0.70456 (0.70316 after acid leaching), and inferred a "high-pressure (upper mantle-lower crust) origin" (i.e. Type **B**).

On balance, several lines of evidence favour classifying feldspar megacrysts as Type **C**:

(a) Feldspar *pheno*crysts (Type **A**) are most unlikely in lamprophyres, because feldspars grow slowly in these volatile-rich media and are known to be suppressed until the latest stages of crystallization in hydrous melts (Yoder & Tilley 1962; Rock 1978).

(b) Given that feldspars normally crystallize last, the homogeneity and high/medium temperature structures of these feldspar megacrysts are difficult to reconcile with the liquidus and solidus temperatures of typical AL (1050°C and 850°C: Rock 1978): some degree of inversion and/or exsolution might be expected if they represented Type **A**.

(c) Eu anomalies in the host-rocks might be expected if these megacrysts had separated directly from the host melt, but these are entirely lacking in AL (Fig.5.6c).

(d) Although insufficient data are available to generalize the extent to which macrocryst and groundmass feldspars in one rock differ, up to *three* megacryst compositions have been found to coexist with *different* groundmass feldspars (Pieri *et al.* 1969).

(e) A crustal feldspathic source (e.g. granitoids, clastic sediments) can be effectively ruled out, because such feldspars would have higher $^{87}Sr/^{86}Sr$ and are known to raise ratios when involved in lamprophyre petrogenesis (Powell & Bell 1970), because granite feldspars would not share the high Fe, Sr and Ba, and because granites are not spatially associated with most macrocryst-bearing AL. A Type **F** origin therefore seems unlikely.

(f) The appearance of feldspar macrocrysts seems linked to magmatic differentiation, in that they can be the *only* inclusions in differentiated AL: Monchique lamprophyres, for example, have low *mg*, Cr and Ni and *resorbed* olivine, but no ultramafic inclusions, indicating fractionation, and yet they do carry orthoclase macrocrysts (Table 6.2).

Given the actual presence of nepheline syenite inclusions in some AL (Table 6.2), and the phonolitic composition of the base in some monchiquites (Section 4.19), perhaps the most likely source for these macrocrysts is *phonolitic differentiate* — a magma known to have consistently low Sr ratios (e.g. Rock 1976a), which is also intimately associated with many AL (Section 8.5), and which has also been invoked as the source of green cores to pyroxenes (Section 6.2). A role for nepheline syenites in AL genesis is therefore considered further in Section 8.4.1.

6.7 Abundance of different inclusion types in different lamprophyres

The relative abundances particularly of crustal- versus mantle-type inclusions in such rocks as alkali basalt, KIL, LL, and the various other lamprophyres may provide supplementary information on relative depths of origin, magma formation in the mantle, mode of irruption, extent of crustal involvement, etc. Some of the otherwise admirable summary in Nixon & Davies (1987) now needs significant revision in the light of the discovery of diamonds and other deep-seated materials in rocks other than KIL and LL (Section 9.1). In so far as generalizations are possible, these are made below:

	Crustal (Type F)	Mantle (Type E)	Discrete Nodule Suite	Diamond
CAL	Very abundant	Very rare (Navajo only)	Very rare (Navajo only)	Not known
AL	Variably abundant	Often abundant and varied	Often abundant and varied	Very rare
UML	Variably abundant	Often abundant and varied	Unrivalled in abundance	Very rare
KIL	Variably abundant	Usually abundant and varied	Usually abundant and varied	Barren to rich
LL	Variably abundant	Uncommon and invariable	Uncommon; often absent	Barren to rich

Although mantle-type inclusions are almost certainly far rarer in LL and (especially) CAL than in KIL, UML or AL, the rarity and lack of variety among LL inclusions is curious and hitherto unexplained, in view of their diamondiferous character. However, more precise rankings are not really possible. For example, the *proportion* of the relatively small number of UML occurrences which bear rich inclusion suites is extremely high (Table 6.2), but it is really too small a sample on which to base comparisons with, say, KIL.

In a number of areas, lamprophyres appear to carry proportionately more mantle material than contemporaneous basalts, taking into account relative host-rock abundances: monchiquites, for example, have yielded some of the most varied and abundant mantle material among Scottish Permo-Carboniferous intrusions (Rock 1983a; Upton *et al.* 1983; many localities in Table 6.2). Again, despite the much greater volumes of alkali basalts erupted, all but one of the known inclusion localities in Greenland are in lamprophyres (Scott Smith 1987). Despite this, the number of inclusion-rich intrusions within individual lamprophyre swarms may be small: Brooks & Rucklidge (1973) found only one among 25 AL dykes in Greenland, and the ratio in Scotland is smaller still.

Perhaps the nearest to a quantitative comparison is afforded by total numbers of inclusion localities worldwide. Table 6.2 records over 80 localities for lamprophyres (many of which are composite, i.e. represent numerous individual intrusions), that is, 2% of the ≈3,500 inclusion localities recorded by Nixon (1987, p.741) in each of alkali basalt and kimberlite hosts. On the assumption that lamprophyres constitute < 2% of the total volume of alkali basalts, but that AL and CAL substantially exceed the volume of kimberlite magmatism (Table 2.1), this suggests that AL are significantly more inclusion-rich than alkali basalts, but considerably less so than kimberlites.

6.8 *P-T* significance of inclusion suites

Accurate *P-T* estimates are hampered by 2 problems: (1) the plethora not only of available geothermometers and geobarometers, but also of equations for calculating *P* and *T* based on given mineral-pairs; (2) strong zoning, which indicates disequilibrium and leads to great uncertainty even in the answers from one equation. These problems are irresolvable within the present scope, so Table 6.2 merely includes various authors' *P-T* estimates for particular inclusions. Not surprisingly, these vary considerably both within and between different lamprophyres. Among the most intriguing, however, are the garnet-peridotites from the Navajo, which suggest the host minettes arose from >120 km depth — not far short of the diamond stability field at 150 km. Since minette and lamproite are so closely related petrologically (Chapters 4–5), and since lamproites are commonly diamondiferous, this may also make minettes potential diamond exploration targets (Section 9.1).

7 Plutonic and Volcanic Equivalents of Lamprophyres (with a contribution by A.E.Wright and D.R.Bowes)

As already shown (Section 3.1), lamprophyres are overwhelmingly hypabyssal, medium-grained intrusions. Although LL commonly exist as volcanic rocks, and AL and CAL do so rarely, KIL and UML almost never do (Section 3.1.7). Again, although relatively coarse-grained (e.g. pegmatoidal) segregations do exist in some dykes (e.g. Macdonald *et al.*1986), no example of a true lamprophyre 'pluton' has ever been identified. Because most writers can see no intrinsic reason for this (R.H.Mitchell 1986, p.43 found the absence of 'plutonic kimberlite' particularly strange), this raises the question of whether *other* named plutonic or volcanic rock-types might be equivalent to lamprophyres. The old suggestion that LL are simply volcanic lamprophyres (Section 1.3.1) can be dismissed — they are merely one *type* of lamprophyre — but other suggestions deserve more careful attention.

It must be stressed that we seek *chemical* but not always *mineralogical* equivalence here: for example, lamprophyres carry minerals such as leucite and melilite which only crystallize in low-*P* regimes, and would not be expected to occur in plutonic rocks.

Table 7.1 Chemical equivalence of the appinite suite and vaugnérite series with calc-alkaline lamprophyres

	Appinite suite					Vaugnérite series								
	Appin			Biella	Av.	Massif Central					Various			Av.
	1	2	3	4	5	6	7	8	9	10	11	12	13	14
SiO2	49.0	48.9	48.3	50.5	52.2	46.7	52.7	55.6	49.0	54.2	51.1	54.7	59.7	51.5
Al2O3	13.8	12.5	12.7	15.5	15.1	17.4	15.6	15.8	17.7	13.3	14.5	18.2	13.3	12.8
Fe2O3*	9.1	9.2	9.9	9.9	8.4	7.9	8.3	7.4	7.9	9.4	9.0	7.5	5.2	7.3
MgO	8.6	14.8	11.6	8.2	6.8	9.3	9.2	7.0	9.4	7.2	8.2	5.3	6.0	7.1
CaO	7.7	7.2	8.4	8.7	6.9	8.4	7.1	5.6	6.0	6.6	5.1	6.5	3.0	6.7
Na2O	2.8	2.6	2.4	2.6	3.4	1.3	2.2	3.5	2.2	2.1	1.9	3.0	2.0	2.0
K2O	2.1	2.2	2.1	1.1	2.1	4.5	2.4	1.6	4.9	4.4	7.3	2.6	7.3	5.6
TiO2	0.8	0.6	1.0	1.5	1.1	1.7	0.8	1.3	1.8	1.1	1.8	1.2	0.8	1.3
P2O5	0.3	0.3	0.3	0.4	0.4	0.9	0.4	0.3	0.5	0.5	—	0.5	0.8	1.0
CO2	3.3	0.4	1.3	—	0.8	—	—		—	—	—	—	—	2.0

For further details of individual examples, see Appendix C, Table C5

[1] Mean of 12 lamprophyres, Appin-Kentallen area, Scotland (Wright & Bowes 1979)
[2] Mean of 19 appinites, type area, Appin-Kentallen area, Scotland (Wright & Bowes 1979)
[3] Mean of 8 kentallenites, Appin-Kentallen area, Scotland (Wright & Bowes 1979)
[4] Appinite, Serie dei Laghi, Biella, Italy (Giobbi Origoni et al.1990)
[5] Average spessartite (Table 5.1)
[6] Vaugnérite, type area, Vaugnéray, Massif Central, France (Sabatier 1980)
[7] Vaugnérite, Estourocs, Massif Central, France (Sabatier 1980)
[8] Lamprophyre, Estourocs, Massif Central, France (Sabatier 1980)
[9] Vaugnérite, Pinay, Massif Central, France (in Sabatier 1980)
[10] Lamprophyre, Pinay, Massif Central, France (in Sabatier 1980)
[11] Type durbachite, Durbach, Schwarzwald, Germany (in Johannsen 1938)
[12] Durbachite, Zvikov, Czechoslovakia (Holub 1977)
[13] Redwitzite, Grafenstein, Czechoslovakia (Troll 1968)
[14] Average minette (Table 5.1)

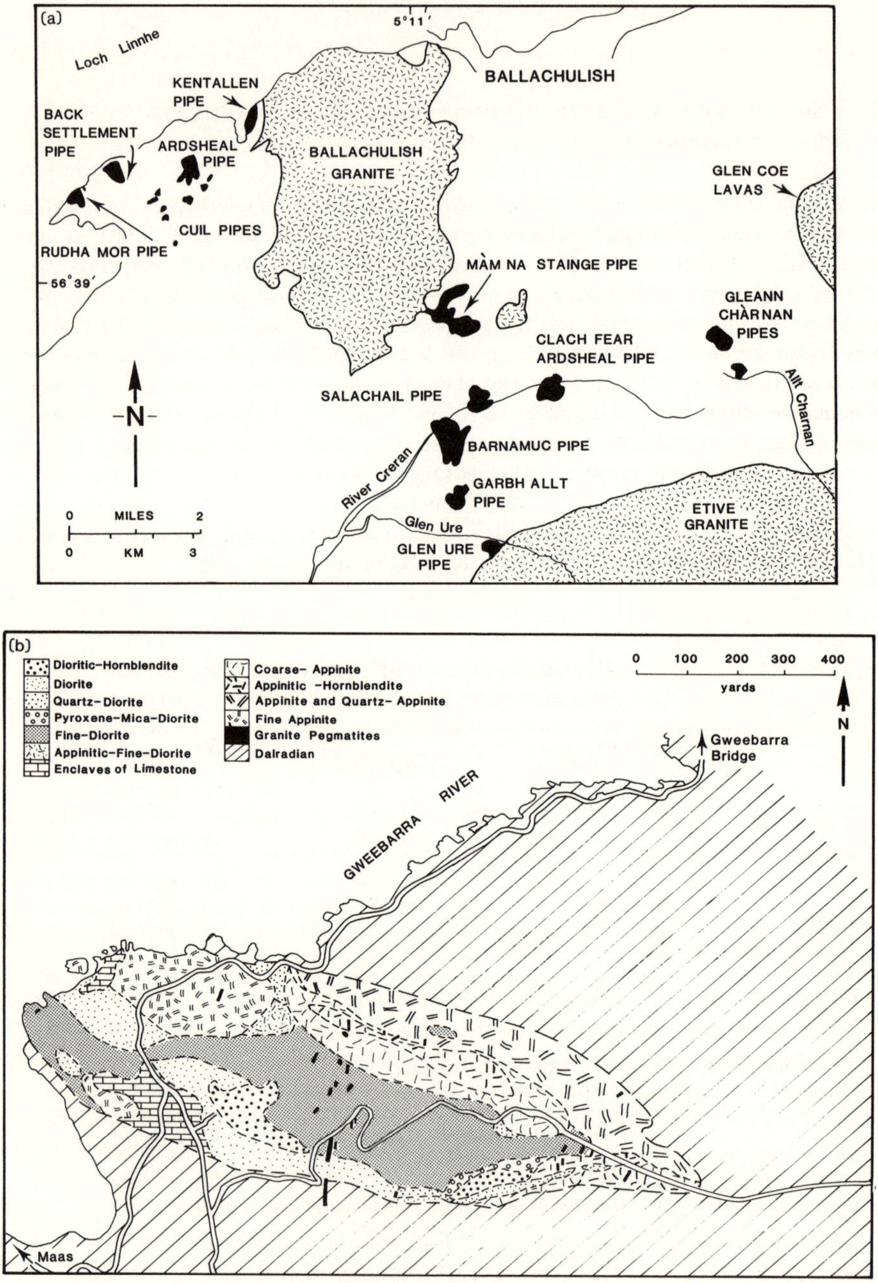

Fig.7.1 *The appinite suite in the Scottish Caledonides.*

(a) Distribution of appinite and kentallenite pipes in the Appin type area, around the Ballachulish pluton (for location and smaller-scale appinite distribution, see Fig.2.6).

(b) Internal structure of an appinite pipe from the Irish Caledonides (partly after French 1966).

7.1 Plutonic and volcanic equivalents of calc-alkaline lamprophyres

Two main rock-groups are considered first, then a miscellany of minor equivalents: (1) the *appinite suite* (appinites, kentallenites, etc: Appendix B; Bowes & McArthur 1976), regarded as plutonic spessartites and vogesites; and (2) the *vaugnérite series* (durbachites, redwitzites and vaugnérites: Appendix B), which may be plutonic minettes and kersantites. The terms *suite* and *series* (*sensu* Rock 1981; Appendix B) indicate the contrast that whereas members of the appinite suite commonly occur together consanguinously, those of the vaugnérite series generally do not. Furthermore, each constituent rock-type has ultramafic, mafic and intermediate variants. Although their names may be obscure to many readers, these rocks appear to be fairly widespread. As they have never previously been reviewed, and as petrologists have long recognized the genetic significance of their intimate association with granitoid plutons (e.g. Pitcher & Berger 1972), they are given significant attention here. Nevertheless, this section should be regarded as condensed and preliminary. Petrogenetic discussion is deferred as usual to Chapter 8. Table C5 (Appendix C) cites sources of supplementary information specifically on these rocks.

7.1.1 The appinite suite: plutonic spessartites and vogesites

The appinite suite was originally described from the Scottish Caledonides, where it is intimately associated with late-tectonic (≈400 Ma) granitoids (Associations A,C: Table 2.2). It displays 4 modes of occurrence (cf. Section 3.1), in approximate abundance as follows: (1) small irregular pipes or vents, clustered by the dozen or even hundred around the related plutons (Fig.7.1a); (2) mafic enclaves and schlieren within the plutons; (3) thick independent dykes or sills; (4) border facies of the plutons. Evidence that so-called "dioritic" enclaves in granitoids are also appinitic is accumulating (e.g. Holden *et al.*1988).

Appinitic rocks are much coarser-grained than associated intrusions of similar size but more felsic composition; this has almost universally been ascribed to high melt volatile contents. The connection with lamprophyres has been established in the field by direct observation of chilled lamprophyric margins and by dykes of lamprophyre originating from intrusions of appinite or kentallenite; it is confirmed by mineralogical and bulk compositional similarity (Fig.7.2–7.3 and see below). The sequence of intrusion in the plutons is usually ultrabasic→acidic, and the appinite pipes tend therefore to be contemporaneous with the earliest plutonic phase. The type rocks of Appin are clustered around the slightly later Ballachulish Granite; this is also the one area where appinites are also closely associated with shoshonitic lavas (Fig.7.1a).

The suite's most distinctive members are appinite itself, and kentallenite, but cortlandtite (Appendix B), diorite, granodiorite, mica±pyroxene-hornblendite, mica-peridotite, mica-pyroxenite, scyelite and syenite also occur, with pyroxene-mica-diorite perhaps dominant. Explosion breccias commonly form *in situ* from country rocks, and may be mobilised into alloclastic intrusive vents (Wright & Bowes 1968; Platten 1983, 1984; Platten & Money 1987). These may in turn develop into breccia dyke-swarms (Platten 1982a).

The apparent rarity of the appinite suite in the Scandinavian and American Caledonides may or may not be real, for appinites have been recognised outside the British Isles primarily by geologists with British connections, who have themselves seen the Scottish appinites first hand. Appinites are consequently 'abundant' in countries such as Australia (e.g. Jennings & Sutherland 1969), and more notably Tanzania, where no less than 22 of the 119 published quarter degree sheets map extensive appinites or kentallenites (Tables C1,

C5). Similar rocks are almost certainly disguised under different names in other countries.

Petrologically, the appinite suite comprises a wide range, but only rocks comparable to lamprophyres are detailed here, the descriptions being condensed largely from Pitcher & Berger (1972) and Hamidullah & Bowes (1987).

Coarse-grained *ultramafic rocks* make up a small but consistent proportion of intrusions,

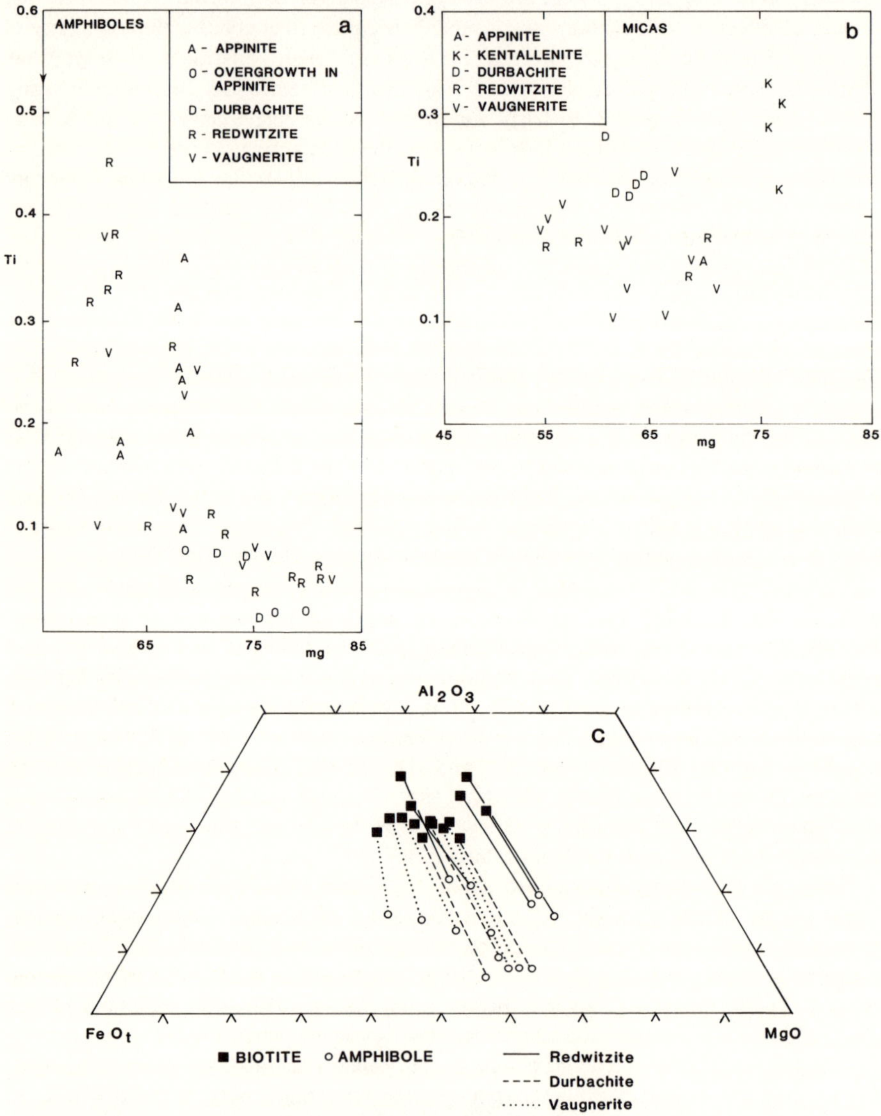

Fig.7.2 *Comparative mineralogy of the appinite suite, vaugnérite series and CAL.*
(a) Amphibole compositions (axes expressed in formula units).
(b) Biotite compositions.
(c) Illustration of the coherence of coexisting biotites and amphiboles.

practically always along the margins. They comprise varying proportions of olivine, clinopyroxene, hornblende and biotite together with rather sodic plagioclase and orthoclase. The characteristic feature of *kentallenite* is the mantling of large olivines (Fo_{83}-Fo_{77}) by orthopyroxene and sometimes clinopyroxene or hornblende, all poikilitically enclosed by phlogopite. The olivines nevertheless appear to have been in equilibrium with the melt. Also occurring as phenocrysts are strongly zoned clinopyroxene (diopside, with Al↑ and Ti↑ core→rim) and Ti-phlogopite (4–6% TiO_2, commonly with a dark, high Ti-Fe core, a lighter, lower Ti-Fe banded zone, and a dark, high Ti-Fe rim). The matrix is made up of smaller grains of olivine (≈Fo_{80}), phlogopite, magnetite and apatite in a coarse interlocking network of feldspar with a variable plagioclase/ anorthoclase ratio (always > 0). *Appinites,* though extremely diverse (notably in colour index), are characteristically composed of very coarse, acicular or stumpy, zoned pargasitic hornblende (commonly with a diopside core, complex overgrowths of actinolite- ferroactinolite showing Ti↓ core→overgrowth), in a matrix of plagioclase and orthoclase. Small grains of corroded or fractured diopside (identical to that coring the amphibole), hornblende, biotite, minor quartz and primary calcite, plus accessory apatite, magnetite and sphene, may occur in the matrix. These rocks differ from ordinary calc-alkaline diorites in the same way that CAL differ from andesites (Sections 1.6.2, 7.2.3) — notably in their relatively Ti- and Mg-rich mafic minerals. They differ from alkali gabbros in the greater abundance of hornblende and biotite, and in the more diopsidic composition of their pyroxene. Crystallization clearly involved early olivine, followed by clinopyroxene-amphibole and finally biotite, with feldspars suppressed until last.

Amphibole is perhaps the most significant mineral of the suite. Complex zoning may involve up to 18 internal bands, varying primarily in *mg* (Fig.7.2a): 4 actinolite bands, 3 darker ferroactinolite bands, with up to 11 further alternations within the middle band on some crystals. However, the number of overgrowth bands is constant within any one pipe, suggesting that a varying physical constraint was active on a different scale in each pipe. Minor phases, such as magnetite, apatite and especially sphene, also seem to have played a significant role in controlling overgrowth composition (notably the drop in Ti from pargasitic core to actinolitic overgrowth). These overgrowths, together with amphiboles in some vaugnérites (next section), are the only known actinolites of definite igneous origin. Si per formula unit scatters from 7.3–7.9 — far in excess of the maximum of 7.45 suggested for igneous amphiboles by Leake (1971), but comparable to hastingsites from some granodiorites (Czamanske & Wones 1973). This peculiarity may imply higher pressure crystallization than the pargasite cores.

Chemically, appinite suite rocks are generally interpreted as the result of differentiation of a basic magma rich in K_2O and volatiles, and the term *shoshonitic* is frequently applied, although the percentage of alkalis is not quite as high as in the type shoshonites. What little isotope work has been done relates to the full range of types at the Garabal Hill Complex, where peridotite, hornblende gabbro, gabbro, pyroxene-mica-diorite, appinitic diorite and granodiorite all have fairly low initial $^{87}Sr/^{86}Sr$ ratios (0.7014–0.7080, mean 0.705), precluding derivation by anatexis of local basement rocks. Contamination is significant only in samples which contain visible metasedimentary xenoliths (these give ≈0.710).

7.1.2 The vaugnérite series: plutonic minettes and kersantites

This series has been reported only from the European Hercynides. However, the recent assertion (Holl *et al.*1989) that the type redwitzite is in fact a *Caledonian* rock metamorphosed in the Hercynian orogeny, raises the possibility that its connection with the Caledonian appinites may be closer than hitherto assumed. Among the most extensive examples are vaugnérite enclaves in the Margaride Granite, France (Table C5), which lie on a lineament extending for 20 km across the pluton, completely independent of the internal compositional zonation. They have autoliths and associated intrusions of biotite-hornblendite, mica- and hornblende-syenite and diorite. From the brief mineralogical descriptions (Appendix B), they appear to share similar parageneses, but average K_2O contents increase in the order appinite→redwitzite→ vaugnérite→durbachite (Fig.7.3a).

Published opinions on the origins of the vaugnérite series echo the now-defunct granite controversy of the 1950s, though hybridization between basic and granitic magmas has proved particularly popular. The pale actinolitic amphibole, abundant evidence of reaction, and common occurrence as enclaves have also been used to support a solid state (e.g. metasomatic) recrystallization origin for both durbachite and vaugnérite. Among strong evidence in favour of an orthomagmatic origin, on the other hand, is the presence of a spessartite dyke cutting one vaugnérite intrusion, the dyke containing phenocrysts and microlites of both Mg-biotite and actinolitic hornblende similar to those in the vaugnérite (Sabatier 1980). There is also evidence, from their spatial distribution along major faults and lineaments, that deep-seated fractures may be a controlling factor in their emplacement.

Parageneses and crystallization histories in all these rocks are complex and variable. Mineralogical features which ally them with lamprophyres include the following:

(1) globular structures, with biotite surrounding quartz and K-feldspar (cf. Section 3.2.4);
(2) salitic pyroxenes, with complex zoning (cf. Section 4.6);
(3) commonly glomeroporphyritic amphiboles showing kaersutitic cores zoned to actinolitic hornblende rims (Fig.7.2a), with complex inclusions and overgrowths;
(4) generally Ti-rich biotites (Fig.7.2b; cf. Section 4.8);
(5) plagioclases which, though dominantly andesine, have cores as basic as An_{80} (cf. Fig.4.6).

Some durbachites differ from normal lamprophyres, however, in their large porphyritic K-feldspars and sometimes in the presence of orthopyroxene. Coexisting biotite and amphibole compositions are highly coherent (Fig.7.2c).

Chemically, the vaugnérite series shows a much wider range in composition than the more closely defined appinite suite, extending from <45% to > 60% SiO_2, with variable but always high MgO and K_2O. K_2O and Mg/Fe ratio clearly distinguish the three major rock-groups (Fig.7.3a) and also show that they are, on the whole, more K- and Mg-rich than the appinite suite, echoing the chemical relationship between minettes on the one hand and vogesites/spessartites on the other (Métais & Chayes 1964b; Section 8.2.1). Isotopic data are completely lacking, but what trace element and REE data are available do confirm the broad similarity of these rocks to minettes and kersantites (Fig.7.3b; cf. Fig.5.6).

7.1.3 Calc-alkaline lamprophyres versus plutonic and volcanic shoshonitic rocks

The intimate association ('B' of Table 2.2) between CAL and Joplin's (1968) shoshonite suite has already been examined (Section 2.4.2). Velde (1971b) and Wimmenauer (1973b) have identified close chemical equivalents to CAL among absarokites, latites, monzonites, murambites and shonkinites (Appendix B). Fig.7.4a shows that some CAL lie within the

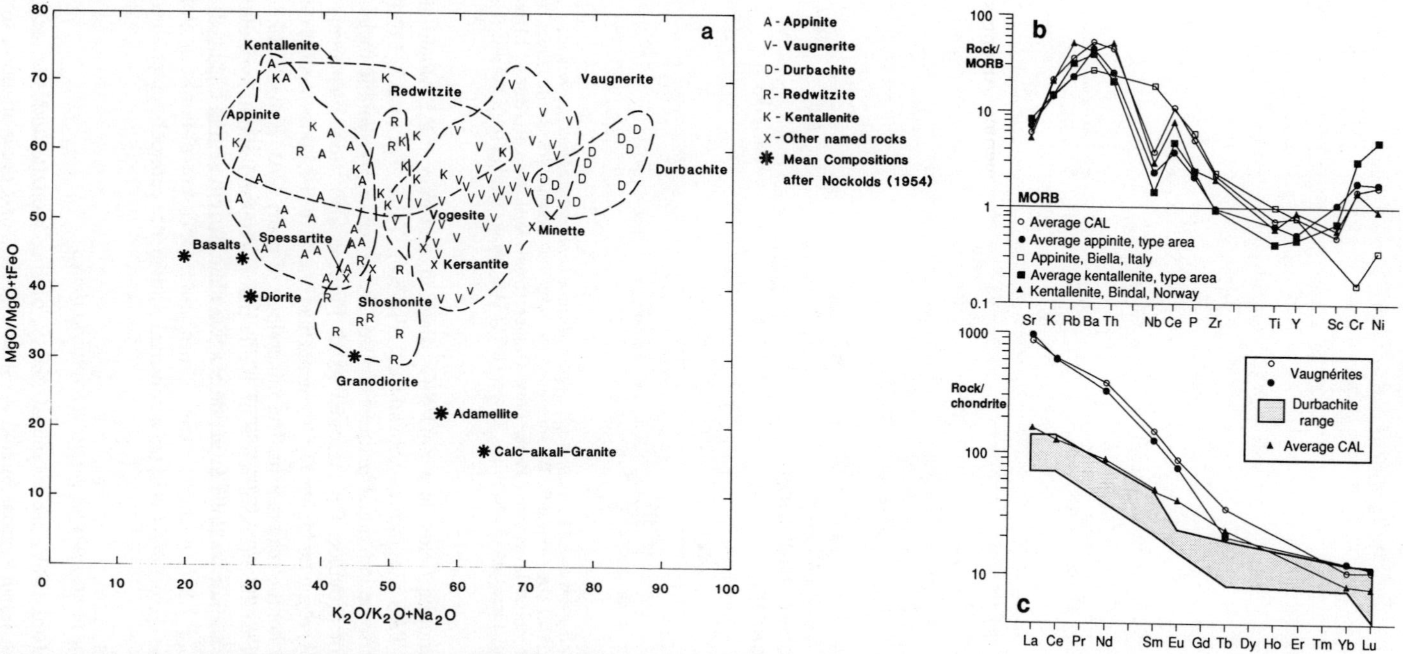

Fig.7.3 *Comparative whole-rock geochemistry of the appinite suite, vaugnérite series and CAL.*
(a) Relationships between the various rock-types on a $K_2O/[K_2O+Na_2O]$ versus *mg* diagram.
(b) MORB-normalized spidergrams (cf. Fig.5.6a). Data from Table 5.1, Wright & Bowes (1979), Sabatier (1980), Giobbi Origoni *et al.*(1988) and Nordgulen & Roberts (1988).
(c) REE plots (cf. Fig.5.6c). Data from Table 5.1, Gil Ibarguchi *et al.*(1984) and Bouska *et al.*(1984).

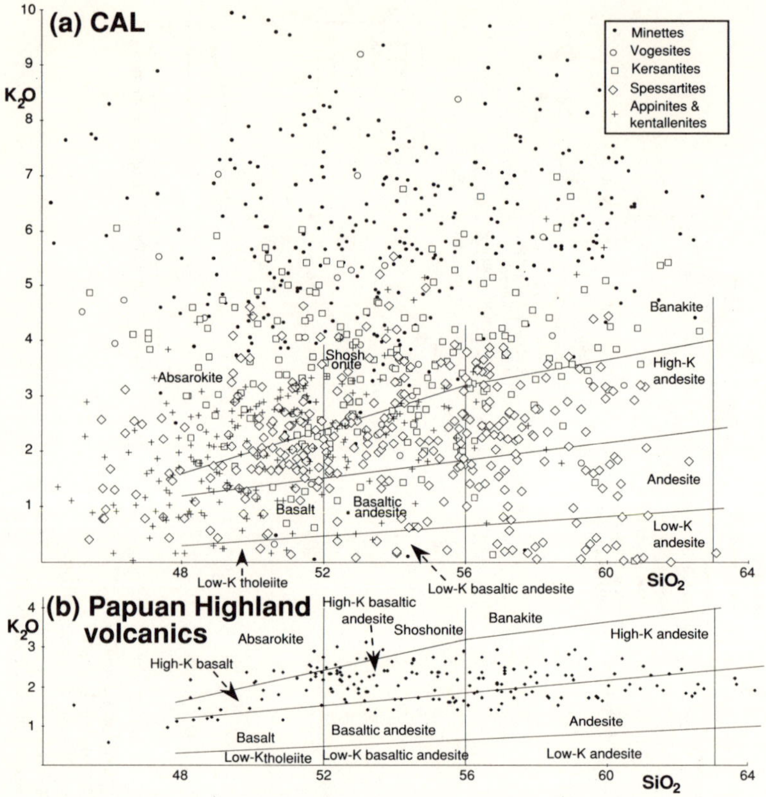

Fig.7.4 *Comparative whole-rock geochemistry of calc-alkaline lamprophyres and shoshonites.* 1590 CAL from LAMPDA (Appendix D) are compared with 180 Tertiary-Recent tholeiitic, calc-alkaline and shoshonitic lavas from Papua New Guinea (from Australian Bureau of Mineral Resources database). Fields after Peccerillo & Taylor (1976). Data recalculated to 100% volatile-free.

absarokite or shoshonite fields on a K_2O–SiO_2 diagram, and many CAL would also be classified via the TAS diagram as shoshonites (Fig.5.1). Table 7.2 (nos.1–11) further shows how similar some individual analyses of CAL are to certain shoshonitic rocks in the same occurrence or province. On the other hand, Fig.7.4a also shows that many CAL (especially minettes) reach far higher K_2O contents than one of the best documented regional expressions of shoshonitic magmatism. The mineralogy of many CAL is also unlike many shoshonites, not least in the occurrence of Ba-Ti-phlogopite and alkali pyriboles, and furthermore, CAL are not confined to the destructive plate margin setting characteristic of shoshonites (Table 2.2). In short, *some* shoshonitic rocks are plutonic or volcanic equivalents of *some* associated CAL, but a universal equivalence cannot be entertained.

7.1.4 Calc-alkaline lamprophyres versus volcanic boninites

Boninites (Crawford 1989) are a type of glassy, olivine-bronzite andesite, originally described from the Japan region. Suzuki & Shiraki (1980) first drew attention to major

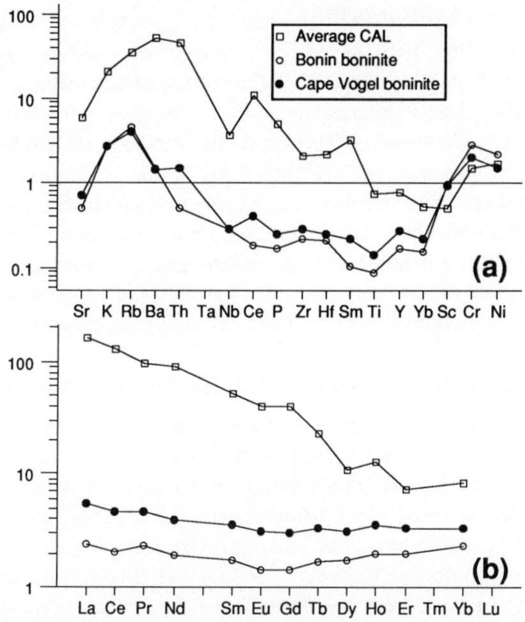

Fig.7.5 *Comparative whole-rock geochemistry of calc-alkaline lamprophyres and boninites.* Average CAL (from Table 5.1) is compared with individual boninites from the type-locality (Chichijima, Bonin Is.) and from Cape Vogel peninsula, Papua New Guinea (Cameron *et al.* 1971; Crawford 1989). Normalizing values and element orders as in Fig.5.6a.

element similarities between the type rocks and certain Japanese spessartites, both being distinguished from normal arc andesites by higher Mg, Cr and Ni (Table 7.2, nos.12-13); the main difference is higher TiO_2 and K/Na in the spessartites. Of course, boninites bear little *mineralogical* resemblance to CAL, and in trace element composition, they are far less enriched in incompatible elements (Fig.7.5). Nevertheless, CAL perhaps better approximate LILE-enriched boninites than LILE-enriched common andesites, since their chemistries clearly merge at the compatible end of the spectrum on Fig.7.5. The appearance of the chapter covering lamprophyric rocks (Foley & Venturelli 1989) in a book on boninites (Crawford 1989) testifies to the increasing recognition of links among these mutually high-MgO, high-SiO_2 rocks.

7.1.5 Minettes versus volcanic olivine-leucitites (ugandites)

Minettes have been reported grading into olivine-leucitites (Chatterjee 1974; Williams *et al.* 1982, p. 231), and the two are also theoretically equivalent (cf. Velde 1971a) thus:

$$\text{[Pyroxene]} + KAlSi_3O_8 + 2KMg_3Si_3AlO_{10}(OH)_2 \equiv \text{[Pyroxene]} + 3KAlSi_2O_6 + 3Mg_2SiO_4 + 2H_2O$$

$$\text{MINETTE} \equiv \text{OLIVINE LEUCITITE (UGANDITE)}$$

However, olivine-leucitites have quite different geochemistry as well as mineralogy from minettes: in particular, their modal leucite/olivine ratios and pyroxene contents are invariably too high, and minette phlogopite/orthoclase ratios are too low, to confirm the above (or similar) equivalences (see Rock 1984 for more detailed comparisons).

7.1.6 Minettes versus volcanic selagites

Pleistocene lavas near Pisa, Italy, have been variously described as "young equivalents of minettes" (Rosenbusch 1923), "selagites", "glimmersyenites" (Johannsen 1938; Barberi & Innocenti 1967), "K-richterite bearing lamproites" (Wagner & Velde 1986a) and "rocks related to lamproites" (Bergman 1987). Despite the presence of K-richterite, all available whole-rock analyses have substantially higher Al_2O_3 and Na/K than most LL, and lack normative *ac*, *ns* and *ks*. Micas are also Al-rich compared with typical LL micas. Like many of the Spanish lamproites, they classify as CAL on discriminant functions (Section 5.1), and as they lie chemically between average minette and lamproite in Table 5.1 (see Table 7.2, nos.8–9), selagites are perhaps best regarded as cocites (Appendix B). Once again, though, they do appear to represent a reasonable volcanic equivalent of *some* minettes.

7.2 Plutonic and volcanic equivalents of alkaline lamprophyres

There is complete gradation between AL and alkali basaltic rocks on all scales, from AL as individual members of regional basaltic dyke-swarms (e.g. coast-parallel swarm in Greenland: Table C2), to local basalt–camptonite gradation within single dykes (e.g. Ramsay 1955; Gallagher 1963). Fig.5.1 further confirms that many AL would be classified on the TAS system as alkali basalts or basanites. Given the higher H_2O and CO_2 in AL, which leads to amphibole and biotite crystallization (Fig.1.3; Table 1.2), there is no problem in equating AL with broad *volatile-enriched* equivalents of alkali basaltic rocks (as in Wimmenauer 1973b).

It is simply a logical extension, therefore, to seek plutonic AL equivalents among amphibole/biotite-rich alkali gabbroic rocks such as berondrites and mafraites (Appendix B), which are commonly found within alkaline intrusions cut by AL dyke-swarms. Even though such rocks have normally undergone extensive accumulation, combined in many cases with metasomatism — so that precise bulk compositional equivalence cannot necessarily be expected — it is not difficult to find major element analyses of AL and gabbroic rocks from such provinces as Iberian, Monteregian or Oslo, which have virtually indistinguishable chemistry (e.g. Table 7.2, nos.14–17), as well as sharing the same mineral assemblages (i.e. combinations of Am or Bi with Ol, Cpx, Pl, Af, Ne, Ac).

7.3 Plutonic and volcanic equivalents of ultramafic lamprophyres

7.3.1 Ultramafic lamprophyres versus volcanic melilitites

Because UML and melilitites are locally associated (e.g. Shrbeny & Machacek 1974), and are the only orthomagmatic rocks to carry melilite, many authors (e.g. Davies 1952) have assumed chemical equivalence *a priori.* The following equations illustrate this in theory:

$$3KMg_3Si_3AlO_{10}(OH)_2+2CaMgSi_2O_6+4CaCO_3 \equiv 4Mg_2SiO_4+3Ca_2MgSi_2O_7+3KAlSiO_4+3H_2O+4CO_2$$
$$3KMg_3Si_3AlO_{10}(OH)_2+CaCO_3(\text{+melilite}) \equiv 4Mg_2SiO_4+3KAlSiO_4+CaMgSi_2O_6+3H_2O+CO_2(\text{+melilite})$$
$$\text{UML} \equiv \text{OLIVINE-NEPHELINE-MELILITITE}$$

In fact, however, MDA on the only large *coeval* melilitite and UML data-set (Shrbeny & Machacek 1974) achieves 100% discrimination (Rock 1986): the UML are poorer in SiO_2 and Al_2O_3 but richer in CaO, H_2O and CO_2 (Table 7.2, nos.31–32). Qualitative comparison for other localities where coeval UML and melilitites coexist, further suggests that although the two types have *broadly* similar compositions, they invariably differ in detail, and particularly in CaO/MgO ratio (Table 7.2, nos.18–26). Globally, too, UML are

Table 7.2 Chemical equivalence of various lamprophyres with related plutonic and volcanic rocks

	1	2	3	4	5	6	7	8	9	10	11	12	13	14	15	16	17
Rock	CM			CM		CM		CK		CS		CS		AM		AC	
SiO2,%	46.0	49.6	47.0	51.8	53.0	51.5	52.9	50.4	52.5	52.8	51.0	56.8	58.0	41.7	42.5	40.7	40.7
Al2O3	12.2	12.7	12.9	14.5	14.0	12.8	13.1	16.2	15.5	14.3	14.4	14.0	13.9	14.8	16.3	14.6	16.1
Fe2O3*	9.0	8.7	9.3	7.6	8.7	7.3	7.0	11.9	10.3	8.6	9.9	8.8	8.6	12.4	13.6	11.8	11.7
MgO	10.4	9.6	7.6	6.9	7.3	7.1	8.0	4.2	5.2	9.2	6.2	7.8	9.6	4.8	4.2	5.0	4.5
CaO	9.0	7.9	9.7	7.4	6.9	6.7	4.3	7.9	6.5	7.3	7.6	8.0	7.5	10.1	10.7	11.6	10.4
Na2O	2.4	1.8	1.9	2.7	2.7	2.0	1.3	2.8	2.6	2.4	2.4	1.7	2.0	4.7	4.8	5.7	5.6
K2O	5.8	4.1	6.5	4.6	5.1	5.6	7.2	3.4	3.3	3.4	5.4	1.0	0.6	2.0	1.9	3.0	3.7
TiO2	0.6	0.9	0.8	0.9	0.7	1.3	1.5	2.0	1.6	1.2	2.3	1.7	0.3	3.9	3.7	4.8	3.9
P2O5	1.1	—	0.9	0.8	0.5	1.0	1.3	0.8	0.8	0.8	0.9	0.2	—	—	1.3	1.4	1.8

For further details of individual examples, see Appendix C, Tables C1–C2

[1] Minette dyke, Arrow Peak, Highwood Mtns., MT, USA (Pirsson 1905)
[2] Average of 4 absarokite lavas, Yellowstone Park, WY, USA (Joplin 1968)
[3] Chilled shonkinite, Shonkin Sag laccolith, MT, USA (Carmichael et al.1974)
[4–5] Average minette and monzonite/shonkinite, 49th parallel area, Canada (Daly 1912, p.357)
[6] Average global minette (from Table 5.1)
[7] Selagite, Montecatini, Tuscany, Italy (Barberi & Innocenti 1967)
[8–9] Kersantite dyke and monzonite of main Åva stock, Åland Is., Finland (Kaitaro 1953)
[10–11] Average CAL dyke and latite lava, Erzgebirge, Germany (Kramer 1976a)
[12] Spessartite, Kasuga-mura, Japan (Suzuki & Shiraki 1980)
[13] Boninite, Bonin Is. (Suzuki & Shiraki 1980)
[14–15] Monchiquite and nepheline-diorite, Montreal area, Canada (Eby 1985b)
[16–17] Camptonite and berondrite, Monchique, Portugal (Rock 1983a)

	18	19	20	21	22	23	24	25	26	27	28	29	30	31	32	33	34
	UL			UL				UL		UL		UL		UP		UP	
SiO2	39.4	34.5	37.3	38.0	37.4	36.8	37.8	27.8	30.6	33.6	35.5	36.7	37.8	33.3	38.0	34.1	42.1
Al2O3	10.8	16.4	9.9	10.7	11.4	2.3	15.5	7.6	9.5	6.8	6.8	7.1	8.4	7.9	9.0	11.7	11.5
Fe2O3*	13.0	9.9	14.5	12.2	10.9	13.2	9.5	15.7	14.0	13.7	12.7	12.6	13.0	11.5	12.3	14.4	13.9
MgO	4.3	3.8	6.4	10.1	8.0	33.2	6.3	11.2	6.4	16.1	11.7	19.1	17.7	17.5	17.8	5.6	3.9
CaO	15.4	15.3	18.4	14.8	20.8	4.1	18.6	28.1	27.3	14.3	16.0	12.3	12.4	15.7	12.3	16.6	11.3
Na2O	7.6	7.8	1.4	5.9	4.7	0.7	6.5	1.5	4.3	1.5	1.6	2.7	0.8	2.1	2.6	5.3	5.3
K2O	4.0	2.9	1.5	2.6	1.7	0.4	2.7	1.3	0.7	3.3	3.3	2.3	1.3	1.9	1.5	2.0	4.4
TiO2	2.7	8.3	3.4	1.9	2.0	1.2	1.6	1.3	1.6	4.7	4.9	2.9	2.9	2.3	2.7	3.4	4.4
P2O5	1.0	0.8	0.9	0.7	0.1	0.2	0.4	1.7	1.8	1.6	1.2	1.1	0.9	1.5	0.9	1.1	0.7

For further details of individual examples, see Appendix C, Table C3

[18–20] Alnöite, turjaite (melilitolite) and melilite, Kisingiri, Kenya (Le Bas 1977)
[21–24] Alnöite, melilitite, meimechite and okaite, Meimecha-Kotui, USSR (Egorov 1970)
[25–26] Alnöite & okaite, Oka, Canada (Gold 1967)
[27] "Katungite" (alnöite), Navajo, USA (Laughlin et al.1989)
[28] Katungite, Uganda (Holmes 1950)
[29–30] Alnöite and ankaratrite, Malaita, Solomon Is. (Allen & Deans 1965)
[31–32] Average polzenite and melilite-nephelinite, Polzen, Czech. (Shrbeny & Machacek 1974)
[33–34] "Bergalite" (polzenite) and melilite-nephelinite, Mt.Elgon, Uganda (Davies 1952)

Each pair or trio compares rocks from the same occurrence or province (except 6–7, 12–13, 27–28)

poorer than melilitites in Si, Al and Na, but richer in Ca, K, CO_2, Y, Th and K/(Na + K); aillikites are compositionally even further removed from melilitites than are alnöites (Rock 1986). Because both local and global comparisons agree, therefore, UML cannot be regarded as volatile-enriched hypabyssal melilitites in the same way that AL are volatile-enriched hypabyssal basaltic rocks (cf. Section 7.2). There *is* gradation (Table 1.3, Section

1.6.3), but it is accompanied by compositional changes, to the extent that UML represent a significantly different magma-type, not only capable of crystallizing greater amounts of amphiboles, carbonates, phlogopite, etc., but also incapable of crystallizing liquidus melilite. Therefore, the 'triangular' equivalence long inferred between alnöites, kimberlites and melilitites (e.g. Taljaard 1937) must be abandoned, as stressed by R.H.Mitchell (1986).

7.3.2 Ultramafic lamprophyres versus plutonic melilitolites

Melilitolites (plutonic melilitic rocks such as okaites, turjaites, uncompahgrites) often accompany UML in carbonatite complexes (Table C3), and share the same paragenesis (Bi, Cpx+Me ± Ne ± Ol). Precise chemical equivalence is not to be expected, given the intensity of accumulation and metasomatism in most examples (Le Bas 1977). Nevertheless, limited analytical data are consistent with a *broad* equivalence (Rock 1986), even though, as in Section 7.2.1, there may be differences particularly in CaO/MgO (Table 7.2, nos.18–34).

7.3.3 Ultramafic lamprophyres versus volcanic meimechites

The type meimechites (Appendix B) are associated with alnöites at Meimecha-Kotui, USSR (Table C3), but rather limited data (cf. Table 7.2, nos.21,23) show that they can no more be regarded as extrusive UML than as extrusive KIL (Rock 1986; R.H.Mitchell 1979,1986).

7.3.4 Ultramafic lamprophyres versus volcanic ankaratrites

Many "ankaratrites" (Appendix B) have the same mineral assemblage as UML. The type-rock and an associated "melilite-ankaratrite" (Johannsen 1938), both lava flows, fall respectively within the present definitions of *ouachitite* and *alnöite*. Other ankaratrites have already been called lamprophyric (e.g. Brousse & Nativel 1963). Many ankaratrites are chemically very similar to closely associated UML (Table 7.2, nos.29–30; see also Mitchell-Thomé 1976, table 54, nos. 20, 31, 39, 43, 44, 49), although they may differ mineralogically (thus Table 7.2 no.30 lacks Me and has a higher Cpx/Bi ratio than no.29). Due to inconsistent usage, the average global chemistry of named ankaratrites is not exactly like UML (Rock 1986), but a *broad* equivalence can once again be entertained.

7.3.5 Ultramafic lamprophyres versus volcanic katungites

Laughlin *et al.*(1989) found extremely close *chemical* equivalence between the katungites (Appendix B) of Uganda and certain alkali-ultramafic dykes of the Navajo (USA) province (Table C6). They therefore also termed the latter "katungites", and inferred heteromorphism (Appendix B) between the two, notwithstanding somewhat differing CaO/MgO ratios (Table 7.2, nos.27,28), and the absence of leucite, kalsilite and glass in the Navajo rocks. The Navajo "katungites" in fact fit the present definition of UML rather better than they fit that of katungite, and are therefore suggested to represent an intrusive UML equivalent of an again *broadly* similar magma-type to the Ugandan katungites. Given additionally that ejected (i.e. plutonic) blocks of rocks actually called alnöites and ouachitites (i.e. UML) are associated with the Ugandan katungites (Holmes 1950), all this tends to confirm the plutonic-volcanic links between the kamafugitic rocks of Uganda and UML already hinted at in Section 1.6.4.

8 Nature, Origin and Evolution of Lamprophyre Melts

Petrogenetic theories for lamprophyres are legion, largely of historical interest, and are reviewed by Rock (1977,1984,1986,1987a,1989a), R.H.Mitchell (1986) and Bergman (1987). This chapter concentrates on: (1) the nature of lamprophyre liquids; (2) genetic unity and diversity; (3) the probability that lamprophyres represent *primary* melts (Appendix B), or parent melts to a range of more evolved (intermediate–felsic) rock-types. Admittedly, little more can be done at present than posing some fairly intractable questions and outlining critical genetic constraints; satisfactory answers will have to await further work.

8.1 Do 'lamprophyre magmas' exist?

Since "an essential requirement for any primary basalt candidate is that the bulk composition be equivalent to that of a liquid" (Rhodes 1981), this is a major consideration. Although much the same question could no doubt be asked about granitic magmas (= silicate melt + restites + precipitates), possible answers take a more extreme form with lamprophyres. Since Bowen (1928) deduced from their abundant phenocrysts that lamprophyres correspond to no fully-liquid magma, they have been suggested instead to form from crystal-laden, supercritical, or hydrothermal *fluids*, with very little actual melt component at all (e.g. Ayres & Higgins 1939; H.G.Smith 1946; Bishop 1964). These fluids might be generated by gas-phase metasomatism rather than by fusion of mantle material, and be emplaced by fluidization rather than by normal intrusion mechanisms (cf. D.K.Bailey 1984, 1987). Remobilisation of cumulates has also been suggested (Barker 1983, p.239).

Even though lamprophyre phenocrysts need not represent liquidus phases, but merely crystals which grew *fastest* in such a volatile-rich medium (Joplin 1966), evidence such as the following can be used to support the crystal-laden condition of lamprophyre liquids:

- The rarity of holohyaline or hypocrystalline selvedges (i.e. chilled margins) may both suggest little liquid is involved (Johnson 1964; Sims & Mudrey 1972; Chatterjee 1974; Section 3.1.1); however, this feature could also reflect relatively long-lived melt flow in fissures (Fabre *et al.*1989).
- Megacrysts may largely fill dyke offshoots (e.g. Weed & Pirsson 1896; Shannon 1920; Kardymowicz 1962; Pitcher & Berger 1972); this cannot happen via crystallization.
- Measured liquidus temperatures for CAL of around 1200℃ (Nemec 1977a; Esperança & Holloway 1987) appear inconsistent with their generally (though not universally) slight contact metamorphic effects (Section 3.1.4), with their late-stage emplacement (along with aplites and pegmatites), and perhaps with their intimate association with granitoids.
- Melt crystallization *ranges* of 250–450 deg.C (Nemec 1977a; Montel & Weisbrod 1986) are far wider than for, say, basaltic liquids, and may imply crystal-laden status, although the abundant volatiles (particularly halogens) may play a role in extending this range.

On the other hand, the lamprophyre combination of [groundmass + phenocrysts + abundant H_2O/CO_2-rich minerals] could be taken as a 'frozen' sample representing [melt + suspended crystals + volatile phase] — indeed a uniquely complete magma system. Lamprophyres would then be "closer to intratelluric magma compositions than aphanitic mafic extrusives,

Table 8.1 Heteromorphism and modal variation in single bodies of lamprophyres

	(a)						Modal » chemical variation (heteromorphism)										(b)	Chemical > modal variation				(c) Both vary		
	1	**2**	**3**	**4**	**5**	**6**	**7**	**8**	**9**	**10**	**11**	**12**	**13**	**14**	**15**	**16**	**17**	**18**	**19**	**20**	**21**	**22**	**23**	**24**
Rock	CK	CM	CS	CK	CV	CK	CS	CK	CK	CS	CK	CS	CV	CK	AC	AM	CS	AC	AC	AM	AM	AM	AC	AC
SiO2,%	45.3	46.7	50.5	50.1	50.2	48.4	51.9	50.0	49.9	50.6	51.6	51.7	44.9	44.0	44.1	43.2	50.5-53.7	47.9–56.7	41.9-43.4	39.4	39.4			
Al2O3	11.2	10.9	13.8	15.0	15.2	14.5	15.0	14.9	15.2	14.5	15.6	15.7	9.9	10.5	16.0	16.1	14.4-15.9	15.6–18.1	15.8-16.4	12.3	8.7	4.1		
Fe2O3*	12.2	11.5	7.9	8.1	7.7	8.0	6.7	7.1	10.3	10.8	11.0	11.1	8.3	9.0	11.1	11.0	9.4-11.0	6.4-13.1	10.5-17.1	12.3	12.1	25.9	18.5	21.6
MgO	7.0	6.8	8.8	8.2	5.8	6.1	8.1	8.8	6.9	6.5	4.8	4.9	12.5	9.6	4.1	4.3	3.6-5.8	1.6–4.3	4.8-6.5	8.6	16.8	21.0	11.3	9.9
CaO	9.6	8.6	8.1	8.0	7.0	7.2	6.1	7.1	8.4	9.1	8.5	7.3	12.5	13.3	10.0	10.2	6.8-7.2	3.7–6.3	7.7-9.2	10.8	8.7	6.3	9.6	11.8
Na20	2.6	2.1	2.0	2.1	3.0	2.7	3.1	2.5	3.0	3.6	3.1	3.2	1.7	2.4	5.3	5.7	3.0-3.5	3.8–5.4	3.5-4.3	2.7	1.0	0.2	1.5	1.7
K20	4.0	4.6	3.3	3.2	4.9	5.1	4.5	3.1	3.6	3.1	3.6	4.0	2.5	2.1	3.4	2.5	1.9-2.5	0.9–4.6	1.8-2.5	5.5	2.7	0.2	1.1	1.5
Ti02	2.3	2.0	1.2	1.2	1.7	1.7	0.2	0.8	1.3	1.4	1.4	1.0	0.7	0.7	3.9	4.0	2.8-3.0	1.2–3.2	3.0-3.3	3.3	2.2	4.1	3.6	6.4
P205	2.3	1.8	0.7	0.6	0.6	0.6	0.4	0.4	0.6	0.3	0.4	0.6	1.0	1.3	0.6	0.9	0.7-1.0	0.3–0.7	0.4-0.6	0.7	0.5	0.3	0.5	0.6
Modal composition																								
Qz,%	•	•			2	•			•	•	•	0.7	tr	3	•	•	•	•	•	•	•	•	•	•
Af	17	24	√	√	60		√	√	10	15	18	15	12	9		√	•	2–12	•	•	•	•	•	•
Pl	24	18	√√√	√√√	•	•	√√√	√√√	39	35	45	37	6	14	√√√	√	33±0.7	45–55	31-42	•	•	√	√√	√√√
Ol	2	•							•	•	•	•	•	•	•	•	•	•	•	√√√	√√	√√√	√√	•
Cpx	8	4			•	•			24	13	20	5	•	•	√√√	√√√	•	1–4	5–8	√	√√	√√√	√√√	√√√
Am	•	•	√√√	√√	21	2	√√√	√√	0.5	22	•	26	22	40	√√	√√	24±1.9	6–19	30	•	√√	√	√√	√√
Bi	24	27	√√	√√√	3	14	√√	√√√	23	13	12	15	44	20	√	√	26±1.4	3–11	trace	•	•	•	•	•
Base	•	•			•	•			•	•	•	•	•	•	•	√√√	•	•		√√√	•	√√√	•	•

For further details of occurrences, see Appendix C, Tables C1,C2

[1–2] Zones in Walsen multiple dyke, Spanish Peaks, USA (Johnson 1964)
[3–4] Zones in Rantirov dyke, Bohemian Mass, Czechoslovakia (Nemec 1975a,1977a)
[5–6] Zones in Hostice dyke, Bohemian Mass, Czechoslovakia (Nemec 1975a,1977a)
[7–8] Zones in Kovadlina dyke, Bohemian Mass, Czechoslovakia (Nemec 1975a,1977a)
[9–12] Zones in Dead River lamprophyre "pluton", USA (Geldon 1972): [9] lower kersantite; [10] lower spessartite; [11] upper kersantite; [12] upper spessartite.
[13–14] Zones in Yellowknife sheet, Canada (Webb & Kerrich 1988)
[15–16] Zones in hybrid lamprophyre body, Monchique, Portugal (Rock 1979)
[17] Range of 10 analyses of Winnboro' multiple dyke, USA (Vogel & Wilbrand 1979)
[18] Range in 8 analyses of Tanegashima sill, Japan (Yagi et al.1975)
[19] Range in 4 analyses of Igdlukasik sill, Gardar province, Greenland (Upton 1965)
[20–21] Edge and centre of dyke, Helsinski, Finland (Eskola 1954)
[22–24] Edge, interior and centre of dyke, Julianeshåb, Greenland (Rivalenti & Sighinolfi 1970)

√ = trace constituent
√√ = minor constituent
√√√ = major constituent

the more fashionable yardstick of parental magmas" (Hughes 1982), because they alone retain a reasonably complete vestige of the original volatile phase.

As discussed below, many lamprophyre mineral assemblages appear to be hybrid, and of diverse sources, including arrested or 'frozen' products of various complex resorption reactions, autometasomatism, etc. It follows that few inferences from natural bulk rocks (analyses or experiments) will have any relevance to those source magmas. In short, the question "do fully liquid lamprophyre magmas exist?" is still largely unanswerable.

8.2 How many distinct lamprophyre magma-types are there?

Whereas most 'common' rock-types (andesite, basalt, etc.) evidently correspond to distinct melt compositions, different petrographically defined lamprophyres (e.g. kersantites and vogesites) may by contrast correspond to *the same* magma-type, via the phenomenon of heteromorphism (Appendix B). Clearly, petrogenetic discussion must revolve around distinct *melt compositions*, not different crystalline rock-types which might happen to crystallize from these; it must thus be resolved at this point in our arguments.

Heteromorphism takes two forms — local and global. Many individual lamprophyre bodies vary between one rock-type and another: for example, between a camptonite centre and a monchiquite margin, or between a porphyry core and a spessartite margin (Table 3.1). *Local heteromorphism* is indicated only where modal variations greatly exceed bulk chemical variations; otherwise, the relationship may be the more common one of mutual differentiation or contamination (Section 8.5). *Global* (*group*) *heteromorphism* is indicated by globally identical compositions of several rock-types. Overall, local heteromorphism seems to be distinctive of many lamprophyre suites, and global heteromorphism certainly exists between certain lamprophyre types, but neither type is universal. This needs to be illustrated by separate discussion of the 5 branches, as follows.

8.2.1 Calc-alkaline lamprophyres

The following equations illustrate that, in theory, the mineral assemblages in minettes, vogesites, kersantites and spessartites can all be chemically equivalent:

$$NaAlSi_3O_8 + 2CaMgSi_2O_6 + KMg_3Si_3AlO_{10}(OH)_2 \equiv KAlSi_3O_8 + NaCa_2Mg_5Si_7AlO_{22}(OH)_2$$
albite + diopside + phlogopite ≡ K-feldspar + edenite
MINETTE ≡ VOGESITE (1)

$$CaMgSi_2O_6 + K_2Mg_5Si_5Al_4O_{20}(OH)_2 + SiO_2 \equiv CaAl_2Si_2O_8 + K_2Mg_6Si_6Al_2O_{20}(OH)_4$$
diopside + eastonite + quartz ≡ plagioclase + phlogopite (+K-feldspar)
MINETTE ≡ KERSANTITE (2)

$$2\,NaAlSi_3O_8 + 2CaMgSi_2O_6 + KMg_3Si_3AlO_{10}(OH)_2 \equiv Na_2CaMg_5Si_8O_{22}(OH)_2 + KAlSi_3O_8 + CaAl_2Si_2O_8$$
albite + diopside + phlogopite ≡ richterite + feldspars
MINETTE ≡ SPESSARTITE (3)

$$KMg_3Si_3AlO_{10}(OH)_2 + 2CaMgSi_2O_6 + NaAlSi_3O_8 \equiv NaCa_2Mg_5Si_7AlO_{22}(OH)_2 + KAlSi_3O_8$$
phlogopite + diopside + albite ≡ edenite + K-feldspar
KERSANTITE ≡ SPESSARTITE (4)

$$NaCa_2Mg_4Si_6Al_3O_{22}(OH)_2 + CaMgSi_2O_6 + SiO_2 \equiv NaCa_2Mg_5Si_7AlO_{22}(OH)_2 + CaAl_2Si_2O_8$$
pargasite + diopside + quartz ≡ edenite + anorthite (+ K-feldspar)
VOGESITE ≡ SPESSARTITE (5)

$$KAlSi_3O_8 + NaCa_2Mg_4Si_6Al_3O_{22}(OH)_2 \equiv KMg_3Si_3AlO_{10}(OH)_2 + NaAlSi_3O_8 + CaAl_2Si_2O_8 + CaMgSi_2O_6$$
K-feldspar + pargasite ≡ phlogopite + feldspars + diopside
VOGESITE ≡ KERSANTITE (5)

Critical thermodynamic data for most participants in the above equivalences are lacking. However, if pargasite has similar properties to tremolite, and phlogopite to fluorphlogopite in Robie *et al.* (1978), then vogesite may be the least thermodynamically favoured [ΔG = –138 (25°C) to –148 (725°C) Kcal/mole, ΔV = +12.6 cm^3/mole, for no.5 proceeding to right]. This may at least partly explain why vogesites are the rarest CAL (Section 3.1).

Local heteromorphism corresponding to equivalences 4 and 5 is clearly seen in 4 dykes varying modally, at virtually constant chemistry, between all 4 CAL rock-types (Table 8.1, nos.1–8). Again, in one lamprophyre stock (Table 8.1, nos.9–12), a kersantite/spessartite lithological boundary cuts across layering, such that kersantites and spessartites both show identical whole-rock and mineral compositions at the same levels in the stock, and the same cryptic variations between different levels. *Global heteromorphism* is suggested by the virtually identical average compositions among the 4 CAL rock-types in Table 5.1, and by MDA (Table 5.2c), which, as Métais & Chayes (1964a,b) found on a smaller data-set, is unable to classify most vogesites or kersantites, or a quarter of spessartites or minettes.

However, minettes do reveal higher K_2O in Table 5.1e and Fig.7.4 (this presumably reflects dominant biotite and K-feldspar). Furthermore, there is evidence that Association **A** minettes (Table 2.2: those with granitoids), and Association **R** minettes (especially those of the Navajo, which bear mantle xenoliths) show significant differences in major element composition (Fig.8.4d; Rock 1984), despite very similar trace element patterns (Fig.8.4c). All this implies that it is convenient and advisable to define *two* CAL magma-types, one of which (termed 'M6' in Table 8.2) crystallizes to Navajo-type and some other minettes, the other ('M7') to most vogesites, kersantites, spessartites and the remaining minettes.

8.2.2 Alkaline lamprophyres

As fully explored by Rock (1977,1979), the relationship between camptonite and monchiquite can be a trivial form of *local* heteromorphism, in which hyalomonchiquite (forming selvedges of camptonite dykes), merely represents a more rapidly cooled melt (e.g. Table 8.1, nos.15–16). Because the high volatile content of AL melt suppresses feldspars to the latest stages of crystallization, selvedges commonly lack essential feldspar

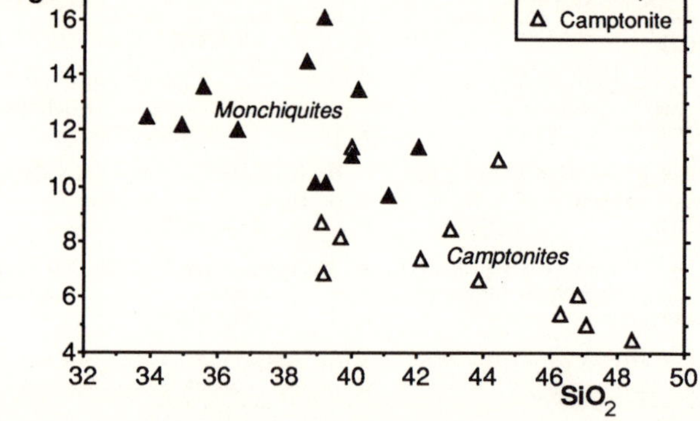

Fig.8.1 *Chemical distinction between camptonites and monchiquites on a province-wide scale.* Data from LAMPDA (Appendix D) for the Orkney dyke-swarm, Scotland, UK (Table C2). The monchiquites approximate nephelinites, and the camptonites approximate basanites in composition.

even when largely crystalline (e.g. Table 8.1, nos.16,20,22) and hence, petrographically, appear more distinct from the centre than would be the case with ordinary basaltic dykes.

Data in LAMPDA also confirm Rock's (1977) demonstration of *global* camptonite-monchiquite hetermorphism (see near-identical averages in Table 5.1b). In some provinces, nevertheless, the two rock-types can be distinct, monchiquites being more Si-poor and Mg-rich (Fig.8.1). This is also underlined by the more undersaturated global normative character of monchiquites (Section 5.2, Table 5.3). Thus camptonites and monchiquites, though sometimes locally heteromorphic and apparently globally very similar, can be distinct on an intermediate (province or occurrence) scale. This paradox is almost certainly a consequence of the definition of monchiquite (Appendix B), since glassy bases ranging from those capable of crystallising plagioclase (and hence potentially heteromorphic with camptonite), to more undersaturated glasses capable of crystallising only feldspathoids, are all encompassed by the name monchiquite. Overall, a complete spectrum exists between the two rock-types, which one magma-type ('M1', Table 8.2) can be used to cover.

8.2.3 Ultramafic lamprophyres

Aillikites and alnöites could theoretically be related by equivalences such as:

$$\underset{\text{ALNÖITE}}{Ca_{1-x}Na_x(Mg,Fe,Al)Si_2O_7 + CO_2} = \underset{\text{AILLIKITE}}{Ca_{1-x}Na_x(Mg,Fe,Al)Si_2O_6 + CaCO_3} \text{ (+ ol, mica, etc)}$$

Yoder (1979) showed that the above reaction will proceed to the right above about 6 kb.

Available analyses are unfortunately too few to test for local heteromorphism via MDA. However, global heteromorphism certainly does not exist, since MDA (Rock 1986) shows aillikites to be lower than alnöites in Si, Al and Na, but higher in Fe, Ti and CO_2 (as in Table 5.1c). Less efficient, though still significant, separation is also obtained on simple bivariate plots (e.g. Fig.5.1d). This difference has two consequences: (1) aillikites are unlikely to be high-*P* variants of alnöites, according to the above equivalence; (2) aillikites as a whole cannot represent carbonated alnöites, since Fe and Ti are higher in aillikites, and CO_2-free recalculation does not wipe out the discrimination. Overall, there seem to be *two* UML magma-types, one of which crystallizes alnöites and polzenites, whereas the other (more carbonatitic) yields aillikites, damkjernites and ouachitites. However, the complete gradation between AL and UML in carbonatite complexes (Section 2.4.8) suggests that 'M1' can be conveniently used to embrace the first group, with only the latter requiring a separate magma-type ('M2').

8.2.4 Kimberlites

The important chemical differences between Group I and Group II KIL have been fully documented by R.H.Mitchell (1986) and Skinner (1989). See also Fig.5.12. Clearly, these two families must correspond to two further, quite distinct melts ('M3' and 'M4').

8.2.5 Lamproites

Trivial, local hetermorphism appears to exist between holocrystalline and glassy lamproites: for example, between orendites/wyomingites and jumillites/verites (Vollmer *et al.* 1984). However, individual lamproite occurrences in Table C4 are each to some extent unique, the Spanish jumillites being the most CAL-like, and the Leucite Hills (USA) wyomingites the most extreme (Section 5.1.2; Fig.5.12). Fig.8.2 reveals globally bimodal MgO

Table 8.2 Lamprophyric magma-types

	Comp.	Xtalline deriv.	Si-sat.	Al-sat.	Na/K	Regimes	Tectonic settings	As	Isotopes	Inclusions(E/F)	Primary?
M1	Basanitic-melilititic	All AC, AM, some UL, UP	Undersaturated (norm.ne,lc)	Saturated (norm.an)	Sodic	Continental and oceanic	Oceanic islands,rifts,triple junctions,subduction zones	M	Tight cluster in or near mantle array (OIB field)	Mantle common; crustal occur	Some
M2	Silico-carbonatitic	UA,UD,UO	Ultra-alkaline (norm.ka,la)	Variable (norm.an/ac)	Sodi--potassic	Continental and oceanic	Oceanic islands, rifts	S	Tight cluster in or near mantle array (OIB field)	Mantle common; crustal occur	Most
M3	Ultrabasic, ultra-alkaline	Group I KIL	Ultra-alkaline (norm.ka,la)	Saturated (norm.an)	Potassic	Continental only	Stable cratons and margins	R	Tight cluster close to CHUR on Sr-Nd; Pb in OIB field	Mantle common; crustal occur	Most
M4	Ultrabasic, ultra-alkaline	Group II KIL, olivine-lamproite	Ultra-alkaline (norm.ka,la)	Saturated (norm.an)	Potassic	Continental only	Stable cratons (S.Africa only??)	R	Tight cluster at eSr +ve, eNd –ve, Pb above growth curve	Mantle common; crustal common	Most
M5	Lamproitic	Most lamproites	Variable (norm.qz/ne)	Perpotassic (norm.ks)	Ultra-potassic	Continental only	Stable cratons, mobile belts, ?subduction-related	R	Several Sr-Nd fields at eSr +ve, eNd –ve; several Pb fields	Mantle occur; crustal occur	Most
M6	Shoshonitic	CM (e.g. Navajo type); some LC	Variable (norm.qz/ne)	Variable (norm.an/ac)	Shosh-onitic	Continental only	Subduction-related, rifts, greenstone belts	R	Tight cluster at slightly +ve eSr, slightly –ve eNd; Pb in OIB field	Mantle occur; crustal common	Some
M7	High-K calc-alkaline	Some CM, most CV,CK,CS,CA	Variable (norm.qz/ne)	Variable (norm.an/ac)	Shosh-onitic	Continental only	Orogenic belts, rifts, island-arcs, greenstone belts	ABC	Broad cluster at slightly +ve eSr, eNd slightly +ve or –ve	Mantle absent; crustal common	Some

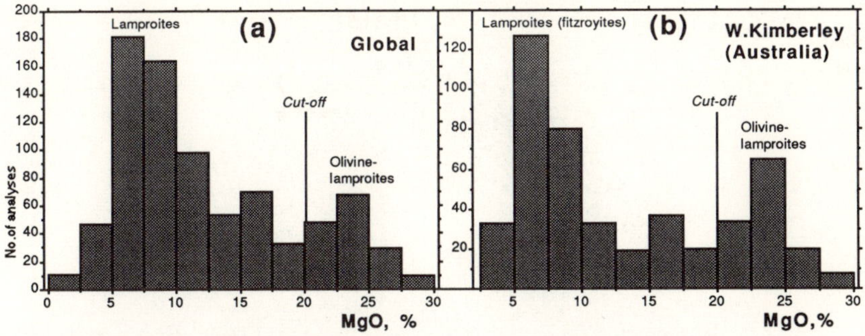

Fig.8.2 *Histograms of MgO contents in lamproites.*
Data (846 analyses in wt.%) from LAMPDA. The consistently bimodal distribution supports the division in Fig.1.2 into *lamproites* and *olivine-lamproites,* with the partition here made for convenience at MgO = 20%: (**a**) global lamproites; (**b**) W.Kimberley lamproites (W.Australia).

distributions, confirming the first-order division in Fig.1.2 into *(leucite)-lamproite* and *olivine-lamproite* families. However, there seems little to choose between Group II KIL and many olivine-lamproites (Section 5.1.2), and several authors have already concluded they may be one and the same (see Bergman 1987, p.169). The overlap between minettes and transitional lamproites (cocites) has also been stressed (Section 5.1.2). To a first approximation, therefore, it seems advisable to distribute lamproites among *three* magma-types in Table 8.2: olivine-lamproites with Group II KIL in 'M4', at least some of the transitional examples within 'M6', and the remainder defined as a final magma-type ('M5'), still broadly coherent though probable more variable than any of the other six.

8.2.6 Seven lamprophyre magma-types

Although there are *five* gradational lamprophyre branches defined on petrographic grounds (Fig.1.2), it thus seems preferable to base genetic discussion on a continuous spectrum of *seven* magma-types (Table 8.2). Each type and its derivatives appears to have broadly similar petrogenesis — more similar, that is, than to the other magma-types. Nevertheless, as with the petrographic branches in Fig.1.4, the series of 7 is considered not only gradational but also circular (cf. Fig.1.4b), with M4 most similar to M3 and M5, and M1 similar not only to M2 but also to M7. The picture is once again one of unity in diversity.

8.3 What parameters can we use to deduce lamprophyres' origins?

Before turning in the next section to consider these 7 magma-types in detail, it must be established quite clearly which parameters can be applied and what they may mean.

8.3.1 Indications of primary magmatic status from whole-rock geochemistry

Primary status is commonly inferred from 'high' *mg*, Sc, Cr, Co and Ni. Absolute values accepted as 'primary' differ greatly between authors, and some low *mg* rocks might be derived directly from localized mantle domains of metasomatized rock-types (e.g. biotite-pyroxenites) which many authors now envisage within a heterogeneous mantle (e.g. Upton *et al.* 1983). Nevertheless, taking the outer limits of various authors' suggestions, melts with *mg#* 65–80%, Sc 15–30, Cr 200-500, Co 25–80 and Ni 90–700 ppm probably

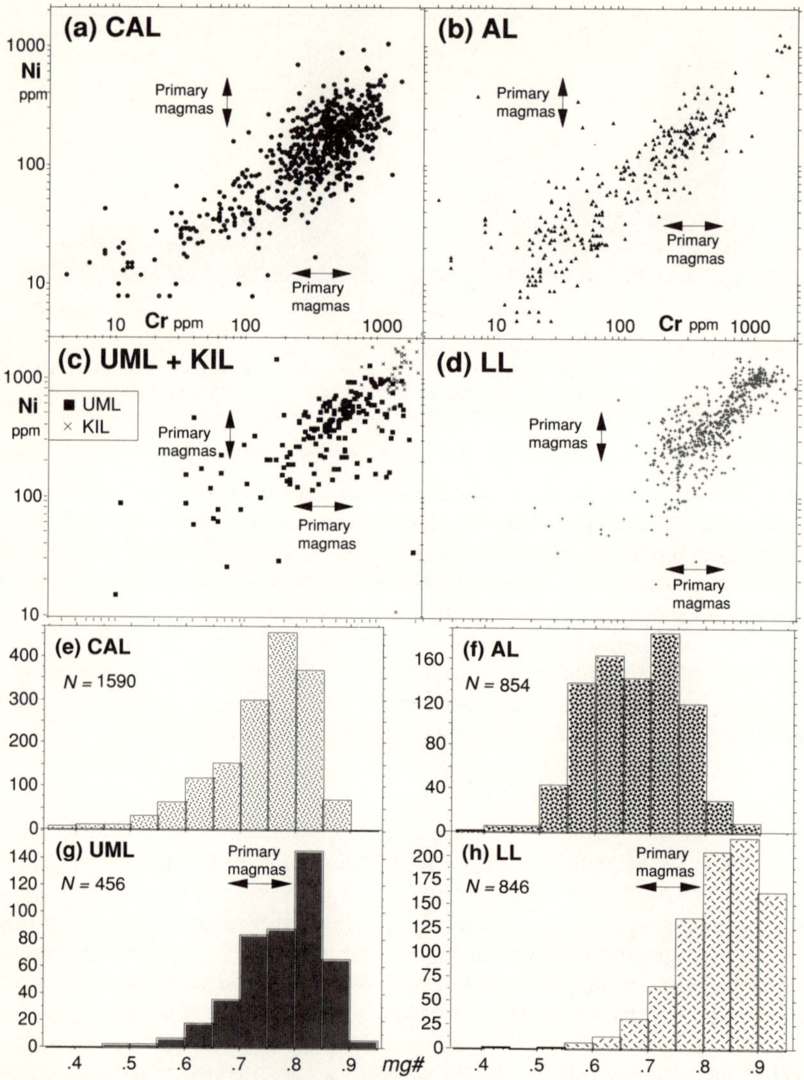

Fig.8.3 *Testing lamprophyres as primary melts.*
Data from LAMPDA (Appendix D). The accepted 'primary' range differs among authors, but the consensus (Rhodes 1981) accepts *mg* approximately 65–80%, Cr 200–500 ppm and Ni 90–700 ppm. Cr and Ni are plotted separately from *mg* because there are many more *mg* determinations available.

represent 1–20% primary melts of mantle lherzolite source (Frey *et al.*1978; Rhodes 1981). High accompanying Fo contents in olivines would tend to support this conclusion.

Another consideration is porphyritic status: do primary magmas have to be cotectic and fully liquid, or can intratelluric magmas be crystal-laden and volatile-rich? Although the general acceptance of heterogeneous sources has introduced new complexities into the old

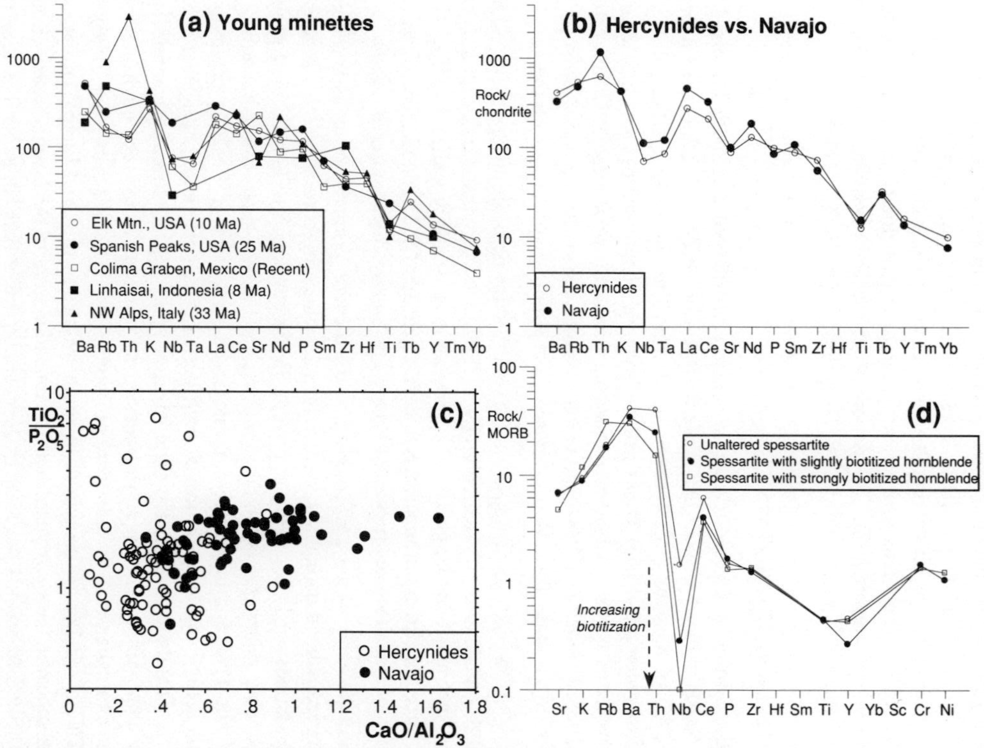

Fig. 8.4 *Geochemical constraints on the origin of calc-alkaline lamprophyres* **(a)** Representative compositions of some Tertiary-Recent minettes, plotted as in Fig 5.6b. **(b)** Averages of Hercynian (Association **A**) and Navajo (Association **R**) minettes, plotted as in **(a)**. **(c)** Discrimination between Hercynian and Navajo minettes, based on major element ratios. **(d)** Growth of a negative Nb anomaly purely by alteration processes in an Archaean CAL from the Yilgarn Block, W. Australia (for raw data and further details see Table Cl and Perring *et al.* 1989b. The three analyses are from the same, small lamprophyre intrusion and differ *only* in the degree of biotitization of their hornblende. All data from LAMPDA (Appendix D).

controversy over basalt genesis (viz. primary picrite versus primary basalt), this does not really alter the basic ground rules about the condition of primary melts.

8.3.2 High LILE and HFSE enrichments

Extreme LILE and HFSE enrichments in lamprophyres (Figs.5.6–5.8) present theoretical problems — which apply in somewhat less extreme form to other alkaline basic rocks (Cullers & Graf 1984) — for they cannot be generated from mantle materials using known partition coefficients and melting models. Three explanations have been forthcoming: (1) extreme fractionation of a garnet-bearing assemblage, such as eclogite, which can account for steep REE profiles (but leads to melt extraction problems due to the extremely low degrees («1%) of melting implied); (2) disequilibrium melting involving a phase rich in REE; (3) a metasomatized mantle source previously enriched in LILE and HFSE.

8.3.3 Coupled negative Ta-Nb-Ti (TNT) anomalies

TNT anomalies form a noticeable feature on spidergrams derived from both M6 and M7 rocks (Fig.5.6a,b), and lead to their distinctive field at low Nb/Pb on Fig.5.9. They appear to occur in CAL of all ages (e.g. Tertiary and Permo-Carboniferous on Fig.8.4b), but Fig.8.4a confirms their presence in *young* minettes whose tectonic settings are fairly clear.

Subduction-related rocks in island-arcs (andesites, etc.) all show these TNT anomalies (e.g. Pearce 1983, Briqueu *et al.*1984, and volume on arc-related alkaline rocks prefaced by Box & Flower 1989). Wide discussion has raised three popular explanations for them:

(1) retention of a Ta-Nb-Ti-rich phase such as perovskite or rutile in the melt residue (although this tends to be negated by Foley's (1989) discovery that rocks with no TNT anomalies, such as olivine-lamproites, are saturated in rutile and ilmenite);
(2) relative source enrichment in adjacent elements (especially La, Th) on Fig.5.6;
(3) assimilation of subducted sediments, which are TNT-poor (Thompson *et al.* 1984).

One question is therefore — are TNT anomalies an indication primarily of tectonic setting, or are they characteristic of particular *rock* assemblages which might occur in other settings? Unfortunately, an answer is not easily arrived at on present data. In Fig.8.4a, the Spanish Peaks and Elk.Mtns. minettes occur in the same state and Cascade-type setting, but only one has a significant TNT anomaly. The Colima Graben minettes on Fig.8.4a occur in a transpressional setting, while the Linhaisai rock lies in a tectonic setting altogether more complex than an island-arc or Cascade-type subduction zone; yet *both* these have anomalies. On the other hand, young lamprophyric rocks in subduction (island-arc) environments which have no anomalies at all include those of Porgera, PNG (Rock & Finlayson 1990). In older rocks, TNT anomalies are ubiquitous in Scottish Caledonian CAL (Table C1), some at least of which appear to have *postdated* subduction and closure of the Iapetus Ocean (Rock *et al.* 1988a). Fig.8.4d sounds yet another *caveat*, suggesting that Nb anomalies may be generated purely by *alteration* processes in ancient lamprophyres.

In short, it is rather early to attach any clear genetic significance to TNT anomalies. One critical comparison would be between the Tertiary CAL and AL of the Japanese island-arc (e.g. Kasuga-mura and Daido-Jima on Figs.2.1–2.2), but unfortunately there seem to be no trace element data yet published for Japanese lamprophyres. For the moment, it is at least clear that TNT anomalies and subduction lack a 1:1 correspondence, since lamprophyres *without* anomalies can occur in subduction settings. However, given that no young lamprophyres *with* anomalies have been noted *outside* subduction settings, it is also a reasonable working hypothesis to regard TNT anomalies as a broad subduction tracer.

8.3.4 *Radiogenic isotopes — crust v. mantle, contamination v. metasomatism*

Table 8.3, comparing previous interpretations of occurrences for which *at least two* isotopes (Sr, Nd or Pb) have been determined, shows that there is no general consensus on what these data mean. Whereas values of $^{87}Sr/^{86}Sr$ and $^{143}Nd/^{144}Nd$ similar to CHUR or

Table 8.3 Published explanations of lamprophyres' combined Sr-Nd±Pb isotopic signatures

Lamp	Occurrence	References	Inferred petrogenesis
AL	Bombay, India	Mahoney et al.(1985)	Chondritic Sm/Nd mantle source, with far less contamination/ mixing of –ve eNd end-member than seen in associated basalts
AL	Wandagee	Jaques et al.(1989b)	Similar to OIB and Group I kimberlite source; variable eSr reflects alteratior since Cpx separates give near-zero eSr, eNd
AL	Alpine swarm, New Zealand	Barreiro & Cooper (1987)	Similar to OIB basalts but may be affected by Dupal isotopic anomaly of southern hemisphere; enriched in Palaeozoic
CAL	Cortlandt, USA	Domenick & Basu (1984)	Nearly chondritic mantle source with possible crustal contamination
CAL	Devon lavas, UK	Thorpe et al.(1986)	Little crustal involvement; inhomogeneous mantle enriched by partial melting associated with subduction of oceanic lithosphere
CAL	Elkhead, USA	Leat et al.(1988)	Aged subcontinental lithospheric mantle source with time-integrated LREE enrichment
CAL	Hercynides	Turpin et al.(1988)	Pb like Hercynian continental crust; mantle source enriched by subduction of crustal sediments
CAL	Limpopo, Zimbabwe	Watkeys & Armstrong (1985)	U-enriched mantle source (no comment on Sr signature)
CAL/ LL	Gondwanas, India	Middlemost et al.(1988)	Metasomatized carbonate-bearing harzburgite; no crustal involvement
KIL	Group II S.African	Fraser et al.(1985)	Subcontinental mantle lithosphere source with high Rb/Sr, low Sm/Nd, chondritic U/Pb for 1–2.5 Ga
LL	Argyle, Australia	Jaques et al.(1989c)	Heterogeneous mantle source with minor crustal contamination?
LL	Gaussberg, Antarctica	Collerson & McCulloch (1983); Nelson et al.(1986)	Mixing of highly enriched, high Rb/Sr, low Sm/Nd subducted oceanic crust source with MORB source-like source; minor crustal involvement or
LL	Holsteinsborg, Greenland	Nelson (1989)	Enriched mantle, isolated from convecting mantle for > 1Ga prior to lamproite emplacement
LL	Leucite Hills, USA	Nelson & McCulloch (1989)	Mantle sources contaminated by high U/Pb sediments subducted into mantle and stored for long periods in subcontinental lithosphere
LL	Leucite Hills, USA	Vollmer et al.(1984)	Two heterogeneous mantle sources with low Sm/Nd, chondritic Rb/Sr; no crustal assimilation; enrichment occurred >1.2 Ga ago
LL	Murcia, Spain	Nelson et al.(1986)	Pb like pelagic oceanic sediments; mantle source contaminated by modern oceanic sediments
LL	Prairie Creek	Alibert & Albarède (1988)	Mantle source with subducted crustal component
LL	Smoky Butte, USA	Fraser et al.(1985)	Two-stage Pb evolution; subcontinental mantle lithosphere source with low Rb/Sr, very low Sm/Nd, U/Pb for 1–2.5 Ga
LL	W.Kimberley, Australia	Nixon et al.(1984);Fraser et al.(1985);Nelson et al.(1986)	3-stage Pb evolution; ancient subcontinental mantle lithosphere source with high Rb/Sr, U/Pb, low Sm/Nd for 1–2.5 Ga possibly mixed with
UML	Coyote Peak, USA	Morgan et al. (1985)	(1) 900Ma: mantle depleted by bas removal; (2) ≈66Ma: mantle melts; (3) melt metasomatizes u.mantle hrz to Ph-weh; (4) 29Ma: weh melts to UML
UML	Oka, Canada	Alibert & Albarède (1988)	Sublithospheric mantle source without significant enrichment

lying in the mantle array can be taken fairly unequivocally to imply a lack of crustal participation, 'high' initial $^{87}Sr/^{86}Sr$ or $^{143}Nd/^{144}Nd$ (radiogenic, positive ε_{Sr} or ε_{Nd}) merely tell us that the analyzed lamprophyre (or its source) has experienced time-integrated enrichment in Rb or Sm, relative to CHUR. This could have occurred in at least 3 ways:

(1) contamination of lamprophyre melt with typically high Rb/Sr or Nd/Sm crustal rocks, during its ascent towards the surface or in a high-level magma-chamber;
(2) deep subduction of crustal rocks to contaminate the mantle itself (Thompson *et al.*1984);
(3) melting of old enriched lithosphere, generated from originally chondritic mantle by LILE-rich fluids rising from deeper in the mantle (Edgar 1983).

Although, in the past, 'high' $^{87}Sr/^{86}Sr$ was commonly taken to mean 'crustal', two arguments have been repeatedly used to suggest that, at least with lamprophyres, strongly positive ε_{Sr} and strongly negative ε_{Nd} remain consistent with essentially mantle derivation:

(1) the highest $^{87}Sr/^{86}Sr$ (up to 0.73, ε_{Sr} up to 250) and lowest ε_{Nd} (down to –25) occur in diamondiferous, mantle inclusion-bearing lamprophyres of very primitive chemistry;
(2) the extremely high Rb, Sr, and LREE in most lamprophyres (Table 5.1; Fig.5.7) mean that crust : magma mixing ratios would have to be inconceivably high (say 5:1) for crustal contamination to affect the isotopic composition of lamprophyres significantly.

8.3.5 Experimental petrology

Although increasingly varied experiments have now been undertaken on lamproites, there are *at most* only one or two direct experimental studies each of CAL, AL, UML or KIL. Indeed, several problems make lamprophyres among the most difficult of igneous rocks for experimentalists to study. Among them, the following remain partly to be addressed:

— *Starting compositions:* It is exceptionally difficult to find natural aphanitic lamprophyres that are sufficiently free of inclusions as to embody useful starting compositions.

— *Pressures*: Evidence such as the presence of diamond (Table 9.1) implies that at least some lamprophyres have experienced pressures in excess of 50 kb — far higher than previously tackled, and until recently not attainable with available experimental apparatus. As a result, many early experiments on lamprophyres were carried out at 1 atm, and are probably of little relevance; few have yet reached more than 30 kb.

— *Volatile fugacities:* To imitate lamprophyres' crystallization conditions must require careful control not only of H_2O *but also* of CO_2, F, Cl and O fugacities; for example, the presence of ferri-diopside (Section 4.6) and high average whole-rock $Fe^3/[Fe^2+Fe^3]$ (Table 5.1) imply unusually oxidizing conditions, while high whole-rock and mineral F, Cl (Table 4.1e,f; Fig.5.8) and primary carbonates (Sections 4.12, 5.7) show that CO_2, F and Cl all play a vital role. Early experiments imitating classic ones on basalts, under dry or H_2O-rich, reducing conditions, are again of limited relevance. Fundamentally different volatile mixtures must distinguish not only the melting conditions of different lamprophyres, but also those of lamprophyres in general from volatile-poor igneous rocks (e.g. Bachinski & Scott 1979; Foley 1989). This in turn means that even the relatively abundant experimental data on lamprophyres' nearest equivalents, such as melilitites, nephelinites, basanites, etc. (e.g. Brey 1978), may have limited applicability.

— *Synthetic systems:* Volatile complexity, coupled with the abundance and critical status of *both* alkalis, TiO_2, P_2O_5, etc., means that even the most complex synthetic systems studied to date probably remain severe oversimplifications of lamprophyric melts.

As a result, even though some lamprophyres have lately been shown to have near-liquidus lherzolitic or harzburgitic assemblages [±Gt], the experimental picture remains incomplete, and sometimes in antithesis to other data, as further discussed below.

8.3.6 Mantle-type inclusions

Type **E** (Chapter 4) inclusions are usually taken to confirm host origins in appropriate parts of the mantle, from relatively shallow (if the inclusions are spinel-lherzolites, to rather deep (if diamondiferous garnet-lherzolite). However, the following *caveats* need to be sounded:

— Similar xenoliths occur in felsic rocks (e.g. phonolites, trachytes: Wright 1966,1969).
— Some ultramafic inclusions in lamprophyres are fairly shallow precipitates (Chapter 6).
— Crystal-laden lamprophyres probably have high yield strengths, enhancing their ability to carry xenoliths (Sparks *et al.* 1977; Spera 1980); xenolith-bearing lamprophyres might conceivably therefore have fractionated even *after* xenolith incorporation. Larsen (1981) argued in this way that one monchiquite had fractionated *despite* its abundant mantle xenoliths and *mg* of 71–74%, because its olivine and Cr-diopside megacrysts had to have crystallized from a melt which was originally still more magnesian.

Perhaps as a result, mantle inclusions do not always marry with whole-rock geochemistry. For example, rich lherzolite suites occur in the Fen damkjernites (Table C3), even though these have *mg** only 50–63 (i.e. below 'primary' values); it must then be decided whether the xenoliths have been retained by fractionated magmas, or the low *mg* is consistent with primary status. Again, therefore, the significance of these inclusions is not clear-cut.

8.3.7 Oceanic versus continental settings

The occurrence of AL and UML on oceanic islands would seem perforce to exclude crustal involvement in their genesis. Even here, however, the argument is not entirely simple. For example, whereas there are several oceanic AL, there is only *one* known oceanic UML, the Malaita alnöite (Table C3), which intrudes an *anomalously thick* area of oceanic crust (the Ontong Java Plateau). Insofar as this may represent a proto-continent (Nixon 1980), the Malaita rock may mirror the proto-oceanic (rift) setting of UML in carbonatite complexes (Association **S** in Table 2.2) rather than a true oceanic setting, and thus suggest that UML require special crustal profiles to be generated outside continental regions.

The confinement of other lamprophyres (CAL, KIL, LL) to continental settings, conversely, is no proof of crustal involvement. KIL and LL in particular are probably characteristic of ancient cratons because the low geothermal gradients there preclude melting within the lithosphere, instead favouring melting at high pressures where volatiles are abundant (see reviews e.g. in R.H.Mitchell 1986, Bergman 1987).

8.3.8 Lamprophyre mineral assemblages — what do they represent?

Bowen (1928) argued that these mineral assemblages are not parageneses, but mixtures of phases formed by the resorption of early-formed minerals into late-stage alkali-rich liquids. Petrologists are now familiar with the notion that kimberlites contain a complex, hybrid assemblage of minerals including all of types **A–F** in Chapters 4,6. Further reactants from resorption, autometasomatism and other low-*T* reactions may also be present (R.H.Mitchell 1986). A similar range, as already hinted, undoubtedly exists in all lamprophyres.

Many peculiarities of lamprophyre mineral assemblages have already been outlined— the rare mineral compositions (Chapter 4), the complex zoning reflecting polybaric crystallization and magma mixing (Sections 4.6–4.8), the peculiar feldspar assemblages (Section 4.9.3). Perhaps the most indicative in the present context is the coexistence of primitive (Mg-rich, i.e. high-T) with evolved (Na-K-Fe-rich, low-T) minerals (Table 4.2); this includes clear incompatibilites, such as minettes with $Fo_{\approx 90}$ olivine phenocrysts in a quartzose matrix. Such features tend to confirm that lamprophyre mineral assemblages represent frozen, incomplete reactions, controlled by *kinetic* rather than *thermodynamic* equilibrium. This also underlies the earlier suggestion that lamprophyre phenocrysts need not necessarily represent liquidus phases, but merely those which grew fastest.

8.4 Petrogenesis of individual lamprophyre magma-types

8.4.1 M1 magma-type (AL/UML): volatile-enriched alkali basaltic to melilititic magmas?

The occurrence of M1-rocks in oceanic areas, together with their 'low' $^{87}Sr/^{86}Sr$ on Fig.5.11 and their almost wholly OIB-like isotopic chemistry on Fig.5.12, sufficiently complement their rich inclusion suites (Table 6.2) to argue convincingly against any direct crustal involvement in their genesis. Several authors in Table 8.3 have indeed argued that AL and UML are the *least* crustally contaminated rocks in their respective suites. The two AL occurrences on Fig.5.12 which lie outside the mantle array are fairly easily accounted for. The more variable ε_{Sr} in the Wandagee monchiquites appears to reflect alteration, since pyroxene separates yield the point near CHUR on Fig.5.12a. The positive ε_{Sr}, negative ε_{Nd}, and much lower $^{206}Pb/^{204}Pb$ of the Loch Roag monchiquite may reflect incidental contamination by its own xenoliths, and/or by the ancient (Lewisian) gneisses through which it was emplaced, for the whole-rock actually lies at the apex of a vertical ε_{Sr}–ε_{Nd} array formed by its rich inclusion suite (Table 6.2). Individual lamprophyre dykes in other suites appear to have experienced similar country-rock interaction: one dyke from the Alpine swarm, for example, has far more radiogenic Sr than others (Barreiro & Cooper 1987).

When added to this evidence implying that M1 melts are of essentially mantle derivation, *mg*, Cr and Ni contents of their crystalline products (AL, some UML) on Fig.8.3, and olivine Fo contents in Fig.4.1, further argue that most of them are primary melts in some sense. Based on the many genetic associations (Table C2) between AL and an established primary melt, namely alkali basalt (Frey *et al.*1978), Wimmenauer (1973b) asserted that "monchiquites, camptonites ... are nothing but basalts, modified by concentration of volatile components", while Upton (1965) similarly concluded that "camptonites result from crystallisation of water-rich basalt ... and the problem of their genesis resolves itself to a search for possible volatile additives". Important questions for this model include the following: (1) can M1 lamprophyres be generated *directly* from mantle source under more volatile-rich conditions than give rise to basaltic magmas?, or (2) does later addition of volatiles to a primary basaltic magma cause it to crystallize as a camptonite?, and if so, (3) *what* volatiles must be added, how are they added, and from whence do they come?

Experimental data on basaltic, basanitic and nephelinitic compositions, starting with Yoder & Tilley (1962), indicate garnet-lherzolite saturation at mantle pressures and thus provide an *indirect* indication of the source material for M1 melts. The experiments also firmly support the idea that *hydrous* basaltic to nephelinitic melts will crystallize as AL and UML, not least because water enhances amphibole/mica and suppresses plagioclase. Rock's (1978) direct experiments on one camptonite successfully reproduced its mineralogy at 2 kbar P_{H_2O}, with a ≈200 deg.C liquidus interval and plagioclase crystallizing only a few degrees above the solidus. Starting with a more nephelinitic composition, or chilling at any stage prior to final crystallization, can therefore explain the genesis of monchiquites, via total suppression of feldspar. Such work also predicts that AL mineral assemblages cannot crystallize from tholeiitic magmas (even volatile-enriched examples), because: (1) amphibole crystallises either subsolidus or only just above the solidus of tholeiite magmas; (2) AL and tholeiites lie on opposite sides of the saturation divide in the basalt tetrahedron.

M1 melt may also give rise to sannaite under elevated fCO_2, which selectively suppresses anorthite component, and may hence lead to dominant *alkali* feldspar (Rock 1976b,1978). Thus in four basalt–AL suites of closely similar chemistry, Ramsay (1955), Woodland (1962) and Gallagher (1963) reported camptonites whereas Carstens (1959) reported *alkali* feldspar-bearing AL ("vogesites"), in which "the absence of the anorthite molecule may be connected with the high CO_2 content of the vogesite magmas".

Turning therefore to *how* and *from whence* the volatile additives come, Upton (1965) inferred that camptonite formed from basalt by addition of volatiles from contemporaneous syenitic magma, and a clear example where a hybridization process of this type has indeed *locally* taken place at crustal levels has since been identified at Monchique (Rock 1979, 1983a). Syenite hybridization might also account for the enigmatic feldspar macrocrysts in AL, which show a strong syenitic signature (Section 6.6.3). Bowen's (1928) classic model, whereby alkali-rich melts react with early olivine and pyroxene to give lamprophyres, amounts to a similar process, and it too has been clearly demonstrated at Monchique (Rock, op.cit.) Another alternative is *auto*enrichment in volatiles merely by crystallization of basaltic melt, as suggested by the occurrence of camptonites as late-crystallizing veins or zones within some alkali basalt dykes (e.g. Ohashi 1980). These, then, are three forms of volatile enrichment which occur *subsequent* to magma generation in the mantle, and they imply that M1 lamprophyres are *secondary* products of volatile enrichment in *primary* basaltic melts.

Secondary volatile addition becomes far less plausible, however, where M1 AL form large regional swarms, or where no syenitic rocks are known. Accumulated data now suggest (cf. Rock 1977) that regional swarms (Association **R**) are no less 'lamprophyric' than central swarms around syenitic plutons (Associations **M,S**), so that even if syenitic magmas may further enhance the volatile contents of lamprophyre melts, control over the volatile content of M1 melts seems likely to be at *mantle* rather than crustal levels. In other words, M1 melts might be direct, *primary* derivatives of unusually hydrous mantle. Tilley & Yoder's (1967) demonstration, that melting of a biotite-pyroxenite could yield a melt capable of crystallizing as alnöite below 6 kb, provides interesting support for this idea.

Most probably, M1 melts can be either primary or secondary: the most primitive, mantle inclusion-rich examples (particularly those in oceanic areas) are most likely to be primary, whereas rocks such as the Elgon, Kaiserstuhl and Tundulu UML, which lack mantle xenoliths, have only accessory modal olivine, and show whole-rock *mg** 40-50, could be secondary fractionates of these primary melts. Those such as the the Avon alnöite, with 66% modal olivine + serpentinite and *mg** = 90, have most probably accumulated olivine.

8.4.2 M2 magma-type (UML): primary carbonatite affililiates?

These show gradation and many similarities with M1 types, and much the same arguments can be advanced as in the previous section for regarding them as primary mantle melts. However, given their equal gradation on the one hand into kimberlites, and on the other into carbonatites, the origin of their extreme CO_2 enrichments needs special consideration.

Models involving the Daly-Shand limestone assimilation hypothesis (Scheumann 1922), or metasomatism or contamination of peridotite or kimberlite (Garson 1962,1966; Vartiainen *et al.* 1978), are now largely of historical interest (see review in Rock 1986). Although no direct experimental data are available for these rocks, more plausible models can be derived by considering fairly extensive data available for other ultra-alkaline rocks. For example, M2 lamprophyres have higher [Ca+Mg]/Si, Ca/Al, K, P, F, H_2O and CO_2 than melilitites (Rock 1986), and probably therefore reflect higher P_{tot}, P_{CO_2}, P_F and smaller percentages of melting than melilitites, based on experimental data such as Brey (1978). The following evidence also suggests that some of them originate at lower *P* and higher *T* than kimberlites: (1) *P/T* figures from mantle xenoliths (Table 6.2); (2) relative rarity of diamonds; (3) preponderance of spinel- over garnet-lherzolites in mantle xenolith suites (Table 6.2); (4) subequal MgO and CaO in aillikites (Table 5.1), suggesting they are generated within the mantle stability field of dolomite, rather than the higher *P* field of magnesite whence many kimberlites may be derived (Brey 1978; Wendlandt 1984; Wyllie 1989). However, magnesite and breunnerite do occur in other UML (Fig.4.8), and the *P-T* regime of several examples may approach that of some KIL, even breaching the diamond field (Eggler 1989).

The most detailed genetic modelling of any M2 rock so far undertaken (Neal & Davidson 1989), yielded a complex scheme that involves melting of a *depleted* (not metasomatized) asthenospheric source, then zone-refining and mantle diapiric upwelling to generate 'proto-alnöite', followed by an assimilation and fractionation (AFC) process involving a subducted derivative of oceanic crust. This particular rock is considered *not* to be a primary melt, notwithstanding its high *mg*, Cr, Ni and rich, varied mantle xenolith and megacryst suite.

The precise relationship between M2 and carbonatite magmatism, and the role of both of these in metasomatizing the mantle itself, are among important topics for further research.

8.4.3 M3 and M4 magma-types (KIL, olivine-lamproites): deep mantle melts?

The latest series of voluminous papers on this hotly debated topic appears in the *4th International Kimberlite Conference Proceedings* (4IKC, 1989). Insofar as a consensus exists, it appears to favour partial melting of subcontinental lithosphere for M5 and of underlying asthenosphere for M4. Experimental studies are now appearing for the first time (Edgar *et al.*1988). Readers are referred to the 4IKC volumes for the full details.

8.4.4 M5 magma-type (LL): multi-reservoir, multi-stage mantle melts?

Reasonably full experiments have been carried out on these (e.g. Barton & Wood 1976; Barton & Hamilton 1978,1979,1982; Arima & Edgar 1983; Edgar & Arima 1983,1985; Foley 1986,1989). The experiments are consistent with primary origin provided that: (1) the mantle source is not normal lherzolite, but has been depleted in garnet and clinopyroxene and enriched in phlogopite (i.e. mica-harzburgite); and (2) melting occurs under higher fH_2O and fF but lower fCO_2 than for M2, M3, or M4 melts. Nevertheless, only ≈ 30% of LL can be primary according to Fig.8.3 and Fig.4.1, and processes such as volatile-phase transfer during ascent may still be required. The numbers of mantle reservoirs and stages involved in M5 genesis (e.g. single versus MORB+enriched; one-stage versus three-stage depletion+enrichment+melting) are also among many debating points (Bergman 1987).

8.4.5 M6 magma-type (Navajo minettes, etc.): slightly contaminated mantle melts?

The Navajo rocks are among the most exhaustively studied of minettes, yet their genesis remains in many respects problematic. Their garnet-peridotite inclusions (Table 6.2) and various geochemical arguments have been used to infer an origin by ≈ 0.3% melting from a metasomatized garnet-peridotite or phlogopite-wehrlite source at low CO_2/H_2O and >130 km depth (Ehrenberg 1979; Carswell 1980; Roden 1981). From melting experiments at 10–20 kb, Esperança & Holloway (1987) further inferred that M6 melts must have ascended rapidly from the mantle at 1,000–1,200°C and chilled quickly without upper crustal modification, since any minette retaining [Ol+Cpx+Ph] as phenocrysts "could not have been significantly reequilibrated at low pressures and temperatures".

Unfortunately, these experiments did not reach high enough pressures to test whether minette could actually be in equilibrium with garnet-peridotite source, and other authors have reached quite different conclusions regarding the extent of upper crustal modification. For example, Menzies *et al.* (1987a) concluded that "most, if not all, of the minettes have been contaminated either with continental crust, lithospheric mantle, or recycled material", thus echoing many earlier papers (e.g. Williams 1936; Powell & Bell 1970; Alibert *et al.* 1986). As a means of resolving these problems, the *contrast* between the Navajo minettes (M6) and the nearby Hopi monchiquites (M1) should now receive more attention: the minettes not only have significantly more variable isotopic signatures (higher ε_{Sr} and lower ε_{Nd}), but also carry a much higher ratio of crustal granite to mantle peridotite inclusions (even though they both intruded the same area in the late Tertiary). Importantly, the critical garnet-lherzolites are restricted to one Navajo volcanic field (Zilditloi), and form <1% of xenoliths even here. This contrast does, therefore, already provide a *prima facie* argument for more crustal involvement in the genesis of M6 than of M1 lamprophyres.

8.4.6 M7 magma-type (CAL): heavily contaminated primary melts?

Given these ambiguities with M6, there is still less chance of resolving the petrogenesis of type M7, for the evidence here is still more complex. The only two available sets of experiments (both on minettes) are unfortunately equivocal. Ruddock & Hamilton (1978a,b) suggested that, in the system Di-Lc-Qtz at 1 atm, liquids with the projected compositions of natural minettes would yield the simplified minette assemblage [Ph + Di + Sa + Qz] on cooling. However, this is not an equilibrium assemblage, so that natural minettes have to be interpreted as random metastable combinations of two liquidus phases [Phl, Di] with fractionated minimum melt [San + Qtz]. Furthermore, inferred crystallization paths would yield petrographical features which are shown by few, if any, natural minettes (Rock 1984).

Nevertheless, compared with M6, most M7 products do show a distinctly more 'crustal' signature: they tend to be richer in Al_2O_3, slightly poorer in MgO, and CaO, more variable in ε_{Nd}, and lower in $^{206}Pb/^{204}Pb$ and $^{206}Pb/^{204}Pb$ (Figs. 5.12, 8.4c). Given also the absence of mantle but abundant crustal inclusions in M7 rocks, and their persistent association with granitic magmatism (i.e. Association **A**), there is surely a good *prima facie* case for a stronger crustal contribution in their genesis. Further support for this has recently been inferred from high ammonium contents in Association **A** minettes (Hall 1988).

Many authors have sought the origin of M7 melts in some form of basalt+continental crust interaction (for historical review see Rock 1984). A case may still be made in specific areas for such a process. For example, the relative distribution in Brittany (France) of Late Carboniferous dolerites (without granitoids) around Trégorrois, and of granite-related CAL around Finisterre (Table C1), can perhaps be explained if the parent melts for the Finisterre CAL were similar to the Trégorrois dolerites, but were modified when rising through the area of granitic plutonism. Some lamprophyres in this area are actually enriched in the very elements in which their resorbed crustal xenoliths are depleted (Cogné 1962).

The overwhelming problem with such a model is that it fails to explain so many other distinctive features of M7 geochemistry (e.g. high F, Ba, Th, P, LREE); moreover, numerous well-documented cases of *in situ* basalt+crust interaction are nowhere known to generate minettes. On the basis that evidence for the 'crust' part of the 'basalt+crust' hypothesis is by far the stronger, Rock (1984) therefore put forward the hypothesis that M7 melts are products not of basalt+crust but of *lamproite* (i.e. M5)+crust interaction. This was based partly on Holmes' (1945) report of olivine-leucite rocks formed by assimilation of granite xenoliths in ugandite lavas, and partly on a geochemical spectrum in the order M5→M6→M7→crust (fig.7 of Rock 1984). This can be interpreted via increasing crustal modification of M5 melts, depending on the mobility and magmatic activity of the crust through which they pass: M5 melts themselves are so reactive to average crust that they only reach the surface unmodified in stable cratonic areas, where they may retain a high mantle/crustal xenolith ratio. In the peripheries of orogenic areas, they suffer sufficient crustal modification to produce the more 'crustal' M6 whole-rock signature, together with a lower xenolith ratio. Within orogens (especially in areas of active granitoid plutonism), M5 melts are thoroughly modified to M7 melts, and the mantle/crustal xenolith ratio reaches zero. The curious 'diaschistic' association (Section 1.1.1) between CAL, pegmatites and

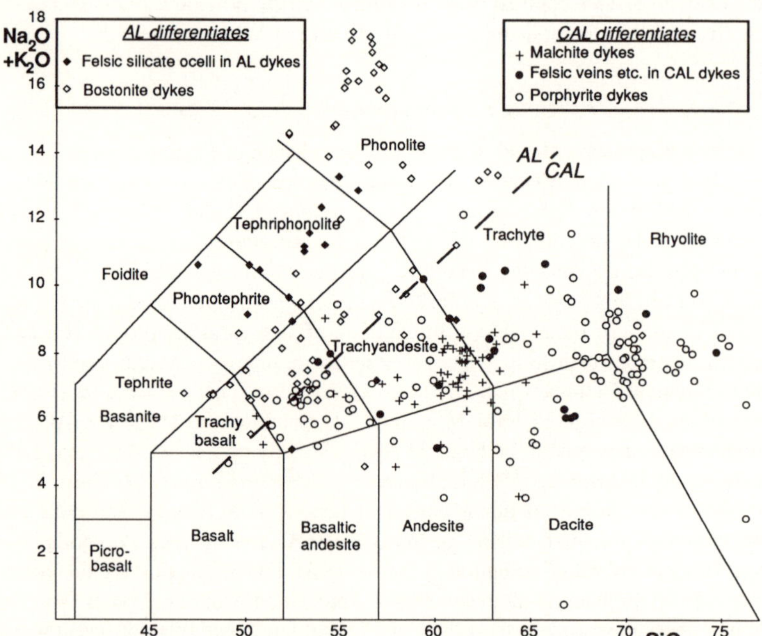

Fig.8.5 *Chemistry and classification of lamprophyric differentiates.*
257 evolved rocks intimately associated with CAL and AL dykes (data from LAMPDA: Appendix D) plotted in the TAS diagram (Le Bas *et al.* 1986). See Appendix B for explanation of nomenclature. Data recalculated to 100% free of volatiles before plotting. "Felsic veins etc." includes patches and schlieren; they and "felsic silicate ocelli" occur within actual lamprophyre dykes (Section 3). The other evolved rocks plotted form separate dykes within the same coeval suites as the lamprophyres.

aplites may also thereby be explained, if modification takes place when the crust is not completely solid, i.e. still at the stage of generating late-stage fluids (cf. Bederke 1947).

A case can still be made for this 'M5+crust' model, but there are also important problems. For example, the Elkhead minettes, USA (Table C1) are intimately associated with rhyolitic rocks, and yet have invariable isotopic compositions which seem to call not for any crustal contribution, but for a uniquely LREE-enriched lithospheric mantle source (Fig.5.12). This is in fact one of an increasing number of cases where acidic rocks appear not to be the *cause* but the *effect* of M7 melts — that is, products of M7 evolution — and it will lead us finally in Section 8.6.1 onto a more general, converse, hypothesis, to explain the M7–granitoid relationship. Before arguing this hypothesis, however, we must first turn to considering the whole question of the evolution of lamprophyric melts.

8.5 Evolution (differentiation) of lamprophyre melts

Several processes have now been shown to cause lamprophyre melts to differentiate (Appendix B) towards more felsic compositions; fractionation of their abundant phenocrysts is only the most obvious. More evolved products of such differentiation can be observed on two scales: (a) *local*, namely relatively felsic globules, ocelli, patches, schlieren or veins within single lamprophyre bodies; (b) *regional*, where separate dykes within a coeval dyke-swarm form a continuous compositional series from lamprophyric to felsic rock-types.

Fig.8.5 shows that these more evolved products cover a wide range. Not surprisingly, however, products of CAL are distinguished by their calc-alkaline chemistry, covering trachybasalt through rhyolite compositions, whereas those of AL are more alkali-rich and dominantly undersaturated, ranging from sodic trachybasalt to phonolite.

8.5.1 Evidence of evolution within single lamprophyre bodies

Strong *prima facie* evidence for essentially *in situ* differentiation is provided by:

(1) aplitic, pegmatitic, syenitic and other mesotype–leucocratic veins, plus complementary mafic schlieren, within CAL dykes (Andrade 1979; Kirchner 1979; Woodward 1970; Roden & Smith 1979; Ayres & Higgins 1939; Chatterjee 1959; Wimmenauer 1973b);
(2) CAL with dark patches, passing into peridotites (Barrow *et al.* 1905; Chatterjee 1974);
(3) AL dykes with foid syenitic globules coalescing toward the top (Fig.3.1).

Table 8.4 compiles evidence of evolution within single bodies (for Harker diagrams, see Rock 1984, fig.9). The Dead River layered stock (Table 8.1, nos.9–12, Table 8.4, nos.5–7; Section 3.1.3) is one of few bodies showing unequivocal crystal fractionation on a largish scale. However, mere association within a single body does not always prove contemporaneity, let alone *in situ* differentiation. One vogesite sill, for example, reportedly grades *downwards* into microgranite (Barrow *et al.*1913, p.84), and components of some composite dykes differ in age by as much as 200 Ma (Rock 1983b). Interpretation of others has proved controversial: Vogel & Wilbrand (1978) and McSween *et al.* (1979) relied on different evidence for their opposing interpretations of the Winnsboro' dyke (Table 3.1).

The consensus in Table 3.1 is nevertheless that most composite CAL–porphyry bodies are products *not* of fractionation but of magma-mixing between lamprophyre and porphyry end-members. Among important evidence is: (a) linearity of bulk chemical variations; and (b) occurrence of mafic macrocrysts in the intermediate members, whose proportions vary inversely with the amount of groundmass, but whose compositions remain invariable and similar to those in the lamprophyres. In some cases, the felsic end-member may carry a differentiate component from the lamprophyre itself (e.g. Macdonald *et al.* 1986), but in others it is generated indirectly (e.g. by anatexis of country-rock), or quite independently.

8.5.2 Evidence of evolution within regional dyke-suites

Some dyke-swarms with reported overall variations from lamprophyres to felsic rocks have later been shown to hide dyke injections of different ages (e.g. Jermanski *et al.*1974). Even where contemporaneity has been proved beyond reasonable doubt, previous interpretations are at variance: Morrison (1918), Solov'yeva (1973) and Perring (1988), for example, attributed Devonian, Jurassic and Archaean CAL→porphyry dyke-series respectively to complex hybridisation processes between CAL and porphyry end-members (mirroring interpretations of composite dykes in the same areas), whereas Nemec (1973b) attributed compositionally very similar Hercynian dyke-series to differentiation of lamprophyre melts.

Detailed studies have revealed a complex picture, and variation diagrams are often ambiguous. For example, Fig.8.6a,b shows apparently continuous chemical variation in the Caledonian CAL→porphyry dyke-suite, suggesting corresponding melt evolution, and yet local examples of this same suite show silica gaps and separate geographical distributions for the CAL and felsic dykes, so mirroring the composite dykes just discussed and again implying *independent* lamprophyre and porphyry melts (e.g. Barnes *et al.*1986).

The only case where the panoply of experimental, mineralogical, geochemical and isotopic data is available to assess these options is once again the Navajo, USA (Table C6), where presumed daughter products of the minettes (Section 8.4.5) include "felsic minettes" ("sanidine trachytes" of Kay & Gast 1973) and potassic latites (e.g. Camp Creek, Arizona). The composional range is largely explicable by fractionation, which must have occurred in the uppermost mantle, because some of the felsic rocks contain abundant peridotite as well as lower crustal inclusions. Given additionally that, in some localities, the mafic minettes have lower $^{87}Sr/^{86}Sr$, higher ε_{Nd}, and carry far fewer mantle xenoliths than the associated felsic minettes, and that an *ultramafic, xenolith-free* ring dyke at Buell Park has lower $^{87}Sr/^{86}Sr$ and higher ε_{Nd} than *xenolith-bearing* mafic minettes elsewhere in the Navajo, the overall data suggest that these felsic rocks formed during ponding of the mafic minette melts near or at the crust-mantle boundary, and then suffered further mechanical contamination by mantle and/or crustal inclusion materials during their subsequent ascent.

8.5.3 Role of liquid immiscibility in the evolution of lamprophyre melts

Liquid immiscibility has a chequered history in petrology (Philpotts 1976). Silicate-*carbonate* immiscibility at least is relatively non-contentious, being well-substantiated by experimental and fluid inclusion data (e.g. Rankin & LeBas 1974; Freestone & Hamilton 1979), and may well therefore explain Type 1 (carbonate-rich) globular structures in lamprophyres (Section 3.2.4). Silicate-*silicate* liquid immiscibility has been championed by Philpotts (1971,1972, 1976), Ferguson & Currie (1971) and Eby (1980,1983) to explain Type 2 (feldspathic) globular structures. They mainly used evidence from Canadian AL, including reported experimental splitting of homogeneous glasses prepared from ocellar lamprophyres into conjugate liquids resembling the original ocelli. They also explained structures such as in Fig.3.1d in terms of a less dense, immiscible felsic melt rising and coalescing within denser lamprophyre melt. The greater abundance of ocelli in AL relative to basalts suggests that, if it does exist at all, this silicate-silicate immiscibility field expands

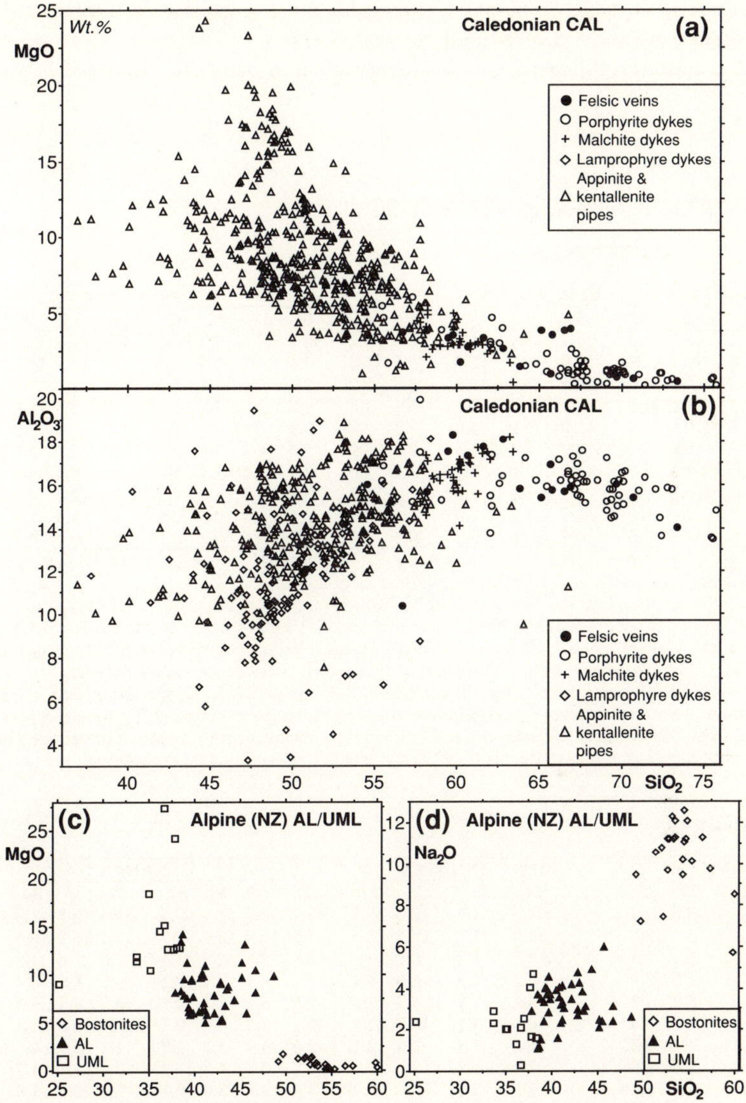

Fig.8.6 *Evolution of lamprophyre melts.*
Illustrated by major element Harker diagrams for two suites of coeval lamprophyres and intermediate–felsic rocks. Data from LAMPDA (Appendix D) and A.F.Cooper (unpubl).
(a,b) 579 Siluro-Devonian (≈ 400Ma) CAL from the UK Caledonides (Table C1; Fig.2.5).
(c,d) 81 Tertiary (≈25 Ma) AL and UML from the Alpine Swarm, New Zealand (Table C2; Fig.2.8).

under higher volatile pressures. However, alternative explanations such as filter-press, segregation of residual liquid into vesicles (cf. amygdales), liquid-filling of bubbles, or replacement of xenoliths, are strongly preferred by Jaffe (1952), Cooper (1979), Carstens (1982) and Foley (1984). Type 3 globular structures have been explained by a plethora of models (Rock 1984, table 6), but there is probably much to be said for their identification with amygdales or varioles (Jaffe 1952; Zimmerle 1977).

Fig.8.6 contrasts apparently continuous variation in one CAL→porphyry dyke-suite

Table 8.4 Examples of magmatic evolution in lamprophyres

(a) *Calc-alkaline lamprophyres*

	MINETTES				KERSANTITES			SPESSARTITES											
	1	2	3	4	5	6	7	8	9	10	11	12	13	14	15	16	17	18	19
SiO2,%	60.2	66.1	48.9	59.5	45.7	49.9	51.7	44.5	53.2	56.2	60.3	63.3	66.1	48.0	55.2	58.2	42.9	53.4	57.6
Al2O3	14.0	14.4	10.1	12.9	7.9	15.2	15.7	12.3	18.0	13.4	14.3	14.5	14.8	10.1	13.9	15.7	17.4	13.2	18.9
Fe2O3	2.2	2.1	4.5	3.0	10.9	9.4	10.1	5.8	5.2	3.4	1.9	1.4	1.3	1.7	1.2	9.6	11.2	4.5	3.8
FeO	2.9	2.1	3.6	1.5	•	•	•	6.4	1.8	2.8	3.8	3.1	2.5	6.9	4.4	2.6	13.2	8.3	3.9
MgO	5.2	2.8	10.0	4.9	13.9	6.9	4.9	11.3	2.1	9.2	6.4	4.9	2.7	14.4	8.4	3.8	8.5	3.6	2.1
CaO	3.9	2.5	9.0	5.3	12.9	8.4	7.3	9.2	8.9	5.3	4.7	3.8	2.7	8.9	5.0	5.1	10.4	5.8	3.7
Na2O	2.2	2.8	1.3	2.5	2.0	3.0	3.2	2.2	3.0	2.9	3.5	3.7	4.0	3.5	4.5	5.3	2.0	5.1	7.4
K2O	6.4	5.9	5.2	7.2	1.9	3.6	4.0	2.2	4.8	2.7	3.1	3.6	4.2	1.1	2.8	4.7	0.3	1.2	0.8
TiO2	1.0	0.7	2.0	0.9	1.7	1.3	1.0	1.3	0.4	0.7	0.6	0.5	0.4	1.4	1.6	1.6	3.1	2.0	0.9
P2O5	1.2	0.5	1.1	0.6	0.5	0.6	0.6	0.4	0.0	0.3	0.1	0.2	0.1	0.2	0.2	0.3	0.2	0.6	0.3

See Appendix C, Tables C1,C6, or Table 3.1 for further details of all examples

[1–2] Cercany composite dyke, Czechoslovakia (Nemec 1973b): [1] marginal malchite; [2] central porphyry.
[3–4] Buell Park, Navajo (Roden & Smith 1979): [3] most mafic and [4] most felsic of separate minette dykes.
[5–7] Dead River lamprophyre "pluton", USA (Geldon 1972): [5] basal pyroxenite; [6] lower and [7] upper kersantite layers.
[8–9] Sandia Mtns., USA (Woodward 1970): [8] spessartite dyke with [9] hornblende-syenite dykelet.
[10–13] Shiviya-Dyrevatka composite dyke, E.Transbaikal, USSR (Solov'yeva 1973): [10] spessartite, 0.85 m from contact; [11] malchite, 1.9m from contact; [12] granodiorite-porphyry, 2.2m from contact; [13] granite-porphyry, centre of dyke.
[14–16] Newmains dyke, SW Scotland (Kennedy & Read 1936): [14] cumulate patch; [15] parent spessartite; [16] syenite schliere
[17–19] Lindesquedrift intrusion, S.Africa (Bisschoff 1972): zones of [17] spessartite [18] syenodiorite, and [19] albite-syenite.

(b) *Alkaline lamprophyres*

	CAMPTONITES				MONCHIQUITES				GENERAL								
	20	21	22	23	24	25	26	27	28	29	30	31	32	33	34	35	36
SiO2	45.6	53.1	43.6	59.7	44.3	53.9	40.3	53.5	43.2	46.9	52.0	60.1	37.4	46.3	51.3	42.3	52.2
Al2O3	14.9	18.4	15.5	18.0	16.8	21.7	15.5	21.6	12.3	12.9	18.1	20.9	12.9	16.9	16.5	14.0	19.2
Fe2O3	4.4	7.5	6.1	1.1	5.7	1.3	6.2	•	6.8	3.8	2.2	1.6	2.6	1.6	1.8	5.9	1.5
FeO	4.8	5.1	7.6	2.9	5.4	2.9	6.6	4.1	7.2	8.1	5.1	0.1	9.0	8.0	6.4	8.1	3.5
MgO	10.4	2.9	5.2	1.8	4.1	1.1	5.3	2.2	7.8	6.1	2.8	0.1	8.0	4.2	3.5	6.0	1.4
CaO	11.0	0.8	9.0	3.8	10.6	5.1	10.3	1.3	9.1	5.7	4.6	0.1	11.4	6.0	4.8	8.8	3.7
Na2O	3.1	6.5	4.6	7.8	5.2	7.5	2.5	2.1	2.6	4.9	3.8	1.7	1.2	3.7	2.5	4.2	7.5
K2O	3.5	2.9	1.6	1.0	2.4	3.7	3.7	10.2	2.1	2.3	4.7	11.3	2.4	2.4	1.7	2.4	3.9
TiO2	0.7	0.4	3.6	1.2	3.6	2.0	4.0	0.7	3.5	3.4	1.0	0.2	3.3	2.3	1.8	3.8	1.6
P2O5	0.0	0.0	1.2	0.4	1.0	0.1	0.6	•	0.7	0.8	•	0.1	0.6	0.8	0.7	1.3	0.3

See Appendix C, Table C2 for further details of all examples

[20–21] Composite dyke, Coldwell complex, Canada (Currie 1980): [20] camptonite centre, [21] nepheline syenite margin.
[22–23] Mt.Johnson, Monteregian Hills, Canada (Eby 1980): [22] camptonite dyke with [23] trachytic ocellus.
[24–25] Ste.Dorothée sill, Montreal, Monteregian Hills, Canada (Philpotts 1972): [24] monchiquite dyke with [25] phonolitic ocellus.
[26–27] Callender Bay, Canada (Ferguson & Currie 1971): [26] monchiquite dyke with [27] phonolitic ocellus.
[28–31] Orkney swarm, Scotland, UK (Rock 1983b): [28] average camptonite dyke; separate dykes of [29] fractionated camptonite; [30] bostonite; [31] potassic bostonite.
[32–34] Gran area, Oslo Province, Norway (Scott & Middleton 1983): separate dykes of [32] most ultrabasic camptonite; [33] intermediate, and [34] most silicic maenite.
[35–36] Alpine swarm, New Zealand (Cooper 1976): (35) AL dyke with (36) phonolitic ocellus.

with markedly discontinous variation in an UML/AL→bostonite suite. Such discontinuities appear to be ubiquitous in the few extensively analyzed AL→bostonite suites (e.g. North Sea and Monteregian–White Mtns provinces in Table C2). Such gaps are also widely present in alkaline basalt→trachyte suites (Macdonald 1974), and have sometimes been interpreted as fortuitous (reflecting sampling problems, differential viscosities, etc). If they are instead taken to indicate a real scarcity or absence of intermediate compositions, this would constitute good evidence against a fractionation relationship, and might suggest an explanation in terms of parent AL splitting via liquid immiscibility into bostonite (represented by the felsic ocelli in AL) and cumulate daughter products. Unfortunately, Fig.8.6 also shows that felsic ocelli in AL are *not* chemically similar to associated bostonite dykes, but largely fill a gap (in the tephriphonolite field) between two clusters of these dykes (broadly trachyandesitic and phonolitic). The bostonites cannot therefore be simply interpreted as products of immiscility represented by the ocelli; they *might* be products of splitting of the ocelli themselves, but of course this runs counter to the textural evidence.

8.6 Lamprophyres as parental melts to other igneous rock-suites

Even though mechanisms may be uncertain, the fact that lamprophyres apparently evolve into intermediate and felsic melts (Section 8.5) does imply that they may represent the *parental* (Appendix B) melts to other rocks. This notion is supported by the following:

(1) *Availability.* In many occurrences such as Borralan, Scotland (Table C1), Lovozero and Khibina, USSR (Table C2), and Sokli, Finland (Table C3) lamprophyres are the *most* primitive or indeed the *only* primitive rocks, associated rocks being carbonatitic, ijolitic or syenitic only. On a larger scale, lamprophyres represent the only primitive melts intruded *throughout* some provinces (e.g. Iberian, Monteregian in Table C2).
(2) *Mantle inclusions.* In other occurrences {e.g. Fen, Norway (Table C3) and Pyrenees, Spain (Table C2)}, lamprophyres are the *only* rocks carrying mantle inclusions.

For these and many other reasons detailed below, lamprophyres have now been equated by many authors with parental melts for the following wide range of igneous suites:

— **M1 magma-type**: *hydrous gabbro–syenite plutons* (e.g. Monteregian Province, Canada: Philpotts 1974); *phonolitic lavas* (Fortaleza, Brazil).
— **M2 magma-type**: *ijolite-syenite-carbonatite complexes* (e.g. Fen, Norway: Griffin & Taylor 1975; Kaiserstuhl, Germany: Keller 1989).
— **M7 magma-type**: *high-K calc-alkaline (peridotite-pyroxenite)-diorite-granodiorite-granite plutons* (e.g. Kasuga-mura, Japan: Suzuki & Shiraki 1980; Cortlandt, USA: Domenick & Basu 1984, Bender et al.1984; UK Caledonides: Macdonald *et al.*1986, Rock & Hunter 1987; UK Hercynides: Leat et al.1987; Aland Is., Finland: Nurmi & Haapala 1986); *andesite–dacite-rhyolite sills* (Elk Head, USA: Leat *et al.*1988); *potassic syenite plutons* (Arkansas, USA: Morris 1987; Biella, Italy: Thompson *et al.*1984; Assynt, Scotland: Thompson & Fowler 1986); *calc-alkaline arc basalts* (Turrialba, Costa Rica: Reagan & Gill 1989).

Examples of each of these possibilities are briefly explored below.

8.6.1 M7 (CAL) magma-type and calc-alkaline suites

The 160+ examples of Association **A** in Table C1 establish beyond doubt that an intimate relationship between CAL and certain types of granitoid plutonism has persisted from about

2.7 Ga to the present. However, protracted discussions have been published concerning the nature of this relationship. Until World War II, it was mostly deemed *diaschistic* (Appendix B), lamprophyre being regarded as one form of late residuum from granitoid magma (with complementary aplite and pegmatite). From about 1950 until recently, many petrologists regarded the association as accidental, the lamprophyres being deemed to have intruded fractures induced by the earlier emplacement of unrelated granitoids (e.g. Carmichael *et al.*1974). This notion can no longer be sustained, because of the unequivocal accumulated evidence that the CAL–granitoid association is both *temporally* and *spatially* intimate:

— dating has time and again proved contemporaneity (Table C1);
— composite dykes evince co-emplacement of the two melts (Table 3.1; Section 3.1.2);
— some lamprophyres are emplaced in time between two phases of one granitoid pluton (e.g. Bailey 1960; Moore & Hopson 1961; Camboly *et al.*1967; Loy 1967);
— other lamprophyres may postdate granites but predate aplites or pegmatites in one suite (Gapeeva 1960; Cantagrel *et al.* 1970; Sims & Mudrey 1972; Suzuki & Shiraki 1980);
— yet more lamprophyres occur in two generations, one predating and one postdating the same granite (numerous cases in Table C1; also Zavaritsky 1935).

An impressive array of data has now led abovementioned authors to suggest that this space-time relationship is also *genetic*: that M7 melts might actually contribute as parents to the granitoids themselves. Among this evidence is the following:

— similarity of chemical composition between dyke porphyries, which may be the products of lamprophyre differentiation (Section 8.5.2), and plutonic granitoids;
— identical initial isotopic ratios of lamprophyres and granitoids;
— close correspondence of lamprophyres with parental melts inferred by granite specialists for the plutons, even if the actual connection has not explictly been made;
— small-scale mimicking within some differentiated lamprophyric dykes of processes inferred to take place within granitoid plutons;
— critical inherited characteristics of granitoids in lamprophyres, and *vice-versa*; for example, lamprophyres decrease in abundance relative to associated felsites towards the Shap granite, UK (Table C1) and share its characteristic orthoclase megacrysts; furthermore, minettes around Shap contrast with kersantites radiating from the nearby Threlkeld granodiorite; Na/K and orthoclase/plagioclase ratios in the lamprophyres correlate with the associated pluton (Piper *et al.* 1978).

Although M7 melts have so far mainly been invoked as parents to calc-alkaline *plutons*, Reagan & Gill (1989) have now extended the suggestion even to 'normal' calc-alkaline *volcanic* rocks. They argue that calc-alkaline basalts may represent mixtures between CAL, representing small-degree melts generated by amphibole breakdown in mantle overlying a subducting slab, and high-degree MORB-like melts, generated by mantle upwelling.

8.6.2 M2 (AL) magma-type and alkaline gabbro-syenite plutons

One of the most intriguing cases is the Monteregian province (Table C2), where lamprophyre dykes gradually change from camptonite in the east via monchiquite in the centre to alnöite in the west; the associated plutons change concomitantly from gabbro-syenite (Mt.Megantic) via essexite-foyaite (Mt.Royal) to ijolite-carbonatite (Oka). The plutonic rocks are cumulates (Eby 1987), and the lamprophyres are among few candidates for primary melts. Philpotts' (1974) inference that camptonite, monchiquite and alnöite were

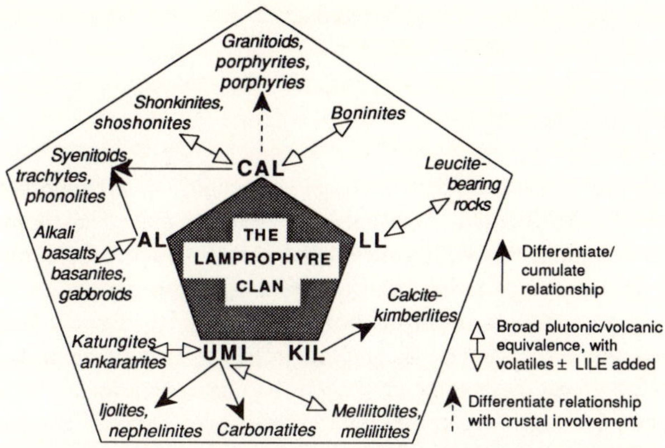

Fig.8.7 *Petrogenetic relationships between lamprophyres and some other igneous rock-types.* The pentagonal arrangement is justified mathematically by Fig.5.2.

parental to the three types of plutonic suites has survived largely unchallenged to date, and indeed has been extended to the voluminous White Mountains plutons: Eby (1985, 1987) for example regards the camptonites as "a reasonable analogue for the magmatic evolution of the mafic series in the White Mountain plutons", and notes for the Mt.Royal pluton that the "initial melt....is also quite similar to the measured composition of the monchiquite".

8.6.3 M3 (UML) magma-type and carbonatite-ijolite-nephelinite complexes

Many UML are doubtless close to the *parent* melts for carbonatite complexes. Indeed, most of the vague names previously ascribed to such parents ("volatile-rich ankaratrite, CO_2-rich melanephelinite, hydrous peridotite", etc. — Heinrich 1966, Tuttle & Gittins 1966; Bell 1989) can be translated as *aillikite*. Among the clearest cases is Fen (Table C3), where the damkjernite (UML) is not only the most abundant rock-type (based on gravity models), but also the sole host to lherzolite xenoliths, whilst possessing the characteristics of both a primary and parental melt (e.g. Griffin & Taylor 1975). M3 melts can probably yield calcitic carbonatites by either fractionation or silicate-carbonate liquid immiscibility, with complementary mafic cumulates or conjugate liquids corresponding to ijolites, pyroxenites, etc. Increasing reports of fenitization around UML dykes even in the absence of separate carbonatites (e.g. Fielding & Jaques 1989) provide further confirmation. Perhaps the main argument against a *global* parental status is that UML have been reported from less than half of the world's carbonatite complexes; however, given the huge recent increase in the number of reported UML, this may yet prove to be a mere information problem.

8.7 Petrogenetic overview

Fig.8.7 proposes a synoptic model for relationships between the 5 lamprophyre branches, and other alkaline and calc-alkaline rocks. If underpinned by future work, it suggests that lamprophyres deserve a status nearer the centre than the periphery of modern petrology.

9 Economic Geology

A decade ago, any suggestion that lamprophyres could be of economic importance would probably have been ridiculed. A turnabout has taken place via mounting evidence that lamprophyres are connected with a medley of precious materials: diamond, gold, sapphire and possibly PGEs. A realization has dawned that lamprophyres are 'ultra-deep' (Section 8.3), and that this makes them economically special. Even confining attention to lamprophyres *other than* kimberlites and lamproites still credits them with economic status. Only the briefest of outlines can be given here, as recent reviews are available in some cases (see below), and the latest developments necessarily involve confidential information.

9.1 Lamprophyres and diamond

The recent history of diamond exploration has been startling. A century-old maxim that diamonds only occurs in kimberlites was buried in the late 1970s with the discovery of the world's richest deposit in the Argyle lamproite (Tables 9.1, C4). Further diamondiferous

Table 9.1 Diamondiferous lamprophyres (excluding kimberlites)

Country	Occurrence	L	Diamond content	Comments	See also
Australia (WA)	Maude Creek	UA	6 microdiamonds from ≈4.6 t of rock	Nearby dykes (Duck Creek, Devils Elbow, Blackfellow Creek) yielded no diamonds	Table C3; Fig.2.8
Australia (WA)	Argyle (AK1), E.Kimberley	LL	Richest known deposit by weight (≈6.8 ct/t)	>35% world production by weight, <10% by value, but rare pinks fetch ≈$1M per carat	Table C4; Figs.2.4/8
Australia (WA)	Nabberu (Bulljah)	UA	Various companies got several microdiamonds	Xenocrysts cover typical kimberlitic indicator minerals (Cr-diopside, etc.)	Table C3; Figs.2.3/8
Australia (WA)	W.Kimberley province	LL	Variable, subeconomic (up to ≈14ct/100t)	Olivine-lamproites are generally richer in diamonds than leucite-lamproites	Table C4; Figs.2.4/8
Australia (WA)	Wandagee	AM	Four microdiamonds (< 0.5mm) recovered	Xenocrysts cover typical kimberlitic indicator minerals (Cr-diopside, picroilmenite, etc.)	Table C2; Figs.2.2/8
Canada (Quebec)	Montreal (Ile Bizard)	UA	10 micros (total 0.06 ct) from 20 cu yd rock	Diamonds presumed genuine by Brummer (1978) but could be laboratory contaminant	Table C3; Figs.2.3/9
India	Chelima	LL	Microdiamonds at low (< 2 ct/100t) grade	Old (ancient?) workings in dykes, nearby alluvial & probable in situ microdiamonds	Table C4; Figs.2.4/13
India	Majhgawan /Hinota	LL	8-15ct/100t (Maj); <1ct/100t (Hinota)	Only currently worked diamond deposit in India; nearby secondary deposits	Table C4; Figs.2.4/13
India	Wajrakarur/ Lattavaram	UA KU	<3ct/100t but variable between the 8 pipes	"Kimberlite" pipes (1,3,4,6) diamondiferous but ?UML pipes (2 & 5) may be barren	Table C6; Figs.2.3/13
Ivory Coast	Bobi, Seguela	LL UL?	Marginally economic		Table C6; Figs.2.3/4
S.Africa	Kimberley	UA	Six extremely small fragmental diamonds	Diamond confirmed in one De Beers dyke but no diamonds in other dykes	Table C3; Fig.2.3
Sweden	Alnö	UA	Three fragmental diamonds in 11.5t UA	"Diamond" figured by Von Eckermann (1967) was garnet; occurrence needs confirmation	Table C3; Fig.2.3
USA (AR)	Prairie Creek	LL	Uncertain but probably low	Mining 1906–1962; area now "Crater of Diamonds" national park but mining planned	Table C4; Figs.2.4/10
USSR	Aldan Shield	LO LL	Accessory	Few details available	Table C4; Fig.2.4
Zambia	Kapamba	LL	Subeconomic; mostly small stones (< 0.1 ct)	Mostly yellow and brown stones with predominant tetrahexahedroid forms	Table C4; Fig.2.4

lamproites have since been discovered (Scott Smith & Skinner 1984b), and other rocks, called "kimberlites" merely because they carry diamonds, have been renamed (Table 9.1).

Continuing exploration has also now revealed microdiamonds in AL and UML (Table 9.1). This must in turn lead to reconsideration of diamond occurrences in the general area of other AL and UML intrusions, not least in Greenland (Scott Smith 1987, p.30), Canada (Brummer 1978), and the Ukraine (Bayrakov 1964), and of several large diamond fields whose source rocks remain enigmatic (notably Kalimantan and Copeton, SE Australia). As an illustration of the topicality of the search, "lamprophyric pipes"— perhaps related to the one known at Gloucester (Table C3) — have recently been hailed in *Australian Business* (December 1989, p.46) and *The Australian* as the long-sought source of the Copeton stones.

All this naturally raises the question: which rocks can in principle carry diamonds? Although experts would differ on their answers, a few generalisations may be made:

— Diamonds are now known not to be entirely confined to pipes, but to occur rarely in subducted eclogite-facies rocks (e.g. Kazakhstan, USSR), ophiolites (e.g. Beni Bousera, Morocco) and other rocks which have been buried to depths within the diamond stability field (> 150 km) and then brought once more to the surface. Economic implications of such occurrences, however, are essentially negative, in showing that not all small alluvial diamond fields warrant an exhaustive search for nearby igneous source pipes.

— Although there are a very few, disputed, reports of diamonds occurring in basaltic hosts (e.g. Sutherland *et al.* 1985), other inclusions accompanying the diamonds appear to have come from depths of no more than 80 km, and the diamonds themselves have peculiar inclusion assemblages and isotopic compositions. This suggests that these basaltic rocks have merely *reworked* diamonds from deeper, hitherto unidentified, primary sources.

— At present, lamprophyres in the sense used here are the only *orthomagmatic* rocks known to host diamonds. Importantly, the *whole clan* fits all the various criteria explorationists have so far delineated for diamond potential: for example, there is a source of carbon in the abundant carbonates, the melts are emplaced rapidly (so preserving metastable diamond), while spinel compositions and xenoliths encompass "favourable" P-T conditions of origin. Again, Eggler (1989) has argued from CMAS projections that *some* UML (e.g. those of Alnö and Montreal) do indeed come from depths > 150 km, which supports the veracity of the persistent reports of trace diamond in these rocks (Table 9.1).

— Nevertheless, evidence in Chapters 4–8 makes it unlikely that UML and AL could host *economic* deposits, or that CAL could host diamonds in more than minute quantities; in summary, the order of economic interest is probably KIL ≈ LL > UML » AL » CAL, with KIL generated at greatest (asthenospheric) depths, but with many LL (i.e. olivine-lamproites) and some UML generated within the diamond stability field.

9.2 Calc-alkaline lamprophyres and gold

A possible connection between gold and lamprophyres has been mooted over many years (e.g. McLennan 1915; Moorhouse 1959, p. 332), and the number of mutual associations in mining camps has led, as Boyle (1979, p.250) puts it, to "an adage among some gold prospectors that lamprophyres are good indicators of the presence of gold deposits". This connection has been fully explored by Rock *et al.* (1987,1988b,c;1989), Rock & Groves (1988a,b), and Wyman & Kerrich (1988a,b;1989a,b). Only a summary need be given here:

— The connection is one between mesothermal (Appendix B) deposits and CAL, and

Table 9.2 Some well-documented gold/calc-alkaline lamprophyre associations
(listed in decreasing order of age)

Area	Age	Comments, including nature of gold-lamprophyre association[a]
Yilgarn Province, Australia	≈2.7 Ga	Hundreds of lamprophyres form swarms in Eastern Goldfields and Murchison Provinces; high Au locally; field/isotopic evidence proves lamprophyres bracket gold mineralization and porphyries[b-f]
Superior Province Canada	≈2.7 Ga	Dozens of associations reported. Moreover, "lamprophyres...uniquely bracket gold mineralization in the Superior Provinces...Neither...granitoids, nor tholeiitic or calc-alkalic magmas display the same degree of temporal or spatial association"[b-f]
Limpopo Belt, Zimbabwe	≈2.7 Ga	Association less intimate than the above: most gold is in marginal belt, lamprophyres in central belt; nevertheless,the two are still contemporaneous[e]
Lupa and Victoria goldfields, Tanzania	≈2.7 Ga	Appinites and associated lamprophyres form wall-rocks to large deposits in several mines (e.g. New Saza); mineralized mica-lamps in Geita deposit (BIF-hosted)
Pine Creek, Australia	≈1.8 Ga	High grade Au-Ag-quartz veins synchronous with granite/syenite/lamprophyre magmatism; some dykes penetrate veins, most don't[e]
Caledonides, UK	≈400 Ma	Fresh lamprophyres carry high Au (up to 523 ppb); gold prospects nearby
Fifield, Australia	≈400 Ma	Pt-Au mineralization associated with numerous plugs of mafic-ultramafic hornblende-rich rocks which closely resemble the Devonian appinite–lamprophyre plugs of the British Caledonides (Chapter 7.1)[e]
Woods Point, Australia	≈400 Ma	95% of gold from reefs actually within lamprophyre dikes in Woods Point swarm, which contains "an abundance of lamprophyric types"[b,c,f]
Tadjikistan, USSR	Palaeozoic	Gold (±rare metals,F,W,Sb,Bi) exactly overlap in age with lamprophyres
Urals, USSR	Permian	Lamprophyres at Magadan (largest USSR Au deposit), Kokpatas[c,f,g]
E.Transbaikal, USSR	Jurassic-Cretaceous	Lamprophyres occur as late stage of magmatism in goldfield; general association of deep lineamenets, lamprophyres and golde
Shandong Peninsula, China	≈100 Ma	Lamprophyres give identical ages and isotopic signatures to gold veins, and have high Au (25-311ppb, mean 150; cf. local granitoids < 2ppb)[b,c,e,f]
N.American Cordillera province	Jurassic–Cretaceous +Tertiary	Good examples at Cripple Creek, Stewart, Willow Creek, etc.; at Rossland, two suites of lamprophyres are associated with two periods of gold mineralization; in Mother Lode area, Au and lamprophyres were both introduced during a hiatus in the Sierra Nevada granitic plutonism, between about 144 and 110 Ma[b,e]; at Bridge River, lamprophyres bracket gold mineralization
Eastern USSR	Early Cretaceous	Au-arsenopyrite mineralization "closely associated with lamprophyre and diorite porphyry dikes"
Azatek, Armenia, USSR	Palaeo-gene	In time,"lamprophyres (are) magmatic formations closest to mineralization"; in space, lamprophyres are commonest where Au-Sb mineralization intense; autometasomatic alteration in lamprophyres resembles Au-related alteration
NW Alps, Italy	≈32 Ma	Gold-quartz veins postdate metamorphic peak by 4–8 Ma but exactly overlap with lamprophyres; lamprophyres are also the most likely parent magmas for the Bialla syenite stock nearby[b,c,e,f]
Linhaisai, Indonesia	Miocene	Lamprophyres occur within 200x30 km corridor of gold prospects of same age, including Kelian, Kasongan, Masuparia, Muro, Muyup
Porgera, Papua New Guinea	Miocene (8–14 Ma)	Gold mineralization intimately associated with Porgera intrusive suite of sills and plugs, which has clear lamprophyric affinities[b,e]

[a] For references and further details, see Tables C1/C5 of Appendix C. In all cases, field and stratigraphical evidence link the lamprophyres and gold within one overall event. In most cases, the lamprophyres belong to a magmatic and mineralization event which also included felsic plutonism and/or volcanism.

[b] Lamprophyres are most numerous around gold occurrences.

[c] Some lamprophyres cut, others are cut by gold veins; the two thus bracket one another in time.

[d] Lamprophyres are foliated by events known to immediately postdate mineralization.

[e] Lamprophyres have been isotopically dated.

[f] Local/regional structural evidence (e.g., lamprophyres parallel Au-bearing veins) suggests genetic relation.

[g] Information obtained from displays at Academician N.Chernyshev Central Geological Research and Survey Museun, Leningrad, USSR.

occurs throughout the geological record, encompassing both of the main (Archaean and Tertiary) episodes of gold deposit formation. Well-documented examples are compiled in Table 9.2.

— Wherever detailed field and/or geochronological evidence is available, the CAL prove to *overlap* gold mineralization in both time and space, and are often the *only* igneous rocks to have been emplaced at precisely the same time as the gold (Table 9.2).
— Nevertheless, because of the tendency for lamprophyres to be emplaced repeatedly in the same areas (Fig.2.14), some gold-lamprophyre associations are purely spatial, and may even coexist with intimate space-time associations in the *same* goldfield (Fig.9.1).
— Overall, the intimate space-time connection between lamprophyres and gold deposits can be supported quantitatively and statistically (Hodgson & Troop 1988), and is as near to a hard scientific 'fact' as can be; debate now centres on whether this connection is primarily *structural* (the gold and lamprophyres merely reflecting similar mantle processes), or *genetic* (the lamprophyres being transporting agents for some of the gold or at least for part of the mineralizing auriferous fluids).
— There is evidence that some lamprophyres may be enriched in gold, although many more accurate data, underpinned by chalcophile and siderophile element determinations, are needed to establish this (Section 5.5). Although high gold in some CAL associated with major gold alteration systems (e.g. McNeil & Kerrich 1986) may partly result from secondary enrichment (even where the rocks appear petrographically fresh), detailed statistical analysis suggests that at least some of the enrichment could still be primary even here (Rock *et al.* 1989). Moreover, high gold contents are also recorded not only in CAL *away from* gold deposits (e.g. SW Scotland), but also in AL, KIL, LL and UML, associated with no gold alteration at all; such enrichments are much more likely to be primary.
— Lamprophyres closely mirror the wall-rock alteration around mesothermal gold deposits, both in their autometasomatic mineralogy and bulk chemistry (e.g. high K, Rb, Ba, CO_2; moderate S; low Cl). This implies that lamprophyric melts are similar to mineralizing fluids in gold alteration systems and, therefore, that they may easily transport gold.
— Deep parts of the mantle (and core) are expected to be intrinsically enriched in gold due to primeval differentiation of the Earth; since lamprophyres come from exceptionally deep, and may have the capacity to transport gold, this might support intrinsic enrichment.
— That CAL are associated with gold (but not diamond), whereas other lamprophyres are associated with diamond (but not gold), may reflect the putative crustal processes involved in CAL genesis (Section 8.4): whereas *all* lamprophyric melts might be Au-enriched at depth, these crustal processes lead to M7 melts giving up gold to crustal hydrothermal systems, whereas other lamprophyres experience no crustal involvement and thus retain their gold.

Overall, the spatial and temporal intimacy of the CAL-gold association makes some CAL gold indicators in their own right. It seems likely that CAL melts contribute fluids and perhaps even gold itself to the mineralizing systems. In any case, the association has major implications for the tectonic setting of mesothermal gold deposits.

9.3 Lamprophyres and miscellaneous mineral deposits

A detailed consideration of relationships between ore mineralization and deep-seated dyke magmatism is well beyond the scope of this preliminary work, but earlier reviews are

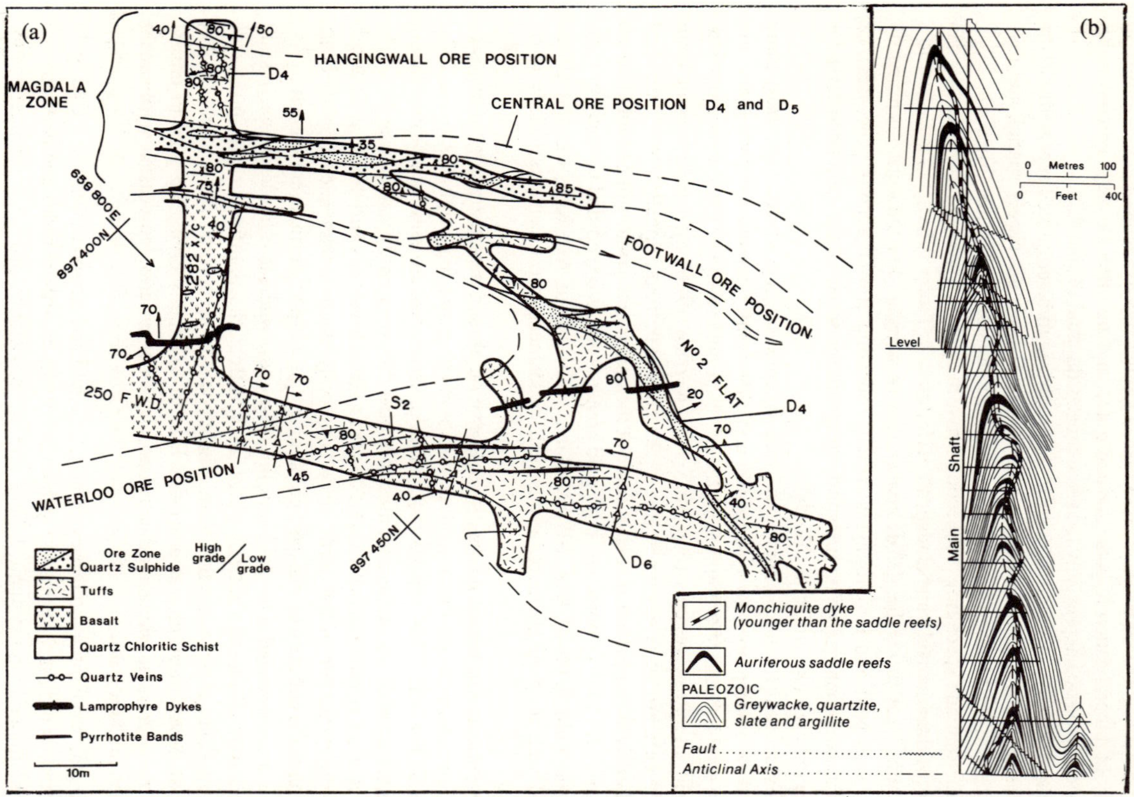

Fig 9.1 *Contrasting gold-lamprophyre associations in the goldfields of Victoria, Australia* Diagrams reproduced with permission from Jones (1988) **(a)** 250 level footwall drive plan of the Magdala goldmine. CAL dykes cut one of the veins here, but are themselves cut by quartz veins; in other areas, veins appear to cut lamprophyre dykes; both are believed to be approximately contemporaneous (Devonian). **(b)** Cross-section of Great Extended Hustler's shaft in the Bendigo goldfield. Here, Devonian auriferous saddle reefs are cut by *Jurassic* AL dykes (Table C2), but this purely spatial-structural association is so consistent that "the presence of a (monchiquite) dyke intersected in a prospecting crosscut (can) be taken as a reliable indicator of the presence of centre country" - that is, rich saddle reefs (Wilkinson 1988a, p.19).

available in Abdullaev (1957), Nemec (1971a) and Shchukin (1974). Table 9.3 merely lists some varied reserves, resources and deposits which show an intimate space-time relationship with *confirmed* lamprophyres.

The odd-man-out in Table 9.3 is the Yogo sapphire deposit, currently being worked and long described as the most valuable gem deposit in the USA. Alkali basaltic rocks in Australia, China, Scotland and Thailand also host significant quantities of sapphire, and a few other lamprophyres, such as the Loch Roag monchiquite (Jackson 1984; Table C2), are known to host even gem quality sapphire crystals, but the Yogo dyke is the only lamprophyre known at present to contain a commercial deposit. The sapphire is undoubtedly xenocrystic, but could represent restite derived from aluminous crustal rocks (cf. the Mull sapphires, NW Scotland) or mantle material.

Remaining examples in Table 9.3 are base metal deposits. The spatial relationship with lamprophyres is commonly indicated by dykes and veins displaying the same structural controls (e.g. coexistence within fractures), while the temporal one — if unconfirmed by direct dating — is usually suggested by mutually cross-cutting relationships which show the lamprophyres bracketing, or only slightly preceding, the veins. Commonly, the lamprophyres postdate one vein set and predate another (as in Seal *et al.*1988). As in Section 9.2, and as with so many mineral deposits, however, a space–time connection is much more easily established than a genetic relationship: many authors have interpreted basically similar geological data either for or against such a relationship (cf. Spurr 1939; Sobolev 1952).

It should be noted that some unconfirmed lamprophyre occurrences in Table C7 are also associated with mineral deposits: for example, Mead *et al.*(1988) reported late lamprophyre dykes in several tungsten mines of the Sonora region, Mexico. These are omitted from Table 9.3, however, simply because the "lamprophyres" are themselves unconfirmed.

Several occurrences in Table 9.3 are from the European Hercynides, where lamprophyric magmatism is commonly interposed in time between granitoid plutonism and later ore deposition. As ore veins are more abundantly associated with lamprophyric than granitic rocks, the space-time relationship between lamprophyres and ores seems stronger in this province than that between granitoids and ores. This might even be taken to warrant expansion of the number of entries in Table 9.3 to include *all* mineralized areas associated with Hercynian granitoids, particularly as the lamprophyres might be parental in part to the granitoids (Section 8.6.1). For example, if Leat *et al.* (1987) are correct that the Cornish minette dykes are involved in the genesis of the Cornubian batholith, UK (Table C1, Fig.2.5), then any component of the well-known Cornish Sn-W mineralization which is not derived from crustal anatexis might logically be attributable to these minettes. The important point here is that lamprophyres are a missing element in the traditional 'granites + mineralization' maxim which should no longer be ignored; it may be at least as reasonable to attribute certain components of mineralizing fluids to deep, mantle-derived, lamprophyric melts as to shallower granitic magmatism.

In general, lamprophyres' enrichment in F, Cl, S, H_2O and CO_2 relative to most igneous rocks, together with their sometimes high temperatures (Section 3.1.4) and explosive emplacement by violent drilling mechanisms (Section 3.1.6), probably confer on them an enhanced ability to dissolve and transport elements which form soluble complexes with one or more of these ligands. Lamprophyre intrusion mechanisms worked out by Currie & Ferguson (1970), which involve repeated explosive drilling by the volatile phase into the country-rocks, followed by resolution of the volatiles into the lamprophyre melt, seem likely to further enhance the dissolution of certain elements from country-rocks.

Table 9.3 Some space-time associations of lamprophyres and miscellaneous mineral deposits

Country	Occurrence	L	Ore mineral(s)	Comments	See also
Canada (NB)	Lake George	CAL (CS)	W-Mo stockwork	In Sb-W-Mo-Au etc deposit; major Sb producer; lamps and ores bracket one another in time	Table C1; Fig.2.9
Canada (NWT)	Keewatin	CAL (CM)	Syngenetic Cu+U+Ag+Pb	Ores "in hydrothermal aureoles adjacent to the dykes"; also epigenetic U+Cu+Pb+Ag+Se+Au±Mo±Zn±Cd±Bi deposits	Table C1; Fig.2.9
Czecho-slovakia	Bohemian Mass	CAL (CK)	Pb-Zn-Ag±Ba (galena-blende)	Lamps interposed in short time interval between granitoids and ores; intimate spatial association	Table C1; Fig.2.5
France	Massif Central	CAL (CM)	U	Mine Henriette, etc.	Table C1; Fig.2.5
Germany	Harz Mtns.	CAL (CK)	Pb-Zn (galena-blende)	Intimate spatial association plus geochemical affinity between CK and ores, but CK may be slightly later	Table C1; Fig.2.5
Italy	NW Alps (Sesio-Lanzo)	CAL (CM)	U–Th (also minor Au veins)	Lamprophyres are radioactive	Table C1; Fig.2.12
Namibia	Aias	AL (AC)	Pb (argentiferous galena)	Wide variety of magmatism occurs, but veins most intimately associated with lamprophyric breccia plugs	Table C2; Fig.2.2
Poland	Swiety Krzyz Mtns.	CAL	Cu-Pb-(Zn-Ni-Co-As)	"Connection between the deposits of nonferrous metals and magmatism is very probable"	Table C1; Fig.2.5
Spain	Sierra de Gadar	CAL ?	Pb (galena–fluorite)	Few details available	—
UK (Scotland)	Appin	CAL (CA)	Cu (chalcopyrite)	High-grade float, low-grade sulphide mineralization in appinite pipes; some similarities to Colorado deposits	Tables C2/5;Fig.2.6
UK (Scotland)	Ardgour (Strontian area)	AL (AC)	Pb-Zn-Ba-Sr (galena, etc.)	Strontianite/Sr type locality; lamps may be somewhat older than ores; further Pb-lamp association in Orkney Is.	Table C2; Fig.2.7
UK (Scotland)	Black Stockarton	CAL	Cu (porphyry copper)	Low-grade mineralization associated with major lamp-porphyry swarms near Criffell granitoid	Table C1; Fig.2.6
USA (AL)	Central York Mtns.	CAL	Syngenetic Be+Sn+fluorite	Space-time relationship is more intimate between ores and lamprophyres than ores and granitoids	Table C1; Fig.2.1
USA (AZ)	Hopi Buttes	AL (AM)	U (uranyl vanadates)	"uranium was introduced by solutions derived from the hydrous phosphate-rich monchiquite magma"	Table C2; Fig.2.10
USA (CO)	Red Mtn.,Urad-Henderson, etc.	CAL (CK)	Mo (porphyry molybdenum)	Mineralization interposed between early bimodal lamp–porphyry and late silicic porphyry magmatism	Table C1; Fig.2.10
USA (MT)	Yogo Gulch	UML (UO)	Sapphire	"...one of the most important gem deposits" in USA; sapphire as xen in lamp; produced 4000 carats (US$3M) in 1984	Table C3; Fig.2.10
USSR	Chukchi peninsula	?	Sn	Few details available	—
USSR	Elbrus	CAL (CS)	Polymetallic	Few details available	—

Such elements include transition metals in which lamprophyres are already *known* to be clearly enriched (Zr, Th, U, etc.). Thus it is possible that lamprophyres could also be enriched in metals such as Mo, Sn, PGE, Au, W, etc, though there are simply no reliable data yet to confirm or deny this. Given furthermore that lamprophyres rise from exceptional depths, they can perhaps not only tap otherwise inaccessible parts of the Earth which may be intrinsically enriched in many elements, but also extract further enrichments in certain elements, on their unusually lengthy but violent journey through the crust. In short, it is conceivable that many lamprophyre melts could reach upper levels of the crust as a cocktail enriched in a medley of rare elements. Their widespread association with mineral deposits might therefore be expected. To date, there are far more known examples of associations with mesothermal gold deposits than with any other type of deposit (cf. Tables 9.2, 9.3), but lamprophyred could well be 'lurking' among the virtually undescribed minor intrusions (and in particular the breccia bodies) associated with many other types of mineral deposit.

A Selective Lamprophyre Bibliography

Compilation

This Bibliography includes most publications which: (a) contain the word *lamprophyre* (or one of the varieties named in Fig.1.2 and Appendix B) either in the title or as a keyword; (b) devote significant attention to lamprophyres in terms of field, petrographical or petrological description; (c) contain analytical data on lamprophyres. Those in which lamprophyres are only very briefly mentioned, and sources prior to and including those compiled by Washington (1917), are not generally included unless giving information which is critical or otherwise unavailable. References to kimberlites are not exhaustively compiled — readers should refer for example to R.H.Mitchell (1986) or to the 7 *International Kimberlite Conference* (IKC) volumes. Abstracts and unpublished works are excluded, unless there is no other published information on the topic.

This Bibliography has been compiled over a period of over 16 years by an exhaustive combination of traditional literature search (aided by *Bibliography and Index of Geology, Bulletin Signalétique, Dissertation Abstracts International* and *Mineralogical Abstracts*), plus usage of the many computerized geological source and reference databases now available (specifically AESIS, GEOARCHIVES, GEOBASE, GEOREF and IGBA). Note that searches from these systems on the free-text strings "minette" and "spessartite" will also raise references to sedimentary rocks and to garnets (e.g. Siehl & Thein 1978). Available compilations dealing with alkaline rocks (e.g. Heinrich 1966; Tuttle & Gittins 1966; Sørensen 1974; Fitton & Upton 1987; Woolley 1987; Bell 1989; *Mem.Geol.Soc.India***15**) and the 7 IKC volumes have also been searched. The rate of acquisition has now slowed to the point where the Bibliography probably includes the vast majority of relevant references; coverage of Russian and Chinese publications is probably weakest.

Ordering

Titles using Roman script are given in their languages; all others are translated into English between [...]
Mac, Mc, St. and similar are listed in strict alphabetical order; 2 or multi-author papers by the same senior author are listed chronologically, then alphabetically. Spaces, apostrophes and similar are ignored.

Symbols

# indicates that some rocks are regarded as different lamprophyres from those named in the source reference (especially camptonites rather than "spessartites"; aillikites rather than "kimberlites" or vice-versa).

¶ indicates that the reference almost certainly hides additional lamprophyres described under other rock-names (e.g. "mafic porphyry", "diabase", "biotite-nephelinite", "K-rich diorite", "mica-peridotite").

? indicates that insufficient information is given to confirm or repudiate the rocks described as "lamprophyres" (or under one variety name); the rocks are therefore regarded as unconfirmed (Table C7).

@ indicates that some or all of the rocks described as "lamprophyres" (or varieties) do not obey the definitions in Streckeisen(1979) or this work; they are thus considered apochryphal (Table C8).

* indicates that chemical data from the source have been incorporated into the database LAMPDA (App.D).

† indicates that the reference is cited in a bibliographical database (e.g. *Bibliography & Index of Geology*) and is therefore believed correct, but has nevertheless proved to be unobtainable.

‡ indicates that the reference is cited by a third party but could not be located in any bibliographical database or obtained directly; hence the veracity of the reference itself is not guaranteed.

****

Abdullaev, K.M. 1957. [*Dykes and ore mineralization*]. In Russian. Moscow.

Abraham, E.M. 1951. Geology of McElroy and part of Boston townships. *Ontario Dept.Mines Geol.Rep.* **59**, 66pp.

* Acejev,B.N. & Harlass,E. 1968.Zum Probleme der Altersstellung von Lamprophyren im westlichen Erzgebirge.*Geologie* **17**,1178-1194.

Acharyya, S.K. 1969. A note on the geology of Darjeeling coalfield and its bearing on the structure of the Eastern Himalayas. *Rec.Geol.Surv.India* **97**, 91-101.

Adams,C.J.D. 1976.Geochronology of the Channel Islands and adjacent French mainland.*J.Geol.Soc.Lond.* **132**,233-250.

Aires Barros, L. 1965. Sobre as rochas feldspatóidicas da região de Zenza do Itombe (Angola). *Garcia de Orta (Lisboa)*, **13**, 425-434.

* Akella,J., Rao,S.P., McCallister,R.H., Boyd, F.R. & Meyer,H.A.O. 1979. Mineralogical studies on the diamondiferous kimberlite of the Wajrakjarur area, southern India. In: Boyd & Meyer (1979a), qv, 172-179.

Albarède,F. & Weisbrod,A. 1981. Hercynian synorogenic lamprophyres from the SE Massif Central (France): evidence for equilibrium of Nd isotopes and REE elements with surrounding granitic rocks. *EOS* **62**, p.1076 (abstr).

Alcock,F.J. 1935.Geology of the Chaleur Bay region.*Mem.Geol.Surv.Can.***183**, 146pp.

* Alderman, A.R. 1929. Magmatic differentiation at Mannum, S.A. *Tr.R.Soc.S.Aust.* **53**, 249-257.

* Alderton,D. 1988.Ag-Au-Te mineralization in the Ratagain complex, northwest Scotland. *Tr.Inst.Ming. Metall.* **97**, B171-180.

‡ Aldrick, D.J., Brown, D.A., Harakai, J.K. & Armstrong, R.L. 1987. Geochronology of the Stewart mining camp, BC. *Ministry of Energy, Mines & Petroleum Resources Geol.Fieldwork 1986, Pap. 1987-1*, 81-92.

* Alibert, C. & Albarède, F. 1988. Relationships between mineralogical, chemical and isotopic properties of some North American kimberlites. *J.Geophys.Res.* **93**, 7643-7671.

Alibert,C., Michard, A. & Albarède, F. 1986. Isotope and trace element geochemistry of Colorado Plateau volcanics. *Geochem. Cosmochim. Acta* **50,** 2735-2750.

† Alietti,A. 1955.Su di una camptonite anfibolica di Val del Fösch presso Mezzavalle in Val di Fassa. *Period. Mineral.* **24**,27-46.

* Allan,J.F. & Carmichael,I.S.E. 1984.Lamprophyric lavas from the Colima graben,Mexico. *Contrib. Mineral.Petrol.* **88**,203-216.

* Allen,J.B. & Deans,T. 1965.Ultrabasic eruptives with alnöitic-kimberlitic affinities from Malaita,Solomon Islands. *Mineral.Mag.* 34,16-34.

* Allen,J.E. & Balk,R. 1954.Mineral resources of the Fort Defiance-Tohatchi Quadrangles.*New Mexico Bur.Mines Res.Bull.***36**,192pp.

Allen,J.R.L. 1968.The Cambrian and Ordovician systems (Post-Cambrian intrusive rocks). In: *The Geology of the East Midlands*(eds.Sylvester-Bradley,P.C. & Ford,T.D.),37-38, Leicester Univ. Press, Leicester.

Allsopp,M.L.,Bristow,J.W.,Skinner,E.M.W.,Scott Smith,B.H. & Danchin,R.V. 1985. Rb-Sr geochronology of some Miocene West Australian lamproites.*Tr.Geol.Soc.S.Afr.***88**,341-346.

? Amodio,L. & Hieke Merlin,O. 1966.I proetti inclusi nelle piroclastiti del Monte Vulture (Lucania). *Mem.Ist.Geol. Mineral.Univ.Padova* **25**,1-49.

Anand, S.N. 1971. Geological and geophysical investigations of diamondiferous volcanic pipe rocks from Lattavaram area, Anantapur district, Andhra Pradesh. *Misc. Publ. Geol. Surv. India* **19**, 128-133.

Ancochea, E. & Nixon, P.H. 1987. Xenoliths in the Iberian Peninsula. In: Nixon (1987), qv, 119-124.

Andersen,T.B. & Jansen,O.J. 1987. The Sunnhordland Batholith, W.Norway: regional setting and internal structure, with emphasis on the granitoid plutons. *Norsk Geol.Tiddskr.* **67**, 159-181.

* Anderson,J.G.C. 1935a.The Arrochar intrusive complex.*Geol.Mag.***72**,263-282.

Anderson,J.G.C. 1935b.The marginal intrusions of Ben Nevis: the Coille Lianachain complex and the Ben Nevis dyke-swarm. *Tr.Geol.Soc.Glasgow* **19**,225-269.

Anderson,J.G.C. 1937a. Intrusions of the Glen Falloch area.*Geol.Mag.***74**,458-468.

Anderson,J.G.C. 1937b.The Etive granite complex.*Q.J.Geol.Soc.Lond.***93**,487-533.

Anderson,J.G.C. & Tyrrell,G.W. 1936.Xenolithic minor intrusions in the Loch Lomond district. *Tr.Geol.Soc. Glasgow* **19**,373-384.

* Andrade,M.M.de. 1950.Um lamprófiro com horneblenda e biotite, da Póvoa de Varzim. *Est.Not.Trab.Serv. Fom. Mineral. (Porto)*, **14**,49-56.

* Andrade,M.M.de. 1954.Contribution a l'étude des roches alcalines d'Angola. 19*th Int.Geol.Congr. (1952) fasc.* **xx**,241-252.

Andrade,M.M.de. 1979.Sobre a ocorrência dum sistema de diques de lamprófiros na Praia de Labruje. *Publ.Mus.Lab. Mineral.Geol.Fac.Ciências Porto* **91**,151-153.

* Andreasson,P.G.,Solyom,Z. & Roberts,D. 1979.Petrochemistry and tectonic significance of basic and alkaline-ultrabasic dykes in the Leksdal Nappe,north Trondheim region,Norway.*Norges Geol.Unders.* **348**,47-72.

André,F. 1981. Un exemple de vaugnéite, cumulat d'une association gabbro-dioritique dans le massif des Ballon (Vosges meridionales). *C.R.Acad.Sci.Paris Ser.II*, **293,** 445-448

Andrews,E.C., Mingaye J.C.H., & Card,G.W. 1907. The geology of the New England Plateau. *Rec.Geol.Surv.N.S.W.* **8**(3),196-251.

Andrews, J.R., Phillips W.E.A. & Molloy M.A. 1978. The metamorphic rocks of part of the north central Ox Mountains inlier of Counties Sligo and Mayo. *J.Earth Sci.R.Dublin Soc.* **1,** 173-194

Anon. 1971. Diamond. *Geol.Surv.IndiaMisc.Publ.* **19**, 269pp.

* Anon. 1988. The mafic dyke swarms of southernmost Sweden; guide. *IGCP-257 Tech.Rep.* **1**, 51-91.

* Anton,L. & Constantinescu,R. 1977.Observatii asupra lamprofirelor din Muntii Fagaras. In Romanian, Engl.abstr. *Rom. Inst.Geol.Geophys. Dari Seama Sedintelor* **64(1)**,53-76.

* Aoki,K. 1981.Chemical composition of potassic basaltic rocks from the Colorado plateau.*Sci.Rep.Tohoku Univ.***xv**,135-139.

Appavadhanulu,K. 1966.Minette and riebeckite-kalisyenite dykes in some upper Cuddapah rocks,Kurnool district,Andhra Pradesh. *Rec.Geol.Surv. India* **94**(2),303-304.

Arima, M. & Edgar, A.D. 1981. Substitution mechanisms and solubility of titanium in phlogopites from rocks of probable mantle origin. *Contrib. Mineral. Petrol.* **77**, 288-295.

*Arima,M. & Edgar,A.D. 1983.A high pressure experimental study on a magnesian-rich leucite-lamproite from the West Kimberley area,Australia: petrogenetic implications.*Contrib.Mineral.Petrol.***84**,228-234.

* Arima, M. & Kerrich, R. 1988. Jurassic kimberlites from Picton and Varty Lake, Ontario: geochemical and stable isotope characteristics. *Contrib. Mineral. Petrol.* **99**, 385-391.

* Arribas,A.San Miguel. 1952.Observations pétrologiques sur les roches lamprophyriques de la Costa Brava Catalane (Bagur). *19th Int.Geol.Congr.Sect.***vi**,77-99.

Arthaud,F. & Matte,P. 1977. Late Palaeozoic strike-slip faulting in Southern Europe and North Africa: results of right-lateral shear zone between Appalachians and Urals. *Bull.Geol.Soc.Am.* **88**,1305-1320.

Arthurton,R.S. & Wadge,A.J. 1981.Geology of the country round Penrith. *Mem.Geol.Surv.G.B. (Sheet 24).*

Asquith,G.B. 1973a.High viscosity "conglomerate" channel deposits in Tertiary lamprophyre sill,Sacramento Mtns.,New Mexico. *Geol.Soc.Am.Abstr.Progr.***5(7)**,p.537 & *Geology*,**1(4)**,149-151.

#* Asquith,G.B. 1973b.Petrography and petrogenesis of Tertiary camptonites and diorites, Sacramento Mountains, New Mexico*New Mexico Bur.Mines Mineral.Resour.Circ.***141**,6pp.

#* Asquith,G.B. 1973c.Flow differentiation in Tertiary lamprophyres (camptonites), Sacramento Mountains, New Mexico. *J.Geol.***81**,643-647.

Assunção, C.F.T.de. 1940. Sobre a petrologia das formações eruptivas do Algarve. *Bol.Mus.Lab.Mineral. Geol.Univ. Lisboa 3rd ser.***7-8**, 47-81.

* Assunção,C.F.T.de & Brak-Lamy,J. 1948.Algumas rochas eruptivas da região de Chaves. *Bol.Soc.Geol. Port.* **8**,107-120.

#* Assunção,C.F.T.de & Canilho,M.H. 1969.A petrografia da ilha de Maio e suas relações com a petrografia do arquipélago de Cabo Verde.*Bol.Mus.Lab.Mineral.Geol.Univ.Lisboa* **11(2)**,161-91.

Atanasiu,I. 1928.Études géologiques dans les environs de Tulghes.*Ann.Inst.Geol.Roumania* **13**,373-511.

* Atkinson,W.J.,Hughes,F.E. & Smith,C.B. 1984a.A review of the kimberlitic rocks of western Australia. In: Kornprobst (1984a), qv, 195-224.

Atkinson,W.J.,Smith,C.B. & Boxer,G.L. 1984b.The discovery and geology of the Argyle diamond deposits. *Proc.Australas.Inst.Ming.Metall.Conf.Darwin*, 141-150.

Auden,J.B. 1949.Dykes in western India. *Rec. Geol. Surv. India* **81(1)**, 223-226.

Aumento,F. & Ade-Hall,J.M. 1973.Deep-drill-1972.Petrology of the Bermuda drill core.*EOS* **54**,p.485 (abstr)

Aumento,F.,Reynolds,P.M. & Gunn,B.M. 1974.The Bermuda seamount: a reactivated section of an older ocean crust.*EOS* **55**,p.455 (abstr.)

Aurousseau,M. & Washington,H.S. 1922.Nepheline syenite of Beemerville*J.Geol.***30**,571-586.

* Ayres,V.L. & Higgins,W.D. 1939.Differentiation in xenolithic lamprophyre dykes at Marquette,Michigan. *J.Geol.* **47**,561-582.

Ayrton, S.N. 1963.A contribution to the geological investigations in the region of Ivigtut, SW Greenland. *Bull Grønlands Geol. Unders.* **37.**

Ayrton, S.N. & Burri, M. 1967.L'evolution du socle Precambrien dans la région de Qagssiarsit, Greenland meridional. *Bull Grønlands Geol. Unders.* **67.**

* Azambre,M.B. 1970.Les monchiquites et autres roches basiques intrusives accompagnant les syénites néphéliniques des Corbières. *C.R.Acad.Sci.Paris* **271**,641-643.

? Babakhodzhaev,S.M. 1956.[Lamprophyres of the R.Varzob-Bolo basin].*Trudy Institut Geologii Akademii Nauk Tadzhikshoy SSR,* **1**,75-85.

Bachinski,S.W. 1986.Reconsideration of 'soda minettes'.*Mineral.Mag.***50**,199-204.

* Bachinski,S.W. & Scott,R.B. 1979.Rare-earth and other trace element contents and the origin of minettes. *Geochim.Cosmochim.Acta* **43**,93-100.

Bachinski,S.W. & Scott,R.B. 1980.Authors' reply.*Geochim.Cosmochim.Acta* **44**,1389-1392.

* Bachinski,S.W. & Simpson,E.L. 1984.Ti-phlogopites of the Shaw's Cove minette: a comparison with micas of other lamprophyres, potassic rocks, kimberlites and mantle xenoliths.*Am.Mineral.***69**,41-56.

* Bachinski,S.W. & Simpson,E.L. 1986.Chemistry and morphology of feldspars of the Shaw's Cove minette, other lamprophyres, and lamproites.Unpubl.posthumous manuscript.

Baechlin,R. 1937. Geologie und Petrographie des M-Tamans-Gebietes (südliches Tessin). *Schweiz.Mineral Petrogr.Mitt.* **17**,1-79.

Bailey,D.K. 1984. Kimberlite: 'The mantle sample' formed by ultrametasomatism. In: Kornprobst (1984), qv, 323-334.

Bailey,D.K. 1987. Mantle metasomatism — perspective and prospect. In: Fitton & Upton (1987), qv, 1-14.

Bailey,E.B. 1958.Some chemical aspects of SW Highland Devonian igneous rocks. *Bull.Geol.Surv.G.B.* **15**,1-20.

Bailey,E.B. 1960. The geology of Ben Nevis and Gle Coe (sheet 53; 2nd Ed). *Mem.Geol.Surv.G.B.*

Bailey,E.B. & Anderson,E.M. 1925. The geology of Staffa, Iona and western Mull (sheet 43). *Mem.Geol.Surv.G.B.*

Bailey,E.B.,Clough,C.T. & Richey,J.E. 1924.Tertiary and post-Tertiary geology of Mull,Loch Aline & Oban. *Mem.Geol.Surv.G.B.*

@ Baillie,P.W. 1973.Structure of the granitic rocks of the Little Mt.Horror area.*Tasmania Dept.Mineral. Tech.Rep.* **15**,32-36.

Baker,B.H. 1953.The alkaline igneous complex of Jombo Hill.*Geol.Surv.Kenya Rep.***24**,32-48.

Balasubrahmanyan, M.N., Murty, M.K., Paul, D.K. & Sarkar, A. 1978. Potassium-argon ages of Indian kimberlites. *J.Geol.Soc. India* **19**, 584-585.

Banerjee,S. 1953.Petrology of the lamprophyres and associated rocks of Raniganj coalfield. *Indian Mineral.* **1**,9-29.

Baratov, R.B., Kukhitov, M.M. & Mushkin, I.V. 1970. [*Volcanic exploration pipes and some features of abyssal structure of the South Gissar.*] In Russian. Dontch, Dushanke, 114pp.

Barber,W.B. 1904.On the lamprophyres and associated igneous rocks of the Rossland mining district,British Columbia. *Am.Geologist* **33**,335-347.

* Barberi,F. & Innocenti,F. 1967.Le rocce selagitiche di Orciatico e Montecatini in Val di Cecina.*Atti Tosc. Soc. Sci. Nat.Mem. Ser.A.*,**74**,139-180.

Bard,J-P. & Gonord,H. 1965.Note sur les lamprophyres micacés d l'est du versant meridionel de la Montagne-noire. *C.R.Soc.Géol.Fr.***5**,165-167.

* Bardet,M.G. 1974.Géologie du diamant.*Mem.Bur.Rech.Géol.Minières (France)*, **83**, 232pp.

Bardet,M.G. & Vachette,M. 1966.Determination of the ages of kimberlites of W.Africa......*3rd Symp.Afr.Geol. Commonwealth Geol. liason office,Rep.GGLO(LR)* **88**,15-17.

Barker,D.S. 1983.*Igneous Rocks*.Prentice Hall,NJ, 417pp.

Barnes,R.P.,Floyd,J.D. & Stone,P. 1988. Big Scare — resurveyed after 110 years. *Rep.Br.Geol.Surv.* **19(2)**, 19-23.

* Barnes,R.P.,Rock,N.M.S. & Gaskarth,J.W. 1986.Late Caledonian dyke-swarms of Southern Scotland: new field, petrological and geochemical data for the Wigtown peninsula, Galloway. *Geol.J.***21**,101-125.

¶* Barnett,R.L.,Arima,M.,Blackwell,J.D.,Winder,C.G.,Palmer,H.C. & Hayatsu, A. 1984.The Picton and Varty Lake ultramafic dikes: Jurassic magmatism in the St.Lawrence platform near Belleville, Ontario. *Can.J.Earth Sci.* **21**, 1460-1472.

* Baroz,F. 1978.Charactères pétrographiques et géochimiques des deux séries volcaniques potassiques du Pentadaktylos (Chypre). *Sci.de la Terre* **20**,297-332.

* Barrabé,L.C. 1952.Les roches intrusives à hornblende brune des Pyrénées.... *19th Int.Geol.Congr.***vi**,9-22.

* Barreiro,B.A. 1983.An isotopic study of Westland dike swarms,South Island, New Zealand. *Carn. Inst.Wash.Yrbk.***82**, 471-475.

* Barreiro, B. & Cooper, A.F. 1987. A Sr, Nd and Pb isotope study of alkaline lamprophyres and related rocks from Westland and Otago, South Island, New Zealand. *Geol.Soc.Am.Spec.Pap.* **215**, 115-125.

Barrington,J. & Kerr,P.F. 1961.Breccia pipe near Cameron,Arizona. *Bull.Geol.Soc.Am.***72**,1661-1674.

Barrington,J. & Kerr,P.F. 1962.Alteration effects at Tuba dike, Cameron,Arizona. *Bull.Geol.Soc.Am.* **73**,101-112.

Barrow,G. & Craig,E.M.C. 1912. The geology of the districts of Braemar, Ballater and Glen Clova (sheet 65). *Mem. Geol.Surv.G.B.*

Barrow, G.,Wilson,G.V. & Craig,E.H.C. 1905. The geology of the country around Blair Atholl, Pitlochry and Abefeldy (sheet 36).*Mem.Geol.Surv.G.B.*

Barrow, G.,Hinxman,L.W.,Craig,E.H.C. & Kynaston,H. 1913. The geology of upper Strathspey, Gaich and the Forest of Atholl (sheet 64). *Mem.Geol.Surv.G.B.*

Barth,T.F.W. 1944.Lamprophyrer av to forskjellige aldre i kystmigmatiten vest for Kristiansand. In Norwegian, Engl. summ. *Norsk.Geol.Tidsskr.***23**,175-185.

Barth,T.F.W. & Ramberg,I.B. 1966.The Fen circular complex. In: Tuttle & Gittins (1966), qv, 225-260.

Bartley, M.W. 1938. The northeastern part of the Schreiber area. *Ontario Dept. Mines* **47(9)**, 29-40.

* Barton,M. 1979.A comparative study of some minerals occurring in the potassium-rich alkaline rocks of the

Leucite Hills, Wyoming,the Vico Volcano,Italy and the Toro-Ankole region,Uganda. *N.Jb.Mineral.Abh.* **137**,113-134.

* Barton,M. & Bergen,M.J.Van. 1981.Green clinopyroxenes and associated phases in a potassium-rich lava from the Leucite Hills, Wyoming.*Contrib.Mineral.Petrol.***77**,101-114.

* Barton.M. & Hamilton,D.L. 1978.Water-saturated melting relations to 5kb of 3 Leucite Hills lavas. *Contrib.Mineral.Petrol.* **66**,41-49.

Barton,M. & Hamilton,D.L. 1979.The melting relationships of a madupite from the Leucite Hills,Wyoming to 30kb. *Contrib.Mineral.Petrol.***69**,133-142.

Barton,M. & Hamilton,D.L. 1982.Water-undersaturated melting experiments bearing upon the origin of potassium-rich magmas. *Mineral.Mag.***45**,267-278.

Barton,M. & Wood,B.J. 1976.Equilibration of orendite with lherzolite mineral assemblages: a combined experimental /theoretical approach.*NERC (UK) Prog.Exptl.Petrol.Publ.Ser.D.* **3**,90-91.

@ Basnett,E.M. 1942.Studies in metamorphism and assimilation in the Wellington district,NSW.II: Dynamic and contact metamorphism of a group of ultrabasic rocks.*J.Proc.R.Soc.NSW* **76**,55-81.

Basta,E.Z.,El-Sharkawy,M.A. & Mikhail,M.A. 1985.Petrographical and petrochemical studies on some Egyptian lamprophyres. *J.Coll.Sci.King Saud Univ.(Riyadh).***16/1**,137-164.

* Basu, A., Rubury, E., Mehnert, H. & Tatsumoto, M. 1984. Sm-Nd, K-Ar and petrologic study of some kimberlites from eastern United States and implication for mantle evolution. *Contrib. Mineral. Petrol.* **86**, 35-44.

Bates,J.A. & Jackson,R.L. 1989 *Glossary of Geology.* 3rd Edn. Am.Geol.Inst., Falls Church, 788pp.

Bautsch,H.J. 1963.Über die Sulfide in den Lamprophyren der Lausitz und ihre genetische Ableitung.*Geologie* **12**,362-364.

Bautsch,H.J. & Rohde,G. 1970.Erzmikroskopische Untersuchungen an einem hydrothermal veretzen Lamprophyrgang des Schneckensteinemassivs. In: *Probleme der Paragenese*,vol.2,Freiberg Forschungsch. Reihe C, **270**,151-159.

* Baxter,A.N. 1987. Petrochemistry of late Palaeozoic alkali lamprophyre dykes from N.Scotland. *Tr.R.Soc.Edinburgh: Earth Sci.* **77**, 267-277.

Baxter,A.N. & Mitchell,J.G. 1984.Camptonite-monchiquite dyke swarms of Northern Scotland: age relationships and their implications.*Scott.J.Geol.***20**,297-308.

Bayrakov,V.V. 1964.Occurrence of eclogitic rock xenoliths in a lamprophyre dyke of the Oktyabr'skiy alkalic stock. *Dokl.Acad.Sci.USSR,Earth-Sci.Sect.***156**,137-140.

* Bea,F. & Corretge,L.G. 1986.Petrography, geochemistry and differentiation models of lamprophyres from Sierra de Gredos, central Spain. *Hercynica* **II**, 1-15.

Bearth,P. 1932. Uber Ganggestine des Malcantone.*Schweiz.Mineral.Petrogr.Mitt* **12**,180-203.

* Bearth,P. 1959.On the alkali massif of the Werner Bjerge in E.Greenland.*Meddr.Grønland* **153(4)**, 63pp.

*Beauseigneur,C. & Rangheard,Y. 1967.Contribution à l'étude des roches éruptives de l'île d'Ibiza (Baléares). *Bull.Soc.Géol.Fr.Sér.*7,**IX**,221-4.

* Beavis,F.C. 1962.The geology of the Kiewa area. *Proc.R.Soc.Vict.***75**, 349-410.

Beckinsale, R.D. & Obradovich, J.D. 1973. Potassium-argon ages for minerals from the Ross of Mull, Argyllshire. *Scott. J. Geol.* **9**, 147-56.

* Bédard, J. 1988. Comparative amphibole chemistry of the Monteregian and White Mountain alkaline suites, and the origin of amphibole megacrysts in alkali basalts and lamprophyres.*Mineral. Mag.* **52**, 91-104.

*Bédard,J., Ludden,J. & Francis,D.M. 1987. The Mégantic intrusive complex, Québec: a study of the derivation of silica-oversturated anorogenic magmas of alkaline affinity. *J.Petrol.* **28**, 355-388.

Bédard, J., Francis, D.M. & Ludden, J. 1988. Petrology and pyroxene chemistry of Monteregian dykes: the origin of concentric zoning and green cores in clinopyroxenes from alkali basalts and lamprophyres. *Can.J.Earth Sci.* **25**, 2041-2058.

Bederke,E. 1947.Zum Probleme der Lamprophyre.*Nachr.Akad.Wiss.Gottingen Math.Phys.* Kl.**1**,53-57.

Beeson,R. 1975.Basic and intermediate dykes in the Philadelphia area,Cape Province. *Ann.Geol.Opname ZAF*,**11**,123-127.

Beger, P.J., 1913. Typenmischung im lamprophyrischen Ganggefolge des Lausitzer Granites. *Ber. Math-Phys. Kl. Königl.-sachs Ges. Wiss.* **65**.

† Beger,P.J. 1916.Beitrage zur Kenntnis der Kalkalkalireihe der Lamprophyre im Gebiete des Lausitzer Granit-Lakkolith. *N.Jb.Beilige-Band* **xl**,583.

* Beger,P.J. 1923.Der Chemismus der Lamprophyre. In: *Gesteins und Mineralprovinzen* (ed.Niggli,P.), Vol.1, *Einfuhrung*, 217-577. Gebrüder Borntrager, Berlin.

Bell, K. 1989. *Carbonatites: genesis and evolution.* Unwin Hyman, London, 618pp.

Bellanger,D. & Buffet,G. 1979.Les filons de kersantite de la zone corticale du massif de Écrins-Pelvoux: un épisode intrusive Carbonifère.*Sci.de la Terre* **7**,p.41.

Bellon,H.,Bizon,G.,Calvo,J.P.,Elizaga,E., Gaudant, J. & Martinez, N.L. 1981. Le volcan du Cerro del Monagrillo (Province de Murcia): âge radiométrique et corrélations avec les sédiments néogènes du bassin de Hellin (Espagne). *C.R.Acad.Sci Paris Ser. II*, **292**, 1035-1038.

Bellotto, S. & Zimmermann, J.L. 1983. Datation par la méthode K.Ar d'un sill lamprophyrique dans la série volcano-sédimentaire minéralisée de l'extrémité SW der Morvan, secteur de Saint Aubain - Gilly-sur-Loire (Saône-et-Loire). *C.R.Acad.Sci.Paris Sér II*, **297**(7), 577-580.

† Belyankin,D.S.,Eremeev,V.P. & Petrov,V.P. 1938.[New data on neointrusions in the basin of the River Urueh].*Tr. Inst.Geol.Sci.Petrogr.Sect.***4**,31pp.

* Belyankin,D.S. & Petrov,V.P. 1940a.[Some new data on teschenite and related zeolite-bearing rocks in the territory of Georgia].In Georgian, Engl. summ. *Mem.Soc.Russe Minéral.Sér.*2.,**69**,276-287.

Belyankin,D.S. & Petrov,V.P. 1940b. [Teschenite-camptonite-monchiquite intrusions from the vicinity of the town of Akhalzykh, Trancaucasia]. In Russian, Engl. summ. *Akad. Sci. USSR Tr.Inst.Geol.Sci. Petrogr. Ser. 7*, **21**, 51-73.

Bender,J.F., Hanson,G.N. & Bence,A.E. 1984. Cortlandt Complex: differentiation and contamination in plutons of alkali basalt affinity. *Am.J.Sci.* **284**, 1-57.

Bennett,G.,Brown,D.D.,George,P.T. & Leary,E.J. 1967. Operation Kapuskasing.*Ontario Dept.Mines Misc. Pap.* **10**.

Benson,W.N. 1913. The geology and petrology of the Great Serpentine Belt of New South Wales. *Proc.Linnaean Soc.N.S.W.* **38**,662-724.

Bentley,M.R.,Maltman,A.J. & Fitches,W.R. 1988. Colonsay and Islay: A suspect terrane within the Scottish Caledonides. *Geology* **16**, 26-28.

* Bergen,M.J.Van, Ghezzo,C. & Ricci,C.A. 1983.Minette inclusions in the rhyodacitic lavas of Mt.Amiata (Central Italy): mineralogical and chemical evidence of mixing between Tuscan and Roman type magmas. *J.Volcanol.Geotherm.Res.* 19,1-35.

* Bergman,S.C. 1987.Lamproites and other potassium-rich igneous rocks: a review of their occurrence, mineralogy and geochemistry. In: Fitton & Upton (1987), qv, 103-190.

* Bergman,S.C.,Dunn,D.P. & Krol,L.G. 1988. Rock and mineral chemistry of the Linhaisai minette and the origin of Borneo diamonds,Central Kalimantan, Indonesia. *Can. Mineral.* **26**, 23-44.

Berkovsky,A.N. 1987.Dyke-swarms of the East European craton: aeromagnetic and geological evidence. In: Halls & Fahrig (1987), qv, 373-377.

* Best,M.G.,Henagel,L.F. & Adams,J.A.S. 1968.Mica peridotite,wyomingite and associated potassic igneous rocks in NE Utah. *Am.Mineral.***53**,1041-8.

* Bettenay, L.F., Rock,N.M.S. & Mather,P.J. 1990. Metamorphosed ultramafic lamprophyres from the Shaw batholith area, Pilbara Block, Western Australia (submitted).

Beuge,P. & Kramer,W. 1977.Lamprophyre Osttueringens und ihre anomalen Quecksilbersehalte im Ergebnis endogener und exogener Anreicherungsprozesse.*Beitr.Geochem. Schr. Geol. Wiss.* **8**,79-99.

Bickford,M.E.,Mose,D.G.,Wetherill,G.W. & Franks,P.C. 1971.Metamorphism of Precambrian granitic xenoliths in a mica peridotite at Rose Dome, Woodson Cty.,Kansas: Part 1,Rb-Sr isotopic studies. *Bull.Geol.Soc.Am.***82**,2863-2868.

Bielski-Zyskind,M.,Wasserburg,G.J. & Nixon,P.H. 1984.Sm-Nd and Rb-Sr systematics of volcanics and ultramafic xenoliths from Malaita,Solomon Islands and the nature of the Ontong Java plateau. *J.Geophys.Res.***89**,2415-2424.

Bigioggero B., Colombo, A., Lanza R. & Sacchi R. 1981.Dati geochimici e paleomagnetici preliminari su filoni nel Sudalpino della zona del Biellese. *Rendiconti Soc.Geol.Ital.* **4 (3)**,271-274.

* Biljon,S.Von. 1940.The Kuboos batholith,Namaqaland.*Tr.Geol.Soc.S.Afr.***42**,123-219.

Billings, M.P. & Fowler-Billings, K. 1975. Geology of the Gorham quadrangle, New Hampshire and Maine. *N.H. Dept. Res. Geol. Dev. Bull.* **6**, 120pp.

Bishop,A.C. 1964.Petrogenesis of hornblende-mica lamprophyre dykes at South Hill,Jersey. *Geol.Mag.* **101**,302-13.

* Bisschoff,A.A. 1972.The dioritic rocks of the Vredefort dome.*Tr.Geol.Soc.S.Afr.***75**,31-45.

Biswas, A.B. & Ray,S. 1952.A biotite-lamprophyre (?mica-peridotite) from the copper belt of Dhalbhum. *Proc. Indian Sci.Congr.39th sess*,**III**,p.177 (abstr.)

Bitschene,P.R. & Mertz,D.F. 1984.Ein neuer Tuffschlot und ein neuer Sandsteingang im Odenwaldkristallin südlich Erbach.*Geol.Jb.***112**,127-137.

Black,L.P. 1977.An Rb-Sr geochronological study of the Proterozoic Tennant Creek Block, central Australia. *J. Aust.Bur.Mineral.Resour.Geol.Geophys.***2**,111-122.

Black,R. & 8 others,1979.Outline of the pan-African geology of Adrar des Aforas (Republic of Mali). *Geol.Rundsch.* **68**,543-64.

Blake,D.A.W. 1955. Oldman river map-area, Saskatchewan. *Mem.Geol.Surv.Can.* **279.**

Blake,D.H. 1980. Volcanic rocks of the Paleohelikian Dubawnt group in the Baker Lake-Angikuni Lake area, district of Keewatin. *Bull.Geol.Surv.Can.* **309**, 39pp.

Bloomfield,K. 1965.Geology of the Zomba area.*Bull.Geol.Surv.Malawi* **16**, 193pp.

* Bloomfield,K. & Garson,M.S. 1965.Geology of the Kirk range-Lisungwe valley area. *Bull.Geol.Surv. Malawi* **17**, 222pp.

¶ Boctor,N.Z. & Yoder,H.S. 1986.Petrology of some melilite-bearing rocks from Cape Province: relationship to kimberlites*Am.J.Sci.***286**,513-539.

Bogatikov,O.A.,Eremeev,H.V.,Mahotkin,I.L.,Kononova,V.A.,Novgoropova,M.I. & Lapyajina,I.P. 1986. [Lamproites of the Aldan and Central Asia]. *Dokl.Acad.Sci.SSSR.***290**(4),154-157.

Bol,L.C.G.M., Bos,A.,Sauter,P.C.C. & Jansen,J.B.H. 1989. Barium-titanium-rich phlogopites in marbles from Rogaland, southwest Norway. *Am.Mineral.***74**, 439-447.

* Bolivar,S.L. 1982a.The Prairie Creek kimberlite,Arkansas. In: *Contributions to the Geology of Arkansas* (ed. McFarland,J.D.) *Ark.Geol.Commission Misc.Publ.***18**,1-21.

Bolivar, S.L. 1982b. *Kimberlite of Elliott County*. Kentucky Geol. Surv.Thesis Series 2, **XI**, 37pp.

Bolivar,S.L. 1987. An overview of the Prairie Creek diamond-bearing intrusion, Arkansas. *Tr.Soc.Ming.Eng. AIME* **280**, 1988-1993.

Bolivar,S.L. & Brookins,D.G. 1979.Geophysical and Rb-Sr study of the Prairie Creek AK kimberlite. In: Boyd & Meyer(1979a), qv, 289-299.

* Bondi,M.,Pirani,R. & Simboli,G. 1968.L'anfibolo bruno delle camptoniti di Predazzo e dei Monzoni. *Mineral. Petrogr.Acta* **14**,71-104

Bonney,T.G. & Houghton,F.T.S.1879.On some mica traps from the Kendal and Sedbergh districts.*Q.J.Geol. Soc.Lond.***35**,165-180.

Boriani,A., Colombo,A., Giobbi Origoni,E. & Pyronel Pagliani,G. 1974. The "Appinite suite" of Massiccio dei Laghi (Northern Italy) and its relationship to the regional metamorphism. *Rend.Soc.Ital.Mineral. Petrogr.* **30**, 893-917.

Boriani,A. & Sacchi,R. 1973. Geology of the junction between the Ivreaa-Verbano and Strona-Ceneri zones. *Mem.Ist.Geol.Mineral.Univ.Padova* **28**,36pp

* Borley,G.D. 1967.Potash-rich rocks from southern Spain.*Mineral.Mag.***36**,364-379.

* Borodin,L.S. & Pyateko,I.K. 1978.General petrological aspects of Palaeozoic alkali magmatism in the Kola peninsula and the rare-earth distribution in alkali-ultrabasic lamprophyre dykes.*Geochem. Int.*124-135.

Borradaile,G.J. 1986. The internal tectonic fabric of minor intrusions and their potential as regional palaeostress indicators. *Geol.Mag.* **123**, 665-671.

* Bose,M.K. 1973.Petrology and geochemistry of the igneous complex of Mt.Girnar, Gujarat, India. *Contrib.Mineral. Petrol.* **39**,247-66.

Bose,M.K. & Nag,S. 1980. Petrology of the alkaline suite of Elchuru, Prakasam district, Andhra Pradesh. *Proc.Indian Acad.Sci.* **89**, 393-401.

* Bouabdli, A., Dupuy, C. & Dostal, J. 1988. Geochemistry of Mesozoic alkaline lamprophyres and related rocks from Tamazert, High Atlas, Morocco. *Lithos* **22**, 43-58.

* Boulton,W.S. 1911.On a monchiquite intrusion in the Old Red Sandstone of Monmouthshire. *Q.J.Geol.Soc.Lond.* **67**,460-476

Bouska,V., Jelínek,E., Pacesová,M., Randa,Z. & Ulrych,J. 1984. Rare-earth elements and other trace elements in the rocks of the Central Bohemian Pluton. *Geol. Zbornik Geol. Carpathica* **35**, 355-376.

* Bowen,N.L. 1922.Genetic features of alnöitic rocks from Ile Cadieux,Quebec.*Am.J.Sci.***3**,1-34.

Bowen,N.L. 1928. *The evolution of the igneous rocks*. Princeton University Press.

Bowes,D.R. 1962.Kentallenite-lamprophyre-granite age relations at Kentallen,Argyll.*Geol.Mag.* **99**,119-122.

* Bowes,D.R.,Kinloch,E.D. & Wright,A.E. 1964.Rhythmic amphibole overgrowths in appinites associated with explosion breccias in Argyll.*Mineral.Mag.***33**,963-973.

Bowes,D.R.,Macdonald,A.S.,Matheson,G.D. & Wright,A.E. 1963.An explosion breccia-appinite complex at Gleann Chàrnan,Argyll. *Tr.Geol.Soc.Glasgow* **25**,19-30.

* Bowes,D.R. & McArthur,A.C. 1976.Nature and genesis of the appinite suite.*Kristalinikum* **12**,31-46.

Bowes,D.R. & Wright,A.E. 1961.An explosion-breccia complex at Back Settlement,nr.Kentallen,Argyll. *Tr.Edinburgh Geol.Soc.***18**,293-314.

Bowes,D.R. & Wright,A.E. 1965.A comparison of the breccia-metagabbro-syenite complex at Fjøne,south central Norway,with some explosion-breccia-appinite complexes in the Caledonian orogenic belt of Scotland. *Norsk.Geol.Tiddskr.***45**,463-472.

Bowes,D.R. & Wright,A.E. 1967.The explosion breccia pipes near Kentallen,Scotland and their geological setting. *Tr.R.Soc.Edinburgh* **67**,109-142.

Bowman,H.N. 1971. *Geology of the Wollongong, Kiama and Robertson 1:50,000 sheets*. Geol.Surv.NSW, 179 pp.

Bowman, N.H., Richardson, S.J. & Dolanski, J. 1979. Narrow 1: 250,000 metallogenic map (SI-55-3), mine data sheets and metallogenic study. *Geol. Surv. NSW Dept. Miner. Resour.* 262-330.

Box,S.E. & Flower,M.F.J. 1989. Introduction to special section on alkaline arc magmatism. *J.Geophys.Res.***94(B4)**, 4467-4468.

Boxer,G.L.,Lorenz,V. & Smith,C.B. 1989. The geology and volcanology of the Argyle (AK1) lamproite diatreme, Western Australia. *Spec. Publ. Geol. Soc. Aust.* **14**, 140-152.

Boyadzhiyev,S., Palivcová,M., Tsvetkova,V. & Hejl,V. 1979. Textures and petrology of the Vetren gabbro (Bulgaria). *Krystalinikum* **14**, 155-170.

Boyd,F.R. & Meyer,H.A.O. 1979a.*Kimberlites,Diatremes and Diamonds.* (Proc. 2nd Int. Kimb. Conf., Vol. 1). Am.Geophys.Union, Washington, 400pp.

Boyd, F.R. & Meyer, H.A.O. 1979b. *The Mantle Sample* (Proc. 2nd Int. Kimb. Conf., Vol. 2). Am.Geophys.Union, Washington, 423pp.

* Boyd,W.W. 1986.Lamprophyre dykes from the Patuxent Range: implications from nodule assemblages for the crustal evolution of W.Antarctica. *Symp.Antarct.Crust.Dev.Cambridge, Aug.'87* (abstr.)

Boyle,R.W. 1979.The geochemistry of gold and its deposits.*Bull.Geol.Surv.Can.***280**,584pp.

Bradshaw,N. 1968.Petrographic examination of seven ultrabasic eruptive rocks and one limestone from Malaita,Solomon Is. *British Solomon Is.Geol.Rec.(1963-7)* **iii**,51-53.

Brammall,A. 1915.The intrusive rock of Marston Jabet,Nuneaton,Warwickshire.*Geol.Mag.***2**,152-158.

Branigan, N.P. 1989. Hybridisation in Middle Proterozoc high level ring complexes, Åland, SW Finland. *Precamb.Res.* **45**,83-96.

Brey,G. 1978. Origin of olivine melilitites: chemical and experimental constraints. *J.Volcanol.Geotherm.Res.* **3**, 61-88.

Brindley,J.C. 1957.The Blackrock granite-breccia,Cty.Dublin: an interpretation.*Sci.Proc.R.Soc. Dublin* **27**,283-286.

Brindley,J.C. 1970.Appinitic intrusions associated with the Leinster granite.*Proc.R.Ir.Acad.***70b**,93-104.

Brindley,J.C.,Gupta,L.N. & Kennan,P.S. 1976.Explosion breccias and related features in the Leinster Caledonian massif. *Proc.R.Ir.Acad.***76b**,337-348.

Briqueu,L., Bougault,H. & Joron,J.L. 1984. Quantification of Nb,Ta,Ti and V anomalies in magmas associated with subduction zones: petrogenetic implications. *Earth Planet.Sci.Lett.* **68**, 297-305.

* Broadhurst,J.R. 1986.Mineral reactions in xenoliths from the Colorado Plateau: implications for lower crustal conditions and fluid compositions. In: *Nature of the lower continental crust* (eds. Dawson,J.B.,Carswell,D.A., Hall,J.& Wedepohl,K.H.) *Spec.Publ.Geol.Soc.Lond.***25**, 331-349.

Brodie,C.G. & Cooper,A.F. 1989.Nodule associations from ouachitite and camptonite lamprophyres,western Otago and South Westland,N.Z. *Spec. Publ. Geol. Soc. Aust.* **14**, 545-559.

* Brøgger,W.C.1894.The basic eruptive rocks of Gran.*Q.J.Geol.Soc.Lond.***50**,15-38.

* Brøgger,W.C. 1921.Die Eruptivgesteine des Kristiangebietes,IV,das Fengebiet in Telemark, Norwegen. *Norsk.Vidensk-Akad.Oslo Skr.Mat.Naturv.Kl.* **9**,408pp.

* Brøgger,W.C. 1931.Die Eruptivgesteine des Oslogebietes.V.Der grosse Hurumvulkan. *Norsk.Vidensk-Akad. Oslo Skr.Mat.Naturv.Kl.* 1930(6),146pp.

* Brøgger, W.C. 1933. Die Eruptivgesteine des Oslogebietes.VI. Ueber verschiedene Gangegesteine des Oslogebietes. *Norsk.Vidensk-Akad.Oslo Skr.Mat.Naturv.Kl.* **2(7)**, 88pp.

Brøgger,W.C. 1934.Die Eruptivgesteine des Oslogebietes.VII.Die chemische Zusammensetzung der Eruptivgesteine. *Norsk.Vidensk-Akad.Oslo Skr.Mat.Naturv.Kl.* **1(1)**,147pp.

Brookins,D.G. 1980. K-Ar age of lamprophyre dike from the Kerr-McGee potash mine, southeastern New Mexico. *Isochron/West* **29**, p.27.

Brooks,C.K.,Fawcett,J.J.,Gittins,J. & Rucklidge,J.C. 1981.The Batjberg complex,east Greenland: a unique ultrapotassic Caledonian intrusion.*Can.J.Earth Sci.***18(2)**,274-285.

Brooks,C.K.,Noe-Nygaard,A.,Rex,D.C. & Rønsbo,J.G. 1978.An occurrence of ultrapotassic dikes in the neighbourhood of Holsteinsborg,W.Greenland.*Bull.Geol.Soc.Denmark* **27**,1-8.

* Brooks,C.K. & Platt,R.G. 1975.Kaersutite-bearing gabbroic inclusions and the late dike swarm of Kangerdlugssuaq, E.Greenland. *Mineral.Mag.***40**,259-283.

* Brooks,C.K. & Printzlau,I. 1978.Magma mixing in mafic alkaline volcanic rocks; evidence from relict phenocryst phases and other inclusions.*J.Volcanol.Geotherm.Res.***4**,315-331.

* Brooks,C.K. & Rucklidge,J.C. 1973.A Tertiary lamprophyre dyke with high pressure xenoliths and megacrysts from Wiedemanns Fjord,E.Greenland.*Contrib.Mineral.Petrol.***42**,197-212.

* Brousse,R. & Nativel,P. 1963.Une picrite-ankaratrite lamprophyrique,près de Saint Micaud. *C.R.Acad.Sci. Paris* **257**,479-481.

* Brown,D.D.,Bennett,G. & George,P.T. 1967.The source of alluvial kimberlite indicator minerals in the James Bay Lowland. *Ont.Dept.Mines Misc.Pap.***7**,33pp.

Brown,J.F. 1975. Potassium-argon evidence of a Permian age for the camptonite dykes: Orkney. *Scott.J.Geol.* **11**,259-262.

Brown,P.E. 1983. Caledonian and earlier magmatism. In: *Geology of Scotland* (ed. G.Y.Craig), 167-204. 2nd Edn. Scottish Academic Press, Edinburgh.

* Brown,P.E.,Brown,R.D.,Chambers,A.D. & Soper,N.J. 1978.Fractionation and assimilation in the Borgtinderne syenite, E.Greenland.*Contrib.Mineral.Petrol.***67**,25-34.

Brownlow, A.H. & Komorowski J.C. 1988.Geology and origin of the Yogo sapphire deposit, Montana. *Econ.Geol.* **83,** 875-880.

Browne,W.R. 1933. An account of post-Palaeozoic igneous activity in New South Wales. *Proc.R.Soc.NSW* **67**,9-95.

Bruce, E.L. 1935. Little Long Lac gold area. *Ont.Dept.Mines Geol. Rep.* **44(3)**, 60 pp.

Brück, P.M. 1976. The andestic and dioritic rocks of west Wicklow and south Dublin. *Bull Geol. Surv. Ireland* **2**,37-51.

* Brueckner,H.K. & Rex,D.C. 1980.K-A and Rb-Sr geochronology and Sr isotopic study of the Alnö alkaline complex, northeastern Sweden.*Lithos* **13**,111-120.

Brummer, J.J. 1978. Diamonds in Canada. *Bull.CIM,* October 1978, 64-79.

* Bryhni,I. & Dons,J.A. 1975.Orbicular lamprophyre from Vestby,S.Norway.*Lithos* **8**,113-122.

Büchi, E von, Bulettik M. & Niggli, E. 1984. Neue Aufschlüsse in schweizerischen Anteil des Schwarzwälder Grundegebirges. *Schweiz. Mineral. Petrogr. Mitt.* **64**,49-65.

Buffière,J.-M. 1964. Les formations cristallines et cristallo-phylliennes du massif du Rochail. *Trav. Lab. Géol. Grenoble* **40**, 45-80.

* Buie,B.F. 1941.Igneous rocks of the Highwood Mountains,part III: Dikes and related intrusives. *Bull.Geol.Soc.Am.* **52**,1753-1808.

Burgess,C.H. 1941.Igneous rocks of the Highwood Mountains,part IV.The stocks. *Bull.Geol.Soc.Am.* **52**,1809-1828.

* Burgess,J.C. & Holliday,D.W. 1979.Geology of the country around Brough-under Stainmore. *Mem.Geol.Surv.G.B.* (Sheets 30/25.31).

Burgess,J.C. & Wadge,A.J. 1974. *Geology of the Cross Fell Area.* HMSO for Br.Geol.Surv,London, 92pp.

Burrett,C.F. & Martin E.L. 1089. Geology and mineral resources of Tasmainia. *Spec.Publ.Geol.Soc.Aust.* **15,** 574pp.

Burri,C. & de Quervain,F. 1934. Uber basische Ganggesteine aus der Ungebung von Brissago (Tessin). *Schweiz.Mineral.Petrogr. Mitt.* **14**, 507-512.

Burri,C. & Parga-Pondal,I. 1935. Beiträge zur Kenntnis einiger jungvulcanischer Gesteine Spaniens. *Schweiz.Mineral.Petrogr. Mitt.* **15**, 266-288.

Burwash, E.M. 1937. Geology of the Lochalsh–Missinaibi area. *Ont.Dept.Mines Geol. Rep.* **44(8)**, 27-38.

? Butakova,E.L. 1974. Regional distribution and tectonic relation of the alkalic rocks of Siberia. In: Sørensen (1974), qv, p.172-189.

Butler,G.T. & Edgar,A.D. 1988. The significance of mantle-derived nodules in the alkaline rocks at Coral Rapids, Ontario. In: Project Lithosphere KS7 Workshop Proceedings, University of Toronto, 119-133.

Butler,G.T., Forsyth, D.M. & Edgar,A.D. 1988. Mantle-derived rocks and their xenoliths at Coral Rapids, Ontario: significance tothe nature of the lithosphere in the KSZ. In: Project Lithosphere KS7 Workshop II Proceedings, University of Toronto, 221-233.

Buturlinov, N.V. 1959. [Lamprophyres of the Shakhta region].*Geologiya i razvdeka* **11**, 50-62.

* Buturlinov,N.V. & Latysh,I.K. 1970.[Gold and silver in magmatic rocks of the Donets Basin]. In Russian. *Geologicheskii Zhurnal (Kiev),***30(5)**, 69-75.

Caby,R. 1972.Preliminary results of mapping in the Caledonian rocks of Canning Land and Wegener Halvø,E.Greenland. *Rapp.Grønlands Geol.Unders.***48**,21-38

Cahen,L., Snelling,N.J., Delhal,J., Vail,J.R., Bonhomme,M. & Ledent,D. 1984. *The geochronology and evolution of Africa.* Clarendon, Oxford, 512pp.

Callegari,E. 1958. Structure di reazione e processi metasomatici nelle rocche femiche di Cima Uzza (Adamello sud-orientale). *Period.Mineral* **27**,99-145.

Calzia,J.P. & Hiss,W.L. 1978. Igneous rocks in northern Delaware Basin, New Mexico and Texas. *Circ. New Mexico Mines & Miner.Resour.***159**, 39-45.

Camboly,J.,Hameurt,J. & Rocci,G. 1967.Relations génétiques entre vaugnérite et kersantite et hypothèse originale sur la génèse des lamprophyres des Vosges.*C.R.Acad.Sci.Paris Sér.D.***264**,25-28.

Cameron,I.B. & Stephenson,D. 1985. *British Regional Geology: The Midland Valley of Scotland* (3rd Edn.) HMSO for Br.Geol.Surv, London.

Cameron,W.E., McCulloch,M.T. & Walker,D.A. 1971. Boninite petrogenesis: chemical and Nd-Sr isotopc constraints. *Earth Planet Sci.Lett.* **65,** 75-89.

* Campbell,I. & Schenk,E.T. 1950.Camptonite dikes near Boulder Dam.*Am.Mineral.***35**,671-692.

* Campbell Smith,W. 1924.Plutonic and hypabyssal rocks of S.Victoria Land. *Br.Antarc.Terra Nova Exped. 1910-13, Geology* **1(6)**,167-227.

* Campbell Smith,W. & Burri,C. 1933.The igneous rocks of Fernando Noronha.*Schweiz.Mineral.Petrogr. Mitt.* **13**,405-434.

Camus,F. 1975. Geology of the El Teniente orebody with emphasis on wall-rock alteration. *Econ.Geol.* **70**, 1341-1372.

* Canilho,M.H.S. 1971.Estudo geológico-petrográfico do maciço eruptivo de Sines. *Bol.Mus.Lab.Mineral. Geol.Fac.Ciências Lisboa* **12**,77-161.

* Cantagrel,J.M.,Valizadeh,M. & Vialette,Y. 1970.Age des granites,granophyres et kersantites de la région de Thiers dans le Massif Central français.*C.R.Acad.Sci.Paris* **270(4)**,600-603.

* Cao, R-L & Zhu, S-H. 1987. Mantle xenoliths and alkali rich host rocks in eastern China. In: Nixon (1987), qv, p.167-180.

* Carmichael,I.S.E. 1967.Mineralogy and petrology of the volcanic rocks from the Leucite Hills,Wyoming. *Contrib.Mineral.Petrol.***15**,24-66.

Carmichael,I.S.E.,Turner,F.J. & Verhoogen,J. 1974*Igneous Petrology*.McGraw-Hill, New York, 739pp.

Carne,J.E. 1908.Geology and mineral resources of the western coalfield.*Mem.Geol.Surv.NSW* **6**, 264pp. See p.147.

Carr,P.F. & Facer,R.A. 1980.Radiometric ages of some igneous rocks from the southern and southwestern coalfields of New South Wales.*Search (Sydney)* **11**,382-383.

Carstens,H. 1958.The origin of feldspar inclusions in the lamprophyres of Kristiansand,Norway. *Norsk.Geol.Tiddskr.***38**,245-252.

* Carstens,H. 1959.Comagmatic lamprophyres and diabases of the south coast of Norway. *Beitr.Mineral.Petrogr.* **6**,299-319.

* Carstens,H. 1962.A post-Caledonian ultrabasic biotite-lamprophyre dike of the Island Ytterøy in the Trondheimsfjord, Norway. *Norges Geol.Unders.***215**,10-21.

Carstens,H. 1982.Spherulitic crystallisation in lamprophyric magmas and the origin of ocelli.*Nature* **297**,493-494.

Carswell,D.A. 1980.Mantle derived lherzolite nodules associated with kimberlite, carbonatite and basalt magmatism: a review. *Lithos* **13**,121-138.

Casati,P. & Bigioggero,B. 1981.Filoni lamprofirici nella Dolomia dello Sciliar della valle Fiscalina. *Riv.Ital.Palaeont. Strat.* **87(2)**,169-175.

* Chalier, M. & Sabourdy,G. 1987. Les lamprophyres du granite hyperalumineux de Saint-Sylvestre (Limousin, Massif Central): caractères pétrologiques et origine. *C.R.Acad.Sci.Paris Sér.II* **305**, 99-104.

* Challis,G.A. 1960.Igneous rocks of the Cape Palliser area.*N.Z.J.Geol.Geophys.***3**,524-542.

* Challis,G.A. 1963.Layered xenoliths in a dyke,Awatere Valley, New Zealand.*Geol.Mag.***100**,11-16.

* Chapman,N.A. 1975a. *Petrology of inclusions from some late Palaeozoic British volcanic rocks.* Ph.D.thesis,Edinburgh Univ.,UK.

Chapman,N.A. 1975b. An experimental study of spinel clinopyroxenite xenoliths from the Duncansby Ness vent, Caithness. *Contrib.Mineral.Petrol.***51**,223-230.

Chapman,N.A. & Powell, R. 1976. Origin of anorthoclase megacrysts in alkali basalts. *Contrib.Mineral. Petrol.* **58**, 29-36.

Chappell,B.W. & White,A.J.R. 1974. Two contrasting granite types. *Pacific Geology* **8**, 173-174.

Chappell,B.W. & White,A.J.R. 1976.Plutonic rocks of the Lachlan mobile zone. *25th Int. Geol. Congr. (Melbourne), Excursion Guide* **13C**, 36pp.

Chatterjee,N.D. 1959.Die Lamprophyre des Spessarts und das Lamprophyrproblem. *Nachr.Ak.Wiss.Gottingen, Math.Phys.*Kl.**(1)**,1-24.

Chatterjee,S.C. 1970.A study of the inclusions in the Girnar complex and their bearing on the age of the intrusion. *Indian Mineral.* **11 (1-2)**,70-77.

Chatterjee,S.C. 1974.*Petrography of the Igneous and Metamorphic Rocks of India.* Macmillan,Bombay, 559pp.

* Chen,J.H. & Moore,J.G. 1979.Late Jurassic Independence dike-swarm in eastern California.*Geology* **7**,129-133.

* Chetverikov,S.D. 1943.A monchikite from Takob*Dokl.Acad.Sci.USSR Earth Sci. Sect.,***38**,37-41.

Chew, Y.N. 1970. *Geology of the Republic of Singapore.* Singapore Public Works Dept., 79pp.

Choubert,G. 1936.Sur quelques filons à facies lamprophyriques du versant lorrain des Vosges et sur leurs enclaves. *Bull.Soc.Géol.Fr.Sér.*5,**5**,629-650.

* Clarke,D.B.,Muecke,G.K. & Pe-Piper,G. 1983.The lamprophyres of Ubekendt Eijland, West Greenland: products of renewed partial melting or extreme differentiation?*Contrib.Mineral.Petrol.***83**,117-127.

* Clarke,D.B. & Pedersen,A.K. 1976.Tertiary volcanic province of W.Greenland. In: Escher & Watt(1976),qv, 364-385.

Clegg,K.E. 1955. Metamorphism of coal by peridotite dykes southern Illinois. *Illinois Geol.Surv.Rep.Inv.* **80**,149-156.

Clegg,K.E. & Bradbury,J.C.,1956.Igneous intrusive rocks in Illinois and their economic significance.*Illinois Geol.Surv. Rep.Inv. 197*,5-19.

* Clement,C.R., Skinner,E.M.W.,Hawthorne,J.B, Kleinjan,L. & Allsopp,H.L. 1979. Precambrian ultramafic dykes with kimberlite affinities in the Kimberley area, S.Africa. In: Boyd & Meyers (1979a), qv, 101-110.

Clement,C.R.,Skinner, E.M.W. & Scott Smith,B.H. 1984. Kimberlite redefined. *J.Geol.* **92**, 223-228.

Codarcea,A.L. 1936.Note préliminaire sur certaines roches éruptives alcalines et sur quelques lamprophyres de la région d'Ogradina (départemente de Severin).*Acad.Roum.Sec.Sci.B*, **18(6-7)**,147-153.

Codarcea,A.L.,Pavelescu,L. & Kissling,A.L. 1965.Contributii la studiud unor camptonite cu katophorit din Poiana Rusca. *Stud.Cerc.Geol.Geofiz.Geogr.Ser.Geol (Bucharest)* **10(2)**,485-90.

* Codarcea,A.L.,Pavalescu,L. & Kissling,A.L. 1967.Contribution a l'étude des camptonites à cataphorite de Poiana Ruscai. *Rev.Roum.Geol.Geophys.Sér.Geol.*11,33-38.See *Mineral.Abstr.* 70-609.

* Cogné,J. 1962.La sizunite et la problème de l'origine des lamprophyres.*Bull.Soc.Géol.Fr.***4**,141-156.

Cogné,J. & Giot,P.R. 1961.Observations à propos d'une lamprophyre micacé à microcline et apatite,rich en enclaves granitiques au flanc sud du Cap Sizun. *C.R.Acad.Sci.Paris* **252**,2569-2571.

Colchester,D.M. 1973.A preliminary note on kimberlite occurrences in South Australia. *J.Geol.Soc.Aust.* **19**,383-386.

* Collerson,K.D. & McCulloch,M.T. 1983.Nd and Sr isotope geochemistry of leucite-beaing lavas from Gaussberg,E.Antarctica. In: *Antarctic Earth Science* (ed. Oliver,R.L.), 676-680. Aust.Acad.Sci.,Canberra.

Collerson,K.D. & Sheraton,J.W. 1986. Age and geochemical characteristics of a mafic dyke swarm in the Archaean Vestfold block, Antarctica: inferences about Proterozoic dyke emplacement in Gondwana. *J. Petrol.* **27**,853-886.

Collins, R.F. 1949. Volcanic rocks of northeastern New Mexico. *Bull. Geol. Soc. Am.* **60**,1017-1040.

†Comin-Chiaramonti. 1981.I tipi filonani nel massiccio granodioritico di Avigna (Alto Adige). Engl. summ. *Studi Trentini Sci. Nat. Acta. Geol.* **58**,39-58.

Condliffe,E. 1976.Melting relationships of selected rocks of dioritic mineralogy.*NERC (UK) Prog.Exptl. Petrol. Publ. Ser. D*, **3**,54-55.

Connor,B.P. 1974.An igneous breccia in the Clara area,Co.Wicklow.*Sci.Proc.R.Soc.Dublin Ser.A.*, **5(6)**,113-115.

Cooper,A.F. 1971.Carbonatites and fenitization associated with a lamprophyric dyke-swarm intrusive into schists of the New Zealand geosyncline. *Bull.Geol.Soc.Am.* **82**, 1327-1340.

* Cooper,A.F. 1979.Petrology of ocellar lamprophyres from western Otago,New Zealand.*J.Petrol.* **20**,139-164.

* Cooper,A.F. 1986.A carbonatitic lamprophyre dike swarm from the Southern Alps, Otago and Westland. In: *Late Cenozoic volcanism in New Zealand*, (ed. Smith,I.M.E.) *Bull.R.Soc.N.Z.* **23**, 313-336.

Cooper, A.F., Barreiro, B.A., Kimbrough, D.L. & Mattinson, J.M. 1987. Lamprophyre dike intrusion and the age of the Alpine Fault, NZ. *Geology* **15**, 941-944.

* Coopersmith,H.G. & Mitchell,R.H. 1989.Geology and exploration of the Rose lamproite,southeastern Kansas,USA. *Spec.Publ. Geol. Soc. Aust.* **14**, 1179-1191.

* Cornen,G. 1981.Des complexes alcalins similaires en domaines océaniques et continental: le Banc de Gorringe comparé à la Serra de Monchique.*C.R.Acad.Sci.Paris* **292**,463-466.

* Cornen,G. 1982.Petrology of the alkaline volcanism of Gorringe Bank(SW Portugal).*Marine Geol.* **47**,101-130.

@ Cortelezzi,C.R. & Rabassa,J. 1976.Contribucion al conocimiento de la geologia del area del Cerro Tandileofu,Tandil, Provincia de Buenos Aires.*6th congr.Geol.Argent.Bahia Blanca Prov. 1975 Librart.Dept. Argentine Sci.Publ.* **1**,475-480.

* Cosgrove,M.E. 1972.The geochemistry of the potassium-rich Permian volcanic rocks of Devonshire. *Contrib.Mineral.Petrol.***36**,155-170.

Coutinho,J.M.V. & Melcher,G.C. 1973.Levantamento geológico e petrográfico na ilha do Monte de Trigo. *Rev.Bras.Geosci.***3(4)**,243-256.

Couturie,J.P. 1977. Le massif granitique de la Margeride (Massif Central Français). *Ann.Fac.Sci.Univ. Clermont-Ferrand* **62**, 319pp.

Cox,K.G., Johnson,R.L., Monkman,L.J., Stillman,C.J., Vail,J.R. & Wood, D.N. 1965. The geology of the Nuantsi igneous province. *Phil.Tr.R.Soc.Lond.***257**, 71-218.

Craig, E.H.C., Wright, W.R. & Bailey, E.B. 1911. The geology of Colonsay and Oronsay, with part of the Ross of Mull (sheet 35). *Mem. Geol. Surv. G.B.*

Crampton,C.B. & Carruthers,R.G. 1914. The geology of Caithness (sheets 110,116). *Mem.Geol.Surv. G.B.*

Crawford,A. & Compston,W. 1973. The age of the Cuddapah and Kurnool systems, southern India. *J.Geol.Soc.Aust. 19*,453-464.

Crawford,A.J. 1989. *Boninites and related rocks*. Unwin Hyman, 465 pp.

* Crawford,A.J., Green,D.A. & Findlay, R.H. 1984. A preliminary petrographical-geochemical survey of dyke-rocks from North Victoria Land, Antarctica. *Geol. Jb. Reihe* B, **60**,153-165.

Crohn,P.W. & Aldershaw,W. 1965. The geology of the Tennant Creek one mile sheet area, NT. *Rep.Aust.Bur.Mineral. Resour.* **114.**

Croneis C, & Billings, M. 1929. New areas of alkaline igneous rocks in central Arkansas. *J. Geol.* **37**,542-561.

Cross,W.1897.Igneous rocks of the Leucite Hills and Pilot Butte,Wyoming.*Am.J.Sci.***4**,115-41.

Cross,W. 1906.Prowserose (syenitic lamprophyre) from Two Buttes,Colorado.*J.Geol.***14**,165-72.

Cross,W. 1914.Dike rocks of the Apishapa quadrangle,Colorado.*U.S.G.S. Prof.Pap.***90-C**,17-31.

Cullers,R.L. & Graf,J.L. 1984.Rare-earth elements in igneous rocks of the continental crust: predominantly basic and ultrabasic rocks. In: *Rare-earth Element Geochemistry* (ed. Henderson.P.), 237-274. Elsevier.

* Cullers,R.L. & Medaris,L.G. 1977.Rare earth elements in carbonatite and cogenetic alkaline rocks: examples from Seabrook Lake and Callander Bay,Ontario.*Contrib.Mineral.Petrol.***65**,143-153.

* Cullers,R.L.,Ramakrishnan,S.,Berendsen,P. & Griffin,T. 1985.Geochemistry and petrogenesis of lamproites,Late Cretaceous age, Woodson Cty.Kansas,USA.*Geochim.Cosmochim.Acta* **49**,1383-1402.

* Currie,K.L. 1975.Geology and petrology of the Ice River alkaline Complex, British Columbia. *Bull. Geol.Surv.Can.* **245**, 68pp.

* Currie,K.L. 1976.The alkaline rocks of Canada.*Bull.Geol.Surv.Can.***239**, 228pp.

* Currie, K.L. 1980.A contribution to the petrology of the Coldwell alkaline complex, Northern Ontario. *Bull.Geol. Surv. Can.***287**,43pp.

Currie,K.L. & Ferguson,J. 1970.The mechanism of intrusion of lamprophyre dikes as indicated by "offsetting" of dikes. *Tectonophysics* **9**,525-535.

* Currie,K.L. & Shafiqullah,M. 1967. Carbonatite and alkaline igneous rocks in the Brent Crater, Ontario. *Nature* **215**, 725-726.

Cwojdzinski,S. 1982.[Pre-granitoid semi-lamprophyre at Rogowek (Klodzko-Zloty stok granitoid massif)]. In Polish, Engl. abstr. *Kwart.Geol.* **25**,31-40.

* Czamanske,G.K. & Atkin, S.A. 1985.Metasomatism, titanian acmite and alkali amphiboles in lithic-wacke inclusions within the Coyote Peak diatreme, Humboldt county, California.*Am.Mineral.***70**,499-516.

Czamanske,G.K.,Lanphere,M.A.,Erd,R.C. & Blake,M.C. 1978.Age measurements of potassium-bearing sulfide minerals. *Earth Planet.Sci.Lett.***40**,107-110.

Czamanske,G.K.,Erd,R.C.,Sokolova,M.N.,Dobrovol'skaya,M.G. & Dmitrieva,M.T. 1979.New data on rasvumite and djerfisherite. *Am.Mineral.***64**,776-778.

Czamanske,G.K.,Leonard,B.F. & Clark,J.R. 1980.Erdite, a new hydrated sodium iron sulfide mineral. *Am.Mineral.* **65**,509-515.

Czamanske,G.K.,Erd,R.C.,Leonard,B.F. & Clark,J.R. 1981.Bartonite, a new potassium iron sulfide mineral. *Am.Mineral.* **66**,369-375.

Czamanske,G.K. & Wones,D.R. 1973. Oxidation during magmatic differentiation, Finmarka Complex, Oslo area, Norway: Part 2, the mafic silicates. *J. Petrol.* **14**, 349-380.

* Dal Piaz,G.V.,Venturelli,G. & Scolari,A. 1979.Calk-alkaline to ultrapotassic postcollisional volcanic activity in the internal northwestern Alps.*Mem.Ist.Geol.Mineral.Univ.Padova* **32**,4-16

Daly,R.A. 1912.Geology of the North American Cordillera at the 49th parallel.*Mem.Geol.Surv.Can.***38**.

@D'Amico,C. 1960.La massa dioritico-quarzifero di Bitti-Onani(Sardegna).*Acta Geol.Alpina* no.**8**, 129-180.

* D'Amico,C. & Guidicini,B. 1961.Le rocce filionane diabische e spessartitico porfiritiche di San Efisio. *Rend.Soc.Mineral.Ital.***17**,245-278.

Daniyelyants,S.Y., & Yakhov,Y.V. 1975.[New data on the relative age of lamprophyres and ores in the Darasun deposit, E.Transbaikal]. In Russian.*Vysshikh Uchebnykh Zavedenii Izvestiya Geologiya i Razvedka* **5**,94-103.

Darton,N.H. & O'Harra,C.C. 1905.Sundance folio.*U.S.G.S.Geol.Atlas* no.**127**.

David,J.W.E. 1950. *Geology of the Commonwealth of Australia.* Arnold, London, 618pp.

Davidson, A. 1972. The Churchill province. In: *Variations in Tectonic Styles in Canada* (eds. Price,R.A. & Douglas,R.J.W.) *Spec.Publ.Geol.Assoc.Can.***11**, 381-433.

* Davies,K.A. 1952.The building of Mt.Elgon.*Mem.Geol.Surv.Uganda* **7**,62pp.

* Davies, H.L. 1983, Wabag, PNG. *Geol. Surv. PNG 1:250,000 geological series explanatory notes*, 84pp.

? Dawes,P. 1976. Precambrian to Tertiary of Northern Greenland. In: Escher & Watt (1976), qv, 248-303.

? Dawes,P. & Soper, N.J. 1970. Geological investigations in northern Peary Land. *Rapp.Grønlands Geol.Unders.***28**,9-15.

Dawson,J.B. 1971. Advances in kimberlite geology. *Earth-Sci.Rev.***7**, 187-214.

Dawson,J.B. 1980. *Kimberlites and their xenoliths*. Springer-Verlag, Berlin, 252pp.

* Dawson, J.B. 1987.The kimberlite clan: relationship with olivine and leucite lamproites, and inferences for upper mantle metasomatism. In: Fitton & Upton (1987), qv, 95-101.

Dawson, J.B. 1989. Geographic and time distribution of kimberlites and lamproites: relationships to tectonic processes. *Spec. Publ. Geol. Soc. Aust.* **14**, 327-342.

Dawson,J.B.,Delaney,J.S. & Smith,J.V. 1978.Aspects of the mineralogy of alnöitic breccia, Malaita, Solomon Islands; comparison with continental kimberlites. *Contrib.Mineral.Petrol.* **67**,189-194.

Dawson,J.B. & Stephens, W.E. 1975. Statistical analysis of garnets from kimberlites and associated xenoliths. *J.Geol.* **83**, 589-607.

* De Almeida,M.A. 1955.Geologia e petrologia do arquipêlago de Fernando de Noronha. *Divn.Geol.Mineral. Dept.Nac.Prod.Mineral. Brazil monograph* **13**.

* De Almeida,M.A. 1961.Geologia e petrologia da ilha de Trindade. *Divn.Geol.Mineral.Dept.Nac. Prod.Mineral. Brazil monograph* **18**.

Deans,T. 1938. Igneous rocks from the Abercorn and Kasama districts, northern Rhodesia. *Geol.Mag.***75**,547-558.

* Deer,W.A. 1950.The diorites and associated rocks of the Glen Tilt complex,Perthshire.II.Diorites and appinites. *Geol.Mag.***87**,181-195.

Deer,W.A. 1953.The diorites and associated rocks of the Glen Tilt complex.III: Hornblende schist and hornblendite xenoliths in the granite and diorite.*Geol.Mag.***90**,27-35.

? Deer,W.A. 1976. Tertiary igneous rocks between Scoresby Sund and Kap Gustav Holm, East Greenland. In: Escher & Watt (1976), qv, 404-429.

Deer,W.A., Howie,R.A. & Zussman,J. 1978. *Rock-forming Minerals.* 2nd Edn. Longmans, London.

* Deines,P. 1970. The carbon and oxygen isotope composition of carbonates from the Oka carbonatite, Quebec, Canada. *Geochem.Cosmochim.Acta* **34**,1199-1225.

Delaney,J.S.,Smith,J.V. & Nixon,P.H. 1979. Model for upper mantle below Malaita, Solomon Islands, deduced from chemistry of lherzolite and megacryst minerals. *Contrib.Mineral.Petrol.***70**,209-218.

Delaney,P.T. & Pollard,D.D. 1981.Deformation of host rocks and flow of magma during growth of minette dykes and breccia-bearing intrusions near Ship Rock,New Mexico.*U.S.G.S. Prof.Pap.***1202**.

* Delor,C.P. & Rock,N.M.S. 1990. Alkaline–ultramafic lamprophyre dykes from the Vestfold Hills, Princess Elizabeth Land, E.Antarctica (submitted).

* Derrick,G.M. 1980.Marraba,Queensland.*Aust.Bur.Miner.Resour.1: 100,000 map commentary.*

Derrick,G.M. & Gellatly,D.C. 1969.New leucite lamproites from the west Kimberley,W.Australia. *Bull.Aust.Bur.Miner.Resour.Geol.Geophys.***125**,103-119.

Desai, S.D. 1964. Geology of the Girnar hills (Gujarat). *J.M.S.Univ.***13(3)**,p. 19.

Deshpande,G.G. & Chakranarayan,A.B. 1973.Occurrence of lamprophyre near Murud-Janjira,Kolaba district,Maharashtra. *Current Sci.***42(11)**,404-405.

Desio,A. 1979.Geologic evolution of the Karakorum. In: *Geodynamics of Pakistan* (eds. Farah,A. & Dejong,K.A.), 111-124. Geol.Surv.Pakistan,Quetta.

* De Sousa,F.P. 1926.La Serra de Monchique.*Bull.Soc.Géol.Fr,Sér.4*,**26**,321-350.

Dessai, A.G. 1985. Ultramafic xenoliths (?) in lamprophyre dykes from Murud-Janjira, Raigarh district, Maharashtra, India. *Current Sci.* **54(23)**, 1236-1238.

* Dessai, A.G. 1987. Geochemistry and petrology of xenolith-bearing lamprophyres from Murud-Janjiri, Raigarh district, Maharashtra, India. *J.Geol.Soc. India* **30(1)**, 61-71.

* Dessai, A.G., Rock, N.M.S., Griffin,B.J. & Gupta,D. 1990. Mineralogy and petrology of some xenolith-bearing alkaline dykes associated with Deccan magmatism south of Bombay. *European J.Mineral* (in press).

*¶ Deutsch,A. 1984.Young Alpine dykes south of the Tauern window(Austria): a K-Ar and Sr isotope study. *Contrib. Mineral.Petrol.***85**,45-57.

* De Villiers,J. & Söhnge.P.G. 1959.Geology of the Richtersweld.*Mem.Geol.Surv.S.Afr.* **48**, 259pp.

Diamond,L.W. & Wiedenbeck,M. 1986.K-Ar radiometric ages of the gold-quartz veins at Brusson,Val d'Ayas,NW Italy: evidence of mid-Oligocene hydrothermal activity in the Northwestern Alps. *Schweiz.Mineral. Petrogr.Mitt.* **66**,385-394.

Didier,J. 1964. Etude pétrographie des enclaves de quelques granites du Massif Central Français. *Ann. Fac. Sci. Clermont-Ferrand* **23**.

@ Dietrichson,B. 1955.Spessartite and pseudotachylite from the thrusting-zone of E.Jotunheim.*Norges Geol.Unders.* **191**,30-55.

* Dimitrijevic,B. 1928. Hemiski karakter iz doline reke Palanke kod Ripnja. *Geol.anali.Balk.pol.knj.IX sv2*,131-132.

Dimroth, E. 1970. Meimechites and carbonatites of the Castignon Lake complex, New Quebec. *N.Jb. Mineral. Abh.***112**,239-278.

Distler, V.V., Ilupin, I.P. & Laputina, I.P. 1987. Sulphides of deep-seated origin in kimberlites, and some aspects of copper-nickel mineralization. *Int. Geol. Rev. 29*, 456-464.

* Dobosi, E. & Horváth, I. 1988. High and low-pressure cognate clinopyroxenes from alkali lamprophyres of the Velence and Buda Mountains, Hungary. *N. Jb. Mineral. Abh.* ***158***, 241-256.

Dobretsov,G.L.,Alyavdina,J.N. & Dobretsova,T.G. 1969.[Kaersutite in camptonites of southern Dzhungaria]. In Russian. *Zap.Vses.Mineral.Obshch.* **98(4)**, 471-475.

Dobretsov,G.L. & Dobretsova,T.G. 1969.[Inclusions of deep-seated origin in camptonites of S.Dzhungaria & their genetic significance]. In Russian. *Akad.Nauk.SSSR Sib.Obt.Inst.Vulkanol. Moscow*, 40-50.

Dolar-Mantuani,L. & Klemen, R. 1940.Ein Spessartit aus dem Pohorje-Gebirge (Jugoslawien). *Zentralb.Mineral.Abt.*A **5**,112-127.

Domenick,R.A. & Basu,A.R. 1982.Age and origin of the Cortlandt complex,New York; implications from Sm-Nd data. *Contrib.Mineral.Petrol.***79**,290-294.

Dons J.A. 1952. Studies on the Igneous Rock Complex of the Oslo Region, Vol XI. Compound volcanic neck, igneous dykes, and fault zone in the Ullern-Husebyåsen area, Oslo. *Norsk.Akad. Oslo Selsk.Skr. Mat.Naturv. Kl.* **XI**, 96pp.

Douglas,J.G. & Ferguson J.A. 1988. *Geology of Victoria.*Victn.Div.Geol.Soc.Aust.**2**,664pp.

* Dressler,B. 1975.Lamprophyres of the north-central Labrador Trough,Quebec.*Neues Jb.Mineral.Mh.*, 268-280.

Droop,G.T.R. 1987. A general equation for estimating Fe^{3+} concentrations in ferromagnesian silicates and oxides from microprobe analyses, using stoichiometric criterial *Mineral.Mag.* **51**, 431-435.

Druecker,M.D. 1985.Mafic dike swarms associated with Mesozoic rifting in eastern Paraguay. In: *Int.Conf.Mafic.Dike Swarms,Toronto,June '85*, 37 (abstr.)

Drysdale,C.W. 1914. Geology of the Thompson River valley, below Kamloops Lake, B.C. *Summ.Rep. Geol.Surv. Can.* 1912, 133-134.

Drysdale,C.W. 1915.Geology and ore deposits of Rossland,B.C.*Mem.Geol.Surv.Can.***77**.

Duncan,R.A.,Hargraves,R.B. & Brey,G.P. 1978. Age, palaeomagnetism and chemistry of melilite basalts in the Southern Cape, South Africa.*Geol.Mag.***115**,317-327.

*Durand-Wackenheim,C. 1983.Sur la composition et l'empreinte métamorphique des filons lamprophyriques du vallon de la Neste de Saux(Hautes-Pyrénées).*C.R.Acad.Sci.Paris* **296**,729-732.

* Durazzo,A.,Taylor,L.A. & Shervais,J.W. 1984.Ultramafic lamprophyre in a carbonate platform environment, Mt.Queglia,Abruzzo,Italy.*N.Jb.Mineral.Abh.***150**,199-217.

* Durgadmath, M.B. 1981. Lamprophyre dykes from Phenai Mata area, Baroda district, Gujarat. *Spec. Publ. Geol. Soc. India* **12**,3-6.

Durrance,E.M. & Kearey,P. 1988. Gravity survey of Colonsay and Oronsay. *Scott J. Geol.***24**,81-88.

*Dvorak,J. 1982.[Stephanian lamprophyres of the Janov-Artmanov anticlinorium in Silesia]. In Czech, Engl. summ. *Sbornik.Geol Ved. (Prague)* **36**,93-113.

Dyer, W.S. 1936. Geology and ore deposits of the Mataschewan area. *Ont.Dept.Mines Geol. Rept.* **44(2)**, 1-58.

Eades,N.W. & Reeves,W.H. 1938. Explanation of the geology of Degree Sheet No. 29 (Singida). *Bull. Geol. Div. Dept. Lands & Mines Tanganyika* **11**, 59pp.

Earp J.R. & Hains,R.A. 1971. *British Regional Geology: The Welsh Borderland.* 3rd Edn. HMSO for Br.Geol.Surv., London.

Eastwood,T., Hollingworth,M.A., Rose,W.C.C. & Trotter,R.M. 1968.Geology of the country around Cockermouth and Caldbeck (sheet 23).*Mem.Geol.Surv. G.B.*

* Eby,G.N. 1975.Abundance and distribution of the rare-earth elements and yttrium in the rocks and minerals of the Oka carbonatite complex,Quebec.*Geochim.Cosmochim.Acta* **39**,597-620

* Eby,G.N. 1980.Minor and trace element partitioning between immisicible ocelli-matrix pairs from lamprophyre dykes and sills, Monteregian Hills petrographic province,Quebec. *Contrib.Mineral.Petrol.* **75**,269-278.

Eby, G.N. 1983.The identification of silicate-liquid immiscibility processes using minor and trace element distributions. In: *The significance of Trace Elements in solving Petrogenetic Problems and Controversies* (ed. Augustithis, S.S.), 27-37. Theophrastus, Athens.

* Eby,G.N. 1985. Age relations, chemistry and petrogenesis of mafic alkaline dykes from the Monteregian Hills and younger White Mountain igneous provinces. *Can.J.Earth Sci.***22**,1103-1111.

* Eby, G.N. 1987. The Monteregian Hills and White Mountains alkaline igneous provinces, eastern North America. In: Fitton & Upton (1987), qv, 433-448.

Edgar,A.D. 1983.Relationship of ultrapotassic magmatism in the western USA to the Yellowstone plume. *N.Jb.Mineral.Abh.* **147**,35-46.

Edgar, A.D. 1989. Barium- and strontium-enriched apatites in lamproites from West Kimberley, Western Australia. *Am.Mineral.* **74**, 889-895.

Edgar, A.D. & Arima, M. 1983. Conditions of phlogopite crystallization in ultrapotassic volcanic rocks. *Mineral. Mag.* **47**, 11-19.

Edgar,A.D. & Arima,M. 1985. F and Cl contents of phlogopites in high pressure melting experiments on ultrapotassic rocks; implications for halogen reservoirs in mantle source regions. *Am. Mineral.* **70**, 529-536.

Edgar, A.D., Arima, M., Baldwin, D.K., Bell, D.R., Shee, S.R., Skinner, E.M.W. & Walker, E.C., 1988. High pressure melting experiments on a SiO_2-poor aphanitic kimberlite from the Wesselton mine, Kimberley, S. Africa. *Am. Mineral.* **73**, 524-534.

Edmonds,E.A.& 9 others. 1968.Geology of the country round Okehampton(sheet 324).*Mem.Geol.Surv. G.B.*

Edmonds, E.A., McKeown, M.C. & Williams, M. 1969. *British Regional Geology: South-West England.* 3rd Edn. HMSO for Br.Geol.Surv., London.

* Edwards,A.B. 1934.Tertiary dykes and volcanic necks of South Gippsland. *Proc.R.Soc.Vic.* **47**,112-132.

Edwards,A.B. 1938.Petrology of the Tertiary Older volcanic rocks of Victoria. *Proc.R.Soc.Vic.***51**,73-98.

Effinova, S.V. 1970.[*Dykes of granitoid complexes, central Kazakhstan*]. In Russian. Nauka,Moscow,237pp.

Eggler,D.H. 1989. Carbonatites, primary melts and mantle dynamics. In: Bell (1989), qv, 561-575.

Eggler,D.H., Dudas,F.O., Hearn,B.C., McCallum,M.E., McGee,E.S., Meyer,H.O.A. & Schultz,D.J. 1987. Lithosphere of the continental United States: xenoliths in kimberlites & other alkaline magmas. In: Nixon (1987), qv, 41-57.

* Egorov,L.S. 1970.Carbonatites and ultrabasic-alkaline rocks of the Maimecha-Kotui region, N.Siberia. *Lithos* **3**,341-359.

* Egorov,O.S. & Karmazin.P.S. 1972.The alkali basaltoid dyke complex of the southern Donetz basin. *Geol.Zh.URSS*, **32**,99-103.

Ehlers, C. & Bergman, L. 1984. Structure and mechanism of intrusion of two postorogenic granite massifs, southwestern Finland. In: *Tectonics Illustrated* (Kroner, A. & Greiling, R., eds.) E. Schweizerbartsche Verlasbuchhandlung, Stuttgart, 173-190.

Ehrenberg,S.N. 1979.Garnetiferous ultramafic inclusions in minette from the Navajo volcanic field. In: Boyd & Meyer (1979b), qv, 330-344.

* Ehrenberg,S.N. 1982a.Rare earth element geochemistry of garnet lherzolite and megacrystalline nodules from minette of the Colorado plateau province.*Earth Planet.Sci.Lett.***57**,191-210.

Ehrenberg,S.N. 1982b.Petrogenesis of garnet lherzolite and megacrystalline nodules from the Thumb,Navajo volcanic field. *J.Petrol.***23**, 507-547.

Ehrenberg,S.N. & Griffin,W.L. 1979.Garnet granulite and associated xenoliths in minette and serpentinite diatremes of the Colorado Plateau.*Geology* **7**,483-7.

@ Eigenfeld,R. 1960.Über den nortigartigen Lamprophyre vom Steinhügel bei Höflas im Munchberger Gneismasse. *Fortschr.Mineral.***38**,162-165.

* Ekren,E.B. & Houser,F.N. 1965.Geology and petrology of the Ute Mountains area,Colorado. *U.S.G.S. Prof.Pap.* **481**, 74pp.

Elders,W.A. 1957.A preliminary note on a xenolith-rich dyke from Lyngen,Northern Norway.*Acta borealis A.Scientia no.12, Tromso museum,*13pp.

Elders,W.A. & Rucklidge,J.C. 1969.Layering and net-veining in hornblende lamprophyre intrusions from the coast of Labrador. *J.Geol.***77**,721-729.

El Goresy,A. & Yoder,H.S. 1974. Natural and synthetic melilite compositions. *Carn.Inst.Wash.Yrbk.* **73**,359-371.

Elsdon,R. & Todd, S.P. 1989. A newly-discovered Caledonian composite spessartite-appinite intrusion from Port-na-Blagh, County Donegal, Ireland. *Geol.J.* **24**, 97-112.

* Embey-Isztin,A. 1972.A study of lamprophyric dyke-rocks of the Velence Hills (Hungary). *Fragm.Mineral.Palaeontol.* **3**,5-24.

Emeleus,C.H. 1964.The Gronnedal-Ika alkaline complex,south Greenland. *Bull.Grønlands.Geol.Unders.***85** & *Meddr.Grønland* **172(3)**.

Emeleus,C.H. & Upton,B.G.J. 1976. The Gardar period in southern Greenland. In: Escher & Watt (1976), qv, 152-181.

@ Emiliani,F. 1958.Studio delle rocce filoniane nel granito a NE di Buddoso(Sassari).*Acta Geol.Alpina* no.7.

? Emmons,R.C.,Reynolds,C.D. & Saunders,D.F. 1953.Genetic and radioactivity features of selected lamprophyres. *Mem.Geol.Soc.Am.* **52**,89-98.

English,R.M. & Grogan,R.M. 1948. Omaha pool mica-peridotite intrusives, Gallatin Cty, Illinois. *Illinois Geol.Surv.Rep.Inv.***130**,189-212.

Erd,R.C. & Czamanske,G.K. 1983. Orichite and coyoteite, two new sulfide minerals from Coyote Peak, Humboldt County, California. *Am.Mineral.***68**, 245-254.

Erickson,R.L. & Blade,L.V. 1963.Geochemistry and petrology of the alkalic igneous complex at Magnet Cove. *U.S.G.S. Prof.Pap.* **425**.

Erlank,A.J. & Finger,L.W. 1970.The occurrence of potassic richterite in a mica nodule from the Wesselton kimberlite,South Africa.*Carn. Inst.Wash.Yrbk.***68**,320-324.

Ervin,C.P. & Mudrey,M.G. 1975.Extension of Vermilion district lamprophyres to western Minnesota from geophysical observations. *Inst.Lake Superior Geol.Tech.Sess.Abstr.Field guides* **21**, p.35.

Ervin,C.P. & Mudrey,M.G. 1976.Extension of a northern Minnesota lamprophyre province by geophysical studies. *J.Geophys.Res.***81**,4917-4922.

Escher, A. & Watt, W.S. 1976. *Geology of Greenland.* Grønlands Geol. Unders, 603pp.

Escorza, C.M. & López Ruiz, J. 1988. Un modelo geodinamico por el volcanismo neogene del suereste iberico. *Estud. Geol.* **44**, 243-251.

* Eskola,P. 1954.Ein Lamprophyrgang in Helsinki und die Lamprophyrprobleme.*Tschermaks Mineral. Petrogr.Mitt.,Ser.*3, **4**,329-337.

Esperança,S. & Holloway,J.R. 1987.On the origin of some mica-lamprophyres: experimental evidence from a mafic minette.*Contrib.Mineral.Petrol.***95**,207-216.

Evans,J.W. 1901.A monchiquite from Mount Girnar,Junagarh.*Q.J.Geol.Soc.Lond.***57**,38-52.

Exley,C.S. & Stone,M. 1982.Petrology of the granites and minor intrusions. In: Sutherland(1982), qv, 293-302.

Eyal,M. & Eyal,Y. 1985.Basic dyke swarms of the Arabo-Nubian Shield. In: *Int.Conf.Mafic.dike swarms,Toronto,June '85*, 40 (abstr.)

Eyles,V.A. 1924.Note on the nature and age of Scottish rocks bearing xenocrysts of anorthoclase. *Geol.Mag.***61**,471-472.

Eyles,V.A. & Blundell,C.R.K. 1957.On a volcanic vent and associated monchiquite intrusions in Monmouthshire. *Geol.Mag.***94**,54-57.

Eyles, V.A., Simpson, J.B. & Macgregor,A.G. 1949. Geology of central Ayrshire (sheet 24). *Mem.Geol.Surv.G.B.*

Fabre,P.,Kast,Y. & Girod,M. 1989. Estimation of flow duration of basaltic magma in fissures. *J.Volcanol. Geotherm.Res.* **37**, 167-186.

Faerseth,R.B. 1978.Mantle-derived lherzolite xenoliths and megacrysts from Permo-Triassic dykes, Sunnhordland, western Norway.*Lithos* **11**,23-35.

* Faerseth,R.B.,Macintyre,R.M. & Natersud,J. 1976.Mesozoic alkaline dykes in the Sunnhordland region, western Norway. *Lithos* **9**,331-345

* Faessler,C. 1962.Analyses of rocks of the province of Quebec.*Quebec Dept.Nat.Res.Geol.Rep.* **103**.

† Fallot,P. & Jérémine,E. 1932.Rémarque sur une varieté nouvelle de jumillite et sur l'extension des laves de ce groupe. *C.R.Congr.Soc.Sav.* 1929,1-13.

Fay, G.H. & Hogarth,D.D. 1969. On the origin of 'reverse' pleochroism of a phlogopite. *Can.Mineral.***10**, 25-34.

@ Fenner,C. 1938.Olivine fourchites from Raymond Fosdick mountains, Antarctica.*Bull.Geol.Soc.Am.* **49**, 367-400.

Fenoglio,M. 1938.Sopra dua "minettes" della Val Nambrone (gruppo dell'Adamello). *Rend.R.Accad.Naz. Lincei Atti. Ser.6*, **28(3-4)**,100-106.

Feraud,G.,Gastaud,J.,Auzende,J-M.,Olivet,J-L. & Cornen,G. 1982.$^{40}Ar/^{39}Ar$ ages for the alkaline vulcanism and the basement of Gorringe Bank, N.Atlantic ocean.*Earth.Planet.Sci.Lett.***57**,211-226.

* Ferguson,J. 1973.The Pilanesberg alkaline province,Southern Africa.*Tr.Geol.Soc.S.Afr.***76**, 249-270.

Ferguson, J., Arculus, R.J. & Joyce, J. 1979. Kimberlite and kimberlitic intrusives of southeastern Australia: a review. *J. Aust. Bur. Mineral. Resour. Geol. Geophys.* **4**, 227-241.

* Ferguson,J. & Currie,K.L. 1971.Evidence of liquid immiscibility in alkalic ultrabasic dikes at Callander Bay, Ontario.*J.Petrol.***12**,561-585.

* Ferguson,J. & Currie,K.L. 1972.Geology and petrology of the alkaline carbonatite complex at Callander Bay. *Bull.Geol.Surv.Can.* **217**,103pp.

Ferguson, J., Martin, H., Nicolaysen, L. & Danchin R.V. 1975. Gross Brukkaros: a kimberlite-carbonatite volcano. *Phys. Chem. Earth* **9**, 219-234.

Ferguson,J. & Sheraton,J. 1979.Petrogenesis of kimberlitic rocks and associated xenoliths of SE Australia. In: Boyd & Meyer (1979a), qv, 140-160.

* Fermoso,M.L. 1967a.El diopsido de las rocas volcánicas de Jumilla.*Estud.Geol.*23,31-33.

* Fermoso,M.L. 1967b.Composicion química de las sanidinas de las rocas lamproíticas españolas. *Estud.Geol.* **23**,29-30.

* Fernández,S. & Hernández-Pacheco,A. 1972.Las rocas lamproíticas de Cabezo Negro,Zeneta. *Estud.Geol.* **28**,267-276.

Ferreira,M.P. & Macedo,C.A.R. 1979.Actividade magmática durante o Mesozóico: I; achega para a datação K-Ar das rochas filonianas básicas intrusivas na zona Centro-Ibérica.*Mem.Not. Inst.Univ.Coimbra* **87**,29-49.

* Ferreira Pinto,A.F. 1968.Lamprófiro com biotite e actinolite de Vilarinho das Furnas. *Mem.Not.Mus.Lab. Mineral.Geol.Univ.Coimbra* **66**,75-80.

* Fettes,D.J.,Mendum,J.R.,Smith,D.I.,Rock,N.M.S. & Macdonald,R. 1989.Geology of the Outer Hebrides. *Mem.Br.Geol.Surv.* (in press).

Fiala,F. 1971.[Diabase volcanism and Ordovician biotite-lamprophyres of the Barrandian basin]. In Czech, Engl.summ. *Sbor.Geol.Ved.G.Ceskosl.* 19,7-97.

Fiala,F. & Chlupac,I. 1973.[Minette dyke in the Devonian of the Barrandian basin]. In Czech. *Cas.Mineral.Geol.* **18/1-4**, 47-55.

* Fielding,D.C. & Jaques,A.L. 1989.Geology,petrology and geochemistry of the Bow Hill lamprophyre dikes, Western Australia. *Spec. Publ. Geol. Soc. Aust.* **14**, 206-219.

* Finlayson, E.J., Rock, N.M.S. & Golding, S.D. 1988. Deformation and regional carbonate metasomatism of turbidite-hosted Cretaceous alkaline lamprophyres (northwestern Papua New Guinea). *Chem.Geol.* **69**, 215-233.

* Fisera,M. 1974.[Bronzite-olivine minette from Brloh in the Pisek area]. In Czech, Engl. summ. *Cas.Mineral. Geol.* 19,71-76.

Fitton, J.G. & Upton, B.G.J. 1987. Alkaline igneous rocks. *Spec.Publ.Geol.Soc.Lond.***30**, 568pp.

¶* Fleming, A.W., Handley, G.A., Williams, K.L., Hills, A.L. & Corbett, G.J. 1986. The Porgera gold deposit, Papua New Guinea. *Econ. Geol.* **81**, 660-680.

#* Flett,J.S. 1900.Trap dykes of the Orkneys.*Tr.R.Soc.Edinburgh* **39**,865-905.

Flett,J.S. 1929.The Inverbeg lamprophyre.*Summ.Prog.Geol.Surv.G.B.(1928),part II*,29-35.

* Flett,J.S. 1935.Petrography. In: *Geology of the Orkneys*, 172-183.*Mem.Geol.Surv.G.B.*

Flett,J.S. & Hill,J.S. 1946. The geology of the country around Lizard & Meneage (sheet 359). *Mem.Geol.Surv. G.B.*

Flinn,D. 1988. The Moine rocks of Shetland. In:Winchester (1988), qv, 74-85.

Foland,K.A.Spera,F.J. & Bergman,S.C. 1980.Strontium isotope relations in megacryst-bearing camptonites from NW Arizona. *Geol.Soc.Am.Abstr. Progr.* **12(2)**,p.36.

Foley,S.F. 1984.Liquid immiscibility and melt segregation in alkaline lamprophyres from Labrador.*Lithos* **17**,127-138.

Foley,S.F. 1986.The oxidation state of lamproitic magmas.*Tschermaks Mineral.Petrogr.Mitt.***34**, 217-238.

Foley,S.F. 1989.The genesis of lamproitic magmas in a reduced fluorine-rich mantle. *Spec. Publ. Geol. Soc. Aust.* **14**, 616-631.

Foley,S.F. & Venturelli,G. 1989. High-K_2O rocks with high-MgO, high-SiO_2 affinities. In: Crawford (1989), qv, p. 72-88.

Foley, S.F., Venturelli, G., Green,D.H. & Toscani,L. 1987. The ultrapotassic rocks: characteristics, classification and constraints for petrogenetic models. *Earth-Sci. Rev.* **24**, 81-134.

@ Forsyth,I.H. & Chisholm,J.I. 1978. Geology of East Fife.*Mem.Geol.Surv.G.B.*

* Fowler,M.F. 1988.Ach'uaine hybrid appinite pipes: evidence for mantle-derived shoshonitic parent magmas in Caledonian granite genesis. *Geology* **16**, 1026-1030.

* Fox,C.S. 1930.The Jharia coalfield.*Mem.Geol.Surv. India* **56**,255pp.

Fozzard,P.M.H. 1965. Brief explanation of the geology, Quarter Degree Sheet 182 Kimamba. *Geol. Surv. Tanzania* 1:125,000 map.

Franks,P.C. 1959.Pectolite in mica peridotite, Woodson County, Kansas.*Am.Mineral.***44**,1082-1086.

* Franks,P.C.,Bickford,M.E. & Wagner,H.C. 1971.Metamorphism of Precambrian granitic xenoliths in a mica peridotite at Rose Dome,Woodson Cty.,Kansas: Part 2,Petrologic and Mineralogic studies. *Bull.Geol.Soc.Am.* **82**,2869-2890.

* Fraser,K.J.,Hawkesworth,C.J.,Erlank,A.J.,Mitchell,R.H. & Scott-Smith,B.H. 1985.Sr,Nd and Pb isotope and minor element geochemistry of lamproites and kimberlites.*Earth Planet.Sci.Lett.***76**,57-70.

Freestone,I.C. & Hamilton,D.L. 1979. The role of liquid immscibility in the genesis of carbonatites — an experimental study. *Contrib.Mineral.Petrol.* **73**, 105-117.

French,W.J. 1966.Appinitic intrusions clustered round the Ardara pluton,Cty.Donegal. *Proc.R.Ir.Acad.* **64b**,303-322.

French,W.J. 1976.The origin of leucodiorites associated with the appinitic intrusions of county Donegal. *Proc.Yorks.Geol.Soc.***41**,107-125.

French,W.J. 1977.Breccia-pipes associated with the Ardara pluton,Cty.Donegal.*Proc.R.Ir.Acad.***77b**,101-117.

* French,W.J. 1978.Lamprophyre dykes associated with the appinitic intrusions of County Donegal.*Sci.Proc.R.Soc. Dublin Ser.A.***6(9)**,97-107.

French,W.J. & Pitcher,W.S. 1959.The intrusion-breccia of Dunmore,co.Donegal.*Geol.Mag.* **96**,69-74.

* Frenzel,G. 1971.Die Mineralparagenese der Albersweiler Lamprophyre.*N.Jb. Mineral.Abh.***115**,164-191.

Frey,F.A.,Green,D.H. & Roy,S.D. 1978. Integrated models of basalt petrogenesis: a study of quartz tholeiites to olivine melilitites from south eastern Australia utilizing geochemical and experimental petrological data. *J.Petrol.* **19**, 463-513.

Frohberg,M.H. 1937.Gold deposits of the Michipicoten area.*Ont.Dept.Mines Rep.* **45**,pt.8(for 1935).

* Furnes,H. & Stillman,C.J. 1985.The petrology and geochemistry of an alkaline lamprophyre sheet intrusion complex in Maio,Cape Verde Republic.*J.Geol.Soc.Lond.***144**,227-242.

* Fúster,J.M.C. 1951.Rocas de Espana central III: granitos y lamprofidos del Puerto de Navacerrada(Segovia). *Estud.Geol.* **14**,431-441.

Fúster,J.M.C. 1955.Transformaciones metasomáticas en los diques diabásicos y lamprofídicos de la Sierra de Guadarrama.*Bol.Real. Soc.Esp.Hist.Natural.***III**,99-127.

Fúster,J.M.C. & Gastesi,P. 1964.Estudio petrológico de las rocas lamproíticas de Barqueros (provincia de Murcia). *Estud.Geol.***20**,299-331.

* Fúster,J.M.C.,Gastesi,P.,Sagredo,J. & Fermoso,M.L. 1967.Las rocas lamproíticas del SE de España. *Estud.Geol.***23**,35-69.

* Fúster,J.M.C. & Ibarolla,E. 1952.Rocas de España Central.IV: pórfidos y lamprófidos magmáticos de la Granja(Segovia). *Estud.Geol.***9**,193-217.

* Fúster,J.M.C. & Pedro,F.de. 1953.Estudio petrológico de las rocas volcánicas lamproíticas de Cabezo Maria.*Estud.Geol.* 19,477-508.

Fyles,J.T.,Harakal,J.E. & White,W.H. 1973.The age of the sulfide mineralization at Rossland, British Columbia. *Econ.Geol.***68**,23-33.

Gabert,G. 1959.Petrologische Beziehungen des Oberharzer Kersantits zu Gang und Tiefengesteinen des Harzes.*Geol.Jb.***75**,79-114.

Gabert,G. 1973. Ober prekambische Olivinmonzonite (Kentallenite) in Zentral-Tanzania (Ostafrika). *Geol. Jb. Reihe B*, **H6**, 51-79.

Gagny,C. 1978. Vaugnérites et durbachites sont des cumulats de magma granitique (l'exemple du magma des Crêtes, Vosges). *C.R.Acad.Sci.Paris* **287**, 1361-1364.

* Gallagher,M.J. 1963.Lamprophyre dykes from Argyll.*Mineral.Mag.***33**,415-30.

Gallagher, M.J. 1964. Rock alteration in some mineralized basic dykes in Britain, *Tr. Inst. Ming.Metall.* **73**, 825-839.

* Gamerith,H. & Kollmer,H. 1973.Untersuchungen an Intrusivgesteinen des östlichen Hindukusch. *Geol.Rundsch.* **62**,161-171.

Ganguly,J. & Bhattacharyya,P.K. 1987. Xenoliths in Proterozoic kimberlites from southern India. Petrology and geophysical implications. In: Nixon (1987), qv, 249-265.

¶Ganju, P.N. & Pant, I.D. 1962. Effects of a mica-peridotite dyke on the Dishergarh seam in the Sodepur colliery, Raniganj coalfield. *Proc. Indian Acad.Sci.* **55B**, 307-319.

* Gapeeva,G.M. 1949.[Camptonites of Toyun]. In Russian. *Zap. Vses. Mineral. Obshch. Ser.2*, **78**, 104-114. See *Mineral. Abstr.* **11**, p.334.

† Gapeeva,G.M. 1950.Kersantite de Taion.*Mem.Soc.Russe Minéral.***79(1)**, 69-71.

Gapeeva,G.M. 1960.[The origin of lamprophyres and their position in the genetic classification of rocks]. In Russian.*Zap.Vses. Mineral.Obshch.***89**,524-544.

Gardiner,C.I. & Reynolds,S.H. 1932.The Loch Doon 'granite' area,Galloway. *Q.J.Geol.Soc.Lond.***88**,1-34.

Gardiner,C.I. & Reynolds,S.H. 1937. The Cairnsmore of Fleet granite and its metamorphic aureole. *Geol.Mag.* **74**,289-300.

* Garson,M.S. 1962.The Tundulu carbonatite ring-complex in southern Nyasaland.*Mem.Geol.Surv.Nyas.***2**

* Garson,M.S. 1965a.Carbonatite and agglomerate vents in the western Shire valley.*Mem.Geol.Surv.Malawi* **3**,167pp.

Garson,M.S. 1965b.Carbonatites in southern Malawi.*Bull.Geol.Surv.Malawi* **15**.

Garson,M.S. 1966.Carbonatites in Malawi. In: Tuttle & Gittins (1966), qv, 33-72.

Garson,M.S. & Campbell Smith,W. 1965.Chilwa Island.*Mem.Geol.Surv.Nyas.***1**.

* Gaskarth, W.J.,Thorpe, R.S. & Macdonald, R. 1989. Caledonian magmatism processes deduced from petrology and geochemistry of minor intrusions in Northern England (in press).

Gaspar,J.C. & Wyllie,P.J. 1982. Barium phlogopite from the Jacupiranga carbonatite. *Am.Mineral.* **59**, 783-789.

Gautneb,H. 1987. *Høy-K dioritter og assosierte bergarter på sydlige Smøla arkipel. Møre og Romsdal.* Unpubl. thesis, Univ. Bergen, 265pp.

Gee, E.R. 1932. The geology and coal resources of the Raniganj coalfield. *Mem. Geol. Surv. India* **61.**

Geijer,P. 1928.Alnöitic dykes from the coast region of Luleå and Lalix in northern Sweden.*Fennia* **50**,1-16.

* Geldon,A.L. 1972.Petrology of a lamprophyre pluton near Dead River. In: *Geol.Minnesota centenary vol.*, 153-159. St.Paul,Minnesota.

* Georgiadès,A.N. 1938.La kersantite du Couvent de St.Grégoire,dans le Mont.Athos.*Akad.Athenon Praktika* **13**,487-491.

* Geraghty,E.P., Carten,R.B. & Walker,B.M. 1988. Timing of Urad-Henderson and Climax porphyry-molybdenum systems, central Colorado, as related to northern Rio Grande Rift tectonics. *Bull.Geol.Soc.Am.* **100**, 1780-1786.

* Gerasimovsky,V.I.,Volkov,V.P.,Kogarko,L.N.,Polyakov,A.I.,Saprikina,T.V. & Balashov,Y.A. 1968. *The geochemistry of the Lovozero alkaline massif.* (Engl.trans). Aust.Nat.Univ.,Canberra, 2 vols.

Gerasimovsky,V.I.,Volkov,V.P.,Kogarko,L.N. & Polyakov,A.I. 1974.Kola Peninsula. In: Sørensen (1974), qv, 206-221.

* Ghose,C. 1949. A petrochemical study of the lamprophyres and associated intrusive rocks of the Jharia coalfield. *Q.J.Geol.Mineral.Metall.Soc. India* **21(4)** 133-147.

? Gibbs,A.K. 1985.Continental mafic intrusives of the Guiana Shield. In: *Int.Conf.Mafic.Dike Swarms, Toronto,June '85*, 46-7 (abstr.)

Gilbert, G. 1948. Rossland Camp. In: *Structural Geology of Canadian Ore Deposits*, 189-195. Can. Inst.Ming.Metall.

Giles,P.S. & Chatterjee,A.K. 1987. Lower crustal xenocrysts and xenoliths in the Tangier dyke, eastern Meguma zone, Nova Scotia. *Rep.Nova Scotia Dept.Mines & Energy* **87-5**, 85-88.

Gil Ibarguchi,G.J.I. 1981. A comparative study of vaugnerites and metabasic rocks from the Finisterre region (NW Spain). *N.Jb.Mineral.Abh.* **143**, 91-101.

Gil Ibarguchi, J.I., Bowden,P. & Whitley J.E. 1984. Rare earth element distribution in some Hercynian granitoids from the Finisterre region, N.W. Spain. *J.Geol.* **92**,397-416.

Gill,J.B. 1982. *Orogenic Andesites and Plate Tectonics*. Springer Verlag, Berlin, 390pp.

Giobbi Origoni,E., Bocchio,R., Boriani,A., Carmine,M. & De Capitani,L. 1990. Late-hercynian mafic and intermediate intrusives of Serie dei Laghi (N.Italy). *Rend.Soc.Ital.Mineral.Petrog.* (in press).

Giobbi Origoni,E., Camerini,R.M. & Caonibo,A. 1975. Metamorfiti & plutoniti pre-granitiche inb assa Valsesia (zone Doccio-Aguma). *Bul.Soc.Geol.Ital.* **94**, 2203-2216.

Girault,J. 1968.Sur une lamprophyre à ovöides de calcite provenant de la région du complexe d'Oka(Canada). *C.R.Acad.Sci.Paris* **266**,649-651.

* Gittins,J.,Hewins,R.H. & Laurin,A.F. 1975.Kimberlitic-carbonatitic dikes of the Saguenay River valley,Quebec, Canada. *Phys.Chem.Earth.***9**,137-148.

Gledhill, T.L. 1927. Michipicoten gold area, District of Algoma. *Ont.Dept.Mines Geol. Rep.* **36**, 1-49.

*Gogineni,S.V.,Melton,C.E. & Giardini,A.A. 1978.Some petrological aspects of the Prairie Creek diamond-bearing kimberlite diatreme, Arkansas.*Contrib.Mineral.Petrol.***66**,251-261.

* Gold,D.P. 1966.The minerals of the Oka carbonatite and alkaline complex,Oka,Quebec. *Int.Mineral.Assoc. Vol.India*, 109-125.

* Gold,D.P. 1967.Alkaline ultrabasic rocks in the Montreal area,Quebec. In: Wyllie(1967), qv, 288-302.

* Gold,D.P. 1972.Monteregian Hills: diatremes, kimberlite, lamprophyres and intrusive breccias W of Montreal. *Int.Geol.Congr.Guidebook* no.**24**, Part **B-10**,35pp. Also in French (32pp.)

Gold, D.P., 1984. A diamond exploration philosophy for the 1980s. *Earth & Mineral Sciences (Penn. State Univ.)* **53 (4)**, 37-42.

Gold,D.P. & Marchand,M. 1969.The diatreme breccia pipes and dykes,and the related alnöite,kimberlite and carbonatite intrusions occurring in the Montreal and Oka areas.Parts I & II. In: Guidebook of the Geology of the Monteregian Hills. *Geol.Assoc.Can.Mineral.Assoc.Montreal*,5-42 and 43-62.

Gon'shakova, V.I., Ruzhitskiy, V.O., Boychuk, M.D., Zaritskiy, A.I. & Streilozov, N.F. 1967. Volcanic pipes and dykes of kimberlite rock on the Russian platform. *Int. Geol. Rev.* **11**, 60-73.

Gordon,M.,Tracey,J.I. & Ellis,M.W. 1958.Geology of Arkansas bauxite region. *US.Geol.Surv.Prof.Pap.* **299**,268pp.

* Goto,K. & Arai,S. 1987. Petrology of peridotite xenoliths in lamprophyre from Shingu, SW Japan:

implications for origin of Fe-rich mantle peridotite. *Mineral.Petrol.* **37**, 137-156.

Grant,N.K. 1966.The Caledonian dykes associated with the north-eastern end of the Newry complex, Cty.Down, Northern Ireland. *Geol.Mag.***103**,44-50.

* Grapes,R.H. 1975.Petrology of the Blue Mountain complex,Malborough,N.Z.*J.Petrol.***16**, 371-428.

* Greenough, J.D., Hayatsu, A. & Papelik, V.S. 1988. Mineralogy, petrology and geochemistry of the alkaline Malpeque Bay sill, Prince Edward Island. *Can. Mineral.* **26**, 97-108.

Gregory,J.W. 1900. Contribution to the geology of British East Africa, Part III. The nepheline syenite and camptonite dykes intrusive in the coast series. *Q.J.Geol.Soc.Lond.***56**,223-229.

Greig,D.C. 1971. *British Regional Geology: The South of Scotland.* 3rd Edn. HMSO for Br.Geol.Surv.,London.

Griffin,W.L. 1973.Lherzolite nodules from the Fen alkaline complex,Norway. *Contrib.Mineral.Petrol.* **38**,135-146.

* Griffin, W.L. & Kresten, P. 1987. Scandinavia — the carbonate connection. In: Nixon (1987), qv, 101-106.

* Griffin,W.L. & Taylor,P.N. 1975.The Fen damkjernite: petrology of a 'central complex kimberlite'. *Phys.Chem.Earth* **9**,163-177.

Grigor'yeva,L.V. & Savitskiy,A.V. 1979.[New data on alkaline lamprophyres of the NE border of the Khibiny massif]. In Russian.*Dokl.Akad.Nauk.SSSR* **248**,1400-1403.

* Grigor'yeva,L.V. & Savitskiy,A.V. 1980.New data on dike intrusions in the southwestern part of the Kola peninsula. *Dokl.Acad.Sci.USSR,Earth-Sci.Sect.***254**,119-122.

Grindley,G.W. 1963. Structure of the Alpine schists of South Westland,Southern Alps, New Zealand. *N.Z.J.Geol.Geophys.***6**, 872-930.

@ Gross,E.B. & Heinrich,E.W. 1966.Petrology and mineralogy of the Mt.Rosa area,El Paso and Teller counties, Colorado. III.Lamprophyres and mineral deposits.*Am.Mineral.***51**,1433-1442.

Grosser,P. 1957.Die Lamprophyre im Gebiet des Hohwaldes bei Neustadt.*Geologie* **6**,p.318 (abstr.)

* Grosser,P. 1966.Differentiation in Lamprophyren der Lausitz.*N.Jb.Mineral.Abh.***105(2)**,133-160.

Grout,F.F. 1932.*Petrography and petrology*.McGraw-Hill, New York, 522pp.

Grove,E.W. 1971. Geology and mineral deposits of the Stewart area, northwestern British Columbia. *Br.Columb. Dept. Mines Petrol. Res.Bull.* **58**, 219pp.

Groves,D.I.,Cocker,J.D. & Jennings,D.J. 1977.The geology, geochemistry and mineralization of the Blue Tier Batholith.*Bull.Geol.Surv. Tasmania* **55**, 171pp.

* Grozdanov,L. 1979.Chemistry and genesis of biotite from lamproites in the vicinity of the village of Svidnya,District of Sofia (Bulgaria).*Dokl.Bulg.Akad.Nauk. (C.R. Acad.Bulg. Sci.)* **32**,341-344.

‡ Grübenmann,U. 1919.Lamprophyrische Ganggesteine im centralen Gotthardgranitgneiss. *Vierteljahrschr. Naturforsch. Ges.Zurich* **84.**

@ Guintrand,Y.,Métais,D. & Thiébaut,J. 1963.Sur une roche de nature lamprophyrique dans la région de Saint Bresson(Haute Saone). *Bull.Soc.Géol.Fr. Sér.5*,**7**,16-19.

Gümbel,C.W.Von.1874.*Die palaeolithischen Eruptivgesteine des Fichtelgeberges*.Franz, München.

Gunn,W.,Clough,C.T. & Hill,J.B. 1897. The geology of Cowal. *Mem.Geol.Surv.G.B.*

* Gupta, A.K., Le Maitre, R.W., Haukka, M.T. & Yagi, K. 1983. Geochemical studies on the carbonated apatite glimmerites from Damodar valley, India. *Proc.Jap.Acad.Sci.* **59B**, 113-116.

* Gupta,A.K. & Yagi,K. 1980.*Petrology and genesis of leucite-bearing rocks*.Springer-Verlag, Berlin, 252pp.

* Gupta,A.K.,Yagi,K.,Lovering,J. & Jaques,A.L. 1986.Geochemical and microprobe studies of diamond-bearing ultramafic rocks from central and south India. *Proc.4th Int.Kimb.Conf.(Perth,WA), Geol. Soc. Aust. Abstr. Ser.* **16**, 27-29.

Gupta Sarma, D., Chetty,T.R.K., Murthy, D.S.N., Rao, R.A.V., Venkatanarayana,B. & Baker,N.R. 1986. Discovery of a new kimberlite pipe in Andhra Pradesh by stream sediment sampling. *J.Geol.Soc. India 27*, 313-316.

Gurney,J.J. 1984. A correlation between garnets and diamonds in kimberlites. *Geol.Dept.& Univ.Extension, Univ.W.Aust.Publ.***8**, 143-166.

Gusev,I.A. & Steblova,V.M. 1974.[Pliocene dikes of camptonite in the eastern Pamirs]. In Russian. *Dokl.Akad.Nauk.Tadzh.SSR* **17(6)**,35-37.

Gwinner,M.P. 1971. *Geologie der Alpen.* Schweizbartsche, Stuttgart, 477 pp.

* Hackmann,V. 1914.Über Camptonitgänge im mittleren Finnland.*Bull.Comm.Géol.Finlande* **42**,1-18.

Hains,B.A. & Horton A. 1969. *British Regional Geology: Central England.* 3rd Edn. HMSO for Br.Geol.Surv.,London.

* Hall,A. 1967.The chemistry of appinitic rocks associated with the Ardara pluton,Ireland. *Contrib.Mineral.Petrol.* **16**,156-71.

* Hall,A. 1982.The Pendennis peralkaline minette.*Mineral.Mag.***45**,257-266.

Hall,A. 1988. Crustal contamination of minette magmas: evidence from their ammonium contents. *N.Jb.Mineral.Mh.*, 137-143.

#* Hall,A.L. 1937.Analyses of rocks, minerals, coals, soils and waters from southern Africa. *Mem.Geol. Surv. S.Afr.* **32**.

* Hallberg, J.A. 1985. *Geology and Mineral Deposits of the Leonora–Laverton area, northeastern Yilgarn Block, Western Australia.* Hesperian Press, Perth, 140pp.

Halliday,A.N., Aftalion,M., Parsons,M., Dickin,A.P. & Johnson,M.R.W. 1987. Syn-orogenic alkaline magmatism and its relationship to the Moine thrust zone and the thermal state of the lithosphere in NW Scotland. *J.Geol.Soc.Lond.* **144**, 611-618.

Halls, H.C. & Fahrig, W.F. 1987. Mafic dyke swarms. *Geol. Assoc. Can. Spec.Pap.***34**,503pp.

Hamidullah,S. & Bowes,D.R. 1987. Petrogenesis of the the appinite suite, Appin District, western Scotland. *Acta Univ.Carolinae Geol.* **4**, 295-326.

Hamilton,R. & Rock,N.M.S. 1990. Geochemistry, mineralogy and petrology of a new find of ultramafic lamprophyres from Bulljah Pool, Yilgarn craton, Western Australia. *Lithos* **24** (in press).

Hanley,G.A. & Bradshaw,P.M.D. 1986.The Porgera gold deposit, Papua New Guinea. In: *Proceedings Gold '86 symposium.* Gold '86, Toronto, 416-424.

Hansen,K. 1980.Lamprophyres and carbonatitic lamprophyres related to rifting in the Labrador sea.*Lithos* **13**,145-152.

Hansen,K. 1981.Systematic Sr-isotopic studies in alkaline rocks from west Greenland.*Lithos* **14**,183-188.

Hansen,K. & Larsen,O. 1974.K-Ar determinations on Mesozoic lamprophyre dykes near Ravns Storø, Fiskenaesset region. *Bull.Grønlands Geol.Unders.***66**,9-11.

Hanson,G. 1929. Bear River and Stewart map areas. *Mem.Geol.Surv.Can.* **159**, 84pp.

Hanson,G. 1935. Portland Canal area, British Columbia. *Mem.Geol.Surv.Can.* **174,** 179pp.

Harcourt, G.A. 1938. The southwestern part of the Schreiber area. *Ontario Dept. Mines Rep.* **47(9)**, 1-18.

Harker,A.1892.The lamprophyres of the north of England.*Geol.Mag.***9**,199-206.

Harker,A. 1912.Lamprophyre dykes in Long Sleddale,Westmorland.*Naturalist* 266-268.

* Harkin,D.A. 1960.The Rungwe volcanics at the northern end of Lake Nyasa.*Mem.Geol.Surv.Tanganyika* **2**,172pp.

Harper,C.T. 1984.Reconnaissance bedrock geology: Nevins-Forsyth area.*Geol.Surv.Saskatchewan Misc. Rep.* **83-4** *(summ. Invest 1983)*, 5-15.

Harper,C.T. 1986.Geology of the Nevins-Forsyth Lakes area.*Geol.Surv. Saskatchewan Open File Rep.***86-4.**

* Harper,L.F. 1915.The southern Coalfield of New South Wales.*Mem.Geol.Surv.NSW* **7**,410pp.

Harpum,J.R. 1959.A camptonite from Lemagrut.*Rec.Geol.Surv.Tanganyika*,**7**, p.93.

Harris, J.F. 1981. Summary of the geology of Tanganyika. Part IV. *Mem. Geol. Surv. Tanganyika* **1**, 143pp.

* Harry,W.T. 1952.An unusual appinitic sill near Killin,Perthshire.*Geol.Mag.***89**,41-48.

Hartley,J. & Leedal,G.P. 1951.A monchiquite vent,Stob a'Ghrianain,Inverness-shire. *Geol.Mag.***88**,140-144.

* Harvey,M. & Joplin,G.A. 1941.A note on some leucite-bearing rocks from N.S.W. with special reference to an ultrabasic occurrence at Murrumburrah. *J.Proc.R.Soc.NSW* **74**,419-442.

† Harvie,R. 1910.On the origin and relations of the Palaeozoic breccia of the vicinity of Montreal. *Tr.R.Soc.Can.Ser.*3, **iv**,249, p. 99.

Hasan, S. 1971. Permian igneous dikes in the southern part of the Baerum Cauldron. *Norsk.Videsnk. Akad.Oslo Skr. Mat.Naturv.Kl.* **XXIII**.

@ Hasan,Zia-Ul. 1969.Petrology of the soda-minette dykes from Haoya, Langesundsfjord, S.Norway. *Norsk.Geol. Tidsskr.***49**,159-170.

Haskell,J.R.,Kennett,S.P.,Prebble,W.M.,Smyth,G. & Willis,I.A.G. 1965.Geology of the middle and Lower Taylor valley of S.Victoria Land.*Tr.R.Soc.N.Z.Geol.Ser.***2**,169-186.

* Haslam,H.W. 1970.Appinite xenoliths and associated rocks from the Ben Nevis igneous complex. *Geol.Mag.* **107**,341-356.

Hatch,F.H.,Wells,A.K. & Wells,M.K. 1972.*Petrology of the Igneous Rocks.* 13th Edn. Murby,London, 551pp.

Haughton,S.H. & Frommurze,H.F. 1930.Notes on the occurrence of argentiferous galena near Aiais,SW Africa. *Tr.Geol.Soc.S.Afr.* **32**,89-102.

Hearn,B.C. 1968.Diatremes with kimberlitic affinities in north central Montana.*Science* **159**,622-625.

* Hearn,B.C. 1989.Alkalic ultramafic magmas in north-central Montana,USA: genetic connections of alnöite, kimberlite and carbonatite. *Spec. Publ. Geol. Soc. Aust.* **14**, 109-119.

Hearn,B.C. & McGee,E.S. 1983.Garnets in Montana diatremes: a key to prospecting for kimberlites. *Bull.U.S.G.S.***1604**.

* Heflik,W., Muszynski,M. & Pieczka, A. 1985. [Lamprophyres from the vicinity of Zawiercie]. In Polish, Engl. summ. *Kwart.Geol.* **29**, 529-550.

Hegemann,F. 1932. Tektonik und Entstehung dioritahnlicher gesteine im nordwestlichen Bayerischen Wald. *N.Jb.Mineral.Beilage Band* **65**, Abt. A, 233-284.

Heinrich,E.W. 1966.*Geology of Carbonatites*.Rand McNally,Chicago, 555pp.

Heinrich,E.W. & Dahlem,D.H. 1967.Carbonatite and alkalic rocks of the Arkansas River area,Fremont City,Colorado: 4. The Pinon Peak breccia pipes.*Am. Mineral.* **52**,817-831.

Heinrich,E.W. & Dahlem,D.H. 1969.Dikes of the McClure Mountain-Iron Mountain alkalic complex, Fremont County, Colorado, U.S.A. *Bull.Volc.* **33**,960-978.

Helmstaedt,H. & Doig,R. 1975.Eclogite nodules from the kimberlite pipes of the Colorado Pleatau: samples of subducted Fransiscan-type oceanic lithosphere.*Phys.Chem.Earth* **9**,95-112.

Helmstaedt,H. & Schulze,D.J. 1979.Garnet clinopyroxenite-chlorite eclogite transition in a xenolith from Moses Rock: further evidence for metamorphosed ophiolites under the Colorado Plateau. In: Boyd & Meyer(1979b), qv, 357-365.

Helwig,J.,Aronson,J. & Day,D.S. 1974.A late Jurassic mafic pluton in Newfoundland.*Can.J.Earth Sci.* **11(9)**,1314-1319.

† Hemmann,M. & Watznauer,A. 1964.Zur Altersstellung der Campto-Spessartite der Lausitz.*Geologie* **13**,482-483.

Henderson,J. 1917. The geology and mineral resources of the Reefton subdivision. *Bull.N.Z.Geol.Surv.* **18**, 232pp.

Hentschel,H. 1934. Über zusammengesetzte Gänge des Meißner Granits im Großenhainer angzug und ihre Metamorphose. *Ber.Math-phys.Kl.Sächs.Akad.Wiss.Leipzig* **86**,215-240.

* Hernández-Pacheco,A. 1965.Una richterita potásica de rocas volcánicas alcalinas,Sierra de las Cabras (Albacete). *Estud.Geol.***20**,265-270.

†Herriot, A. 1977. The pre-Tertiary dykes of Catacol, Arran.*Trans. Leeds Geol. Assoc.* **9(7)**,10p.

* Heyl,G.R. 1936.Geology and mineral resources of the Bay of Exploits area.*Bull.Nfdld. Dept.Nat.Res.***3**, 66pp.

Hibsch,E. 1925.Über die camptonitische Fazies basaltischer Gesteine als Spaltungsvorgang basaltischer Schmelzflüsse. *Tschermaks Mineral.Petrogr.Mitt.***38**,262-267.

Hieke,O. 1945. I giacimenti di contatto del Monte Costone.*Mem.Ist.Geol.Mineral.Univ.Padova* **15**,46pp.

Hill,J.B. & Kynaston,H. 1900. On kentallenite and its relations to other igneous rocks in Argyllshire. *Q.J. Geol.Soc.Lond.* **56**, 531-558.

Hill, J.B.,Peach,B.N. & Clough, C.T. 1905. The geology of mid-Argyll (sheet 37). *Mem.Geol.Surv.G.B.*

Hills,E.S. 1952.The Woods Point dyke-swarm. In: *Sir Douglas Mawson Anniversary Volume* (eds. Glaessner, M.F. & Rudd,E.A.), 87-100. Univ.Adelaide,Adelaide.

Hinde, S. 1988. Central Deborah. In: Jones (1988), qv, 28-31.

Hinxman,L.W., Carruthers,R.G. & Macgregor,M. 1923. The geology of Corrour and the Moor of Rannoch (sheet 54). *Mem.Geol.Surv. G.B.*

Hinxman,L.W. & Wilson,J.S.E. 1902. The geology of lower Strathspey (sheet 85). *Mem.Geol.Surv.G.B.*

Hirayama, K. 1957. [The so-called spessartite and granophyres in the neighbourhood of Shiratorihonmachi]. In Japanese. *Rep.Geol.Surv.Japan* **140**.

* Hjelle,A. 1962.Ultrabasik biotitførende lamprofyrgang ved Skabberud,Stange. In Norwegian. *Norges.Geol.Unders.* **215**,30-36.

Hjelmquist,S. 1930. Kullait van Dalby. *Geol. Foren. Förh.* **52**,247-268.

Hjelmquist,S. 1939. Some post-Silurian dykes in Scania and problems suggested by them. *Sver.Geol Unders. Ser.C*, 430, Årbok **33**(10).

* Hochman,M.B. 1980.Geochemical investigations in the Mt.Bundey area,NT. In: *Uranium in the Pine Creek Geosyncline* (eds. Ferguson,J. & Goleby,A.B.),335-342. Int. Atomic Energy Authority,Vienna.

Hodder,R.W. 1961. Alkaline rocks and niobium deposits near Nemegos, Ontario. *Bull.Geol.Surv.Can.* **70**,75pp.

Hodgson, C.J. & Troop, D.G. 1988. A new computer-aided methodology for area selection in gold exploration: a case study from the Abitibi greenstone belt. *Econ. Geol.* **83**, 952-977.

Holden,P.,Halliday,A.N., Stephens,W.E. & Henney,P. 1990. Mass transfer between mafic enclaves and silicic magmas with particular reference to mobile versus immobile elements (submitted).

* Holgate,N. 1950.The Glen Banvie igneous complex of Perthshire.*Q.J.Geol.Soc.Lond.***106**,433-460.

Holl,P.K., Drach, V.von, Muller-Sohnius,D. & Kohler,H. 1989. Caledonian ages in Variscan rocks: Rb-Sr and Sm-Nd isotopic variations in dioritic intrusives from the North Western Bohemian maassif, West Germany. *Tectonophysics* **157**,179-194.

* Holmes,A. 1915.A contribution to the petrology of N.W.Angola.*Geol.Mag.***2**,322-328.

Holmes,A. 1945. Leucitized granite xenoliths from the potash-rich lavas of Bunyaruguru, SW Uganda. *Am.J.Sci.***243A**, 313-332.

Holmes,A. 1950.Petrogenesis of katungite and its associates.*Am.Mineral.***35**,772-792.

Holub,F.V. 1977. Petrology of inclusions as a key to petrogenesis of the durbachitic rocks from Czechoslovakia. *Tsch.Mineral.Petrogr. Mitt.*, **24**, 133-150.

Holub,F.V. 1978. Príspêvek ke geochemii durbachitickych hornin. *Acta Univ.Carolinae Geol.* **3-4**, 351-364.

Hopgood,A.M. & Bowes,D.R. 1987. Structural succession and tectonic history of the gneiss-amphibolite--granulite-mantle peridotite association near the eastern margin of the Moldanubian Zone, Central European Hercynides. *Acta Univ.Carolinae Geol.* **1**, 51-88.

Hopkins, P.E. 1924. Ontario gold deposits. *Ont.Dept.Mines Geol. Rep.* **30**, 73 pp.

Hopson,C.A. 1957.Origin of lamprophyres associated with the Chelan batholith, Washington. *Bull.Geol.Soc.Am.***68**, p.1747 (abstr.)

Horne,J. 1923.The geology of the lower Findhorn and lower Strath Nairn (sheets 84/94). *Mem.Geol.Surv.G.B.*

Horne,J. & Hinxman L.W. 1914. The geology of the country around Beauly and Inverness (sheet 83). *Mem.Geol.Surv. G.B.*

Horne,R.R. & Thompson,M.R.A. 1967.Post-Aptian camptonite dykes in SE Alexander Island. *Bull.Br. Antarc.Surv.* **14**,15-24.

* Horváth,I. & Odor,L. 1984.Alkaline ultrabasic rocks and associated silicocarbonatites in the NE part of the Transdanubian Mountains (Hungary). *Mineral.Slovaca (Bratislava)***16**,115-119.

Houston,J.R.,Bates,R.G.,Velikanje,R.S. & Wedow,H. 1958.Reconnaissance for radioactive deposits in southeastern Alaska,1952. *Bull.U.S.G.S.* **1058-A**,30pp.

* Hovorka,D. 1967.[Porphyrites and lamprophyres of the crystalline complex of Tatraveporid]. In Slovak,Engl.summ. *Sbor.Geol.Viek.Zk.Ceskosl.***8**,51-78.

‡ Hovorka,D.,Chovan,M. & Michalek,J. 1982.[Olivine kersantite in granodioritic enclosing rocks at Dubrava Mountains, Lower Tatras]. In Slovak.*Mineral.Slovaca (Bratislava)***14(1)**,85-90.

* Howard, W.V. 1922. Some outliers of the Monteregian Hills. *Tr.R.Soc.Canada Ser.3*, **16**, 47-95.

Howard,W.V. 1926. Devonian volcanic rocks near Dalhousie, New Brunswick. *Bull.Geol.Soc.Am.***37**, 475-496.

Howell, F.H. & Molloy,J.S. 1960. Geology of the Braden orebody, Chile. *Econ.Geol.* **55**, 863-905.

* Howie,R.A. 1963.Kaersutite from the Lugar sill and from alnöite breccia,Alnö Island. *Mineral.Mag.* **33**,718-719.

Hubbard, F. & Branigan, N. 1987. Late Svecofennian magmatism and tectonism, Åland, Southwest Finland. *Precamb.Res.* **35**,241-256.

* Hubberten,H-W,Katz-Lehnert,K.& Keller,J. 1988. Carbon & oxygen isotope investigation in carbonatites and related rocks from the Kaiserstuhl,Germany.*Chem.Geol.***70**,257-274.

Hudson,S.N. 1937.The volcanic rocks and minor intrusions of the Cross Fell inlier. *Q.J.Geol.Soc.Lond.* **93**,368-404.

Hughes,C.J. 1982.*Igneous petrology*.Developments in Petrology, **7**. Elsevier,Amsterdam, 551pp.

* Huhma, A. 1981. Youngest Precambrian dyke rocks in north Karelia, East Finland. *Bull. Geol. Soc. Finland* **53(2)**, 67-82.

Hunt,G.M. & Engelhardt,R.T.,1973. Carbon and oxygen isotope ratios of carbonate from ultramafic dikes of Western Kentucky. *Geol.Soc.Am.Abstr.Progr.* **5**, p.324.

* Hunter,D.R. & Reid,D.L. 1987. Mafic dyke swarms in southern Africa. In: Halls & Fahrig (1987), qv, 445-456.

Hunter,M. & Rosenbusch,H. 1890. Ueber Monchiquit, ein Camptonitisches Ganggestein aus der Gefolgschaft der Elaeolithsyenit. *Tschermaks Mineral.Petrogr.Mitt.* **11**, 445-466.

* Hunter, R.H. & Upton, B.G.J. 1987. The British Isles — a Palaeozoic mantle sample. In: Nixon (1987), qv, 107-118.

‡ Hurrle,H. 1976.Ocelle- und Mandelbildung der ultrabasischen Basalte im kalisalzlager Buggingen und im kristallin des Schwarzwaldes.*J-H.Geol.Landesant.Baden-Wurttemberg* **18**,19-37.

Hurst, M.E. 1936. Gold deposits in the vicinity of Red Lake. *Ont.Dept.Mines Geol. Rep.* **44(6)**, 1-52.

Hutchison,C.S. 1964.A gabbro-granodiorite association in Singapore Island.*J.Geol.Soc.Lond.***120**,283-298.

* Hutton,C.O. 1940. Limburgite from Nevis Bluff, Kawarau Gorge, central Otago. *Tr.R.Soc.NZ* **73**, 58-67.

Hyndman,D.W. 1972.*Petrology of Igneous and Metamorphic Rocks*. McGraw-Hill,New York, 533pp.

‡ Ignatev,N.A. 1936.[Kersantite dykes from the Artem shaft,Vlasovsky mine,Donbass]. In Russian, Engl. summ. *Perm.Univ. Sci.Mem.* **2(3)**,237-261.

IMA (1978) Nomenclature of amphiboles. *Mineral.Mag.* **42**, 533-563; *Am.Mineral.* **63**, 1023-1052.

IMA (1988) Nomenclature of pyroxenes. *Mineral.Mag.* **52**, 535-550; *Am.Mineral.* **73**, 1123-1133.

* Issler,R.S. & Roisenberg,A. 1972.Lamprófiro das proximidades do Cerro Tupana,quadrícula de Arroio São Sape,Rio Grande do Sul. *Pesquisas* **1**,43-9 and *22nd Geol.Congr.Brasil,resumo* 7(1968).

#* Ito,M. 1986.Kimberlites and their ultramafic xenoliths from western Kenya.*Tschermaks Mineral. Petrogr.Mitt.* **34**,193-216.

? Ivanushko,O.S. 1973.[Structural position and genesis of lamprophyre dykes in the Obitochnaya River basin]. In Russian,Engl.summ.*Dopov.Akad.Nauk.Ukr.SSR.Ser.B,* **10**,879-885.

? Ivanyuk, B.O. 1986. Conjugation of gold and tin-tungsten mineralization in the northeastern USSR. *Dokl. Acad. Sci. USSR, Earth Sci. Sect.* **291**, 62-65.

Jackson, B. 1984.Sapphire from Loch Roag, Isle of Lewis, Scotland. *J. Gemnol.* **19**,336-432.

* Jaffe,H.F. 1952.Amygdalar camptonite dykes from Mount Jo,Mount Marcy Quadrangle,Essex County, New York.*Am.Mineral.***38**,1065-1077.

Jahn,B. 1973.A petrogenetic model for the igneous complex in the Spanish Peaks region. *Contrib.Mineral.Petrol.* **41**,241-58.

* Jahn,B.,Sun,S.-S.,& Nesbitt,R.W. 1979.REE distribution and petrogenesis of the Spanish Peaks igneous complex,Colorado. *Contrib.Mineral.Petrol.***70**,281-98.

Jahns,R.M. 1938. Analcite-bearing intrusives from South Park,Colorado.*A.J.Sci.,5th ser.***36**,8-26.

James,E.W. 1989. Southern extension of the Independence dike swarm of eastern California. *Geology* **17**,587-590.

* Janse,A.J.A. 1969. Gross Brukkaros: a probable carbonatite volcanic in the Nama plateau of south West Africa. *Bull.Geol.Soc.Am.* **80**, 573-586.

Janse, A.J.A. 1971. Monticellite-bearing porphyritic peridotite from Gross Brukkaros, South West Africa. *Tr.Geol.Soc.S.Afr.* **74**, 45-55.

* Janse, A.J.A., Downie, I.F., Reed, L.E. & Sinclair, I.G.L. 1989. Alkaline intrusions in the Hudson Bay lowlands, Canada: exploration methods, petrology and geochemistry. *Spec. Publ. Geol. Soc. Aust.* **14**, 1192-1203.

Janse,A.J.A. & Sheahan,P.A. 1987. Bibliochrony of igneous rocks of Arkansas with particular emphasis on diamonds. *Spec.Pap.Geol.Soc.Am.* **215**, 249-292.

Jaques,A.L. & Foley,S.F. 1985.The origin of Al-rich spinel inclusions in leucite from the leucite lamproites of western Australia.*Am.Mineral.***70**,1143-1150.

* Jaques,A.L.& Perkin,D.J. 1984.A mica pyroxene ilmenite megacryst-bearing lamprophye from Mt.Woolooma, northeastern New South Wales.*Bur.Aust.Bu.Miner.Res.J.Geol.Geophys.***9**,33-41.

* Jaques,A.L.,Lewis,J.D.,Smith,C.B.,Gregory,G.P.,Ferguson,J.,Chappell,B.W. & McCulloch,M.T. 1984a. The diamond-bearing ultrapotassic (lamproitic) rocks of the W.Kimberley region,western Australia. In: Kornprobst (1984a), qv, 225-234.

Jaques,A.L.,Webb,A.W.,Fanning,C.M.,Black,L.P.,Pidgeon,R.T.,Ferguson,J.,Smith,C.B. & Gregory,G.P. 1984b. The age of the diamond-bearing pipes and associated leucite lamproites of the West Kimberley region,western Australia. *J.Aust.Bur.Miner.Resour.Geol.Geophys.***9**,1-7.

Jaques,A.L.,Creaser,R.A.,Ferguson,J. & Smith,C.B. 1985.A review of the alkaline rocks of Australia. *Tr.Geol.Soc.S.Afr.***88**,311-334.

* Jaques,A.L.,Lewis,J.D. & Smith,C.B. 1986.The kimberlites and lamproites of western Australia. *Bull.Geol.Surv.W.Aust.* **132**, 268pp.

* Jaques, A.L., Haggerty, S.E., Lucas, H. & Boxer, G.L. 1989a. Mineralogy and petrology of the Argyle (AK1) lamproite pipe, Western Australia. *Spec. Publ. Geol. Soc. Aust.* **14**, 153-169.

* Jaques,A.L.,Kerr,I.D.,Lucas,H.,Sun,S-S. & Chappell,B.W. 1989b.Mineralogy and petrology of picritic monchiquites from Wandagee,Carnarvon Basin,Western Australia. *Spec. Publ. Geol. Soc. Aust.* **14**, 120-139.

* Jaques, A.L., Sun, S-S. & Chappell, B.W. 1989c. Geochemistry of the Argyle (AK1) lamproite pipe. *Spec. Publ. Geol. Soc. Aust.* **14**, 170-188.

Jehu,T.J. 1922.The Archaean and Torridonian formations and the later intrusive igneous rocks of Iona.*Tr.R.Soc. Edinburgh* **53**,165-187.

Jehu,T.J. & Craig,R.M. 1923-7.Geology of the Outer Hebrides.*Tr.R.Soc.Edinburgh* **53**,419-441,615-641; **54**,467-489; **55**,457-488.

Jelinek, E., Soucek, J., Turdy, J. & Ulrych, J. 1989. Geochemistry and petrology of alkaline dyke rocks of the Roztoky volcanic centre, Ceske Stredehori Mountains, CSSR. *Chem.Erde* **49**, 201-217.

* Jennings,D.J. & Sutherland,F.L. 1969.Geology of the Cape Portland area with special reference to the Mesozoic appinitic rocks.*Tech.Rep.Dept.Mines Tasmania* **13**,45-82.

Jensen,H.I., 1907. The geology of the Nandewar Mountains. *Proc.Linnaean Soc.N.S.W.* **32**, 842-913.

Jensen, L.S. 1978. Archean komatiitic, tholeiitic, calc-alkalic and alkalic volcanic sequences in the Kirkland Lake area. In: *Toronto '78 Field Trip Guidebook* (eds. A.L.Currie & W.O.Mackasey). *Geol.Assoc.Can. Guidebook,* 237-259.

* Jérémine,E. 1927.Dolérites et lamprophyres de La Hague.*C.R.Congr. Soc.Sav.*, 65-68.

* Jérémine, E. 1954. Étude pétrographique des roches éruptives et métamorphiques du massif de Bou Agrao. *Notes Serv. Geol. Maroc.* Tome II, **74**, 119-147.

† Jérémine,E. & Fallot,P. 1929.Sur la présence d'une varieté de jumillite aux environs de Calasparra (Murcia). *C.R.Acad.Sci.Paris* **18**,800.

* Jérémine,E. & Sandréa,A. 1955.Sur la composition chimique d'espichellite et sa position dans la famille des lamprophyres.*C.R.Acad.Sci.Paris* **241**,891-893.

Jermanski,J.,Depciuch,T. & Pendias,H. 1974.[Igneous vein rocks of the Chocianow area in the fore-Sudetic block]. In Polish,Engl.summ.*Kwart.Geol.***18(4)**,690-698.

* Johannsen,A. 1938.*A Descriptive Petrography of the Igneous Rocks,vol.IV.*Chicago Univ.Press.

@ Johnson,M.R.W. & Dalziel,I.W.D. 1966.Metamorphosed lamprophyres and the late thermal history of the Moines. *Geol.Mag.***103**,240-249.

Johnson,R.B. 1961.Patterns and origins of radial dike-swarms associated with the W Spanish Peaks and Dike Mountain,Colorado. *Bull.Geol.Soc.Am.***72**,579-590.

* Johnson,R.B. 1964.Walsen composite dyke near Walsenburg,Colorado.*U.S.G.S. Prof.Pap.***501-B**,69-73.

Johnson,R.B. 1968.Geology of the igneous rocks of the Spanish Peaks region,Colorado.*U.S.G.S. Prof.Pap.***594-C**,1-47.

Johnson,R.W. & Milton,C. 1955. Dike rocks of Central-Western Virginia. *Bull.Geol.Soc.Am.* **66**, p. 1959.

Johnstone,G.S. 1989. *British Regional Geology: The Northern Highlands.* 4th Edn. HMSO for Br.Geol.Surv., London.

Johnstone G.S. 1990. *British Regional Geology: The Grampian Highlands.* 3rd Edn. HMSO for Br.Geol.Surv., London.

Johnstone,G.S. & Wright,J.B. 1951.The camptonite-monchiquite suite of Loch Eil.*Geol.Mag.***88**,148.

* Johnstone,M.H. 1952.The geology of the Hamersley Siding area. *J.R. Soc.W.Aust.* **36**, 45-75.

* Jones, A.P. & Smith, J.V. 1983.Petrological significance of mineral chemistry in the Agathla Peak and The Thumb minette,Navajo volcanic field.*J.Geol.***91**,643-656.

* Jones, A.P. & Smith, J.V. 1985. Phlogopite and associated minerals from Permian minettes in Devon, south England. *Bull. Geol. Soc. Finland* **57**, 89-102.

Jones, D.G. 1988. Central Victorian gold deposits. *Bicentennial Gold '88 Excursion Guidebook, Univ. W. Aust. Geol. Dept. & Extension Publ.* **13**, 77pp.

Joplin,G.A. 1957.Basic and ultrabasic rocks near Happy Jacks and Tumut Pond in the Snowy Mountains of NSW. *J.Proc.R.Soc.NSW.* **91**,120-141.

Joplin,G.A. 1959.On the origin and occurrence of basic bodies associated with discordant bathyliths. *Geol.Mag.* **96**,361-373.

* Joplin,G.A. 1965.Chemical analyses of Australian rocks. Part i. *Bull.Aust.Bur.Mineral.Resour.Geol. Geophys.* **65**.

Joplin,G.A. 1966.On lamprophyres.*J.Proc.R.Soc.NSW* **99**,37-44.

* Joplin,G.A. 1968.The shoshonite association — a review.*J.Geol.Soc.Aust.***15**,275-294.

Joplin,G.A. 1971. *A Petrography of Australian Igneous Rocks.* Angus & Robertson,Sydney.

* Joplin,G.A. 1973.Chemical analyses of Australian rocks,Part iii: igneous and metamorphic, supplement, 1961-1969. *Bull.Aust.Bur.Mineral.Resour.Geol.Geophys.***146.**

Judd,J.W. 1885. On the Tertiary and older peridotites of Scotland. *Q.J.Geol.Soc.Lond.***41**, 354-418.

Junner,N.R. 1920. The geology and ore-deposits of the Walhalla–Wood's Point auriferous belt. *Proc.Aust.Inst. Ming. Metall.* **39**, 127-258.

† Juskova,G.E. & Mal'kov,B.A. 1969.[Fluorine in the lamprophyres and metagabbros of northern Timan]. In Russian. *Refrativny Zurnal,Ser.Geol.Moskow* **3**,10.

Just,G., & Kramer,W. 1989. Syenitoide und uhre Beziehungen zu lamprophyrische Bildungen in der Fichtelgebirgisch -Erzgeburgischen Antiklinalzone. *Z.Geol.Wiss.Berlin* (in press).

* Kaiser,E. 1923.Kaoliniserung als Verwitterungsworgange in des narmiburuiste Sudwestafrikas. *Zeit.Krist.***58**,125-146.

* Kaitaro,S. 1953.Geologic structure of late Pre-Cambrian intrusives in the Ava area,Åland Is. *Bull.Comm.Geol. Finlande* **162**.

Kaitaro,S. 1956.On central complexes with radial lamprophyric dykes.*Bull.Comm.Geol.Finlande* **172**, 55-65.

Kaminsky, F.V. 1976. [Alkali basaltoid breccia of Onoga peninsula.] In Russian. *Isv. Akad. Nauk. SSSR Earth Sci. Sect.* **7**, 50-59.

* Kaminskiy, F.V., Frantsesson, Ye. V. & Khvostova, V.P. 1974. First information on platinum metals (Pt, Pd, Rh, Ir, Ru, Os) in kimberlitic rocks. *Dokl. Acad. Sci. USSR Earth-Sci.Sect.* **219**, 190-193.

† Kanchaveli,A.L. 1964.[Lamprophyre dykes associated with the Tchiatour quartz porphyry]. *Soobshch. Akad.Nauk.Gruz.SSR* **36(1)**,115-118.

Kaplan,G.,Faure,D.,Elloy,R. & Heilammer,R. 1967.Contribution a l'étude de l'origine des lamproites. *Bull.Centre Rech.PAV-SNPA***1(1)**,153-159.

* Kapp,H. 1960.Zur Petrologie der Subvulkane zwischen Mester Vig und Antarctic Havn.*Meddr.Grønland* **153(2)**,203pp.

Kapustin,Y.L. 1974.[Dyke-series of picrite-alnöite in ultrabasic-alkalic rock massifs]. In Russian.*Soviet.Geol.***8**,43-59.

Kapustin, Y.L. 1981. Damkjernites — dike equivalents of carbonatites. *Int. Geol. Rev.* **23**, 1326-1334.

* Karamata,S. 1955.Petroloska studija magmatskih i kontakno-metamorfnih stena Boranje. *Glasnik Prir.Mus.s.z.Ser.A.knj* **6**, 1-130.

* Kardymowicz,J. 1962.[Petrological studies of lamprophyres in the Swiety Krzyz Mountains(Central Poland)]. In Polish,Engl.summ. *Kwart.Geol.***6**,271-311.

Kardymowicz,J. 1963.[Biotite of the Swiety Krzyz Mountain lamprophres]. In Polish,Engl.summ. *Pr.museum zierni,Warsaw,***7**,43-68.

* Kawano,G. & Kishira,K. 1940. [Dyke swarms of spessartite and granophyre through granite at Shiratori, Kagawa Prefecture]. In Japanese. *J.Jap.Assoc.Min.Petrol.Econ.Geol.***23**, 165-177.

* Kay,R.W. & Gast,P.W. 1973.The rare-earth content of alkali-rich basalts.*J.Geol.***81**,653-682.

Kay, S.M., Snedden, W.T., Foster, B.P. & Kay, R.W. 1983. Upper mantle and crustal fragments in the Ithaca kimberlite. *J. Geol.* **91**, 277-290.

* Keiling,K. 1961.Die minette im Plauenschen Grunde bei Dresden.*Jb.Staatl.Mus.Mineral.Geol.*, 75-110.

Keller,J. 1989. Extrusive carbonatites and their significance. In: Bell (1989), qv, 70-88.

Kelterburn,P. 1923. Geologische und petrographische untersuchungen in Malcantone (Tessin).*Verh.Natf. Ges.Basel.* **34**, 128-232.

Kemp,J.F.1897.The Leucite Hills of Wyoming.*Bull.Geol.Soc.Am.***8**,169-182.

* Kemp.J.F. & Billingsley,P. 1921.Sweetgrass Hills,Montana.*Bull.Geol.Soc.Am.***32**,437-478.

Kemp,J.F. & Knight,W.C. 1903.Leucite Hills of Wyoming.*Bull.Geol.Soc.Am.***14**,305-336.

Kemp,J.F. & Marsters,V.F.1893. The trap dykes of the Lake Champlain region. *Bull.U.S.Geol.Surv.* **107**, 11-62.

Kempster,R.M.F., Clarke,D.B., Reynolds,P.H. & Chatterjee,A.K. 1989. Late Devonian lamprophyric dykes in the Meguma zone of Nova Scotia. *Can.J.Earth Sci.***26**,611-613.

Kennan,P.S. 1979. Plutonic rocks of the Irish Caledonides.*Spec.Publ.Geol.Soc.Lond.* **8**, 705-711.

* Kennedy,W.Q. & Read,H.H. 1936.The differentiated dyke at Newmains,Dumfries-shire and its contact and contamination phenomena. *Q.J.Geol.Soc.Lond.***92**,116-144.

* Kepezhinskas,V.V.,Kepezhinskas,P.K. & Usova,L.A. 1984.Origin of camptonites of the Agardag complex of the Sangilan Highlands of Tuva.*Soviet Geol.Geophys.***25(4)**,51-56.

Kesse,G.O. 1971.Contribution to the geology of the area around Ober-Heisling.*Jb.Geol.Bundersrepub.Austria Sonderband* **17**,121-138.

@ Key,C.H. 1977.Origin of appinitic pockets in the diorites of Jersey,Channel Islands.*Mineral.Mag.* **41**,183-192.

* Kicula,K. & Wieser,T. 1970.[PreCambrian sediments and lamprophyres in the borehole Opatkowice-1]. In Polish,Engl.summ. *Pol.Tow.Geol.Rocz.(Ann.Soc.Geol.Pologne)* **40(1)**,111-129.

Kidwell,A.L. 1947. Post Devonian igneous activity in southwestern Missouri. *Missouri Geol.Surv.Water Res.Rept.Inv.* **4**,77pp.

King,A.F. & McMillan,N.J. 1975.A mid-Mesozoic breccia from the coast of Labrador.*Can.J.Earth Sci.* **12**,44-51.

* King,B.C. 1937.The minor intrusions of Kirkcudbrightshire.*Proc.Geol.Assoc.***48**,282-306.

* Kirchner,J.G. 1979.Petrographic significance of a carbonate-rich lamprophyre from Squaw Creek,Northern Black Hills,S.Dakota. *Am.Mineral.***64**,986-992.

Kirkley, M.B., Stuart Smith, H. & Gurney, J.J. 1989. Kimberlite carbonates — a carbon and oxygen stable isotope study. *Spec. Publ. Geol. Soc. Aust.* **14**, 264-281.

Klemm,W. 1923.Über die Minetten,Vogesite und Kersantite des Odenwaldes. *Notizbl.Hess.Geol.Landesamt. Folge* **6**,5-27.

Klemm,W. 1928.Über einen gemischten Augitminette-Kersantitgang und der Steinmauer bei Heppenheim. *Notizbl.Ver.Erdkde Darmstadt* **11**,7-24.

† Klimov, G.K. 1986. [Difference between camptonites and lamprophyric essexites of central Tadzhikstan).*Dokl. Akad. Sci. Tadzhikskoy SSR* **29(5)**,302-306.

Klitchenko,M.A. & Suprichov,V.A. 1974.[Ferrierite,gismondine and analcime from monchiquite and camptonite of the southern Donbass]. In Ukranian,Engl.summ.*Dopov.Akad.Nauk. RSR,ser.B,* **8**, 687-90.

‡Knezevic,V. 1956. O pojavama minete na planini Rudniku. *Vesnik zav.Geol.istr.NRS knj XII*, 195-199.

* Knezevic,V. 1976.[Lamprophyres from Cer Mountain and other localities in Serbia and Bosnia in

systematics of igneous rocks]. In Serbian,Engl.summ.*Zapisnici srpsko geoloskodrustvo*. 1975-6,41-50.
Knill,D.C. 1969.The Permian igneous rocks of Devon.*Bull.Geol.Surv.G.B.* **29**,115-38.
Knill,D.C. 1982.Permian volcanism in SW England. In: Sutherland(1982), qv, 329-332.
* Knopf,A. 1936.Igneous geology of the Spanish Peaks region.*Bull.Geol.Soc.Am.***47**,1728-1784.
* Knopf,D. 1970.*Les kimberlites et les roches apparentées de Côte d'Ivoire*.Sodemi-Abidjan.
Koch,I. 1938.Die Kuselite des Saar-Nahe-Gebietes.*N.Jb.Beilage Band* **73,Abt.A,H3**,419-494.
Kochhar, N. 1987. Alkaline magmatism in Delhi Supergroup. *J. Geol. Soc. India* **30**, 537-539.
* Koenig, J.B. 1956. The petrography of certain igneous dikes of Kentucky. *Bull. Geol. Surv. Kentucky Ser. IX*, **21**.
Kogarko, L.N. 1987. Alkaline rocks of the eastern part of the Baltic shield (Kola Peninsula). In: Fitton & Upton (1987), qv, 531-544.
* Koghan,B.S. 1976.[Phlogopite camptonites of southern Primorye]. In Russian. *Isv.Akad.Nauk.SSSR Geol. Ser.* **1**,139-143.
¶ Korn,H. & Martin,H. 1954. The Messum igneous complex in South-West Africa. *Tr.Geol.Soc.S.Afr.* **57**,83-122.
Kornprobst, J. 1984. *Kimberlites and Related Rocks.* (Proc.3rd Int.Kimb.Conf., 2 vols.) Elsevier Developments in Petrology **11 a/b.**
Korobeynikov,A.F. 1989.Behaviour of gold in magmatic and metasomatic processes. *Geochem.Int.*171-179.
† Kowalski,W.M. 1966.Lamprophyres from Zloty Stok.*Bull.Acad.Polon.Sci.Ser.Geol.Geogr.* **14(2)**,87-90.
Kramer,W. 1966.Ein gemischter Quartzporphyr-Lamprophyrgang bei Brand-Erbisdorf südlich Freiburg und seine Einschlüsse. *Geologie* **15**,1033-1041.
* Kramer,W. 1976a.Genese der Lamprophyre im Bereich der Fichtelgebirgisch-Erzgebirgisch Antiklinalzone. *Chem.Erde* **35**,1-49.
Kramer,W. 1976b.Zur tektonischen Stellung und Bedeutung von Lamprophyren und Doleriten im Saxothuringikum-Lugikum. *Nova Acta Leopoldina* **45**,37-43.
Kramer,W. 1976c.Zur Petrologie und metallogenetischen Bedeuutung der Dolerite (Lamprophyre) des Lausitzer Massivs. *Z.Geol.Wiss.(Berlin)* **4**,975-994.
Kramer,W.,1988.Magmengenetische Aspekte der Lithosphären-entwicklung. *Akademie-Verlag, Series in Geol.Sci.* **26**,135pp.
Kramer,W. & Rosler,H.J. 1976.Ein Beitrag zur Genese der Lamprophyre.*Z.Geol.Wiss.Berlin* **4**,667-683.
* Kramer,W. & Seifert,W. 1984.Xenolithe,Lamprophyre under Kruste-Mantel-Beziehungen. *Forsch.Reihe.C (Leipzig)*,**389**,38-49.
Kranck,E. 1928.On turjaite and the ijolite stem of Turja,Kola.*Fennia* **51**,1-104.
Kranck,E. 1939.Bedrock geology of the seaboard region of Newfoundland Labrador. *Bull.Nfld Geol.Surv. 19*, 44pp.
* Kranck,E. 1953.Bedrock geology of the seaboard of Labrador between Domino and Hopedale. *Bull.Geol.Surv.Can.***26.**
* Kresten,P. 1976.Chrome-pyrope from the Alnö complex.*Geol.Fören.Förh.*.**98**,179-180.
Kresten,P. 1980.The Alnö complex;tectonics of dyke emplacement.*Lithos* **13**,153-158.
* Kresten,P.,Ahman,E. & Brunfelt,A.O. 1981.Alkaline ultramafic lamprophyres and associated carbonatite dykes from the Kalix area,northern Sweden.*Geol.Rundsch.***70**,1215-1231.
Kresten,P. & Brunfelt,A.O. 1979.Lamprophyres and carbonatites from the Kalix-Lulea archipelago.*Lithos* **13**,216-217.
* Kresten,P. & Edelman,N. 1975.A boulder of ouachitite from Inderskar,Åland Is.*Bull.Geol.Surv.Finland* **47**,167-169.
Kresten,P. & Nairis,H.J. 1982.Alnö diamonds.*Geol.Fören.Förh.(Stockholm).***104**,210.
Kresten,P. & Persson,L. 1975.Discrete diopside in alnöite from Alnö Island.*Lithos* **8**,187-192.
† Krishnan, M.S. 1925. The petrography of the rocks from the Girnar and Osham hills, Kathiawar. *Rec.Geol.Surv. India* **58**,407-415.
Krist,E. 1967.[Geologic and petrographic characteristics of spessartite-kersantite dike in the lower Tatras.]In Slovak,Germ.summ. *Acta Geol.Geogr.Univ.Comenianae Geol.***12**,63-75.
* Kubo,K. & Satoh,H. 1984.[Petrology of lamprophyre in the Urakawa area,Hokkaido,Japan]. In Japanese, Engl. summ.*J.Geol.Soc.Japan* **90**,717-731.
Kubovics,I., Sólymos, K.G. & Szabó, C. 1985. Petrology and geochemistry of ultramafic xenoliths in mafic rocks of Hungary and Burgenland (Austria). *Geol.Zborn.(Bratislava)*,**36**, 433-450.
Kubovics,I., Szabó,C. & Sólymos,K. 1989. Lamprophyre dykes in NE Transdanubia,Hungary: the Alcsútdoboz-2 borehole (in press).
† Kuchukova,M.S. 1973.[Problems of the age and mineralogy of the lamprophyres of Tchakyl-Kalyan Mountains]. *Uzbek.Geol.Zh.***7(3)**,41-48.

† Kudryavtseva,N.G.,Filatov,Y.I. & Antip'yev,G.A. 1967.[Post-ore lamprophyres of the Kadaya deposit]. In Russian. *Vyssh.Ucheb.Zaved.Isv.Geol.Razved.***5**,24-27.

* Kuehner,S.M.,Edgar,A.D. & Arima,M. 1981.Petrogenesis of the ultrapotassic rocks from the Leucite Hills, Wyoming. *Am.Mineral.***66**,663-677.

† Kumar,J.C.N. 1968.[Spessartite and diorite-porphyry dikes in the Elbrus polymetallic deposit]. In Russian.*Vyssch.Ucheb.Zaved.Isv.Geol.Razved.***3**,161-165.

* Kushmuradov,O.K. 1970.[Notes on the gold content of sedimentary,metamorphic and granitoid rocks of the Northern Nuratau Mountains]. In Russian.*Usbeskskii geologischeskii Zhurnal* **14(1)**,8-11.

† Kuzebnyy,V.S. & Mar'in,A.M. 1973. [Lamprophyre-like rocks in the granitoids of Rudny Altai]. *Tr.Akad.Nauk.Kaz.SSR Inst.Geol.***33**,55-69.

Kynaston,H. & Hill,J.B. 1908. The geology of the country near Oban and Dalmally (sheet 45). *Mem.Geol.Surv.G.B.*

Lacroix,A. 1917. La composition chimique de la vaugnérite et la position de cette roche dans la systématique. *Bull.Soc.Fr.Minéral.Crist.* **40,** 158-162.

* Lacroix,A. 1922.*Minéralogie de Madagascar.*A.Challamel,Paris,3 vols.

* Lacroix,A. 1928.La constitution lithologique des iles volcaniques de la Polynesie Australe. *Mem.Acad. Sci.Paris* **59**,1-80.

* Lacroix,A. 1933a.Contribution à la connaissance de la composition chimique et minéralogique des roches éruptives de l'Indochine. *Bull.Serv.Géol. Indochine* **20(3)**,98-99.

Lacroix,A. 1933b.Les roches éruptives potassiques leucitiques ou non, du Tonkin occidental. *C.R.Acad.Sci. Paris* **197**,625-627.

* Lapadu-Hargues,S. 1959.Note sur la fraidronite de la Lozère. *Bull.Serv.Carte.Géol.Fr.* **57,no.261,** 395-409 (107-121).

Lapointe,P.L. 1979.Paleomagnetism of Notre Dame Bay lamprophyre dykes,Newfoundland,and the opening of the N.Atlantic ocean. *Can.J.Earth Sci.***16(9)**, 1823-1831.

Larouzière,F.D.de,& Bordet,P. 1983.Sur la genèse de certains types de lamproïtes du bassin de Mazarron. *C.R.Acad.Sci.Paris Ser.*II,**296(14)**,1071-1076.

Larsen,E.S.,Hurlbut,C.S.,Buie,B.F. & Burgess,C.H. 1941.Igneous rocks of the Highwood Mountains,part vi. Mineralogy. *Bull.Geol.Soc.Am.***52**,1841-1856.

* Larsen,J.G. 1981.Medium pressure crystallisation of a monchiquite magma—evidence from megacrysts of Drever's block, Ubekendt Eijland,W.Greenland.*Lithos* **14**,241-262.

Larsen,J.G. 1982.Mantle-derived dunite and lherzolite nodules from Ubekendt Ejland,West Greenland Tertiary Province.*Mineral Mag.***46**,329-336.

Larsen,L.M. 1980.Lamprophyric and kimberlitic dykes associated with the Sarfartôq carbonatite complex,SW Greenland. *Rapp.Grønlands Geol.Unders.***100/79**,65-69.

Larsen,L.M.,Rex,D.C. & Secher,K. 1983.The age of carbonatites,kimberlites and lamprophyres from SW Greenland: recurrent alkaline magmatism during 2500 million years.*Lithos* **16**,215-221.

Larsson,W. 1943.Zur Kenntnis der alkalinen ultrabasischen Ganggesteine des Kalixgebietes,Nordschweden. *Sveriges Geol.Unders.***456c.**

* Laughlin,A.W.,Aldrich,M.J.,Shafiqullah,M. & Husler,J. 1985-6.Tectonic implications of the age, composition and orientation of lamprophyre dykes,Navajo volcanic field,Arizona.*Earth Planet.Sci.Lett.***76**, 361-374 & **80**,418-420.

Laughlin,A.W.,Charles,R.W. & Aldrich,M.J. 1989.Heteromorphism and crystallisation paths of katungites, Navajo volcanic field,Arizona,USA. *Spec. Publ. Geol. Soc. Aust.* **14**, 582-591.

* Laukkanen,J. 1987. [The lamprophyres of Middle Finland]. In Finnish,Engl.abstr. *Rep. Investigations Geol. Surv. Finland* **76**,91-98.

Laukkanen,J. & Mäkipää,H. 1983.On the chemical classification of lamprophyric rocks. *Geologi* (Helsinki) **35(3)**,54-55.

Lawson, K.D. 1959.Geology of Boston and Pacaud township. *Ontario Dept.Mines Geol.Rep.* **64**, 55 pp.

Lazarenkov,V.G. 1976.[Alkalic lamprophyres from the Los massif,Guinea.]. In Russian, Engl. summ. *Geol.Geofiz. Akad.Nauk.SSR.Sib.Otd.***5**,120-127.

Leake, B.E. 1971. On aluminous and edenitic hornblendes. *Mineral. Mag.* **38**, 389-407.

Leake, R.C. & Cooper, C. 1983. The Black Stockarton Moor subvolcanic complex, Galloway. *J. Geol. Soc. Lond.* **140,** 665-676.

*Leat,P.T.,Thompson,R.N.,Morrison,M.A.,Hendry,G.L. & Trayhorn,S.C. 1987. Geodynamic significance of post-Variscan intrusive and extrusive potassic magmatism in SW England.*Tr.R.Soc.Edinburgh: Earth Sci.* **77**, 349-360.

*Leat,P.T.,Thompson,R.N.,Morrison,M.A.,Hendry,G.L. & Dickin, A.P. 1988. Silicic magma derived by fractional-crystallization from Miocene minette, Elkhead Mountain, Colorado. *Mineral. Mag.* **52**, 577-586.

Leavy,B.D. & Hermes,O.D. 1979.Mantle xenoliths from southeastern New England. In: Boyd & Meyer (1979b), qv, 374-381.

* Le Bas,M.J. 1977.*Carbonatite-Nephelinite Volcanism.* Wiley, New York, 347pp.

Le Bas, M.J. 1984. Oceanic carbonatites. In: Kornprobst (1984), qv, 169-180.

Le Bas, M.J. 1987. Nephelinites and carbonatites. In: Fitton & Upton (1987), qv, 53-83.

Le Bas M.J., Le Maitre,R.W., Streckeisen, A. & Zanettin,B. 1986. A chemical classification of volcanic rocks based on the total alkali-silica diagram. *J.Petrol.* **27,** 745-750.

Le Bas, M.J., Mian,I. & Rex,D.C. 1987. Age and nature of carbonatite emplacement in North Pakistan. *Geol.Rdsch.* **76(2),** 317-323.

* LeCheminant,A.N., Miller, A.R. & LeCheminant,G.M. 1987. Early Proterozoic alkaline igneous rocks, district of Keewatin: petrogenesis and mineralization. In: *Geochemistry and Mineralization of Proterozoic Volcanic Suites* (eds. Beckinsale, R.D. & Pharoah, T.D.) *Spec.Publ.Geol.Soc.Lond.***33**, 412-437.

Lee,G.W. & Bailey, E.B. 1925. The pre-Tertiary geology of Mull,Loch Aline and Oban. (sheets 35, 43, 44, 45, 52). *Mem.Geol.Surv. G.B.*

Lee,W.T. & Mertie,J.B. 1922. Geology of the Raton-Brilliant-Kochler Region. *USGS Atlas Folio* **214**.

Leedal,G.P. 1951.Faulted Permian dykes in the Highlands.*Geol.Mag.***88**,60-64.

Leelanandam,C. & Ratnakar,J. 1980.Ocellar lamprophyres from the Purimetla alkaline pluton,Prakasam district,Andhra Pradesh. *Q.J.Geol.Ming.Metall.Soc. India* **52(2),**77-79.

Leelanandam, C. & Srinivasan, T.P. 1986. Occurrence of ocellar lamprophyres in the Settupalle alkaline pluton, Prakasam district, Andhra Pradesh. *Current Sci.* **55**, 474-476.

* Lees,G.J. 1974.Petrochemistry of the mica lamprophyres(minettes) of Jersey.*Proc.Ussher Soc.***3**,149-155.

Lehmann,E. 1977.The mineralogical variations in spessartites of the Asswan area,Egypt. *N.Jb.Mineral. Abh.***129**,66-74.

Leitchl, C.H.B., Dawson, K.M. & Godwin, I. 1988. Late Cretaceous-Early Tertiary gold mineralisation: a galena lead isotope study of the Bridge River mining camp, southwestern British Columbia. *Bicentennial Gold '88 (Melbourne, Australia), Geol. Soc. Aust. Abstr. Ser.* **23**, 448-451.

* Le Maitre,R.W. 1975.Volcanic rocks from Edel.No.1 petroleum exploration well,offshore Cameron basin, western Australia. *J.Geol.Soc.Aust.***22**,167-174.

Le Maitre,R.W. 1982. *Numerical Petrology.* Elsevier Developments in Petrology **8**, 281pp.

Le Maitre,R.W. 1989. *A classification of igneous rocks and glossary of terms. (Recommendations of the International Union of Geological Sciences Subcommission on the Systematics of Igneous Rocks).* Blackwell, Oxford, 193pp.

Leonardi,P. 1967. *Le Dolomiti, geologia dei Monti Tra Isarco e Piave.* Cura Consiglio Nazionale Ricerche, Trento, 3 vols, 1019 pp.

* Lepvrier, C. & Velde, D. 1976. A propos des intrusions tertiaires de la marge nord africaine entre Cherchel et Ténès (Algérie). *Bull. Soc. Géol. Fr.* **7**, 991-998.

* Leroy, J. & Sonet, J. 1976. Contribution a l'étude géochromologique des filons de lamprophyres recoupant le granite a deux micas de Saint-Sylvestre (Limousin — Massif Central français). *C.R. Acad. Sci. Paris, Ser. D* **283**, 1477-1480.

Leutwein,F.,Sonet,J. & Zimmermann,J.L. 1972.Dykes basiques du massif Armoricain septentrional: contribution a leur étude géochronologique.*C.R.Acad.Sci.Paris* **275**,1327-1330.

* Lewis, J.D. & Davy, R. 1981. Felsic dykes of the Mount Edgar batholith. *Ann.Rep.Geol.Surv. W.Aust. (1980)*, 103-108

Lewis,R. 1982.Petrology and mineralogy of the monticellite alnöite associated with the Omaha oil field, Gallatin Co.,Illinois. *Geol.Soc.Am.Abstr.Progr.***14(5)**, p.265.

? Lichak,I.L.& Piskorskaya,Y.K. 1968.[Lamprophyres of Volhynia]. In Russian. *Geol.Zh.* **28**, 66-73.

Lindgren, W. & Ransome,F.L. 1906. Geology and gold deposits of the Crippel Creek district, Colorado. *Prof.Pap.U.S.G.S.* **54**, 516pp.

* Lis,J. 1970.[The geochemistry of certain elements in the Karkonosze granitoid massif,Sudetes]. In Polish,Engl.summ. *Inst.Geol.Biul.Polska* **4(224)**,5-101.

Löffler,K. 1961.Petrologische Studien an einem Gangkreuz Lausitzer lamprophyre bei Nieder-Friedersdorf. *Ber.Dt.Geol.Ges.***6**,72-84.

* Lokka,L. 1934.Neure chemische Analysen von Finnischen Gesteinen.*Bull.Comm.Geol.Finlande* **105**.

Long,C.B. & Max,M.D. 1977. Metamorphic rocks in the SW Ox Mountains Inlier, Ireland; their structural compartmentation and place in the Caledonian orogen. *J.Geol.Soc.Lond.***133**, 413-432.

* López Ruiz,J. 1970a.Estudio petrográfico y geoquímico del complejo filoniano de Fuerteventura (Islas Canarias). *Estud.Geol.***26**,173-208.

López Ruiz,J. 1970b.Le complexe filionien de Fuerteventura.*Bull.Volc.***33,**1166-1185.

López Ruiz,J. & Rodríguez Badiola,E. 1980.La region volcanica Neogena del sureste de España. *Estud.Geol.* **36**,5-63.

* Love,J.D. 1970.Cenozoic geology of the Granite Mountains area, Wyoming.*U.S.G.S.Prof.Pap.* **495-C.**

* Loy,W. 1967. Intrusion d'une phase lamprophyrique entre deux phases granitiques dans le batholite de Trégastel. *C.R.Acad.Sci.Paris Sér.*D,**264,**2265-2267.

Lucchini,F. & Mezzetti,R. 1969.Megacristalli di sanidino nella camptonite di Imana(Predazzo). *Mineral. Petrogr.Acta* **15,**185-190.

* Lucchini,F.,Mezzeti,R. & Simboli,G. 1969.The lamprophyres of the Predazzo-Monzoni area: Camptonites. *Mineral.Petrogr.Acta* **15,**109-145.

* Lucchini,F.,Simboli,G.,Zenatti,M., Barbieri,M., Nicoletti, M. & Petrucciani, C. 1983. Petrochimica e dati radiometrici K/Ar e Rb/Sr dei «filoni basici» di Corvara in Bodia e dei lamprofiri di Val Fisealina (Dolomiti Orientali). Revisione classificativa ed implicazione genetiche nel quadro del magmatismo filoniani delle Alpi. *Mineral.Petrogr.Acta* **27**,233-249.

* Luhr,J.F. & Allan,J.F., Carmichael,I.S.E., Nelson,S.A. & Hasenara,T. 1989. Primitive calc-alkaline and alkaline rock types from the western Mexican volcanic belt. *J.Geophys. Res.* **94,** 4515-4530.

* Luhr,J.F. & Carmichael,I.S.E. 1981.The Colima volcanic complex,Mexico.Part II.Late-Quaternary cinder cones. *Contrib.Mineral.Petrol.***76,**127-147.

* Luhr,J.F.& Kyser, T.K. 1989.Primary igneous analcime: the Colima minettes. *Am. Mineral* **74,** 216-223.

Lumbers,S.B. 1971. Geology of the North Bay area. *Geol.Rep.Ont.Dept.Mines* **94,** p.51.

Lumsden,G.I., Tulloch,W. Howells,M.F. & Davis,A. 1967. The geology of the neighbourhood of Langholm (sheet 11). *Mem.Geol.Surv. G.B.*

* Lupanova,N.P. 1934.[Basaltic hornblende and augite from monchikite of Khibina Mtns]. In Russian. *Trav. Inst.Petrogr.Acad.Sci.URSS.***6,**53-64. See also *Mineral.Abstr.***vi** (137),p.420.

Ma,H-Y. 1948.On the occurrence of agmatite in the Rogart migmatite area,Sutherland: a study in granitization. *Geol.Mag.***85**,1-18.

Maaløe,S. 1987.The generation and shape of feeder dykes from mantle sources. *Contrib.Mineral.Petrol.* **96**,47-55.

Macdonald, A.J. 1986. Ore Deposit Models #12. The platinum group element deposits: classification and genesis. *Geoscience Canada*, **14(3)**, 155-166.

Macdonald,R. 1974. The role of fractional crystallization in the formation of alkaline rocks. In: Sørensen (1974), qv, 442-458.

* Macdonald,R.,Rock,N.M.S.,Rundle,C.C. & Russell,O.J. 1986.Relationships between late Caledonian lamprophyric and acidic magmas in a differentiated dyke, SW Scotland. *Mineral.Mag.***50**,547-557.

* Macdonald,R.,Thorpe,R.S., Gaskarth, J.W. & Grindrod, A.R. 1985.Multi-source origin for Caledonian lamprophyres of northern England. *Mineral.Mag.***49**,485-494.

MacGregor,A.G. 1939.Exhibition of Scottish monchiquitic rocks containing large apatites. *Q.J.Geol.Soc. Lond.* **95**,cxviii.

MacGregor,A.G. 1955.Xenolithic monchiquite.*Geol.Mag.***92**,82-83.

MacGregor,A.G. & Kennedy,Q.C. 1931.The Morvern-Strontian 'granite'.*Summ.Prog.Geol.Surv.G.B.part* **II,** 105-119.

* Macgregor,M. 1937. The western part of the Criffell-Dalbeattie igneous complex. *Q.J.Geol.Soc.Lond.***93,** 457-486.

Mackasey, W.O. 1976. Geology of Walters and Leduc townships, district of Thunder Bay. *Ontario Dept. Mines Geol. Rep.* **149**, 1-60.

MacKenzie,W.S.,Donaldson,C.H. & Guilford,C. 1982.*Atlas of Igneous Rocks and their Textures*. Longmans, London, 148pp.

MacKenzie,W.S. & Guildford, G. 1980. *Atlas of Rock-forming Minerals in Thin Section*. Longmans, London, 98pp.

MacKevitt,E.M. 1955.Geology and ore deposits of the Bokan Mountain Uranium Thorium area, southeastern Alaska. *Bull.US Geol.Surv.***1154,**125pp.

MacLean, A. 1956. Geology of Lebel Township. *Ontario Dept Mines Geol. Rep.* **150**, 63 pp.

* MacNevin,A.A. 1977.Diamonds in NSW.*Geol.Surv.NSW Mineral.Resour.Rep.***42,**125pp.

Madden-McGuire, D.J., Silberman, M.L. & Church, S.E. 1988. Geological relationships, K-Ar ages and isotopic data from the Willow Creek mining district, Southern Alaska. *Bicentennial Gold '88 (Melbourne, Australia), Geol. Soc. Aust. Abstr. Ser.* **23**, 368-370.

* Madhavan, V. & Khurram, M.Z.A.K. 1989. The alkaline gneisses of Khariar, Kalahandi district, Orissa. *Mem.Geol.Soc.India* **15**, 265-289.

* Madhavan, V. & Leelanandam, C. 1988. Petrology of the Elchuru alkaline pluton, Prakasam district, Andhra Pradesh, India. *J. Geol. Soc. India* **31**, 515-537.

Madhavan, V., Mallikharjuna Rao,J., Subrahmanyam,K., Krishna, S.G. & Leelanandam,C. 1989. Bedrock geology of the Elchuru alkaline pluton, Prakasam district, Andhra Pradesh. *Mem.Geol.Soc.India* **15**, 189-205.

* Mahoney,J.J., MacDougall,J.D., Lugmair,E.W., Gopalan,K. & Krishnamurthy,P. 1985. Origin of contemporaneous tholeiitic and K-rich alkalic lavas: a case study from the northern Deccan Plateau,India. *Earth Planet.Sci.Lett.***72**,39-53.

* Majer,V. 1965.[Lamprophyres from around Kriva Palanka: II: lamprophyre (minette) and its relationships to the cenotype extrusives in east Macedonia (Yugoslavia)]. In Slav, German abstr.*Acta Geol.(Zagreb)* **4**,225-250.

Majer,V. & Karamata,S. 1983.[Possibility of new genetic interpretation of K-rich lamprophyres and basalts in Yugoslavia]. In Slav.*Geol.Vjesnik (Zagreb)* **36**,181-188.

? Makeyev,B.V. & Yefimov,V.F. 1972.[Some petrologic characteristics of lamprophyres in one of the tin ore areas in the central Chukchi peninsula]. In Russian.*Sov.Geol.***12**,105-112.

Mal'kov,B.A. 1967.Differentiation in dykes of camptonite and essexite porphyry.*Dokl.Acad.Sci.USSR, Earth-Sci. Sect.***177**,1174-1177.

* Mal'kov,B.A. 1970.Differentiation in monchiquite dykes.*Dokl.Acad.Sci.USSR, Earth-Sci.Sect.* **194**,155-158.

* Malpas,J.,Foley,S.F. & King,A.F. 1986.Alkaline mafic and ultramafic lamprophyres from the Aillik Bay area,Labrador.*Can.J.Earth Sci.***23**,1902-1918.

† Manea, A.Z. 1983.Petrology of the magmatites in the Sana Liliana-Valea Galbena region (Bihor manif). *Ann. Inst. Geol. Geophys. Roumania* (Bucharest) **61**,250-213.

Manzoni,M. & Nanni,T. 1977.Palaeomagnetism of Ordovician lamprophyres from Taylor Valley,Victoria Land,Antarctica. *Pure Appl.Geophys.***115(4)**,961-977.

* Marco,L.de. 1958.Su alcuni filoni lamprofirici radioattivi del complesso Sesia-Lanzo. *Comit.Nat. Ric. Nucl. Stud.Ric.Div.Geol.Mineral.Ital.***1**,3-30.

Marsden,M.A.H. 1988.Middle Devonian igneous activity. In: Douglas & Ferguson (1988), qv, 149-151.

* Marston,R.J. 1971.The Foyers granitic complex,Inverness-shire,Scotland.*Q.J.Geol. Soc.Lond.***126**,331-368.

* Martens,J.H.C. 1924. Igneous rocks of Ithaca, New York and vicinity.*Bull.Geol.Soc.Am.***35**,305-320.

* Mason,B. 1958.The intrusive rocks of the Kaikoura Mountains, N.Z.*Tr.R.Soc.N.Z.***85**,247-262.

* Mason,B. 1961.Some analysed rocks from Westland,N.Z. *N.Z.J.Geol.Geophys.***4**,347-351.

* Mata, J. & Munhá, J. 1986. Geochemistry of some late-Variscan calc-alkaline lamprophyres from the Ossa-Morena zone (Elvas, central Portugal). *Comun. Serv. Geol. Port.* 72, 9-31.

*¶ Mathias, M. 1956. The petrology of the Messum complex, SW Africa. *Tr.Geol.Soc.S.Afr.***59**, 23-57.

Mathias, M. 1974. Alkaline rocks of southern Africa. In: Sørensen (1974), qv, 189-202.

Mathur, K.K.,Debey,V.S. & Sharma, N.L. 1926. Magmatic differentiation in Mount Girnar. *J.Geol.* **34**, 289-307.

* Matos Alves, C.A. 1964. Estudo geológico do maciço eruptivo de Sintra. *Rev.Fac.Ciências Lisboa Ser.2.* **C12**, 123-289.

Matshanow, D. 1964. Über zonare Gänge des Almalyk-Bezirkes. *Usbek.Geol.J.***6**,54-61.

* Mauger, R.L., 1988a. Ocelli: transient disequilibrium features in a Lower Carboniferous minette near Concord, North Carolina. *Can. Mineral.* **26**, 117-132.

* Mauger, R.L., 1988b. Geochemical evidence for sediment cycling from North Carolina (USA) minettes. *Can. Mineral.* **26**, 133-142.

Maxey,L.R. 1976.Petrology and geochemistry of the Beemerville carbonatite-alkalic rock complex, New Jersey. *Bull.Geol.Soc.Am.***87**,1551-9 and supplementary material **76-20**.

McArdle,P. 1974.A Caledonian lamprophyre swarm in SE Ireland.*Sci.Proc.R.Soc.Dublin Ser.*A **5**,117-122.

* McArdle,P. & O'Connor,P.J. 1987. The distribution, geochemistry and origin of appinites and lamprophyres associated with the East Carlow deformation zone, Ireland. *Bull.Geol.Surv.Ireland* **4**, 77-88.

McBirney,A. 1963. Breccia pipe near Camerona, Arizona: discussion. *Bull.Geol.Soc.Am.***74**, 227-232.

McCall,G.J.H. 1958.Geology of the Gwasi area.*Rep.Geol.Surv.Kenya* **45.**

* McCallien,W.J. 1927.Preliminary account of the post-Dalradian geology of Kintyre.*Tr.Geol.Soc.Glasgow* **18**, 40-126.

* McCulloch,M.T.,Jaques,A.L.,Nelson,D.R. & Lewis,J.D. 1983.Nd and Sr isotopes in kimberlites and lamproites from Western Australia: an enriched mantle origin.*Nature* **302**,400-403.

McDougall,I. & Wellman,P. 1976.Potassium-argon ages for some Australian Mesozoic igneous rocks. *J.Geol.Soc.Aust.* **23**,1-9.

McDowell,F.W.,Roden,M.F. & Smith,D. 1986.Comments on: "Tectonic implications of the age, composition and orientation of lamprophyre dykes,Navajo volcanic field, Arizona", by Lauglin et al. *Earth Planet.Sci.Lett.***80**, 415-417.

McGetchin,T.R., Nikhanj,Y.S. & Chodos,A.A. 1973. Carbonate-kimberlite relations in the Cone Valley diatreme, San Juan county, Utah. *J.Geophys.Res.***78**, 1854-1869.

McGetchin,T.R. & Silver,L.T. 1970. Compositional relations — minerals from kimberlite and related rocks in the Moses Rock dike, San Juan County,Utah. *Am Mineral.***55**,1738-1771.

McGetchin,T.R. & Silver,L.T. 1972. A crustal-upper mantle model for the Colorado Plateau based on observation of crystalline rock fragments in the Moses Rock dyke. *J.Geophys Res.* **77**,7022-7037.

* McHone, J.G. 1978. Distribution, orientation and ages of mafic dikes in central New England. *Bull. Geol. Soc. Am.* **89,** 1645-1655.

McHone, J.G. 1984. Mesozoic igneous rocks of northern New England and adjacent Quebec. *Geol. Soc. Am. Map & Chart Series MC-49*, 1:690,000 map + 5pp. text.

McHone, J.G. & Butler,J.R. 1984.Mesozoic igneous provinces of New England and the opening of the North Atlantic Ocean. *Bull.Geol.Soc.Am.***95**,757-765.

* McHone,J.G. & Corneille,E.S. 1980.Alkalic dykes of the Lake Champlain valley.*Vermont Geol.* **1**, 16-21.

McHone, J.G., Ross, M.E. & Greenough, J.D. 1987. Mesozoic dyke swarms of eastern North America. In: Halls & Fahrig (1987), qv, 279-288.

* McHone,J.G. & Trygstad,J.C. 1981.Mesozoic mafic dykes of southern Maine.*Maine Geol.* **2**.

McIntyre,D.B. 1954.The discovery of the Moine Thrust.*Proc.Geol.Assoc.***65**,203-223.

* McIver,J.R. 1981.Aspects of ultrabasic and basic alkaline intrusive rocks from Bitterfontein, South Africa. *Contrib.Mineral.Petrol.* **78**,1-11.

McIver,J.R. & Ferguson,J. 1979.Kimberlitic, melilititic, trachytic and carbonatite eruptives at Salpetre Kop,Sutherland, South Africa. In: Boyd & Meyer (1979a), qv, 111-128.

McKelvey,B.C. & Webb,P.N. 1961. Geological investigations in southern Victoria Land,Antarctica. *N.Z.J.Geol. Geophys.***5**, 143-162.

McLennan, J.F. 1915. Quartz veins in lamprophyre intrusions. *Eng. Ming. J.* **99**, 11-13.

* McMahon,B.H. & Haggerty,S.E. 1979.The Oka carbonatite complex: magnetite compositions and the related role of titanium in pyrochlore. In: Boyd & Meyer(1979a), qv, 382-392.

* McNeil,A.M. & Kerrich,R. 1986.Archean lamprophyre dikes and gold mineralisation,Matheson,Ontario: the conjunction of LIL-enriched magmas,deep crustal structures and Au concentration.*Can.J.Earth Sci.* **23**,324-343.

McSween,H.Y.,Coish,R.A. & Norman,M.D. 1979.Coexisting acid and basic melts: geochemistry of a composite dyke; a discussion. *J.Geol.***87**,209-14.

? Mead,R.D., Kesler,S.E., Foland,K.A. & Jones,L.M. 1988. Relationship of Somoran tungsten mineralization to the metallogenic evolution of Mexico.*Econ.Geol.* **83**,1943-1965.

Meighan,I.G. & Neeson,J.C. 1979. The Newry igneous complex, county Down. *Spec.Publ.Geol.Soc. Lond.* **8**, 717-722.

Meinhold,K.D. 1970. Petrographie, Metamorphose, Tektonik und stratigraphische Stellung der Konse-Serie in Zentral-Tanzania (Ostafrika). *Geol.Jb.*, **H.91**, 137pp.

Mendez,V. & Villar,L.M. 1977.Edad de los filones ultrabasicos(alnöitas) de Rio Piedras,Sierras subandinas. *Asoc.Geol.Argentina Rev.***32(1)**,77-78.

Mendum,J.R. & Tonkin,P.C. 1976. Geology of the Tennant Creek 1:250,000 sheet area, Northern Territory. *Rec.Aust.Bur.Miner.Resour.* 1976/88. BMR Microform MF96.

Menzies, M.A., Arculus, R.J., Best, M.G., Bergman, S.C., Ehrenberg, S.N., Irving, A.J., Roden, M.F. & Schulze, O.J. 1987a. A record of subduction processes and within-plate volcanism in lithospheric xenoliths of the southwestern USA. In: Nixon (1987), qv, 59-74.

* Menzies,M.A.,Halliday,A.N.,Palacz,Z.,Hunter,R.H.,Upton,B.G.J.,Aspen,P. & Hawkesworth,C.J. 1987b. Evidence from mantle xenoliths for an enriched lithospheric keel under the Outer Hebrides.*Nature* **325**,44-47.

* Menzies, M.A., Halliday, A.N., Hunter, R.H., Macintyre, R.M. & Upton, B.G.J. 1989. The age, composition and significance of a xenolith-bearing monchiquite, dike, Lewis, Scotland. *Spec. Publ. Geol. Soc. Aust.* **14**, 843-851.

Menzies, M.A. & Hawkesworth, C.J. 1987. *Mantle Metasomatism.* Academic Press, London, 463pp.

* Métais,D. 1960.Caractères pétrographiques et chimiques des kersantites de la rade de Brest. *C.R.Acad.Sci.Paris* **251**,2202-2203.

* Métais,D. 1961.Les roches filoniennes basiques de la rade de Brest: kersantites et dolérites. *Bull.Soc.Géol.Fr. Sér.3*, **7**,387-396.

Métais,D. & Chayes,F. 1963.Varieties of lamprophyre.*Carn. Inst.Wash.Yrbk.***62**,156-157.

Métais,D. & Chayes,F. 1964a.Kersantites and vogesites:a possible example of group heteromorphism. *Carn. Inst.Wash.Yrbk.***63**,196-199.

Métais,D. & Chayes,F. 1964b.Classification of lamprophyres-a possible petrographic application of multigroup discriminant function analysis.*Carn. Inst.Wash.Yrbk.***63**,182-185.

* Métais,D.,Ravier,I. & Duong,P.K. 1962. Nature et composition chimique des micas de deux lamprophyres. *Bull.Soc.Fr.Minéral.Crist.***85**,321-328.

Meyer,H.O.A. 1976. Kimberlites of the continental United States: a review. *J.Geol.* **84,** 377-403.

* Meyer,H.O.A. & Mitchell,R.H. 1988. Sapphire-bearing ultramafic lamprophyre from Yogo, Montana: a ouachitite. *Can. Mineral.* **26**, 81-88.

* Meyer,H.O.A. & Villar,L.M. 1984.An alnöite in the Sierras Subandinas,northern Argentina. *J.Geol.* **92**, 741-751.

† Michalewa,I.A. 1962.Zur genese der lamprophyre in Sudosten des Gorayj Altais (W.Sibirien). *Geol.I.Geofis.* **8**,110-113.

Michel-Levy,A. & Lacroix,A. 1887. Sur le granite a amphibole de Vaugnéray (Vaugnérite de Fournet). *Bull.Soc.Fr.Minéral.Cristall.* **10**, 27-31.

Michon,E. 1987. Les vaugnérites de l'Est du massif central français: apport de l'analyse statistique multivariée in l'étude géochimique des éléments majeurs.*Bull.Soc.Geol.France* **8** LIII, 591-600.

Middlemost,E.A.K. 1967.Petrology of the Bremen granite-syenite complex, South West Africa. *Tr.Geol.Soc. S.Afr.* **70**,117-134.

Middlemost,E.A.K. 1987. *Magmas and Magmatic Rocks.* Longmans, London, 266pp.

* Middlemost, E.A.K., Paul, D.K. & Fletcher, I.R. 1988. Geochemistry and mineralogy of the minette-lamproite association from the Indian Gondwanas. *Lithos* **22**, 31-42.

* Milashev, V.A. 1988. *Explosion Pipes.* Springer, Berlin, 260pp.

Miles,K.R. 1948.Some Western Australian lamprophyres*J.R.Soc.W.Aust.***31**,1-15.

Miller,T.P. 1972. Potassium-rich alkaline intrusive rocks of western Alaska. *Bull.Geol.Soc.Am.***83**, 2111-2127.

@ Minguzzi,C. 1940.Richerche petrografiche sopra alcuni lamprofiri porfiriti e dioriti della val d'Ultimo (Alto Adige). *Stud.Trentini Sci.Nat.An.***21(1-2)**,113-158.

Miser,M.D. & Purdue,A.M.,1929. Geology of the DeQueen and Calder Gap quadrangle, Arkansas. *Bull.U.S. Geol.Surv.* **808**,195pp.

Miser,M.D. & Ross,C.S.,1923a. Diamond-bearing peridotite in Pike Country,Arkansas USA. *Bull.U.S. Geol.Surv.* **735**,279-322.

Miser,M.D. & Ross,C.S.,1923b.Peridotite dykes,Scott Country,Arkansas*Bull.U.S.Geol.Surv.***735**,271-278.

Mitchell,J.G. 1971.K-Ar ages of mica lamprophyres from southern E. Greenland.*Rapp.Grønlands Geol. Unders.* **45**,53-5.

Mitchell,J.G. & Roberts,D. 1986.Ages of lamprophyres from Ytterøy and Lerkehaug,near Steikjer,central Norwegian Caledonides.*Norsk.Geol.Tidsskr.***66**,255-262.

Mitchell,R.H. 1970.Kimberlite and related rocks — a critical reappraisal*J.Geol.***78**,686-704.

Mitchell,R.H. 1979.The alleged kimberlite-carbonatite relationship: additional contrary mineralogical evidence. *Am.J.Sci.* **279**,570-589.

Mitchell,R.H. 1980.Pyroxenes of the Fen alkaline complex*Am.Mineral.***65**,45-54.

* Mitchell,R.H. 1981.Titaniferous phlogopites from the leucite lamproites of the W.Kimberley area,western Australia. *Contrib.Mineral.Petrol.***76**,243-251.

Mitchell,R.H. 1983.The Ile Bizard intrusion,Montreal,Canada — kimberlite or lamprophyre? *Can J.Earth Sci.* **20**,1493-1496.

* Mitchell,R.H. 1985.A review of the mineralogy of lamproites.*Tr.Geol.Soc.S.Afr.***88**,411-438.

Mitchell, R.H. 1986. *Kimberlites: mineralogy, geochemistry and petrology.* Plenum, New York, 442pp.

Mitchell, R.H. 1987. Mantle-derived xenoliths in Canada. In: Nixon (1987), qv, 33-40.

Mitchell,R.H. 1989.Aspects of the petrology of kimberlites and lamproites: some definitions and distinctions. *Spec.Publ.Geol.Soc.Aust.* **14**, 7-45.

Mitchell,R.H. & Brunfelt,A.O. 1975.Rare earth element geochemistry of the Fen alkaline complex,Norway. *Contrib.Mineral.Petrol.***52**,247-259.

Mitchell,R.H. & Crocket,J.H. 1972.Isotopic composition of strontium in rocks of the Fen alkaline complex. *J.Petrol.***13**,83-97.

* Mitchell,R.H. & Janse,A.J.A. 1982.A harzburgite-bearing monchiquite from Wawa,Ontario. *Can.Mineral.* **20**,211-216.

Mitchell,R.H. & Lewis,R.D. 1983.Priderite-bearing xenoliths from the Prairie Creek mica peridotite, Arkansas. *Can.Mineral.* **21**,59-64.

Mitchell, R.H. & Meyer, H.A.O. 1989. Mineralogy of micaceous kimberlites from the New Elands and Star Mines, Orange Free State, South Africa. *Spec.Publ.Geol.Soc.Aust.* **14**, 83-96.

Mitchell,R.H.& Platt,R.G. 1984.The Freemans Cove volcanic suite: field relations,petrochemistry and tectonic setting of nephelinite-basanite volcanism associated with rifting in the Canadian Arctic archipelago.*Can.J.Earth Sci.***21**,428-436.

* Mitchell,R.H., Platt,R.G. & Downey,M. 1987. Petrology of lamproites from Smoky Butte, Montana. *J.Petrol.* **28**, 645-678.

Mitchell,R.S. 1985. *Dictionary of rocks.* Van Nostrand Reihold, Amsterdam, 228pp.

* Mitchell-Thomé,R.C. 1976.*Geology of the Middle Atlantic Islands.*Gebrüder Borntraeger,Berlin, 382pp.

Mogarovskiy,V.V. 1986.Trace elements in alkalic basaltoids and lamprophyre of the Southern Tien Shan and Pamirs. *Dokl.Acad.Sci.USSR Earth Sci.Sect.***291**,170-172.

Mogarovskiy,V.V.,Klimov,G.K. & Mel'nichenko,A.K. 1973.[Basaltic and alkali basaltic dikes of the Zeravshan-Hissar structural zone, Tien Shan]. In Russian.*Isv.Akad.Nauk.SSSR Ser.Geol.***1**, 131-133.

* Mohr,P.A. 1970.Catalog of chemical analyses of rocks from the intersection of the African, Gulf of Aden and Red Sea Rift systems.*Smithsonian Contrib.Earth Sci.***2**.

* Mokhtari,A. & Velde,D. 1987.Sector-zoned kaersutite in camptonites from Morocco. *Mineral.Mag.* **51**,151-156.

* Mokhtari,A. & Velde,D. 1988. Xenocrysts in Eocene camptonites from Taourirt, Northern Morocco. *Mineral Mag.* **52,** 587-602.

Mokhtari,A., Wagner,C. & Velde, D. 1985. Presence of late crystallizing ferriannite-rich annite in basic eruptive rocks from Morocco. *N.Jb.Mineral. Mh.* 1985, 513-520.

Monese,A. 1969.I lamprofiri del Lago della Vacca(Adamello meridionale). *Mem.Accad.Patavina Sci.Lett.Arti atti.* **80(2)**,165-187.

Montel,J.M. 1988. Première découverte d'une vaugnérite à orthopyroxène: pétrographie, géochimie et conséquences pour la genèse des vaugnérites. *C.R.Acad.Sci. Paris Sér.II,* **306(4),** 985-990.

Montel,J.M. & Weisbrod,A. 1986. Characteristics and evolution of "vaugneritic magmas": an analytical and experimental approach, on the examples of the Cevennes Medians (French Massif Central). *Bull.Minéral.,* **109**, 575-587.

@ Moody,C.L. 1949.Mesozoic igneous rocks of the Northern Gulf coastal plain.*Bull.Am. Assoc.Pet.Geol.* **33**,1410-1428.

Moore, A.E., 1988. Olivine: a monitor of magma evolutionary paths in kimberlites and olivine kimberlites. *Contrib. Mineral. Petrol.* **99**, 238-248.

¶ Moore,A.E. & Verwoerd,W.J. 1985.The olivine melilitite-"kimberlite"-carbonatite suite of Namaqualand and Bushmanland,S.Africa.*Tr.Geol.Soc.S.Afr.***88**,281-294.

Moore,D. & Ashton,P.R. 1968.Lamprophyre dyke in the Ingletonian.*Tr.Leeds Geol.Assoc.***7**,269-274.

Moore, E.S. 1937. Geology and ore deposits of the Ramore area. *Ontario Dept.Mines Geol.Rep.***45(6)**, 1-37.

Moore,J.G. & Hopson,C.A. 1961.The Independence dike-swarm in eastern California.*Am.J.Sci.***259**,241-259.

Moorhouse,S.J. & Moorhouse,V.E. 1988. The Moine Assemblage in Sutherland. In: Winchester (1988), qv, 54-73.

Moorhouse,W.W. 1959.*The study of rocks in thin section.*Harper & Row,New York, 514pp.

* Morgan,J.W.,Czamanske,G.K. & Wandless,G.A. 1985.Origin and evolution of the alkalic ultramafic rocks in the Coyote Peak diatreme,Humboldt County,California.*Geochim.Cosmochim.Acta* **49**,749-759.

* Morris,E.M. 1987. The Cretaceous Arkansas alkalic province; a summary of petrology and geochemistry. *Spec.Pap.Geol.Soc.Am.* **215**, 217-234.

Morris,J.H., Predergast,T.,Synnott,P.,Delahunty,R.,Crean,E. & O'Brien,C. 1986. The geology of the Monaghan-Castleblayney district, Co.Monaghan: a provisional summary. *Bull.Geol.Surv.Ireland* **3**,337-349.

Morrison,J. 1918.The Shap minor intrusions.*Q.J.Geol.Soc.Lond.***74,**116-144.

* Morrison,M.A.,Hendry,G.L. & Leat,P.T. 1987.Regional and tectonic implications of parallel Caledonian and Permo-Carboniferous lamprophyre dyke swarms from Lismore,Ardgour. *Tr.R.Soc.Edinburgh Earth-Sci.,* **77,** 279-288.

* Morrison,M.A.,Thompson,R.N.,Gibson,I.L. & Marriner,G.F. 1980.Lateral chemical heterogeneity in the Palaeocene upper mantle beneath the Scottish Hebrides.*Phil.Tr.R.Soc.Lond.Ser.A*,**297**,229-244.

Mosebach,R. 1934.Einsprenglingsartige Orthoklaskristalle in Dioritischen und Lamprophyrischen Gesteinen des kristallinen Spessarts.*Senckenbergiana* **16**,38-48.

Mudrey,M.G. & Geldon,A.L. 1973.A lower Precambrian lamprophyre pluton near Ely, Minnesota. *19th Ann. Inst.Lake Superior Geol.Tech.Progr.Abstr.Madison,Wisconsin*,p.25.

Mueller, A.G., Harris, L.B. & Lungan, A., 1988. Structural control on greenstone-hosted gold mineralization by transcurrent shearing: a new interpretation of the Kalgoorlie mining district, Western Australia. *Ore Geol. Rev.* **3**, 359-387.

Muir, T.L. 1982a. Geology of the Hemlo area, district of Thunder Bay. *Ontario Geol.Surv.Rep.* **217**, 1-65.

Muir, T.L. 1982b. Geology of the Heron Bay area,district of Thunder Bay.*Ontario Geol.Surv.Rep.***218**,1-89.

* Mukherjee,K.K. 1961.Petrology of the lamprophyres of the Bokaro coalfield, Bihar. *Q.J.Mineral. Metall. Soc.India* **33(1)**, 69-87.

@ Mulas Sanchez,J. 1963a.Lamprofido mangeritico en el Rio Espinaredo.*Not.Comun. Inst.Geol.Mineral. España* **72**,289-298.

@ Mulas Sanchez,J. 1963b.Un lamprofido interesante.*Not.Comun. Inst.Geol.Mineral.Esp.***69**,193-206.

* Müller,M. 1984.Die lamprophyre des Schwarzwaldes.*Fortschr.Mineral.Beiheft* 2, **62**,106-111.

Murphy,F.C. 1987.Late Caledonian granitoids and timing of deformation in the Iapetus suture zone of eastern Ireland.*Geol.Mag.***124**,135-142.

Murphy,J.B., Keppie,J.D. & Hynes,A.J. 1990. Geology of the Antigonish Highlands. *Mem.Geol. Surv. Can.* (in press).

Murthy,M.V.N. 1948.Camptonite dykes from Inishowen,Cty.Donegal.*Tr.Geol.Soc.Glasgow* **21**,205-206.

Murthy,S.R.N. 1977. Petrochemistry and origin of the kimberlites of Wajrakarur and Lattavaram. *Rec.Geol.Surv. India* **109** (2) 148-160.

Murty,Y.G.K., Rao,V.B., Guptasarma,D., Rao,J.M.,Rao,M.N. & Bhattacharji,S. 1987. Tectonic, petrochemical and geophysical studies of mafic dyke swarms around the Proterozoic Cuddapah Basin,South India. In: Halls & Fahrig (1987), qv, 303-316.

† Mushkin,I.V. 1966.[Basaltoids and alkaline lamprophyres from the Karataou River region]. In Russian. *Uzbek.Geol.Zh.* **10**,28-36.

* Mushkin,I.V. & Kotunov,A.Y. 1976.[Principal features of the petrology of monchiquite-camptonite dykes of Bukantan]. In Russian.*Uzbek.Geol.Zh.***1**,91-94.

* Mushkin,I.V. & Larin,N.M. 1979.[Alnöite of middle Ferghana]. In Russian.*Dokl.Akad.Nauk.SSSR* **244**,439-441.

Mykura,W. 1976. *British Regional Geology: Orkney and Shetland.* HMSO for Br.Geol.Surv., London.

Mykura,W. & Phemister,J. 1976. The geology of Western Shetland. *Mem.Geol.Surv. G.B.*

Naeser,C. 1971.Geochronology of the Navajo-Hopi diatremes: Four Corners area. *J.Geophys.Res.* **76**, 4978-4985.

Nag,S. 1983.Geochemical study on Proterozoic mafic alkaline rocks of Elchuru,Prakasam district,Andhra Pradesh,India. *N.Jb.Mineral.Abh.***147**,217-227.

† Nag,S., Smith,T.E. & Chakraverty, P.S. 1983. Alkaline basic intrusives of Elchuru, Prakasam district, Andhra Pradesh. *Proc.Indian. Acad. Sci. Earth Planet Sci.* **92(1)**,73-80.

Nagabhushanam, B. & Venkatanarayana,B. 1985. Geology and geochemistry of kimberlites of Wajrakarur area, Anantapur district, Andhra Pradesh. *Geophys.Res.Bull.NGRI, Hyderabad* **23 (1)**, 43-54.

Nakamura,N. 1974. Determination of REE, Ba, Mg, Na and K in carbonaceous and ordinary chondrites. *Geochim. Cosmochim. Acta* **38**, 757-775.

Nambiar,A.R. 1987.Alkaline magmatism in parts of West Khasi Hills and East Garo Hills district, Meghalaya. *Rec.Geol.Surv.India.***115**,25-45.

* Nambiar,A.R. 1988.Petrology of lamprophyres from parts of East Garo Hills and West Khasi Hills district, Meghalaya.*J.Geol.Soc.India.***32**,125-136.

Nambiar,A.R. 1989.Ocellar structures of late stage melt segregation in alkaline lamprophyres from Meghalaya.*Current Sci.* (in press)

* Nathan,S. 1968.Experimental crystallisation of a lamprophyre glass.*N.Z.J.Geol.Geophys.* **11**,1191-1197.

Neal, C.R. 1988. The origin and composition of metasomatic fluids and amphiboles beneath Malaita, Solomon Islands. *J. Petrol.* **29**, 149-180.

Neal, C.R. & Davidson, J.P. 1989. An unmetasomatized source for the Malaitan alnöite (Solomon Islands): petrogenesis involving zone refining, megacryst fractionation and crustal assimilation. *Geochim. Cosmochim. Acta* **53**, 1975-1990.

Neal,C.R. & Nixon,P.H. 1985.Spinel-garnet relationships in mantle xenoliths from the Malaita alnöite,Solomon Islands.*Tr.Geol.Soc.S.Afr.***88**,347-354.

Nehru, C.E. & Reddy, A.K. 1989. Ultramafic xenoliths from Vajrakarur kimberlites, India. *Spec.Publ.Geol. Soc.Aust.* **14**, 745-758.

Nel,L.T. & Jansen,H. 1957. The geology of the country around Vereeniging (sheet 62). *Geol.Surv.S.Afr. Sheet Explanation*, 90 pp.

Nelson,D.R. 1989. Isotopic characteristics and petrogenesis of the lamproites and kimberlites of central west Greenlandl. *Lithos* **22**, 265-274.

Nelson, D.R. & McCulloch, M.T. 1989. Enriched mantle components and mantle recycling of sediments. *Spec.Publ.Geol.Soc.Aust.* **14**, 560-570.

* Nelson,D.R.,McCulloch,M.T. & Sun,S.-S. 1986.The origins of ultrapotassic rocks as inferred from Sr,Nd and Pb isotopes.*Geochim.Cosmochim.Acta*.**50**,231-245.

Nemec,D. 1966.Plagioclase albitization in the lamprophyric and lamproid dykes at the eastern border of the Bohemian mass. *Beitr.Mineral.Petrogr*.**12**,340-353.

Nemec,D. 1968.Fluorine in lamprophyre and lamproid rocks.*Geochim.Cosmochim.Acta* **32**,523-529.

Nemec,D. 1970.Lamprophyrische und Lamproide Ganggesteine im Sudteil der böhmisch-mährischen Anhöhe(CSSR).*Tschermaks Mineral.Petrogr.Mitt. Ser*.3,**14**,235-284.

Nemec,D. 1971a.On the relationship of lamprophyres to hydrothermal ore veins in the Bohemian-Moravian Heights. *Geol.Rundsch*.**60**,718-724.

Nemec,D. 1971b.Phosphorus in lamprophyre and lamproid rocks.*Contrib.Mineral.Petrol*.**34**,236-250.

* Nemec,D. 1972a.Micas of the lamprophyres of the Bohemian Massif.*N.Jb.Mineral.Abh*.**117**,196-216.

* Nemec,D. 1972b.Gangesteine aus der Umgebung von Raabs.*Verh.Geol.B.-A*.**2**,247-262.

* Nemec,D. 1973a.Amphibole lamprophyrischer Ganggesteine der Böhmischen Masse. *Cas.Mineral.Geol*. **18**, 31-46.

* Nemec,D. 1973b.Differentiation series of minettes in the central Bohemian pluton.*J.Geol*.**81**,632-642.

* Nemec,D. 1973c.Paragenetische Analyse der Ganggesteine der Minettengruppe.*Chem.Erde* **32**,1/2,80-97.

* Nemec,D. 1974.Petrochemistry of the dyke rocks of the central Bohemian pluton. *N.Jb.Mineral.Mh*. **H5**,193-209.

* Nemec,D. 1975a.Petrographie der lamprophyrischen und lamproiden Ganggesteine im Nordostteil der Böhmischen Masse(CSSR). *Z.Geol.Wiss.Berlin* **3**,23-36.

* Nemec,D. 1975b.Petrochemie und Genese der lamprophyrischen und lamproiden Ganggesteine im Nordostteil der Böhmischen Masse.*Z.Geol.Wiss.Berlin* **3**,37-52.

Nemec,D. 1975c.Lamprophyrische und lamproide Ganggesteine im Nordteil der böhmisch-mährischen Höhe. *Verh.Geol.B.-A*., **2-3**,223-268.

Nemec,D. 1975d.Thallium in mica lamprophyres.*Acta Univ.Carolinae geologica* **1**,61-73.

* Nemec,D. 1977a.Differentiation of lamprophyre magma.*Kristalinikum* **13**,73-87.

Nemec,D. 1977b.Barium in den Ganggestein der plutonischen Abfolge der Böhmischen Masse(CSSR). *Z.Geol.Wiss.Berlin* **5**,1011-1019.

Nemec,D. 1977c.Beryllium in Ganggesteinen der plutonischen Abfolge.*Z.Geol.Wiss.Berlin* **5**,329-338.

* Nemec,D. 1978a.Ganggesteine auf der Storung von Trebic.*Z.Geol.Wiss.Berlin* **6**,1211-1217.

Nemec,D. 1978b.Obliquity of potassium feldspars in the differentiation series of minettes.*Tschermaks Mineral.Petrogr.Mitt*.**25**,63-69.

* Nemec, D. 1988a. The amphiboles of potassium-rich dyke rocks of the southeastern border of the Bohemian Massif. *Can. Mineral.* **26**, 89-96.

* Nemec, D. 1988b. Origin of syenite porphyries in the Central Bohemian pluton by magma mixing. *N.Jb.Mineral.Abh.* **159**, 59-71.

* Nemtsovich,V.M. 1976.Agardag complex of alkalic basaltoids,Southeastern Tuva.*Dokl.Acad.Sci.USSR Earth-Sci. Sect.* **227**,148-149.

@ Netelbeck,T.A.F. 1959.Lamprophyres in the ultrabasics south of Cozani.*Bull.Soc.Geol.Greece* **4(1)** 6-10.

* Nicholls,G.D. 1950.The Glenelg-Ratagain igneous complex.*Q.J.Geol.Soc.Lond*.**106**,309-344.

Nielsen,T.F.D. 1987a.Tertiary alkaline magmatism in East Greenland: a review. In: Fitton & Upton (1987), qv, 489-516.

Nielsen,T.F.D. 1987b.Mafic dyke-swarms in Greenland: a review. In: Halls & Fahrig (1987), qv, 349-360.

Niggli,P. 1923.*Gesteins und Mineralprovinzen*. Gebrüder Borntrager, Berlin.

Nikishov,K.N.,Kovak'sky,V.V. & Martinshev,V.K. 1972.Alkalic ultrabasic rocks (alnöites,kimberlites and carbonatite) of the NE Siberian platform.*24th Int.Geol.Congr*.**14**,51-55.

* Nikishov,K.N.,Poberezhskij,V.A. & Makhotko,V.F. 1979.Composition of olivine and monticellite from monticellite-alnöites of Siberia.*Dokl.Acad.Sci.USSR, Earth-Sci. Sect*.**244**,116-121.

Nixon,P.H. 1980.Kimberlites in the South-West Pacific.*Nature* **287**,718-720.

Nixon, P.H. 1987. *Mantle xenoliths*. Wiley, New York, 844pp.

Nixon,P.H. & Boyd,M.R, 1979.Garnet-bearing lherzolites and discrete nodules from the Malaita alnöite, Solomon Islands, and their bearing on oceanic mantle composition and geotherm. In: Boyd & Meyer (1979b), qv, 400-423.

Nixon, P.H., Boyd, F.R. & Lee, D.C. 1987. Western Australia — xenoliths from kimberlites and lamproites. In: Nixon (1987), qv, 281-286.

Nixon,P.H.& Coleman,P.J. 1978.Garnet-bearing lherzolites and discrete nodules from the Malaita alnöite,Solomon Islands, and their bearing on the nature and origin of the Ontong Java plateau. *Bull.Aust.Soc.Expl. Geophys.* **9**,103-106.

Nixon, P.H. & Davies, G.R. 1987. Mantle xenolith perspectives. In: Nixon (1987), qv, 741-756.

* Nixon,P.H.,Mitchell,R.H. & Rogers,N.W. 1980.Petrogenesis of alnoitic rocks from Malaita,Solomon Is. *Mineral.Mag.***43**,587-596.

Nixon, P.H. & Neal, C.R. 1987. Ontong Java Plateau: deep-seated xenoliths from thick oceanic lithosphere. In: Nixon (1987), qv, 335-346.

* Nixon,P.H.,Rex,D.C. & Condliffe,E. 1986.A note on the age and petrogenesis of lamprophyre dykes of the Cautley area, Yorkshire Dales.*Proc.Leeds Geol.Assoc.***10(4)**,40-52.

* Nixon,P.H.,Thirlwall,M.F.,Buckley,F. & Davies,C.J. 1984.Spanish and western Australian lamproites: aspects of whole-rock geochemistry. In: Kornprobst (1984a), qv, 285-296.

Nobel,F.A.,Andriessen,P.A.M.,Hebeda,E.H.,Priem,H.N.A. & Rondeel,H.E. 1981.Isotopic dating of the post-Alpine Neogene volcanism in the Betic cordillera,southern Spain.*Geol.Mijnbouw* **60**,209-214.

* Nockolds,S.R. 1941.The Garabal Hill-Glen Fyne igneous complex.*Q.J.Geol.Soc.Lond.***96**,451-510.

Nockolds,S.R.,Knox,R.W.O'B. & Chinner,G.A. 1978.*Petrology for students*.Cambridge Univ. Press, 435pp.

Nordgulen,O. & Mitchell,J.G. 1988. Kentallenite (olivine-monzonite) in Bindal, Central Norwegian Caledonides. *Bull.Norges.Geol.Unders.* **413**, 51-60.

Norrish,K. 1951.Priderite, a new mineral from the leucite-lamproites of the West Kimberley area,Western Australia. *Mineral.Mag.***29**,496-501.

Novgorodova,M.T., Galuskin,Y.V., Boyarskaya,R.V. & Mokhov,A.V. 1987. Accessory minerals in lamproite-like rocks from central Asia. *Int.Geol.Rev.* **29**, 295-306.

* Nowakowska,A. & Teisseyre,A.K. 1971.[The Carboniferous and Tertiary volcanic rocks in the northern margin of the Intrasudetic Basin(Central Sudetes)]. In Polish,Engl.summ.*Geol.Sudetica* **5**,211-236.

‡ Nuez,J.de la,Ubanell,A.G. & Villaseca,C. 1981.Diques lamprofídicos norteados con facies brechiodales eruptivas en la region de la Paramera de Avila (Sistema Central-Espanol).*Cuad.Lab.Xeol.Lax.***3**,53-73.

Nurmi, P.A. & Maapala, I., 1986. The Proterozoic granitoids of Finland: granite types, metallogeny and relation to crustal evolution. *Bull.Geol.Soc.Finlande* **58**,203-233.

Nystuen,J.P. 1975.Plutonic and subvolcanic intrusions in the Hurdal area,Oslo region.*Norges.Geol.Unders.* **317**,1-21.

* Oberhänsli,R. 1986.Geochemistry of meta-lamprophyres from the central Swiss Alps. *Schweiz.Mineral. Petrogr.Mitt.* **66**,315-342.

Oberhänsli, R. 1987. Mineralogy and Alpine metamorphism of meta-lamprophyres from the Central Swiss Alps. *Schweiz. Mineral. Petrogr. Mitt.* **67**, 321-338.

Obolenskaya,R.V. & Firsov,L.V. 1966.Lower Mesozoic mica lamprophyres of the Chuysk complex in the Altay Mountains. *Dokl.Acad.Sci.USSR, Earth-Sci.Sect.* **170(4)**,110-2.

O'Brien,H.E.,Irving,A.J. & McCallum,S. 1988. Complex zoning and resorption of phenocrysts in mixed potassic mafic magmas of the Highwood Mountains, Montana. *Am.Mineral.* **73**, 1007-1025.

O'Connor,P.J. 1974. Some Caledonian appinitic intrusions in E.Carlow,I reland. *Irish Naturalist J.* **18**,103-108.

Ödman,O. 1930. Volcanic rocks of Mt.Elgon in British East Africa. *Geol.Fören.Förh.* **52**, 455-536.

* Ohashi,F. 1980.An alkali olivine basalt and its related rocks from the Setogawa group,Shizuoka Prefecture. *J.Geol.Soc.Japan* **86(12)**,799-813.

@ Oji,Y. & Takeshita,H. 1970.Kersantite inclusion in a calc-alkaline andesite from N.Yatsugataka volcano, central Japan. *J.Jap.Assoc.Mineral.Petrol.Econ.Geol.***63(1)**,1-15.

Okrusch,M. & Jamal-Allil,S. 1979.Studies in the Abha crystalline complex,Asir,Saudi-Arabia. *N.Jb.Mineral.Abh.* **135**,148-179.

Owen,J.V.,Greenough,J.D.,Hy,C. & Ruffman,A.,1988. Xenoliths in a mafic dyke at Popes Harbour, Nova Scotia: implications for the basement to the Meguma Group.*Can.J.Earth Sci.***25**,1464-1471.

Padovani, E.R. & Tracy, R.J. 1981. A pyrope-spinel (alkremite) xenolith from Moses Rock dike: first known North American occurrence. *Am. Mineral.* **66**, 741-745.

Palivcová,M. & Knotek,M. 1975. [The textural variability of the Pecerady gabbro and its causes]. In Czech. *StudieCeskoslov.AkadVed.* **12**, 35-68.

Palm,Q.A. 1954. Vaugnérites et amphibolites, deux types de rockes amphiboliques dam les Cévennes a la hauteur de Largentière (Ardèche). *Bull.Soc.Géol.Fr.* **IV**, 627-641.

Palmieri, J.H. & Arribas A. 1975. El Complejo alcalino potasico de Sapukal (Paraguay oriental). In: Congr. Ibero-Americano de Geologia economica. *Dept. Argentina Sci. Publ. Buenos Aires,* **IV**, 267-300.

Pankhurst,R.J., Andrews,J.R., Phillips,W.E.A., Sanders,I.S. & Taylor,W.E.G. 1976. Age and structural setting of the Slieve Gamph Igneous Complex, Co. Mayo, Eire. *J.Geol.Soc.Lond.* **132**, 327-334.

Pareek,H.S. 1966.Petrological characteristics of Barakar coal seams metamorphosed by lamprophyre sill in the Jharia coalfield,Bihar.*Proc. Indian Acad.Sci.Sect.B*,**63**,261-70. Also *Econ.Geol.* **59**, 926-929 (1964).

Parga-Pondal,I. 1935. Quimismo de las manifestaciones magmáticas Cenozóicas de la Peninsula Iberica. *Trab.Mus.Nac. C.Nat.Ser.Geol.* **39.**

Parker,R.L. 1926. Ueber Lamprophyre im Granite von Baveno. *Schweiz.Mineral.Petrogr.Mitt.***6**,102-114.

Parker,R.L.& Sharp,W.N. 1970.Mafic-ultramafic rocks and associated carbonatites of the Gem Park complex,Custer and Fremont Counties,Colorado.*US Geol.Surv.Prof.Pap.***649**,24pp.

* Parrot,J-F. 1974.Le secteur de Tamimah: étude d'une séquence volcano-sédimentaire de la région ophiolitique du Baër-Bassait (NE Syrie).*Cah.ORSTOM Sér.Géol.***6(2)**,127-146.

Parsons, G.E. 1948. Bidgood Kirkland Mine. In: *Structural Geology of Canadian ore deposits*, 653-658.Can. Inst. Ming.Metall. Spec.Vol.

* Part,G.M. 1950.Volcanic rocks from the Cape Verde Islands.*Bull.Br. Mus.Nat.Hist.(Min).***1(2)**.

† Pascual,C.M.M. & Bea,B.F. 1986.Estudio de los ocelos en las camptonites de la Sierra de Gredos, España central. *Bol.Geol.Mineral.* **92(2)**,94-110.

* Pathak,J.C. 1984.Basic and lamprophyric rocks of Pat region,west of Lohardaga,Bihar. *Geoviews (Secunderabad)*, **12(4)**,89-100.

* Paul, D.K., Crocket, J.H. & Nixon, P.H. 1979, Abundances of Pd, Ir and Au in kimberlites and associated nodules. In: Boyd & Meyer (1979b), qv, 272-279.

* Paul,D.K. & Potts,P.J. 1981.Rare-earth abundances and origin of some Indian lamprophyres. *Geol.Mag.* **118(4)**, 393-399.

* Paul,D.K.,Potts,P.J.,Rex,D.C. & Beckinsale,R.D. 1977.Geochemical and petrogenetic study of the Girnar igneous complex, Deccan volcanic province,India.*Bull.Geol.Soc.Am.***88**,227-234.

Paul, D.K. & Sarkar, A. 1984. Petrogenesis of some Indian lamprophyres. *Spec. Publ. Geol. Surv. India* **12**, 45-54.

Pawlowska,K. 1958. [New data on lamprophyres from the region of Iwaniska in the Swiety Krzyz Mtns]. In Polish,Engl.summ.*Kwart.Geol.***2**,688-705.

Peach,B.N.,Kynaston, H. & Muff, M.B. 1909. The geology of the seaboard of mid-Argyll (sheet 36). *Mem Geol.Surv. G.B.*

Peach B.N.,Horne J.,Woodward, H.B. & Clough, C.T. 1910. The geology of Glenelg, Lochalsh and the south-east part of Skye (sheet 71). *Mem.Geol.Surv.G.B.*

Peach,B.N.,Gunn,W.,CloverC.T. & Hinxham,L.W. 1912.The Geology of Ben Wyvis (sheet 93). *Mem.Geol.Surv.G.B.*

Peach,B.N.,Horne,J, Hinxman, L.W. & Crampton, C.B. 1913a. The geology of central Ross-shire (sheet 82). *Mem.Geol.Surv.G.B.*

Peach,B.N.,Horne,J.,Gunn,W. & Clough, C.T. 1913b. The geology of the Fannich Mountains (sheet 92). *Mem.Geol.Surv.G.B.*

Peacock J.D.,Berridge N.G.,Harris A.L. & May F. 1968. The geology of the Elgin district (sheet 95). *Mem.Geol.Surv.G.B.*

Pearce, J.A. 1983. Role of the sub-continental lithosphere in magma genesis at active continental margins, In: *Continental Basalts and Mantle Xenoliths* (eds. Hawkesworth, C.J. & Norry, M.J.), 230-249. Shiva, Orpington.

Pearson,R.C. 1959.Metamorphosed lamprophyre and related dikes,northern Sawatch range,central Colorado. *Bull.Geol.Soc.Am.***70**, p.1784 (abstr.)

Peccerillo,A. & Taylor, S.R. 1976. Geochemistry of Eocene calc-alkaline volcanic rocks from the Kastamonu area, northern Turkey. *Contrib.Mineral.Petrol.* **58**, 63-81.

* Pelczar,A. 1973.[Minette and vogesite from the Borzeta IG-1 borehole].*Kwart.Geol.***17(4)**,940-42 (abstr.)

Pellicer,M.J. 1973.Estudio petrológico y geoquímico de un nuevo yacimiento de rocas lamproíticas situado en las proximidades de Aljorra (Murcia).*Estud.Geol.***29**,99-105.

Pendias,H. & Ryka,W. 1974.[Magmatic alkaline rocks in the central part of west Pomerania]. In Polish,Engl.summ.*Kwart.Geol.***18**,1-16.

* Permingeat,F. 1954.Les filons de minettes,satellites du granite d'Azegour,Haut-Atlas. *Not.Mem.Serv. Geol. Maroc.***122**,x,81-8.

Perring,C.S., 1988. Petrogenesis of the lamprophyre "porphyry"suite from Kambalda, Western Australia. *Geol.Dept.Univ. Extension W.Aust. Publ.***12**,277-294.

Perring, C.S., Barley, M.E., Cassidy, K.F., Groves, D.I., McNaughton, N.J. & Rock, N.M.S. 1989a. The association of linear orogenic belts, mantle-crustal magmatism and Archean gold mineralization in the Eastern Yilgarn Block of Western Australia. *Econ.Geol.Monograph* **6**, 571-584.

* Perring, C.S., Rock, N.M.S., Golding, S.D. & Roberts, D.E. 1989b. Criteria for the recognition of

metamorphosed or altered lamprophyres: a case study from the Archaean of Kambalda, Western Australia. *Precamb.Res.***43,** 215-237.

Peyronnet,P.de. 1984a.Caractères géochimiques des roches filonniennes du Crétacé des Pyrénées.Y a-t-il une continuité du sillon houiller? *C.R.Acad.Sci.Paris, SérII,***298**, 403-406.

Peyronnet,P.de. 1984b.Les roches filoniennes des massifs hercyniens français: éléments de comparison avec les autres roches plutoniques. *C.R.Acad.Sci.Paris SérII,* **299**, 951-956.

Peyronnet,P. & Lameyre,J. 1984. Les rockes filoniennes des massifs hercyniens français. Éléments de comparison avec des autres roches plutoniques. *C.R.Acad.Sci.Parris. Sér II,* **299,** 951-956.

Phemister,J. 1952.The riebeckite-bearing dykes of Shetland.*Mineral.Mag.*29,359-373.

Phillips,W.J. 1973. Interpretation of crystalline spheroidal structures in igneous rocks. *Lithos* **6**, 235-244.

* Phillips,W.J. 1956.Minor intrusive suite associated with the Criffell-Dalbeattie granodiorite complex. *Proc.Geol.Assoc.* **67**,103-121.

* Philpotts,A.R. 1972.Density,surface tension and viscosity of the immiscible phase in a basic alkaline magma. *Lithos* **5**,1-18.

* Philpotts,A.R. 1974.The Monteregian Hills. In: Sorensen (1974), qv, 293-310.

* Philpotts,A.R. 1976.Silicate liquid immiscibility: its probable extent and petrogenetic significance. *Am.J.Sci.***276**,1147-1177.

? Piepoli,P. & Collari,N. 1939.Su alcune concentrazioni di magnetite in una spessartite dell'Ogliastra (Sardegna). *Period.Mineral.* **10(3)**,301-320.

* Pieri,R.de,Vecchi,G.de & Quarensi,S. 1969.Su alcuni megacristalli feldspatici di una camptonite. *Atti Mem.Acad.Patavina Sci.Lett.Arti* **82**,189-219.

Piper,J.D.A. 1981a.Palaeomagnetic study of the late Precambrian west Greenland kimberlite-lamprophyre suite: definition of the Hadrynian track.*Phys.Earth Planet. Int.***27**,164-186.

Piper,J.D.A. 1981b.Palaeomagnetism of pseudotachylites from the Ikertôq shear belt and their relationships to the kimberlite-lamprophyre province,central west Greenland.*Bull.Geol.Soc.Denmark* **30**,51-61.

Piper,J.D.A., McCook,A.S., Watkins,K.P., Brown,G.C. & Morris,W.A. 1978. Palaeomagnestism and chronology of Caledonian igneous episodes in the Cross Fell inlier and northern Lake District. *Geol.J.* **13**,73-92.

Pirajno,F. 1982.Lamprophyre dykes in the Victoria range sector of the Karamea batholith,NZ. *N.Z. J.Geol.Geophys.* **25**,499-502 and **27**, 399-401.

Pirsson,L.V.1895.Complementary rocks and radial dykes.*Am.J.Sci.,3rd Ser.* **50**,116-121.

Pirsson,L.V.1896.On the monchiquites or analcite group of igneous rocks.*J.Geol.***4**,680-690.

Pirsson,L.V. 1905.Petrography and geology of the igneous rocks of the Highwood Mountains,Montana. *Bull.U.S.G.S.* **237**.

Pitcher,W.S. & Berger,A.R. 1972.*Geology of Donegal.* (Chapter 7; the appinite suite: basic rocks genetically associated with granite).Wiley,New York,435pp.

Pitcher,W.S. & Read,H.H. 1952.An appinitic intrusion-breccia at Kilkenny, Maas Co., Donegal. *Geol.Mag.* **89**, 328-336.

Pivec, E., Povondra, P., Rutsek, J. & Tulrych, J. 1986. [Petrology and geochemistry and the Osecná intrusion: Jested foothills, northern Bohemia.] In Czech,Engl.abstr.*Acta Montana UGC CSAU* **74**, 23-31.

* Platt,R.G. & Mitchell,R.H. 1979.The Marathon dykes. I: Zr-rich titanian garnets and manganoan magnesian ulvöspinel-magnetite spinels.*Am.Mineral.***64**,546-550.

* Platt,R.G. & Mitchell,R.H. 1982a.The Marathon dykes: ultrabasic lamprophyres from the vicinity of McKellar harbour,NW Ontario.*Am.Mineral.***67**,907-916.

Platt,R.G. & Mitchell,R.M. 1982b.Rb-Sr geochronology of the Coldwell complex,northwestern Ontario, Canada. *Can.J. Earth Sci. 19*,1796-1801 & **21**,p.126.

Platt,R.G.,Mitchell,R.H. & Holm,P.M. 1983.Marathon dikes:Rb-Sr and K-Ar geochronology of ultrabasic lamprophyres from the vicinity of McKellar harbour,northwestern Ontario,Canada.*Can.J.Earth Sci.***20**,961-967.

Platten,I.M. 1982a. A late Caledonian breccia swarm in Glen Creran, near Glen Coe in the Grampian Highlands.*Geol. Mag.* **119**, 169-180.

Platten,I.M. 1982b. Partial melting of feldspathic quartzite around late Caledonian minor intrusions in Appin, Scotland. *Geol. Mag.* **119**, 413-419.

Platten,I.M. 1983. Partial melting of semipelite and the development of marginal breccias around a late Caledonian minor intrusion in the Grampian Highlands of Scotland. *Geol.Mag.* **120**, 37-49.

Platten,I.M. 1984. Fluidized mixtures of magma and rock in a late Caledonian breccia dyke and associated breccia pipes in Appin, Scotland. *Geol.J.* **19**, 209-226.

Platten,I.M. & Money,M.S. 1987 Formation of late Caledonian subvolcanic breccia pipes at Cruachan Crinn, Grampian Highlands, Scotland. *Tr.R.Soc.Edinb.Earth Sci.***78**, 85-103.

Podwysocki,M.H. & Dutcher,R.R. 1971.Coal dykes that intrude lamprophyre sills,Purgatore River valley. *Econ.Geol.***66**,267-80. Also as *Spec.Pap.Geol.Soc.Am.***121**, p.140 (1969).

Poitevin,E. & Cooke,H.C. 1946.Camptonite dykes from Sherbrooke district, Quebec.*Tr.R.Soc.Canada Ser.*3, **40**, 87-92.

Popov, A.S. 1967. Palaeozoic volcanism of the Kola peninsula. *Dokl. Acad. Sci. USSR* **174**, 32-34.

Popov,V.S. 1972.[Globular structure of lamprophyres]. In Russian. *Zap.Vses.Mineral.Obshch.***106**,370-379.

* Powell,J.L. & Bell,K. 1970.Strontium isotopic studies of alkalic rocks: localities from Australia,Spain and the western USA. *Contrib.Mineral.Petrol.***27**,1-10.

* Powell,J.L., Hurley,P.M. & Fairbairn,H.W. 1966. The strontium isotopic composition and origin of carbonatites. In: Tuttle & Gittins (1966), qv, 365-378.

* Praegel,N-O. 1981.Origin of ultramafic inclusions and megacrysts in a monchiquite dyke at Streap, Inverness-shire, Scotland. *Lithos* **14**,308-322.

Prest, V.K. 1957. Geology of Hislop township. *Ontario Dept.Mines Ann.Rept. for 1956,* **65 (5)**, 51pp.

* Prider,R.T. 1939.Some minerals from the leucite-rich rocks of the W.Kimberley area,W.Australia. *Mineral.Mag.***25**,373-387.

* Prider,R.T. 1960.The leucite lamproites of the Fitzroy basin,W.Australia*J.Geol.Soc.Aust.* **6**,71-118.

Prider,R.T. 1965.Noonkanbahite,a potassic batisite from the lamproites of western Australia. *Mineral.Mag.* **34**,403-405.

* Prider,R.T. 1982.A glassy lamproite from the West Kimberley area,western Australia. *Mineral.Mag.* **45**,279-282.

Prider,R.T. & Cole,W.F. 1942.The alteration products of olivine and leucite in the lamproites from the W.Kimberley area. *Am.Mineral.***27**,373-384.

Priem,H.N.A.,Verschure,R.H.,Boelrijk,N.A.,Hebeda,E.H. & Thorkildsen,C.D. 1968.Rb-Sr and K-Ar age measurements on phlogopitic biotite from the ultrabasic lamprophyre dike on the island of Ytterøy*Norsk.Geol.Tidsskr.***48**,319-321.

Pryce,M.W.,Hodge,L.C. & Criddle,A. 1984.Jeppeite a new K-Ba-Fe-titanate from the Walgidee Hills, W.Australia. *Mineral.Mag.***48**,263-266.

Pye, E.G. 1976. Geology and mineral deposits of the Guv River area. *Ontario Dept.Mines Open File Rep.* **5152** (1956, publ. 1976).

? Quartino,B.J. 1961.Acerca de una tinguaita de Iruya (Provincia de Salta) y de la rocas alkalinas del noroeste Argentino. *Rev.Asoc.Geol.Arg.***XV (3 & 4),**145-157.

? Quartino,B.J.& Llambias,E. 1964.Synopsis on rocks with alkaline affinities in the Argentine Republic. *22nd.Int.Geol. Congr.***16,**1-16.

Quick, D. 1988. The Stawell Goldfield. In: Jones (1988), qv, 61-68.

* Raeside,R.P. & Helmstaedt,H. 1982.The Ile Bizard intrusion,Montreal,Canada — kimberlite or lamprophyre? *Can.J.Earth Sci. 19*,1996-2011 and **20**,1496-8.

Raggatt, H.G. 1937. Geological survey of the Condobolin-Trundle District. *NSW Dept. Mines Ann.Rep.for 1936*, 92-95.

* Ramasamy,R. 1984.Vogesite from carbonatite complex of Tiruppattur,Tamil Nadu,India*J.Geol.Soc. India* **25(5),** 307-310.

* Ramsay,J.G. 1955.A camptonitic dyke-suite at Monar,Ross-shire and Inverness-shire. *Geol.Mag.* **92**,297-308.

Rankin,A.H. & Le Bas,M.J. 1974. Liquid immiscibility between silicate and carbonate melts in naturally occuring ijolite magma. *Nature* **250**, 206-209.

@ Rao,M.G. 1973.Alkali lamprophyres from the Garo Hills,Assam*Rec.Geol.Surv. India* **105(2)**,121-124.

* Rao,N.K.L.,Rao,S.K.V. & Chakrapani Naidu,M.G. 1966. Lamprophyre from Chelama,Kurnool district, Andhra Pradesh. *Indian Mineral.***7(1-2)**,8-12.

Rao,R.B.N. 1939.Hornblende lamprophyre dykes near Yachenahalli, Mysore district*Rec.Mysore Geol.Dept.* **37**,117-120.

Rao, S.P. & Phadtre, P.N. 1966. Kimberlite pipe rocks of Wajrakarur area, Anatapur district, Andhra Pradesh. *J.Geol.Soc. India* **7**, 118-123.

Rao,S.S. 1964.The geology of the igneous complex of the Girnar Hills, Gujarat State, India. *22nd Int.Geol.Congr.* **7**,42-60.

Rao,S.S. 1968.Petrology of the lamprophyres of the Girnar Hills,Kathiawar,India*J.Geol.Soc. India* **9**,82-87.

Rao,T.V.S., Rao,Y.J.B., Sivaraman,U. & Gopalan,K. 1989. Rb-Sr age and petrology of the Elchuru alkaline complex: implications to alkaline magmatism in the Eastern Ghat mobile belt. *Mem.Geol.Soc.India* **15**, 207-223.

* Ratcliffe,N.M. 1981.Cortlandt-Beemerville magmatic belt: a probable late Taconian alkalic cross-trend in the central Appalachians.*Geology* **9(7)**,329-336.

* Ratcliffe,N.M., Armstrong,R.L., Mose,D.G., Seneschal,R., Williams,N. & Baiamonte,M.J. 1983. Emplacement history and tectonic significance of the Cortlandt complex,related plutons and dike swarms in the Taconide zone of New York, based on K-Ar and Rb-Sr investigations.*Am.J.Sci.***282**,358-390.

Ratnakar,J. & Leelanandam,C. 1989. Petrology of the alkaline plutons from the eastern and southern peninsula India. *Mem.Geol.Soc.India* **15**, 145-176.

* Ravich,M.G.,Solov'ev,D.S. & Federov,L.V. 1985. *Geological structure of MacRobertson Land (East Antarctica).* Russian Translation Series No.24. Balkema, Rotterdam, 247pp.

Razdorozhryi,V.F. 1974.[Spessartites from the Shakhty District of the Donets basin]. In Ukranian, Engl. summ. *Dopov.Akad.Nauk.Ukr.RSR.Ser.B.***5**,425-427.

Read,H.H. 1923.Geology of the country around Banff, Huntly & Turriff (sheets 86,96). *Mem.Geol.Surv.G.B.*

Read,H.H. 1926.Mica-lamprophyres of Wigtown.*Geol.Mag.***63**,422-429.

Read,H.H. 1931.The geology of Central Sutherland (sheets 108-9). *Mem.Geol.Surv.G.B.*

Read,H.H.,Ross,G. & Phemister,J. 1925. The geology of the country around Golspie (sheet 103). *Mem.Geol.Surv.G.B.*

* Read,H.H.,Phemister,J. & Ross,G. 1926.Geology of Strath Oykell and lower Loch Shin (sheet 102). *Mem.Geol.Surv.G.B.*

Reagan,M.K. & Gill,J.B. 1989. Coexisting calcalkaline and high-niobium basalts from Turrialba volcano, Costa Rica: implications for residual titanates in arc magma sources. *J.Geophys.Res.* **94(B4),** 4619-4633.

* Reddy, Ajit Kumar T. 1986.Petrology and geochemistry of Vajrakarur kimberlites. *Rec.Geol.Surv. India* **115**, part 5, 54-66.

* Reddy, Ajit Kumar T. 1987.Kimberlite and lamproite rocks of Vajrakarur area, Andhra Pradesh. *J.Geol.Soc. India* **30**, 1-12.

* Reed,J.C.,Marvin,R.F. & Mangum,J.H. 1970.K-Ar ages of lamprophyre dykes near Great Falls,Maryland. *U.S.G.S. Prof.Pap.***700C**,145-149.

Reed,J.C. & Reed,J.C. 1969. Gold veins near Great Falls, Maryland. *Bull.U.S.Geol.Surv.***1286**,21 pp.

Reid, C. 1906. The geology of the country around Newquay (sheet 346). *Mem.Geol.Surv.G.B.*

Reid, C. 1910. The geology of the country around Padstow (sheets 335-6). *Mem.Geol.Surv.G.B.*

Reid, C. 1911.The geology of the country around Tavistock and Launceston (sheet 337).*Mem.Geol.Surv.G.B.*

Reid, C. 1912. The geology of the country around Dartmoor (sheet 338). *Mem.Geol.Surv.G.B.*

Reid,D.L. & Barton,E.S. 1983. Geochemical characterization of granitoids in the Namaqaland geotraverse. *Geol.Soc.S.Afr.Spec.Publ.***10**, 67-82.

* Restaino,S. 1934.Contributo allo studio chimico di una roccia eruttiva della Punta delle Pietro Nere (Foggia). *Rend. R.Accad.Sci.Fis.Mat.(Soc.R.Napoli) Ser.*4,**4(1-6)**,167-169.

Rex,D.C. 1970.Age of a camptonite dyke from SE Alexander Island.*Bull.Br.Antarc.Surv.* **23,**p.103.

* Reynolds,D.L. 1931.Dykes of the Ards Peninsula,Cty.Down.*Geol.Mag.***68**,97-111 and 145-165.

Reynolds,D.L. 1936.Demonstrations in petrogenesis from Kiloran Bay,Colonsay; 1: Transfusion of quartzite. *Mineral.Mag.***24,**367-407.

Reynolds,D.L. 1938.Transfusion phenomena in lamprophyre dykes.*Geol.Mag.***75**,51-75.

Reynolds,P.H. & Aumento,F. 1974.Deep drill 1972.Potassium-argon dating of the Bermuda drill core.*Can.J.Earth Sci.* **11**,1269-1273.

Rhodes, J.M. 1981. Characteristics of primary basaltic magmas. In: *Basaltic Volcanism on the Terrestrial Planets.* Pergamon, New York, 409-432.

Rice,C.M. & Davies,B. 1979.Copper mineralisation associated with an appinite pipe in Argyll,Scotland. *Tr. Inst.Ming.Metall.*,**88**,B154-160.

Richards,J.P. 1988. Geochemistry of alkaline intrusives at the Porgera gold deposit, Papua New Guinea. *Ann.Rept.Research School Earth Sci.Aust.Nat.Univ. 1988*, p.121.

Richey,J.E. 1939.The dykes of Scotland.*Tr.Edinburgh Geol.Soc.***13**,393-435.

Richey,J.E. & Thomas,H.H. 1930. The geology of Ardnamurchan, north-west Mull and Coll (sheets 51-52). *Mem.Geol.Surv.G.B.*

Rimsaite, J.H.Y. 1967. Studies of rock-forming micas. *Bull.Geol.Surv. Can.* **149,**82pp.

* Rivalenti,G. & Sighinolfi,G.P. 1970.Geochemistry and differentiation phenomena in a basic dyke of the Frederikshåb district, South West Greenland.*Atti.Soc.Tosc.Sci.Nat.Mem.Ser.A.***77**, 358-80.

Robertson,J.A. 1966. Salter Township. *Ont.Dept.Mines Prelim.Geol.Map.* **P378.**

Robertson,J.A. 1970. Geology of the Massey area. *Ont.Dept.Mines open file Rep.* **5043.**

* Robey,J.V.A.,Bristow,J.W.,Marx,M.R.,Joyce,J.,Danchin,R.V. & Arnot,F. 1989.Alkalic ultrabasic dikes near Norseman,Western Australia. *Spec.Publ.Geol.Soc.Aust.* **14**, 382-391.

Robie, R.A., Hemingway,B.S. & Fisher,J.R. 1978. Thermodynamic properties of minerals and related substances at 298.15°K and 1 bar pressure and at higher temperatures. *Bull.U.S.Geol.Surv.***1452.**

* Rock,N.M.S. 1976a.Comparative strontium isotopic composition of alkaline rocks: new data from southern Portugal and East Africa.*Contrib.Mineral.Petrol.***56**,205-28.

Rock, N.M.S. 1976b. The role of CO_2 in alkali rock genesis.*Geol.Mag.***113**, 97-113.

Rock,N.M.S. 1977.The nature and origin of lamprophyres: some definitions, distinctions. and derivations. *Earth-Sci.Rev.***13**,123-169.

Rock,N.M.S. 1978.The effect of CO_2 on the liquid relations of alkali basalts and its relevance to the nephelinite problem: a preliminary investigation. *NERC (UK) Prog.Exptl.Petrol.Publ.Ser.D.***11**, 164-170.

* Rock,N.M.S. 1979.Petrology and origin of the type monchiquites and associated lamprophyres of Serra de Monchique,Portugal. *Tr.R.Soc.Edinburgh* **70**,149-170.

Rock,N.M.S. 1980.Rare-earth elements and the origin of minettes;critical comment on a paper by Bachinski & Scott. *Geochim.Cosmochim.Acta.***44,**1385-1388.

Rock,N.M.S. 1981. How should igneous rocks be grouped? *Geol.Mag.* **118**, 449-461.

Rock,N.M.S. 1982a.Petrography and age of agglomeratic vents near Toscaig, Applecross, Scotland. *Proc.Geol.Assoc.***93**,305-8.

*Rock, N.M.S. 1982b. The Late Cretaceous alkaline igneous province in the Iberian Peninsula, and its tectonic significance. *Lithos* **15**, 111-131.

*Rock,N.M.S. 1982c. Chemical mineralogy of the Monchique alkaline complex, Portugal. *Contrib.Mineral. Petrol.* **81**, 64-79.

* Rock,N.M.S. 1983a.Alguns aspectos geológicos, petrológicos e geoquímicos do complexo eruptivo de Monchique. *Comun.Serv.Geol.Port.* **69**,325-372.

* Rock,N.M.S. 1983b.The Permo-Carboniferous camptonite-monchiquite dyke-suite of the Scottish Highlands & Islands: distribution,field and petrologic aspects. *Inst.Geol.Sci.(UK)Rep.***82/14**, 36pp.

Rock,N.M.S. 1984.Nature and origin of calc-alkaline lamprophyres:minettes,vogesites,kersantites & spessartites. *Tr.R.Soc.Edinburgh: Earth Sci.***74**,193-227.

Rock,N.M.S. 1986.The nature and origin of ultramafic lamprophyres: alnöites and allied rocks. *J.Petrol.* **27**, 155-196.

Rock,N.M.S. 1987a.The nature and origin of lamprophyres: an overview.In: Fitton & Upton(1987), qv, 191-226.

Rock,N.M.S. 1987b.A global database of analytical datra for alkaline syenitoid, trachytoid and phonolitoid rocks. *Modern Geol.***11**, 51-68.

Rock, N.M.S. 1988a. 'Lamprophyre' minor intrusions of Colonsay — a comment.*Geol. Mag.* **125**, 307-310.

Rock, N.M.S. 1988b.Major late Caledonian and Hercynian shear movements on the Great Glen Fault — discussion. *Tectonophysics* **154**, 171-175.

Rock,N.M.S. 1988c.Summary statistics in geochemistry: a study of the performance of robust estimates. *Math.Geol.***20**,243-275.

Rock,N.M.S. 1988d. Numerical Geology.*Springer-Verlag Lecture Notes in Earth Sciences* **18**, 427pp.

Rock,N.M.S. 1989a.Kimberlites as varieties of lamprophyres: implications for geological mapping, petrological research and mineral exploration. *Spec.Publ.Geol.Soc.Aust.***14**, 46-59.

Rock,N.M.S. 1989b. Comment on "Colonsay and Islay: a suspect terrane within the Scottish Caledonides". *Geology* **17**, 98-99.

Rock,N.M.S. 1989c. CIPW: a terminal-interactive FORTRAN 77 program for tabulating extended CIPW norms from large batches of analyses. *COGS Computer Contrib.***5 (1)**, 17-41.

* Rock, N.M.S. & Barley,M.E. 1989. Calc-alkaline lamprophyres from the Pilbara Block, Western Australia. *J.R.Soc.W.Aust.* **71**, 7-13.

* Rock, N.M.S. & Finlayson,E.J. 1990. Petrological affinity of intrusive rocks associated with the giant gold deposits at Porgera, Papua New Guinea. *J.South-East Asian Earth Sci.*(in press).

* Rock,N.M.S. & Groves,D.I. 1988a.Do lamprophyres carry gold as well as diamonds? *Nature* **332**, 253-255.

Rock, N.M.S. & Groves, D.I. 1988b. Can lamprophyres resolve the genetic controversy over mesothermal gold deposits? *Geology* **16**, 538-541.

* Rock,N.M.S. & Hunter,R.H. 1987.Late Caledonian dyke-swarms of northern Britain: spatial and temporal intimacy between lamprophyric and granitic magmatism around the Ross of Mull pluton,Inner Hebrides. *Geol.Rundsch.***76**, 805-826.

Rock,N.M.S. & Leake,B.E. 1984. The International Mineralogical Association amphibole nomenclature scheme: computerization and its consequences. *Mineral.Mag.* **48**, 211-227.

Rock,N.M.S. & Paul, D.K. 1989. 'Lamprophyres', 'lamproites' and 'kimberlites' in India: a bibliography and preliminary reappraisal. *Mem.Geol.Soc.India* **15**, 291-311.

Rock,N.M.S. & Rundle,C.C. 1986.Lower Devonian age for the "Great (basal) Conglomerate", Scottish Borders. *Scott.J.Geol.* **22**,285-288.

Rock,N.M.S. & Wheatley,M.R. 1989. Computers in mineral exploration: establishing aknowledge base in the search for diamonds. *Aust.Inst.Ming.Metall.MINPET Symposium 1989, Sydney*, 15-24.

Rock, N.M.S. & 9 others. 1983. Geology of the Lubuksikaping quadrangle, Sumatra. *Geological Development and Research Centre, Bandung, Java,* 60 & 1:250,000 scale map.

* Rock,N.M.S.,Cooper,C. & Gaskarth,J.W. 1986a.Late Caledonian subvolcanic vents and associated dykes in the Kirkcudbright area,Galloway,SW Scotland.*Proc.Yorks.Geol.Soc.***46**,29-38.

* Rock,N.M.S.,Gaskarth,J.W. & Rundle,C.C. 1986b.Late Caledonian dyke-swarms of Southern Scotland; a regional zone of primitive K-rich lamprophyres and associated vents.*J.Geol.***94**,505-522.

* Rock,N.M.S.,Duller,P.,Haszeldine,R.S. & Groves,D.I. 1987.Lamprophyres as potential gold exploration targets. *Geol.Dept.& Univ.Extension, Univ.W.Aust. Publ.***11**,271-286.

* Rock,N.M.S.,Gaskarth,J.W.,Henney,P.A. & Shand,P. 1988a.Late Caledonian dyke-swarms of northern Britain: Some tectonic and petrogenetic implications of distribution and chemical variations. *Can.Mineral.* **26**, 3-22.

* Rock, N.M.S., Griffin, B.J., Paul,D.K. & Hergt,J.M. 1990. Lamproites, olivine-lamproites and minettes from the Gondwana coalfields, India (submitted).

Rock, N.M.S., Groves, D.I. & Ramsay, R.R. 1988c. Lamprophyres: a girl's best friend? *Geol.Dept.& Univ.Extension, Univ.W.Aust. Publ.***12**, 295-308.

Rock, N.M.S., Groves, D.I., Perring, C.S. & Golding, S.D. 1989. Gold, porphyries and lamprophyres: what does their association mean? *Econ.Geol.Monograph* **6**, 601-617.

* Rock, N.M.S., Hallberg, J.A., Groves, D.I. & Mather, P.J. 1988d. Archaean lamprophyres in the Yilgarn Block, Western Australia: new indications of their widespread distribution and significance. *Geol.Dept.& Univ.Extension, Univ.W.Aust. Publ.***12**, 245-275.

* Roden,M.F. 1981.Origin of coexisting minette and ultramafic breccia,Navajo volcanic field. *Contrib.Mineral.Petrol.* **77**,195-206.

* Roden,M.F. & Smith,D. 1979.Field geology,chemistry and petrology of Buell Park minette diatreme, Apache county,Arizona. In: Boyd & Meyer (1979a), qv, 364-381.

* Roden,M.F.,Smith,D. & McDowell,F.W. 1979.Age and extent of potassic volcanism on the Colorado Plateau. *Earth Planet.Sci.Lett.* **43**,279-284.

Rogers,A.N. 1922. The geology of the country around Heidelberg. *Geol.Surv.S.Afr.Sheet Explanation*, 84pp.

@ Rogers,J.J.W. & Longshore,J.D. 1960.Differentiation of a lamprophyre sill in the northern La Plata Mountains, Colorado. *Am.Mineral.***45**,774-782.

* Rogers,N.W.,Bachinski,S.W.,Henderson,P. & Parry,S.J. 1982.Origin of potash-rich basic lamprophyres: trace element data from Arizona minettes.*Earth Planet.Sci.Lett.***57**,305-312.

Rohde,G. 1972.Ueber Pentlanditentmischungen in Pyrrhotinen an Lausitzer Lamprophyren. *Dtsch.Ges.Geol. Wiss.Ber. Reihe B.Mineral Lagerstättenforsch.***16**,265-269.

Rohde,G. & Ullrich,H-J. 1969.Über einige Erzminerale in Pyrrhotinparagenesen verschiedener Lausitzer Lamprophyre. *Dtsch.Ges.Geol.Wiss.Ber.Reihe B.Mineral.Lagerstättenforsch.* **14**,315-326.

Rombouts,L., 1987. Geology and evaluation of the Guinean diamond deposits. *Ann.Soc.géol.Belgique,* **110**,241-259.

Rosenbusch,H. 1897. *Mikroscopische Physiographie.* 3rd Edn. Schweizerbart, Stuttgart.

* Rosenbusch,H. 1923. *Elemente der Gesteinlehre.* Schweizerbart, Stuttgart.

* Ross,C.S. 1926a.A nepheline-hauyne alnöite from Winnett,Montana.*Am.J.Sci.***11**,218-227.

#* Ross,C.S. 1926b.A Colorado lamprophyr of the verite type.*Am.J.Sci.***12**,217-229.

† Rosseykin,L.V. & Razhmanov,A.M. 1971.[Relation of lamprophyre dikes to the ore mineralisation in the Maykurian deposit]. In Russian.*Dokl.Akad.Nauk.Tadzh.SSR* **14**,44-47.

Roux,J. & Hamilton,D.L. 1976. Primary igneous analcite: an experimental study. *J.Petrol.* **17**, 244-257.

Rubinowski,Z. 1962.[The lamprophyres of the Daleszyce region and mineralisation symptoms connected with them]. In Polish,Engl.summ.*Kwart.Geol.***6**,245-270.

* Rucklidge,J.C.,Brooks,C.K. & Nielsen,T.F.D. 1980.Petrology of coastal dykes at Tugtilik,SE Greenland. *Greenland Geosci.***3**,19pp.

Ruddock,D.I. & Hamilton,D.L. 1978a.The system $KAlSi_2O_6$-$CaMgSi_2O_6$-SiO_2 at 4kb.*NERC (UK) Prog. Exptl.Petrol.Publ.Ser.D.***11**,25-27.

Ruddock,D.I. & Hamilton,D.L. 1978b.Stability of carbonate in a simple potassium-rich rock model. *NERC (UK) Prog.Exptl.Petrol.Publ.Ser.D.***11**,28-31.

Russell,M.J. & Smythe,D.K. 1978.Evidence for an early Permian oceanic rift in the Northern North Atlantic.In: *Petrology and Geochemistry of Continental Rifts* (E.R. Neuman & I.B. Ramberg,eds.),173-179. Reidel, Dordrecht.

Rust, G.W. 1937. Preliminary notes on explosive volcanism in southeastern Missouri. *J.Geol.***45**, 48-75.

Ryka,W. 1974.[Diabase-lamprophyre association of the NE margin of the upper Silesia coal basin]. In Polish,Engl.summ. *Pol. Inst.Geol.Biul.***278**,35-69.

Sabatier,H. 1978. Rémarques préliminaires sur quelques vaugnérites du Massif Central Français. *C.R. Acad.Sci.Parris.* **286,** 9-11.

Sabatier, H. 1980, Vaugnérites et granites: une association particulière de roches grenues acides et basiques. *Bull. Minéral.* **103,** 507-522.

* Sabine,P.A. 1953.The petrography and geological significance of the post-Cambrian minor intrusions of Assynt. *Q.J.Geol.Soc.Lond.***109**,137-171.

Sabine,P.A. 1963.The Strontian granite complex,Argyllshire.*Bull.Geol.Surv.G.B.***20**,6-42.

Sabourby,G. 1975.Apport de la géochimie la connaissance de la petrogénèse des granitoids de Cévennes méridionales, Massif Central Français*Ann.Sci.Univ.Clermont-Ferrand.***52**,1-278.

* Sadashivaiah,M.S. & Durgadmath,M.B. 1976.Camptonite dykes from Phenai Mata area,Baroda district, Gujarata State, India*J.Karnatak Univ.Sci.***21**,39-46. See *Mineral. Abstr.* **78**-3535.

Saether,E. 1957.The alkaline rock province of the Fen area in southern Norway. *Det.Kgl.Norsk.Vidensk. Selsk.Skr.* No.**1**.

Sage, R.P. 1985. Chipman Lake area. *Ontario Geol.Surv.Study* **44**, 40pp.

Sage, R.P. 1987a. Borden Township carbonatite complex. *Ontario Geol.Surv.Study* **33**, 62pp.

* Sage, R.P. 1987b. Nemegosenda Lake alkalic rock complex. *Ontario Geol.Surv.Study* **34**, 132pp.

Sage, R.P. 1987c. Big Beaver House carbonatite complex. *Ontario Geol.Surv.Study* **51**, 71pp.

* Sage, R.P. 1987d. Shenango Township alkalic rock complex. *Ontario Geol.Surv.Study* **35**, 119pp.

Sage,R.P. 1988a. Hecla-Kilmer alkalic rock complex. *Ontario Geol.Surv.Study* **38**,38pp.

Sage,R.P. 1988b. Firesand River carbonatite complex. *Ontario Geol.Surv.Study* **47**, 81pp.

Sage,R.P. 1988c. Seabrook Lake carboantite complex. *Ontario Geol.Surv.Study* **31**, 45pp.

Sage,R.P. 1988d. Sturgeon Narrows and Squaw Lake alkalic rock complexes.*Ontario Geol.Surv.Study* **49**, 117pp.

Sage,R.P. 1988e. Killala Lake alkalic rock complex. *Ontario Geol.Surv.Study* **45**, 120pp.

Sage,R.P. 1988f. Lackner Lake alkalic rock complex. *Ontario Geol.Surv.Study*,**32**, 141pp.

* Saha,A.K.,Sankaran,A.V. & Bhattacharya,T.K. 1973.Geochemistry of the newer dolerite suite of intrusions within the Singhbhum granite — a preliminary study*J.Geol.Soc. India* **14(4)**,329-346.

Sahama,T.G. 1974. Potassium-rich alkaline rocks. In: Sørensen (1974), qv, 96-109.

? Sanbonsugi,M. 1938.[On some monchiquitic rocks in Hokkaido].In Japanese*J.Geol.Soc.Japan* **45(541)**,769-770 (abstr.)

* Sainsbury,C.L. 1969. Geology and ore deposits of the Central York Mountains, Western Seward Peninsula, Alaska. *Bull.U.S.Geol.Surv.***1287**, 101pp.

Sandford,B.V. & Norris,A.W. 1975. Devonian stratigraphy of the Hudson platform. *Mem.Geol.Surv.Can.* **379**, 372pp.

San Miguel,M. 1935.Una erupción de jumillita en la Sierra de las Cabras (Albacete). *Bol.R.Soc.Esp.Hist. Nat.* **35**,147-154.

San Miguel,M.,Almela,A.&Fúster,J.M. 1951.Sobre un volcán de veritas recientemente descubierto en el Mioceno de Barqueros (Murcia).*Estud.Geol.***7**,411-429.

San Miguel,M.& Pedro,F.de. 1945.Afloriamientos de fortunita en Puebla de Mula (Murcia).*Notas y Comun.* **33**,9-24.

Sanyal,S.P. 1961.A differentiated lamprophyre sill from Sudamdih,Jharia coalfield. *Indian Mineral.***15**,p.290.

* Sanyal,S.P. 1964.Petrology of certain lamprophyres from the Jharia coalfield,Bihar, with a discussion on the differentiation of the Sudamdih sill.*Misc.Publ.Geol.Surv. India* **8**,27-44.

* Sarcia,J.A. & Sarcia,J.A. 1956.Les gites d'uranium du Nord-Linousin.II:La Mine Henriette.*Sciences de la Terre (Nancy)* **iv(3-4)**,253-305.

* Sarkar,A.,Paul,D.K.,Balasumbrahmanyan,M.N. & Sengupta,N.R. 1980.Lamprophyres from Indian Gondwanas: K-Ar ages and chemistry. *J.Geol.Soc. India* **21**, 188-193.

Sarkisyan,G.A. 1973.[Lamprophyre dikes of the Azatek gold-antimony deposit]. In Russian. *Dokl.Akad. Nauk.Arm.SSR* **57**,84-88.

‡ Sarma, B.S.P. 1983. A report on ground magnetic survey over Chelima dyke (Cuddapah Basin). *Tech.Rep.Nat.Geophys.Res. Inst.Hyderabad*, 83-210.

Sathe,R.V. & Desai,S.D. 1968.On lamprophyre-nepheline syenite composite intrusions of Mt.Girnar, Saurashtra. *J.Sci.Technol.Poona Univ.***34**,77-80.

Sathe,R.V. & Oka,S.S. 1975.Petrogenesis of lamprophyres of Mt.Girnar, Saurashtra.*Q.J.Geol.Mineral. Metall. Soc. India* **47**,61-67.

Satterly, J. 1952. Geology of Munro Township. *Ontario Dept. Mines Geol. Rep.* **60**, 60 pp.

Scheibner, E. 1976. *Explanatory notes on the tectonic maps of New South Wales.* NSW Geol. Surv., Sydney, 283pp.

Scheumann,K.H. 1913.Petrographische Untersuchungen an Gesteinen des Polzengebietes in Nord-Böhmen. *Abh.Königl.Sachs.Ges.Wiss.Math.-Phys.Kl.***32**,607-776.

Scheumann,K.H. 1922.Zur Genese alkalisch-lamprophyrischer Ganggesteine. *Zentralb.Mineral.Geol.Paläont.* **16**,495-545.

Scheumann,K.H. 1952.Die Bedeutung der Lamprophyrischen Differenziation fur die Herkunft granitischer Plutonit. *19th Int.Geol.Congr.***6**,163-168.

Schmidt,D.L. & Ford,A.B. 1969.Geology of the Pensacola and Thiel Mountains.*Geol.Antarctic map folio series,* Folio **12**, plate v. Am.Geographical Society.

* Schmitt,H.H.,Swann G.A. & Smith D. 1974. The Buell Park kimberlite pipes, northeastern Arizona. *Geol.Soc.Am.Rocky Mountain Sec.Guidebook no.***27**,672-698.

* Schnetzler,C.C. & Philpotts,J.A. 1968.Partition coefficients of rare-earth elements and Ba between igneous matrix material and rock forming mineral phenocryst materials. In: *Origin and Distribution of the Elements* (ed. L.H.Ahrens), 929-938. Pergamon, Oxford.

Schrader,F.C. 1909.Mineral deposits of the Cerbat Range,Black Mountains and Grand Washington Cliffs. *Bull.U.S.G.S.***397**.

Schulz,A.R. & Cross,W. 1912.Potash-bearing rocks of the Leucite Hills, Sweetwater County, Wyoming. *Bull.U.S. Geol.Surv.* **512**,39pp.

Schulze, D.J. 1987. Megacrysts from alkalic volcanic rocks. In: Nixon (1987), qv, 433-452.

* Schulze,D.J.,Smith,J.V. & Nemec,D. 1985.Mica chemistry of lamprophyres from the Bohemian massif, Czechoslovakia. *N.Jb.Mineral.Abh.***152(3)**,321-334.

Schulze,E-G. 1968.Zur Petrographie,Petrochemie und Spurenelement-Verteilung einiger Gesteingänge in nördlichen Oberharz. *Tschermaks Mineral.Petrogr.Mitt.Ser.*3,**12**,403-438.

* Scott,B.H. 1979.Petrogenesis of kimberlites and associated potassic lamprophyres from central W.Greenland. In: Boyd & Meyer (1979a), qv, 190-205.

* Scott,B.H. 1981.Kimberlite and lamproite dykes from Holsteinsborg,W.Greenland.*Greenland Geosci.***4**,24pp.

Scott Smith, B.H. 1987. Greenland. In: Nixon (1987), qv, 23-32.

Scott Smith,B.H. 1989.Lamproites and kimberlites in India. *N.Jb.Mineral.Abh.* **161(2)**,193-225.

* Scott Smith,B.H. & Skinner,E.M.W. 1984a.A New look at Prairie Creek,Arkansas. In: Kornprobst (1984a), qv, 255-284.

Scott Smith,B.H. & Skinner,E.M.W. 1984b. Diamondiferous lamproites. *J.Geol.***92**, 433-438.

* Scott Smith,B.H.,Skinner,E.M.W. & Loney,P.E. 1989.The Kapamba lamproites of the Luangwa valley, eastern Zambia. *Spec. Publ. Geol. Soc. Aust.***14**, 189-205.

* Scott,P.W. 1980.Zoned pyroxenes and amphiboles from camptonites near Gran,Oslo region. *Mineral.Mag.* **43**,913-917.

* Scott,P.W. & Middleton,R. 1983.Camptonite and maenite sills near Gran,Hadeland,Norway. *Norges Geol.Unders.* **389**,1-26.

@ Scribano,V. 1976.Notizie geologica-petrografiche sui filoni ed altre piccole intrusioni dell'Alpe Lusia(Moena). *Studi Trentini Sci.Nat.Ital.***53(5)**,9-19.

*Seal,R.R., Clark,A.H. & Morrissy,C.T. 1988. Stockwork tungsten (scheelite)-molybdenum mineralization, Lake George, Southeastern New Brunswick. *Econ.Geol.* **83**, 1259-1282.

Searle, M.P. 1984. Alkaline peridotite, pyroxenite and gabbroic intrusions in the Oman Mountain, Arabia. *Can.J.Earth Sci.* **21**,396-406.

* Sederholm,J.S. 1926.On migmatites and associated Precambrian rocks of SW Finland:ii:the region around Barosundsfjord. *Bull.Comm.Geol.Finlande* **77**. Also book published by Oliver & Boyd, Edinburgh (1967).

Sedlacek,M. 1939.Ein Spessartitgang bei Engabrunn (Niederdonau).*Tschermaks Mineral.Petrogr.Mitt.* **52(4-5)**, 278-281.

Seifert,A. & Troger,E. 1937.Ein Polzenitvorkommen im Rumburger granit. *Zentralb.Mineral.Abt.A.* 173-181.

Sen, S.N. & Rao,C.N., 1971. Chelima dykes — a source for diamonds in Kurnool district, Andhra Pradesh. *Geol. Surv. India Misc. Publ.* **19**, 92-94.

‡ Sen, S.N. & Rao, N.K.L. 1970. Chelima dykes. *Proc.2nd Upper Mantle Symp.,Proj.Session 5 on Combined Geosurveys in Dharwar and Cuddapah Basin (Hyderabad),* 435-439.

Sethna,S.F. 1989. Petrology and geochemistry of the acid, intermediate and alkaline rocks associated with the Deccan basalts in Gujarat and Maharashtra. *Mem.Geol.Soc.India* **15**, 47-61.

Seymour,H.J. 1900.Occurrence of blue amphibole in kersantite from Co.Down.*Geol.Mag.***vii**, 257-260.

Shafiqullah,M.,Tupper,W.M. & Cole,T.J.S. 1968. K-Ar ages on rocks from the crater at Brent,Ontario. *Earth Planet. Sci.Lett.* **5**,148-152.

* Shand,S.J. 1923.The alkaline rocks of the Franspoort Line, Pretoria district.*Tr.Geol.Soc.S.Afr.* **25**,81-100.

Shannon,E.V. 1920.Petrography of some lamprophyric dyke-rocks of the Coeur d'Alene mining district,Idaho. *Proc.U.S. Nat.Mus.***57**,475-495.

@ Shannon,W.G. 1924.The petrography and correlation of the igneous rocks of the Torquay promontory. *Geol.Mag.* **61**,193-210.

‡ Sharma, N.L. 1939. Pseudomorphs after leucite in the "Mica-trap" of Jharia coalfield. *Science* **8 (9).**

Sharma,N.L. & Subrahmanyam,N.V.R. 1952.A preliminary note on the petrogenesis of some mica-traps of Ramnagar area, Raniganj coalfield.*Q.J.Geol.Mineral.Metall.Soc. India* **24**,133-140.

†Shatalov,M.M. 1981. [First finding of a composite dyke in the western Azov Sea region]. *Dopov.Akad. Nauk Ukrainskoy RSR Ser.B.Geol.Khim.Biol. 1981, No.***7**, *34-37.*

Shchukin,S.I. 1974.[Lamprophyres and ores]. In Russian.*Geol.Rud.Mestorozhd.***16:6**,97-101.

* Shee,S.R.,Bristow,J.W.,Bell,D.R.,Smith,C.B.,Allsopp,H.L. & Shee,P.B. 1989.The petrology of kimberlites, related rocks and associated mantle xenoliths from the Kuruman Province, South Africa. *Spec. Publ. Geol. Soc. Aust.* **14**, 60-82.

Sheppard,S., Crookes, R.A.,Groves,D.J.,Simpson,P.G. & McNaughton, N.J. 1990. The early Proterozoic Tom's Gully gold-silver deposit, N.T.*Abstr.Ser.Geol.Soc.Aust.***25**,159-160.

* Sheraton,J.W.& Cundari,A. 1980.Leucitites from Gaussberg,Antarctica. *Contrib.Mineral.Petrol.* **71**, 417-427.

* Sheraton,J.W. & England,R.N. 1980.Highly potassic mafic dykes from Antarctica.*J.Geol.Soc.Aust.***27**, 129-135.

Sheraton, J.W., Thomson, J.W. & Collerson, K.D. 1987. Mafic dyke swarms of Antarctica. *In:* Halls & Fahrig (1987), qv, 419-432.

Shervais,J.W., Taylor,L.A. & Laul,L.C. 1987. Magma mixing and kimberlite genesis: mineralogic, petrologic and trace element evidence from eastern USA kimberlites.*Spec.Pap.Geol.Soc.Am.* **215**, 101-114.

Sheraton,J.W., Black,L.P., McCulloch,M.T. & Oliver,R.L. 1990. Age and origin of a compositionally varied mafic dyke-swarm in the Bunger Hills, East Antarctica. *Chem.Geol.* (in press).

Sherwin,L. & Holmes.G.G. 1986. *Geology of the Wollongong and Port Hacking 1:100,000 sheets 9029,9129.* Geol.Surv.NSW, 179pp.

Shoemaker,E.M. 1956. Occurrence of uranium in diatremes on the Navajo and Hopi reservations, New Mexico and Utah. *USGS Prof.Pap.* **300**, 179-185.

Shoemaker,E.M.,Roach,C.H. & Byers,F.M. 1962.Diatremes and uranium deposits in the Hopi Buttes,Arizona. In: *Petrologic Studies* (eds. Engel A.E.J. *et al.*) *Geol.Soc.Am.Buddington Vol.*, 327-355.

* Shrbeny,O. & Machacek,V. 1974.Microelements in melilitic rocks of northern Bohemia.*Cas.Mineral.Geol.* **19**,15-25.

Sial, A.N. 1987. The tertiary alkaline province of Fortaleza, State of Ceará, Brazil: oxygen isotopes and REE-geochemistry. *Geochimica Brasiliensis,* **1(1)**, 41-51.

* Siddiqui,F.A.,Chaudhry,M.N. & Shakoor,A. 1969.Geology and petrology of the feldspathoidal syenites and the associated rocks of the Koga area,Chamla valley,Swat,West Pakistan.*Geol.Bull.Punjab.Univ.***7**,1-29.

Sieber,J. 1956.Gibt es bonenanzeigende Pflanzen des Lamprophyres?*Chem.Erde* **18(1-2)**,27-30.

Siehl,A. & Thein,J. 1978.Geochemische Trends in der Minette.*Geol.Rundsch.***67**,1052-1077.

* Simboli,G. 1958.Alcuni lamprofiri nella parte orientale del cristallino di Cima d'Asta. *Rend.Soc.Mineral. Italiana* **14**,316-325.

† Simic,V.,Kostic,A. & Milojkovic,R. 1965.[Lamprophyre dykes of Klisura village on Selicevica Mountain]. In Bulgarian. *Glasn.Prirod.Muz.Beogradu a*,**19-20**,21-29.

* Simpson,E.S. 1926. *Minerals of Western Australia.* 3 vols. Hesperian Press, Perth.

* Simpson,E.S. 1954.The Okonjeje igneous complex. *Tr.Geol.Soc.S.Afr.* **57**,125-172.

Sims,P.K. & Mudrey,M.G. 1972.Syenitic plutons and related lamprophyres. In: *Geology of Minnesota*, 140-152. Minnesota Geol.Surv.

* Singewald,J.T. & Milton,C. 1930.An alnöite pipe,its contact phenomena and ore deposition near Avon, Missouri. *J.Geol.***38**,54-66.

Singh,K.N. 1981.On the studies of the lamprophyres from the area between Damchini and Jemunia Nalas in parts of the Jharia coalfield, district Dhanbad, Bihar. *Indian Mineral.* **35**,p.41.

Skeats,E.W. 1914.On a volcanic agglomerate containing glaciated pebbles at Kangaroo Gully,Bendigo. *Proc.R.Soc.Vic.* **26**,373-385.

Skerl, A.C. & Oates, F. 1938. The geology of the north Ilunga area. *Short Pap. Geol. Div. Dept. Lands & Mines Tanganyika,* **18**, 35pp.

* Skhirtladze,N.I. 1942.[Camptonite-monchiquite dyke rocks of the Kutais-Tkvibuli districts.] In Georgian. *Bull.Acad.Sci.Georg.URSS* 3,803-810.

Skinner, E.M.W. 1989. Contrasting Group I and Group II kimberlite petrology: towards a genetic model for kimberlites. *Spec. Publ. Geol. Soc. Aust.* **14**, 528-544.

Skinner,E.M.W. & Scott,B.H. 1979. Petrography, mineralogy and geochemistry of kimberlites and associated lamprophyre dykes near Swartruggens, western Transvaal. Extended Abst.Kimberlite Sympos II,Cambridge.

Skinner,E.M.W.,Smith,C.B.,Bristow,J.W.,Scott Smith,B.H. & Dawson,J.B. 1985.Proterozoic kimberlites and lamproites and a preliminary age for the Argyle lamproite pipe,Western Australia. *Tr.Geol.Soc.S.Afr.* **88**,335-340.

† Skuridin,V.A.,Sotnikov,V.I. & Nikitina,Y.I. 1972.[Petrology and mineralogy of complex dikes in the Shakhtaminskiy deposit of the copper-molybdenum ore formation]. In Russian. In: [*Geology and genesis of Siberian formations*],168-183.Izd.Nauka,Moscow.

* Sliwinski,S. 1960.[Lamprophyres and volcanic breccias in Devonian limestone of Dziewski near Siewierz]. In Polish, Engl.abstr.*Przegr.Geol.Polska* **8(12)**,655-657.

* Smith,C.B., Gurney,J.J., Skinner,E.M.W., Clement,C.R. & Ebrahim,N. 1985a. Geochemical character of southern African kimberlites: a new approach based on isotopic constraints. *Tr.Geol.Soc.S.Afr.* **88**, 267-280.

Smith, C.B., Alsopp M.L., Kramers, J.D., Hutchinson, G., & Roddick, J.C. 1985b. Emplacement ages of Jurassic–Cretaceous South African Kimberlites by the Rb-Sr method in phlogopite and whole rock samples. *Tr. Geol. Soc. S. Afr.* **88**, 249-266.

Smith,C.B., Atkinson W.J. & Tyler,E.W.J. 1989. Diamond exploration in Western Australia, Northern Territory and South Australia. *Australas. Inst.Ming.Metall Monograph,***13.**

Smith,D. 1979.Hydrous minerals and carbonates in peridotite inclusions from the Green Knobs and Buell Park kimberlitic diatremes on the Colorado Plateau. In: Boyd & Meyer(1979b), qv, 345-356.

Smith,D. 1987. Genesis of carbonate in pyrope from ultramafic diatremes on the Colorado Plateau, southwestern United States. *Contrib.Mineral.Petrol.* **97**, 389-396.

Smith, D. & Boyd, F.R. 1989. Compositional heterogeneities in minerals of sheared lherzolite inclusions from African kimberlites. *Spec. Publ. Geol. Soc. Aust.* **14**, 709-724.

Smith,D. & Ehrenberg,S.N. 1984.Zoned minerals in garnet peridotite nodules from the Colorado Plateau: implications for mantle metasomatism and kinetics.*Contrib.Mineral.Petrol.***86**,274-285.

Smith,D.I. 1979.Caledonian minor intrusions of the N Highlands of Scotland. In: *Caledonian of the British Isles —Reviewed* (eds.Harris A.L.et al.) *Spec.Publ.Geol.Soc.Lond.***8**,683-98.

* Smith,H.G. 1916.The Luergecombe Mill lamprophyre and its inclusions.*Q.J.Geol.Soc.Lond.***72**,77-83.

Smith,H.G. 1929.Some features of Cornish lamprophyres.*Proc.Geol.Assoc.***40**,260-268.

Smith,H.G. 1930.Some features of lamprophyres near Sedbergh.*Proc.Geol.Assoc.***41**,336-342.

Smith,H.G. 1933.Some lamprophyres of the Channel Islands.*Proc.Geol.Assoc.***44**,121-130.

Smith,H.G. 1936a.The South Hill lamprophyre,Jersey.*Geol.Mag.***73**,87-91.

Smith,H.G. 1936b.New lamprophyres and monchiquites from Jersey.*Q.J.Geol.Soc.Lond* .**92**,365-381.

Smith,H.G. 1939.New lamprophyres at Grève de Lecq,Jersey.*Geol.Mag.***76**,165-169.

Smith,H.G. 1946.The lamprophyre problem.*Geol.Mag.***83**,165-171.

* Smith,J.G. 1973.A Tertiary lamprophyre dyke province in south eastern Alaska.*Can.J.Earth Sci.***10**,408-420.

Smith,J.W. & de Ferranti, R.Z. 1968. Report on the Porgera goldfield, Western Highlands, Papua New Guinea. *Geol.Surv.PNG data file* **12**AM.

Smith, P.M. 1986. Dupert, a structurally controlled gold deposit, northwestern Ontario, Canada. In: *Gold '86* (ed. Macdonald, A.J.), 197-212. Gold '86, Toronto.

Smith,R.P. 1987. Dyke emplacement at Spanish Peaks, Colarado. In: Halls & Fahrig (1987), qv, 47-54.

Smithson,S.B.,Fikkan,P.R. & Toogood,D.J. 1970.Early geologic events in the ice-free valleys,Antarctica. *Bull.Geol.Soc.Am.* **81**,207-210.

* Snaveley,P.D. & Wagner,H.C. 1961.Differentiated gabbroic sills and associated alkalic rocks in the central part of the Oregon coast range,Oregon.*Researches U.S.G.S. 1961,short papers in the geologic and hydrologic sciences, articles* **293-435**,D156-160.

* Sobolev, A.V.,Sobolev, N.V.,Smith, C.B.& Debussy, J. 1989. Fluid and melt compositions in lamproites and kimberlites based on the study of inclusions in olivine.*Spec. Publ. Geol. Soc. Aust.* **14**, 220-240.

Sobolev,V.S. 1952.[Important characteristic of femic minerals of lamprophyres as connected with their genesis]. In Russian.*Mineral.Sborn.Lvov. Geol.Obsch.***6**,147-152.

Sobolev,V.S.,Bazarova,T.J. & Yagi,K. 1975.Crystallisation temperature of wyomingite from Leucite Hills. *Contrib. Mineral.Petrol.* **49**,301-308.

† Soellner,J. 1913.Ueber Bergalith, ein neues melilitreiches Ganggestein aus dem Kaiserstuhl. *Mitt.Bad.Geol. Landesanst.* **7**,415-466.

Soellner,J. 1939.Ueber den Chemismus der lamprophyrischen und mondhaldeitischen Ganggesteine des Kaiserstuhls. *Zentralb.Mineral.Abt.A*,**11**,300-319 & **12**,321-338.

Sokolowski,I.S. 1970. *Geology of Poland*, Vol.I. Wydawnictwa Geologiczne, Warsaw, 651pp.

Solomin,Y.S. 1969.[Criteria for age determination in lamprophyres and lamprophyric rocks, eastern Transbaikal.] In Russian.*Vyssh.Uchebn.Zaved.Isv.Geol.Razved.***10**,23-32.

* Solov'yeva,N. 1973.Derivation of the minor intrusion series of E.Transbaikal. *Int.Geol.Rev.* **15(7)**, 837-848.

Sørensen,H. 1974.*The alkaline rocks*.Wiley,New York, 622pp.

* Sørensen,H.,Hansen,J. & Bondesen,E. 1969.Preliminary account of the geology of the Kvanfjeld area of the Ilimaussaq intrusion,S.Greenland.*Rep.Geol.Surv.Greenland.***28**,40pp.

Spangenberg,K. 1951.Ein Ganggefolge mit Natronvormacht aus dem Peridotit der Chromerzlagerstätte von Tampadel am Zobten. *N.Jb.Mineral.Abh.***82**,247-316.

Sparks,R.S.J., Pinkerton,H. & Macdonald, R. 1977. The transport of xenoliths in magma. *Earth Planet.Sci.Lett.* **35**, 234-238.

* Sparks,R.S.J. & Wadge,G. 1975.Geological and geochemical studies of the Sintra alkaline igneous complex, Portugal. *Bull.Volc.***39**,385-406

Speight,J.M. & Mitchell,J.G. 1979.The Permo-Carboniferous dyke-swarm of northern Argyll and its bearing on dextral displacement on the Great Glen Fault. *J.Geol.Soc.Lond.***136**,3-12.

Spera,J. 1980. Aspects of magma transport. In: *Physics of Magmatic Processes* (ed. Hargreaves,R.B.), 265-323. Princeton Univ.Press.

Spurr,J.E. 1939. Diaschistic dykes and ore deposits. *Econ.Geol.* **34,** 41-48.

Staatz,M.H. 1983.Geology and description of thorium and uranium deposits in the southern Bear Lodge Mountains, Northeastern Wyoming.*US Geol.Surv.Prof.Pap.***1049-D**,52pp.

Stansfield,J. 1923a.Nomenclature and relations of the lamprophyres.*Geol.Mag.***60**,550-554.

* Stansfield,J. 1923b.An extension of the Monteregian petrographical province to the west and north-west. *Geol.Mag.***60**,433-453.

Stark,J.T.,Johnson,J.H.,Behre,C.H.,Powers,W.E.,Howland,A.L.& Gould,D.B. 1949.Geology and origin of South Park, Colorado.*Mem.Geol.Soc.Am.***33,**188pp.

Steele,K.F. & Wagner,G.H. 1979.Relationship of the Murfreesboro kimberlite and other igneous rocks of Arkansas,USA. In: Boyd & Meyer(1979a), qv, 393-399.

Stefanini,G. 1934. Il complexo eruttivo di Orciatico e Montecatini in provincia di Pisa. *Mem.Atti.Soc.Tosc. Sci.Nat.* **44**, 224-300.

Stefanova,M. 1966.[Petrochemical pecularities of the Svidnja potassium alkaline rocks.] In Bulgarian, Engl. summ.*Isz. Bulg. Akad. Nauk. Geol. Inst.* **15**,191-203.

Stefanova, M. & Boyadzhiyeva, R. 1974. [Genetic and chemical characterization of apatite from the biotite-katophorite lamproites near Svidnja,Sofia]. In Bulgarian, Engl. summ. In: *Minerogenezis* (ed. Aleksiev, E.), 477-285. Isv. Bulg. Akad. Nauk, Sofia.

Stefanova,M. & Boyadzhiyeva,R. 1975.[On the geochemistry of niobium and tantalum in potassium-alkaline lamproitic rocks from the village of Svidnja,Sofia.] In Bulgarian, Engl. summ. *Geokhim.Mineral.Petrol.* **3**,16-30.

Stefanova, M. & Pavlova, M. 1974. [Geochemistry and isotopic composition of lead from potassium-alkaline rocks of lamproite character]. In Bulgarian, Engl. summ. In: *Minerogenezis* (ed. Aleksiev, E.), 333-348. Isv. Bulg. Akad. Nauk, Sofia.

Steiner,H. 1984.Mineralogische,petrographische,geochemische und isotopengeologische Untersuchungen an einem Meta-Lamprophyre und seinem granodioritischen Nebengestein (Matorello-Gneis) aus der Maggia-Decke. *Schweiz.Mineral.Petrogr.Mitt.***64**,261-71.

Stemmerik,L. & Sørensen,M. 1980.Upper Permian dykes in southern Scoresby Sund,E.Greenland. *Rapp.Grønlands Geol.Unders.***100**,108.

* Stewart,J.W. 1970.Precambrian alkalic-ultramafic/carbonatite volcanism at Qagssiarssuk,SW Greenland. *Bull.Grønlands Geol.Unders.***84,**70pp & *Meddr.Grønland* **186(4).**

* Stigzelius,H. 1944.Über die Erzgeologie des Viljakkalagebietes im sub-westlichen Finland. *Bull.Comm. Geol.Finlande* **134.**

Stillwell,F.L. 1912.Preliminary notes on monchiquite dykes of Bendigo coalfield.*Proc.R.Soc.Vic.* **25**, 1-14.

Stobbe,H.R. 1949. Petrology of volcanic rocks of northeastern New Mexico. *Bull.Geol.Soc.Am.* **60**,1041-1093.

Stockley, G.M. 1936. Geology of south and south-eastern regions of the Musoma District. *Short Pap.eol. Div.Dept.Lands & Mines Tanganyika* **13**, 46pp.

@ Stockley, G.M. 1948. Geology of north, west and central Njombe District, Southern Highland Province. *Bull. Geol.Div.Dept.Lands & Mines Tanganyika* **18**, 67pp.

Storetvedt,K.M. 1967.Magnetic properties of an ultrabasic biotite-lamprophyre dyke from the island of Ytterøy. *Norsk.Geol.Tidsskr.***47**,171-184.

Storetvedt,K.M. 1987. Major late Caledonian and Hercynian shear movements on the Great Glen Fault. *Tectonophysics* **143**, 253-268.

Storetvedt, K.M., Carmichael, C.M., Hayatsu, A. & Palmer, H.C. 1978. Palaeomagnetism and K/Ar Results from the Duncansby Volcanic Neck, N.E. Scotland: superimposed magnetizations, age of igneous activity, and tectonic implications. *Phys. Earth Planet. Int.* **16**, 379-392.

Stracke,K.J.,Ferguson,J. & Black,L.P. 1979. Structural setting of kimberlites in south-east Australia. In: Boyd & Meyer (1979a), qv, 71-91.

* Streckeisen,A. 1954.Das Nephelinsyenit Massiv von Ditro.*Schweiz.Mineral.Petrogr.Mitt.* **32**,251-309 & **34**,336-409.

Streckeisen,A. 1976. To each plutonic rock its proper name. *Earth-Sci.Rev.***12**,1-33.

Streckeisen,A. 1979.Classification and nomenclature of volcanic rocks,lamprophyres,carbonatites and melilitic rocks. *N.Jb.Mineral.Abh.***134**,1-14; also *Geology* **7**,331-335; *Geol.Rundsch.* **69**,194-207.

Streckeisen,A. & Hunziker J.C. 1974. On the origin and age of the nepheline syenite massif of Ditro. *Schweiz Mineral. Petrog.Mitt.* **54**,59-77.

* Strong,D.F. & Harris,A. 1974.Petrology of Mesozoic alkaline intrusives of central Newfoundland. *Can.J.Earth Sci.* **11(9)**,1208-1219.

Stuckless, J.S. & Irving,A.J. 1976. Strontium isotope geochemistry of mageacrysts and host basalt from SE Australia. *Geochim.Cosmochim.Acta* **40**, 209-214.

Subrahmanyam,N.V.R. & Nandan,K.R. 1961.A note on the petrology of apatite-biotite-monchiquite from Giridih coalfield,Bihar. *J.Geol.Soc. India* **2**,39-45.

Suda, C.E., Brookins, D.G. & Dellavalle, R. 1982. Uranium and other trace element geochemistry of the Hopi Buttes volcanic province, northeastern Arizona. *Econ.Geol.***77**,1210-1220.

* Sukheswela,R.N. & Aavasia,R.K. 1972.Carbonatite-alkalic complex of Panwad-Kawant,Gujarat, and its bearing on the structural characteristics of the area. *Bull.Volc.***35**,564-578.

* Sukheswela,R.N. & Poldevaart,A. 1958.Deccan basalts of the Bombay area,India. *Bull.Geol.Soc.Am.* **69**,1475-1494.

@ Sukheswela,R.N. & Sethna,S.F. 1962.Deccan traps and associated rocks of the Bassein area.*J.Geol.Soc. India* **3**,125-146.

Suppel, D.W. & Barron, L.M. 1986. Platinum in ultrabasic intrusive complexes at Fifield; a preliminary report. *NSW Geol. Surv. Quarterly Notes,* **65**, 1-8.

Suppel, D.W., Warren, A., Watkins, J.J., Chapman, J., Tenison Woods, K. & Barron, L.M. 1986. A reconnaissance study of the geology and gold deposits of the West Wylong-Temora-Adelong district. *NSW Geol. Surv. Quarterly Notes* **64**, 1-23.

Süssmilch,C.A. 1922. *An Introduction to the Geology of New South Wales.* 3rd Edn. Angus & Robertson, Sydney.

Süssmilch,C.A. & Stone,W.G. 1915. Geology of the Jenolan Caves district. *J.R.Soc.NSW* **49**, 332-384.

Sutcliffe,R.H. 1981. Geology of Fletcher Lake area, district of Thunder Bay.*Ont.Geol.Surv.Rept.***251**,65pp.

* Sutherland,D.S. 1982. *Igneous Rocks of the British Isles.* Wiley, New York, 645pp.

* Sutherland,F.L. 1973.The shoshonitic association in the upper Mesozoic of Tasmania.*J.Geol.Soc.Aust.* **19**,487-496.

* Sutherland,F.L. & Corbett,E.B. 1974.The extent of upper Mesozoic igneous activity in relation to lamprophyric intrusions in Tasmania.*Pap.R.Soc.Tasmania* **107**,175-190.

Sutherland, F.L., Hollis, J.D. & Raynor, L.R. 1985. Diamonds from nepheline mugearite? A discussion. *Mineral. Mag.* **49**, 748-751.

Sutton,J.S. 1970.The Termon granite and associated minor intrusives,Cty.Mayo,Ireland. *Sci.Proc.R.Soc. Dublin, Ser.A*,**3(28)**,293-302.

* Suvanapradip,N. & Thiébaut,J. 1964.Pétrologie de la minette de Termes.*Ann.Sci.Univ.Besançon, 2nd Sér.Géol.Fasc.***18.**

* Suzuki,K. & Shiraki,K. 1980.Chromite-bearing spessartites from Kasuga-mura,Japan and their bearing on possible mantle origin andesite.*Contrib.Mineral.Petrol.***71**,313-322.

Sviridov,V.V. & Makhajeva,T.V. 1972.[Linear correlation between major oxides in monchiquites of the Donets Basin]. In Ukranian, Engl. summ. *Dopov.Akad.Nauk.Ukr. RSR, Ser.B* **5**, 410-413.

Svisero,D.P.,Meyer,H.A.O.,Haralyi,N.L.E. & Hasui,Y. 1983.A note on the geology of some Brazilian kimberlites. *J.Geol.***92**, 331-338.

Szabó, C. 1985 Xenoliths from Cretaceous lamprophyre of Alsútdoboz-2 borehole, Transdanubian Mountains, Hungary. *Acta Mineral. Szeged.* **27**, 39-50.

Szabó,C. 1986.Mineralogy,petrology and geochemistry of ultramafic nodules in lamprophyre dykes of Alcsútdoboz-2 borehole (Bakonyium,Hungary): their origin and genetic implications. *Ann.Univ. Scientiarium Budapestinsis de Rolando Eötvös Nominatae Sect.Geol.***26**,31-32.

Szczepanowski,W.J. 1962.[Geochemical investigation of vulcanites in the Swiety Krzyz Mountains,part i:lamprophyres]. In Polish,Engl.summ.*Kwart.Geol.***6**,313-29.

Taljaard, M.S. 1937. South Africa melilite basalts and their relations. *Tr.Geol.Soc.S.Afr.* **39**, 281-315.

@ Tane,J-L. 1963.La génèse des lamprophyres et des laves spilitiques dans le massif de Pelvoux. *C.R.Acad.Sci.Paris* **256(11)**, 2451-2452.

@ Tane,J-L. 1974.Quelques observations nouvelles relatives à la genèse des laves spilitiques. In: *Spilites and Spilitic Rocks* (ed. Amstutz,G.C.),349-358.Springer-Verlag,Berlin.

Tarnowska,M. 1967.[Mineralisation and contact metamorphism associated with lamprophyres in the Wszachow 1,2 & Iwaniska-3 bores]. In Polish. *Kwart.Geol.***11(2)**,462-463 (abstr.)

Tarnowska,M. 1968.[Morphostructural features of lamprophyres from the Iwaniska-Wszachow region, Swietokrzyskie Mtns]. In Polish, Engl.summ. *Kwart.Geol.***12(4)**,1104-1105 (abstr.)

* Tarnowska,M. 1969.[New sub-Quaternary outcrops of lamprophyres in the Swietokrzyskie Mtns]. In Polish, Engl.summ. *Kwart.Geol.***13(4)**,751-775.

Tarnowska,M. 1974.[Lamprophyre in the middle Devonian of Gory Swietokrzyskie Mtns]. In Polish, Engl.summ. *Pol. Inst. Geol.Biul.***275**,63-93.

Tarr, W.A. & Keller,W.D., 1933. A post-Devonian igneous intrusian in SE Missouri. *J.Geol.***41**, 815-823.

Taube,A. 1984.The geology of the Woodcutters lead-zinc-silver deposit and its environment,Rum Jungle, NT, Australia. *Aust. Inst.Ming.Metall.Conf.,Darwin*,347-356.

Taylor,B.J., Burgess I.C., Land,D.H., Mills,D.A.C., Smith,D.B. & Warren,P.T. 1971. *British Regional Geology: Northern England.* 4th Edn. HMSO for Br.Geol.Surv., London.

Taylor,K. & Rushton,A.W.A. 1971.The pre-Westphalian geology of the Warwickshire coalfield. *Bull.Geol. Surv.G.B.* **35**,38-43.

Taylor,W.E.G. 1968. The Dalradian rocks of Slieve Gamph, western Ireland. *Proc.R.Ir.Acad.* **67B**, 63-82.

Teale,E.O.,Eades,N.W. & Oates,F. 1935. The eastern Lupa Goldfield. *Bull.Geol.Surv.Tanganyika* **8**, 59pp.

* Teixeira,C. & Assunção,C.F.T.de,1958.Rochas básicas de fácies gabróica e dolerítica intrusivas nos granitos da Beira. *Rev.Fac.Ciências Lisboa,2nd Ser.C* **vi**,81-123.

Tessensohn, F. & Roland,N.W. 1987. German Antarctic North Victoria Land Expedition, 1982/3. *Geol.Jb.Reihe B*, **66**, 321 pp.

Thompson, R.N. 1982. Magmatism of the British Tertiary volcanic province. *Scott.J.Geol.* **18**, 49-107.

* Thompson,R.N. & Fowler,M.B. 1986.Subduction-related shoshonitic and ultrapotassic magmatism: a study of Siluro-Ordovician syenites from the Scottish Caledonides. *Contrib.Mineral.Petrol.***74**,507-522.

* Thompson,R.N.,Morrison,M.A.,Hendry,G.L. & Parry,S.J. 1984.An assessment of the relative roles of crust and mantle in magma genesis: an elemental approach.*Phil.Tr.R.Soc.Lond.***A310**,549-590.

Thompson,T.B.,Pierson,J.R. & Lyttle,T. 1980.Petrology & petrogensis of the Bokan Granite complex, southeastern Alaska. *Bull.Geol.Soc.Am.***93**,898-908.

Thomson, J.E., Charlewood, G.H., Griffin, K., Hawley, J.E., Hopkins, K., Macintosh,C.G., Ogrizio, S.P., Perry, O.S. & Ward, W. 1950. Geology of the main ore zone at Kirkland Lake. *Ontario Dept.Mines Ann. Rep. for 1948* **57(5)**, 55-103.

Thomson J.E. & Griffis, A.T. 1944. Geology of Gauthier Township, east Kirkland Lake area. *Ontario Dept. Mines Geol. Rep.*, **50**, 29pp.

Thonon,P. 1973.Quelques roches filoniennes de la rade de Brest,I & II.*Penn. Bed.* **9(72)**,17-24 & **10(80)**,6-16.

Thorpe,R.S.,Cosgrove,M.E. & van Calsteren,P.W.C. 1986.Rare earth element, Sr- and Nd-isotope evidence for petrogenesis of Permian basaltic and K-rich volcanic rocks from south-west England. *Mineral.Mag.***50**,481-489.

* Thy,P. 1982.Richterite-arfvedsonite-riebeckite-actinolite assemblage from MARID dikes associated with ultrapotassic magmatic activity in central west Greenland.*Terra Cognita* **2(3)**,247-250.

* Thy,P., Stecher, O. & Korstgård, J.A. 1987. Mineral chemistry and crystallization sequences in kimberlite and lamproite dykes from the Sisimiut area, central W. Greenland. *Lithos* **20**, 391-418.

* Tidmarsh,W.G. 1932.The Permian lavas of Devon.*Q.J.Geol.Soc.Lond.***88**,712-773.

Tilley,C.E. & Yoder,H.S. 1967.Pyroxenite-facies conversion of volcanic and subvolcanic melilite-bearing assemblages. *Carn. Inst.Wash.Yrbk.* 1966,257-259.

Tilton,G.R. 1983.Evolution of depleted mantle:the lead perspective.*Geochim.Cosmochim.Acta* **47**, 1191-1197.

Tingey,R.J.,McDougall,I. & Gleadow,A.J.W. 1983.The age and mode of formation of Gaussberg, Antarctica. *J.Geol.Soc.Aust.***30**,241-246.

Tolman,C. & Landes,K.L. 1939.Igneous rocks of the Mississippi valley lead-zinc districts. *Spec.Pap.Geol. Soc.Am.***24**, 71-103.

* Tomic,J. 1929. Litolska serija trahit-kajanita iz oblasti iz medu Bregalnice i Vardara u Juznoj Srbiji. *Belgrade Glas. SKA* **137**,55-94.

* Tomkeieff,S.I. & Marshall,C.E. 1940. The Killough-Ardglass dyke-swarm. *Q.J.Geol.Soc.Lond.***96**, 321-338.

Tomkeieff,S.I. 1983. *Dictionary of Petrology.* Wiley, New York, 680pp.

Tompkins,L. 1987. Exploration for kimberlites in the southwest Goiás region; Brazil: mineral chemistry of stream sediment samples. *J. Geochem.Expl.* **27**, 1-28.

Tonika,J. 1979. The Mutênín ferrodiorite ring intrusion, west Bohemia. *Krystalinikum* **14**, 195-208.

Tröger,W.E. 1932.Über einen Riebeckit-führenden Lausitzer Lamprophyr.*Sitzungsber. U.Abh.Naturwiss ges.Isls. Dresden.* 1931,159-167.

Tröger,W.E. 1935.*Spezielle Petrographie der Eruptivgesteine.*Verl.Deut.Mineral.Gesell.,Bonn, 360pp.

Tröger,W.E. 1939.Über theralith und monchiquit.*Zentralb.Mineral.Geol.Palaeont.Abt.A.*80-94.

Troll,G. 1968. Gleiderung der redwitzitischen Gesteine Bayerns nach Stoff- und Gefüge Merkmalen (Teil. 1), *Abh.Bayer.Akad. Wiss., Mathem.-naturw. Kl.* **N.F.133.**

* Trommsdorf,A. 1962a.Über Lamprophyre aus den nördlichen Kalkalpen. *Tschermaks Mineral.Petrogr.Mitt. Österreich* **8(2)**,281-325.

† Trommsdorf,A. 1962b.Lamprophyre im wetterstein und Karwendelgebirge.*Naturwissenschaften Dtsch.* **49(8)**,179-180.

? Trowell,N.F. 1983.Geology of the Squaw Lake-Sturgeon Lake area district of Thunder Bay. *Ontario Geol.Surv.Rep.* **227**,114pp.

Trubelja,F. & Paskvalin,L. 1962. Lamprofirska zica iz okolice sase kod Srebrenice u Bosni. *Geol.Glasnik (Sarajevo)* br. **6**,61-64.

Trzcienski,W.E. & Marchildon,N. 1989. Kyanite-garnet-bearing Cambrian rocks and Grenville granulites from the Ayer's Cliff, Quebec, Canada, lamprophyre dike suite: deep crustal fragments from the northern Appalachians. *Geology* **17**, 637-640.

Tucker,D.H. & Collerson,K.D. 1972.Lamprophyric intrusions of probable carbonatitic affinity from S.Australia. *J.Geol.Soc.Aust.* **19**,387-391.

Tukey,J.F. 1977. *Exploratory Data Analysis.* Addison-Wesley, Reading, 688pp.

Tulloch,A.T. 1984.Lamprophyre dykes in the Victoria Range sector of the Karamea batholith,NZ: comment. *N.Z.J.Geol.Geophys.***27**,339-400.

Turner,F.J. 1932.Tinguaites and camptonites from the vicinity of Haast Pass.*Tr.N.Z. Inst.***62**,215-229.

* Turpin, L., Velde, D. & Pinte, G. 1988. Geochemical comparison between minettes and kersantites from the West European Hercynian orogen: trace element and Pb-Sr-Nd isotope constraints on their origin. *Earth Planet. Sci. Lett.* **87**, 73-86.

Tuttle,O.F. & Gittins,J. 1966.*Carbonatites.*Wiley,New York, 591pp.

@ Tyrrell,G.W. 1912. The late Palaeozoic alkaline igneous rocks of the west of Scotland.*Geol.Mag.***9**,69-80 & 120-131.

@ Tyrrell,G.W. 1928.A further contribution to the petrography of the late-Palaeozoic igneous suite of the west of Scotland. *Tr.Geol.Soc.Glasgow* **18**,259-294.

Tyrrell,G.W. & Neilson,A.T. 1938.Igneous rocks from the neighbourhood of Mount Jombo and the Sabaki River. Chapter ix in *Monographs Geol.Dept.Hunterian Mus.Glasgow* **v**, 108-116.

@ Ujike,O. 1979.Petrochemistry and mineralogy of a K_2O-rich mafic dyke in Shodo-Shima,Kagawa prefecture,Japan. *J.Jap.Assoc.Mineral.Petrol.Econ.Geol.***74**,57-67.

Ulmer,P. Callegari, E. & Sanderegger,U.C. 1983. Genesis of the mafic and ultramafic rocks and their genetical relation to the tonalite-trondhjemitic granitoids of the southern part of the Adamello batholith. *Mem.Soc.Geol.Ital.* **26**, 171-222.

* Ulrych,J. 1983.[Compositional sector-zoning in clinopyroxene of monchiquite and rongstockite from the Ceske Stredehori Mountains]. In Czech,Engl.summ. *Acta Univ.Carolinae Geol.(Prague)* **1-2**,117-131.

* Ulrych, J., Pivec, E. & Rutsek, J. 1986. Spinel zonation in melilite rocks of the Ploucnice river region, Czechoslovakia. *N. Jb. Mineral. Abh.* **155**, 129-146.

* Ulrych, J., Pivec, E., Povondra, P. & Rutsek, J. 1988. Perovskite from melilite rocks, Osecna complex, northern Bohemia, Czechoslovakia. *N. Jb. Mineral. Mh.*, 81-95.

Ulrych,J., Povondra,P.,Rotsek,J. & Pivec,E. 1988b.Melilitic and melilite-bearing subvolcanic rocks from the Ploucnice river region, Czechoslovakia. *Acta Univ.Carolinae Geol. (Prague)*,**2**,195-231.

* Upton,B.G.J. 1965.The petrology of a camptonite sill in south Greenland.*Meddr.Grønland* **169(11).**

@ Upton,B.G.J. 1973.*Field excursion guide to the Carboniferous volcanic rocks of the Midland valley of Scotland.* Geol.Soc.Edinburgh, 46pp.

Upton,B.G.J. 1974.The alkaline province of south-west Greenland. In: Sørensen(1974), qv, 221-238.

Upton,B.G.J. 1982.Carboniferous to Permian volcanism in the stable foreland. In: Sutherland (1982), qv, 255-76.

Upton,B.G.J., Aspen,P. & Chapman,N.A. 1983.The upper mantle and deep crust beneath the British Isles: evidence from inclusions in volcanic rocks.*J.Geol.Soc.Lond.***140**,105-121.

* Upton,B.G.J. & Emeleus,C.H. 1987. Mid-Proterozoic alkaline magmatism in southern Greenland: the Gardar province In: Fitton & Upton (1987), qv, 449-471.

*¶ Upton,B.G.J.,Emeleus,C.H. & Beckinsale,R.D. 1984.Petrology of the Nortern East Greenland Tertiary Flood Basalts: evidence from Hold with Hope and Wollaston Foreland. *J.Petrol.***25**,151-184.

* Upton,B.G.J. & Thomas,J.E. 1973.Precambrian potassic ultramafic rocks: South Greenland. *J.Petrol.***14**, 509-534.

* Urry,W.D. & Holmes,A. 1941.Age determinations of Carboniferous basic rocks of Shropshire and Colonsay.*Geol.Mag.***78**,45-61.

* Uto, K., Hirai, H., Goto, K. & Arai, S. 1987. K-Ar ages of carbonate- and mantle-nodule bearing lamprophyre dikes from Shingu, central Shikoku, southwest Japan. *Geochem. J.* **21**, 283-290.

† Uvad'ev,L.I. 1980.[Riphean alkali magmatism in the Kandalakish graben].*Dokl.Akad.Nauk.SSSR.* **254(16)**,1435-6.

Val'ter,A.A. & Yeremenko,G.K. 1973.[Genesis and composition of augite megacrysts from camptonite dikes of the eastern Azov region]. In Russian,Engl.summ.*Dopov.Akad.Nauk.Ukr.SSR Ser.B,* **10**,873-879.

* Val'ter,A.A. & Yeremenko,G.K. 1974.Megacrysts of alkalic feldspar in camptonite of the Azov region. *Dokl.Acad.Sci.USSR Earth Sci.Sect.***217**,129-132.

Vaniman,D.,Laughlin,A.W. & Gladney,E.S. 1985.Navajo minettes in the Cerros de las Mujeres,New Mexico.*Earth Planet.Sci.Lett.* **74**,69-80.

* Van Schmus,W.R. 1971.Ages of lamprophyre dikes and of the Mongowin pluton,north shore of Lake Huron,Ontario. *Can.J.Earth Sci.***8**,1203-1209.

* Van Zijl,P.J. 1962.The geology,structure and petrology of the alkaline intrusions of Kalkfeld and Okorusus. *Ann.Univ.Stellenbosch,Ser.A,* **37**, 237-340.

* Vardabasso,S. 1928.I lamprofiri della provincia petrografica di Predazzo.*An.R.Scuola Ingegn.Padova* **4**,1-24.

* Vartiainen,H.,Kresten,P. & Kafkas,Y. 1978.Alkaline lamprophyres from the Sokli complex,northern Finland. *Bull.Geol.Soc.Finland* **50(1-2)**,59-68.

Vecchi,G.de.,1966.I filoni basici ed ultrabasici dell'Altipiano di Tonezza.*Mem. Inst.Geol.Mineral.Univ. Padova* **25**,58pp.

Vejnar,Z. 1975. Hornblendes and problems of recrystallisation of gabbroic rocks. *Lithos* **8**, 59-68.

#* Velde,D. 1967.Sur un lamprophyre hyperalcalin potassique. La minette de Sisco(Ile de Corse). *Bull.Soc.Fr. Minéral. Crist.* **90**,214-223.

Velde,D. 1968a.A new occurrence of priderite.*Mineral.Mag.***36**,867-870.

Velde,D. 1968b.Les transformations de l'olivine dans les lamprophyres et lamproites.*Bull.Soc.Géol.Fr.* **10(5)**, 601-612.

#* Velde,D. 1969.Les micas des lamprophyres: kersantites,minettes et lamproites. *Bull.Soc.Fr.Minéral.Crist.* **92**, 203-223.

* Velde,D. 1970.Remarques sur la nomenclature des minettes.*C.R.Sommaire séances Soc.Géol.Fr.***6**,222-223.

* Velde,D. 1971a.Les lamprophyres à feldspath alcalin et biotite: minettes et roches voisines. *Contrib.Mineral.Petrol.* **30**,216-239.

* Velde,D. 1971b.Les kersantites: étude des lamprophyres à plagioclase et biotite. *Bull.Soc.Fr.Minéral.Crist.* **94**,411-426.

* Velde,D. 1971c.A note on an analcite-bearing lamproite from Devonshire.*Geol.Mag.***108**,201-204.

* Velde,D. 1975.Armalcolite-Ti phlogopite-diopside-analcite-bearing lamproites from Smoky Butte,Montana. *Am.Mineral.***60**,566-573.

* Velde,D. & Tournon,J. 1970.La camptonite de San Feliú de Buxalleu.*Bull.Soc.Fr.Minéral.Crist.* **93**, 482-487.

Velde,D. & Yoder,H.S. 1976.The chemical compostion of melilite-bearing eruptive rocks. *Carn.Inst.Wash. Yrbk.***75**, 574-580.

Velde,D. & Yoder,H.S. 1977. Melilite and melilite-bearing igneous rocks.*Carn.Inst.Wash.Yrbk.***76**,478-485.

Vendl,M. 1926.Gangesteine aus dem Nephelinsyenitstock von Ditro.*Math.Termegzettud ertesito Budapest* **43**,215-243.

* Venturelli,G.,Capedri,S.,Battistini,G.di,Crawford,A.,Kogarko,L.N. & Celestini,S. 1984a.The ultrapotassic rocks from southeastern Spain. *Lithos* **17**,37-54.

* Venturelli,G.,Thorpe,R.S.,Dal Piaz,G.V.,Del Moro,A. & Potts,P.J. 1984b.Petrogenesis of calc-alkaline, shoshonitic and associated ultrapotassic Oligocene volcanic rocks from the northwestern Alps,Italy. *Contrib.Mineral.Petrol.***86**, 209-220.

* Venturelli, G., Mariani, E.S., Foley, S.F., Capedri, S. & Crawford, A.J. 1988. Petrogenesis and conditions of formation of Spanish lamproitic rocks. *Can. Mineral.* **26**, 67-79.

* Vila,J-M.,Hernandez,J. & Velde,D. 1974.Sur la présence d'un filon de roche lamproïtique recoupant le flysch de type Guerroch entre Azzaba et Hamman-Meskoutine,dans l'est de Constantinois. *C.R.Acad.Sci. Paris Ser.D* **278**,2589-2592.

‡ Villar,L.M.,Pezzutti,N.E. & de Svetliza,S.S. 1981.Las kimberlites micaceas de Los Alisos,Sierras Subandinas de Salta. Significado geológico-económica.*Actas 8th Congr.Geol.Argentina* **IV**,903-919.

Villar Favre,J.F.,Gonzalez,R. & Toselli,A. 1973.Los lampróﬁros intrusivos en el granito de la Cuesta de Miranda. *Acta Geol.Lilloana* **12(3)**, p.37.

† Villaseca,C.,Lopez Garcia,J.A.,Nuez,J.de la & Ubanell,A.G. 1983.Contribuición al estudio de los diques camptoníticos heteromorfos com subfacies ocelares y de diatrema asociados, Sierra de la Paramera de Avila. *Rev.Materiales Procesos Geol.* **1**,103-118.

* Villaseca,C. & Nuez,J. 1986.Diques camptoníticos en el Sistema Central Español. *Estud.Geol.***42**, 69-78.

* Vincent,E.A. 1953. Hornblende-lamprophyre dykes of basaltic parentage from the Skaergaard area, E. Greenland. *Q.J.Geol.Soc.Lond.***109**,21-50.

Vishnevskii, S.A., Dolgov, Y.A. & Sobolev, N.V. 1986.Lamproites of the Talakhtakh diatreme on the eastern slope of the Anabar Shield. *Sov.Geol.Geophys.* **27(8)**,15-24.

* Visser,J.N.J. 1964.Analyses of rocks, minerals and ores.*Geol.Surv.S.Afr. Dept.Mines Handbook* **5**.

* Viterbo,C. & Zanettin,B. 1959.I filoni lamprofirici dell'Alto Baltoro.*Mem.Acad.Patavina Class Sci.Mat. Nat.***71**,3-39.

†Vladykin,N.V. 1985. [First discovery of lamproites in the USSR]*Dokl.Akad.Nauk SSSR* **280(7)**,718-722

Vlasov,K.A.,Kuz'menko,M.Z. & Es'kova,E.M. 1966. *The Lovozero alkali massif.* Oliver & Boyd, Edinburgh, 627pp.

* Vogel,T.A. & Wilbrand,J.T. 1978.Coexisting acid and basic melts:geochemistry of a composite dyke. *J.Geol.* **86**,353-371.

* Vollmer,R.,Ogden,P.,Schilling,J.G.,Kingsley,R.H. & Waggoner,D.G. 1984.Nd and Sr isotopes in ultrapotassic volcanic rocks from the Leucite Hills,Wyoming.*Contrib.Mineral.Petrol.***87**,359-368.

Von Eckermann,H. 1928. Håmrongite,a new Swedish alkaline mica lamprophyre. *Fennia* **50 (13)**.

* Von Eckermann,H. 1948.The alkaline district of Alnö Island.*Sveriges Geol.Unders.***36**.

* Von Eckermann,H. 1958.The alkaline and carbonatitic dykes of the Alnö formation on the mainland NW of Alnö Island. *Kungl.Svenska vetteriskapsakademiem handlinger fjarde serien band* **7(2)**,61pp.

* Von Eckermann,H. 1963.Contributions to the knowledge of the alkaline dikes of the Alnö region. *Ark.Mineral.Geol.* **3(12)**, 259-275 & 521-535.

Von Eckermann,H. 1966a.Progress of research on the Alno carbonatite. In: Tuttle & Gittins(1966), qv, 3-32.

Von Eckermann,H. 1966b.The age relations between the alnöite dykes at Alnö and the sovite pegmatites,and the possibility of inflammable gas beneath the alkalic diatremes.*Arkiv.Mineral.Geol.***4(5)**,paper 12,369-375.

Von Eckermann,H. 1967.A comparison of Swedish, African and Russian kimberlites. In: Wyllie(1967), qv, 302-11.

* Von Eckermann,H. 1968.The chemical and optical properties of some minerals of the Alnö alkaline rocks. *Arkiv.Mineral.Geol.***5**,93-210

von Sauer, A. 1893. Der Granitit von Durbach im nördlichen Schwarzwalde und seine Grenzfazies von Glimmersyenit (Durbachit). *Mitt.Bad.Geol.Landesanstalt* **2**, 231-275.

Vyalov,O.S.& Sobolev,V.S. 1959.Gaussberg,Antarctica. *Int.Geol.Rev.***1(7)**,30-40.

* Wade,A. & Prider,R.T. 1940.The leucite-bearing rocks of the W.Kimberley area,W.Australia. *Q.J.Geol.Soc.Lond.* **96**,39-98.

Wade, F.B. & Oates, F. 1938. Explanation of Degree Sheet No. 52 (Dodoma). *Short Pap. Geol. Div. Dept. Lands and& Mines Tanganyika* **17**, 58pp.

* Wagner,C. & Velde,D. 1985.Mineralogy of two peralkaline,arfvedsonite-bearing minettes.A new occurrence of Zn-rich chromite. *Bull.Minéral.* **108**,173-187.

* Wagner,C. & Velde,D. 1986a.Lamproites in North Vietnam: a re-examination of cocites*J.Geol.***94**,770-76.

* Wagner,C. & Velde,D. 1986b.The mineralogy of K-richterite-bearing lamproites. *Am.Mineral.***71**,17-37.

* Wagner,C. & Velde,D. 1987. Aluminous spinels in lamproites: occurrence and probable significance. *Am.Mineral.* **72**, 689-696.

* Wagner,P.A. 1912.On an interesting dyke intrusion in the upper Waterbeg system.*Tr.Geol.Soc.S.Africa* **15**,26-30.

Wagner,P.A. 1914. *The Diamond Fields of South Africa.* Transvaal Leader, Johannesberg.

* Wahlgren,C.H. & Kahr,A.M. 1977.Mica-lamprophyres in western Varrnland,SW Sweden. *Geol.Foren. Forh.* **99**,291-295.

Wakhaloo, S.N. 1967. On the nature of volcanic eruption and of differentiation of the Girnar igneous complex, Junagarh, Kathiawar peninsula, India. *Proc.Sympos.Upper Mantle Project, Hyberabad*, 430-449. Nat.Geophys.Res. Inst.India.

* Waldman,M.A.,McCandless,T.E. & Dummett,H.T. 1987.Geology and mineralogy of the Twin Knob #1 lamproite, Pike County,Arkansas. *Spec.Pap.Geol.Soc.Am.***215**, 205-216.

Waldmann,L. 1935.Über eine Minette mit Ågirin und Alkalihornblende im Nordmähren. *Verhandl.Geol. Bundesanstalt*, 141-144.

Walker,F. 1921.A monchiquite dyke in Lauderdale.*Tr. Geol.Soc.Edinburgh* **11**,349-350.

* Walker,F. 1927.The igneous geology of Ardsheal Hill,Argyllshire.*Tr.R.Soc.Edinburgh* **55**,147-157.

* Walker,F. 1949.The paragenesis of a hornblende lamprophyre from Bellville.*Tr.Geol.Soc.S.Afr.* **32**, 189-194.

Walker,G.P.L.& Leedal,G.P. 1954.The Barnesmore complex,Co.Donegal.*Sci.Proc.R.Soc.Dublin* **26**, 207-243.

Walker,G.P.L. & Ross,J.V. 1955.A xenolithic monchiquite dyke near Glenfinnan, Inverness shire.*Geol.Mag.* **91**, 463-472.

* Walker,K.R. & Mond,A. 1971.Mica lamprophyre (alnöite) from Radok Lake, Prince Charles Mountains,Antarctica. *Rec.Aust.Bur.Mineral.Resour.Geol.Geophys.* **1971/108.**

Walker,T.L. & Parsons,A.L. 1927.The contact phenomena of the nepheline syenites of Port Coldwell,Ontario. *Univ. Toronto Studies Geol.Ser.***24**,28-32.

Wallace,R.C. 1975.Mineralogy and petrology of xenoliths in a diatreme from South Westland,N.Z. *Contrib.Mineral.Petrol.***49**,191-199.

Walter,P. 1950. Das ostende des basischen GertGesteinzugeszuges Ivrea-Verbano und de angrenzenden Teile der Tessiner Wurzelzone. *Schweiz.Mineral.Petrogr.Mitt.* **30,** 1-144.

Walton,B.J. 1965.Sanerutian appinitic rocks and Gardar dykes and diatremes,N of Narssarssuaq. *Bull.Grønlands Geol.Unders.***57**.

Walton,B.J. 1966.Carbonatite-lamprophyre dykes of Mesozoic age.*Bull.Grønlands Geol.Unders.* **62**,37-38.

* Walton,B.J. & Arnold,A.R. 1970.Plutonic nodules in lamprophyric carbonatite dykes from Frederikshaab,SW Greenland. *Meddr.Grønland* **190(5)**,26pp,and *Bull.Grønl.Geol.Unders.* **91**,1-25.

Warren, J.R., 1956. A study of magnetic anomalies associated with ultrabasic dykes in the Western Kentucky fluorspar district. *Bull.Kentucky Geol.Surv.* **19**, SerIX, 38pp.

* Washington,H. 1917.Chemical analyses of igneous rocks.*U.S.G.S.Prof.Pap.***99**.

Waters,A. 1927.Concerning the differentiation of a lamprophyric magma at Corbaley Canyon,Washington. *J.Geol.***35,**158-170.

Waters,F.G. 1987.A suggested origin of MARID xenoliths in kimberlites by high pressure crystallisation of an ultrapotassic rock such as lamproite.*Contrib.Mineral.Petrol.***95**,523-533.

* Watkeys,M.K. & Armstrong,R.A. 1985.The importance of being alkaline - deformed late Archaean lamprophyric dykes,Central zone,Limpopo Belt.*Tr.Geol.Soc.S.Afr.***88**,195-206.

@ Watson,J. 1975. The Lewisian complex. *Spec.Rept.Geol.Soc.Lond.***6**, 15-29.

Watson,J. 1984. The ending of the Caledonian orogeny in Scotland. *J.Geol. Soc. Lond.* **141**, 193-214.

* Watson,K.D. 1955.Kimberlite at Bachelor Lake, Quebec.*Am. Mineral* **40**,565-579.

* Watson,K.D. 1957.Hornblende lamprophyre dykes from SW Lesuer township,Ontario.*Can.Mineral.***6**,15-30

Watson,K.D. 1967a. Kimberlite pipes of northeastern Arizona. In: Wyllie (1967), qv, 261-268.

Watson,K.D. 1967b.Kimberlites of eastern North America. In: Wylie (1967), qv, 312-322.

Watson,K.D., Bruce,G.S.W. & Halladay,L.B. 1978. Kimberlitic dyke in Keith township, Ontario. *Can.Mineral.* **16**, 97-102.

* Watt,W.S. 1966.Chemical analyses from the Gardar igneous province,S.Greenland. *Rep.Geol.Surv. Greenland* **6**,92pp.

* Watterson,J. 1968.Plutonic development of the Ilordleq area,part 2:late kinematic basic

dykes.*Meddr.Grønland* **185(3)**; also *Grønlands Geol.Unders.* **70**, 104pp.
Watznauer,A. 1964.Der heutige Stand des Lamprophyrproblems in der Lausitz.*Geologie* **13 (H6/7)**,813-820.
* Webb, D.R. & Kerrich, R. 1988. An Archean ultramafic lamprophyre, Yellowknife: implications for tectonics and source regions. *Contrib.Geol. Northwest Territories* **3**, 115-122.
Wedepohl, K.H. 1978. *Handbook of geochemistry*. Springer-Verlag, Berlin.
Weed,W.H. & Pirsson,L.V.1896.Geology of the Castle Mountains mining district. *Bull.U.S.G.S.* **139**.
Weller,S. & St. Clair,S. 1928. Geology of St.Genevieve Country, Missouri. *Missouri.Bur.Geol.Mines 2nd Ser.* **22**, 352pp.
Wellman,P. 1973.Early Miocene potassium-argon age for the Fitzroy lamproites of W.Australia. *J.Geol.Soc. Aust. 19*(**4**), 471-474.
Wellman,P. & Cooper,A. 1971.Potassium-argon ages of some New Zealand lamprophyre dikes near the Alpine fault. *N.Z. J.Geol.Geophys.***14(2)**,341-350.
Wellman,P.,Cundari,A. & McDougall,I. 1970.Potassium-argon ages from leucite-bearing rocks from NSW. *J.Proc.R.Soc.NSW* **103**,103-107.
@ Wells,A.K. & Bishop,A.C. 1955.An appinitic facies associated with certain granites in the Channel Isles. *Q.J.Geol. Soc.Lond.* **111**,143-166.
* Wells,R.C. 1937.Analyses of rocks and minerals from the laboratory of the USGS, 1914-36.*Bull.U.S.G.S.* **878**,134pp.
* Wendlandt,R.F. 1977.Barium-phlogopite from Haystack Butte,Highwood Mountains,Montana. *Carn. Inst.Wash.Yrbk.* **76**,534-539.
Wendlandt,R.F. 1984. An experimental and theoretical analysis of partial melting in the system $KAlSiO_4$-CaO-MgO-SiO_2-SiO_2 and applications to the genesis of potassic magmas, carbonatites and kimberlites. In: Kornprobst (1984), qv, 359-370.
Westoll,N.D.S. & Miller,J.A. 1969.The age of some kentallenite intrusions in Argyll.*Scott.J.Geol.***5**,11-14.
Whittingham,J.K. 1959. Brief explanation of the geology, Quarter Degree Sheet 215 Iringa. *Geol.Surv. Tanganyika* 1:125,000 map.
* Wierzcholowski,B. 1977.[Dyke-rocks of the Klodzo-Zloty stock granitoid massif]. In Polish,Engl.summ. *Geol.Sudetica* **12(2)**,7-28.
* Wierzcholowski,B. 1979.[Petrological and chemical study of the minette from Stojkow,Middle Sudetes]. In Polish, Engl.summ.*Arch.Mineral.(Warsaw)***35(2)**,67-77.
Wilkinson, H.E. 1988a. Bendigo geology. In: Jones (1988), qv, 17-21.
Wilkinson, H.E. 1988b. Bendigo reef nomenclature. In: Jones (1988), qv, 22-27.
Wilkinson,J.F.G. 1977. Analcime phenocrysts in a vitrophyric analcimite — primary or secondary? *Contrib.Mineral.Petrol.* **64**, 1-11.
Wilkinson, P. 1986. *SYSTAT: the system for statistics*. Systat Inc., Evanston, IL.
Willemse,J. 1937.On the old granite of the Vredefort region and some of its associated rocks. *Tr.Geol.Soc.S.Afr.* **40**,43-119.
Williams,D. 1923.The Cronkley mica lamprophyres.*Proc.Liverpool Geol.Soc.***13**,323-334.
Williams,G.H. 1888. The peridotites of the Cortlandt series on the Hudson River near Peekskill. *Am.J.Sci.*, **3rd Ser.**, **31**, 26-31.
Williams,H.1890.The geology of Arkansas.*Ann.Rep.Geol.Surv.Arkansas for 1890*,vol.II.
* Williams,H. 1936.Pliocene volcanoes of the Navajo-Hopi country.*Bull.Geol.Soc.Am.***47**,111-172.
Williams,H.,Turner,F.J. & Gilbert,C.M. 1982.*Petrography*. (2nd Edn.)Freeman,San Fransisco, 626pp.
Williams,M.Y. & Dyer,W.S. 1930.Geology of southern Alberta and SW Saskatchewan. *Mem.Geol.Surv.Can.* **168**,78-80.
Willmann,K. 1920. Die Redwitzite, eine neue Gruppe von granitischen Lamprophyren. *Z.Dtsch.Geol. Gesellschaft* **71 (1/2)**,1-33.
Wilshire,H.E., Meyer, C.E., Nakata,J.K., Calk,L.C., Shervais J.W., Nielson,J.E. & Schwarzman, E.C. 1989. Mafic and ultramafic xenoliths from volcanic rocks of the Western United States. *USGS. Prof. Pap.* **1443**, 179pp.
Wilson, A.E. & Gilligan,A. 1924. The basic dykes and sills of the Howgill Fells. *Proc.Yorks.Geol.Soc.* **21**,59-70.
Wilson, H.E. 1972. *Regional Geology of Northern Ireland*. HMSO, Belfast.
Wilson, J.S.G. 1886. Explanation of sheet 87. *Mem.Geol.Surv. G.B.*
* Wilson,N.W. 1965.Geology and mineral resources of part of the Gola forests, south-eastern Sierra Leone. *Bull.Geol.Surv.Sierra Leone* **4**.
Wimmenauer, W. 1966. The eruptive rocks and carbonatites of the Kaiserstuhl, Germany. In: Tuttle & Gutins (1966), qv, 183-204.
Wimmenauer,W. 1972.Die Lamprophyre des Schwarzwaldes.*Fortschr.Mineral.***50**,34-37.

Wimmenauer,W. 1973a.Granites et lamprophyres.*Bull.Soc.Géol.Fr.***15(3)**,195-198.

* Wimmenauer,W. 1973b. Lamprophyre,Semilamprophyre und anchibasaltische Ganggesteine. *Fortschr. Mineral.* **51**,3-67.

Wimmenauer,W. 1974.The alkaline province of central Europe and France. In: Sørensen(1974), qv, 238-271.

* Wimmenauer,W. 1976.Lamprophyres and associated rocks in the dike suite of the Dikeos monzonite,Kos. *Ann.Geol.Pays.Hell.***28**,427-435.

* Wimmenauer,W. & Hahn-Weinheimer,P. 1966.Geochemische und petrographische Kriterien für die Herkunft von Kersantiten und Minetten.*N.Jb.Mineral.Mh.***7**,201-220.

Winchester,J.A. 1988. *Later Proterozoic Stratigraphy of the Northern Atlantic Regions.* Blackie, Glasgow, 279pp.

@ Windom,K.E. & Boettcher,A.L. 1980.Mantle metasomatism and the kimberlite-lamprophyre association: evidence from an eclogite nodule from Roberts Victor mine,S.Africa.*J.Geol.* **88**, 705-712.

* Witkind,I.J. 1969.Clinopyroxenes from acidic,intermediate and basic rocks,Little Belt Mountains,Montana. *Am.Mineral.* **54**, 1118-1138.

* Witkind,I.J. 1970.Composite dikes in the Little Belt Mountains,Central Montana.*U.S.G.S.Prof.Pap.* **700-C**,82-88.

* Witkind,I.J. 1973.Igneous rocks and related mineral deposits of the Barker quadrangle,Little Belt Mountains,Montana. *U.S.G.S. Prof.Pap.***752**,1-58.

Wolff,J.E. 1929.Mount Monadnock,Vermont-a syenite hill.*J.Geol.***37**,1-16.

@? Wolff,J.E. 1938.Igneous rocks of the Crazy Mtns.,Montana.*Bull.Geol.Soc.Am.***49**,1569-1625.

* Woodland,B.G. 1962.Lamprophyric dykes of the Burke area,Vermont.*Am.Mineral.***47**,1094-1110.

Woods,M.J. 1976.Fractionation and origin of the Highwood Mountain volcanics.*Northwest Geol.***5**,1-9.

* Woodward,L.A. 1970.Differentiation trends of spessartite dykes,Sandia Mountains,New Mexico. *J.Geol.* **78**,741-745.

Woolley, A.R. 1987. *Alkaline Rocks and Carbonatites of the World. Part 1: North and South America.* British Museum (Natural History), London, 216pp.

* Woolley, A.R. & Jones, G.C. 1987. The petrochemistry of the northern part of the Chilwa alkaline province, Malawi. In: Fitton & Upton (1987), qv, 335-356.

* Woussen,G. 1970.La géologie du complex ignée du Mont Royal.*Can.Mineral.* **10**,432-451.

Wright,A.E. & Bowes,D.R. 1968.Formation of explosion breccias.*Bull.Volc.***32**,15-32.

* Wright,A.E. & Bowes,D.R. 1979.Geochemistry of the appinite suite. In: *Caledonian of the British Isles — Reviewed* (eds.Harris A.L.et al.) *Spec.Pub.Geol.Soc.Lond.* **8**,699-704.

Wright,J.B. 1966. Olivine nodules in phonolite. *Nature* **210**, 519.

Wright,J.B. 1969. Olivine nodules in trachyte. *Nature* **223**, 285-286.

Wroblewski,T. 1974.[Geologic and structural characterisation of the location of acidic lamprophyre in Sierakow near Daleszyce]. In Polish,Engl.summ.*Pol. Inst.Geol.Biul.***275**,95-112.

Wyllie, P.J. 1967. *Ultramafic and Related Rocks.* Wiley,New York.

Wyllie, P.J. 1989. The genesis of kimberlites and some low-SiO_2, high-alkali magmas. *Spec. Publ. Geol. Soc. Aust.* **14**, 603-615.

Wyman, D.A. & Kerrich, R. 1988a. Alkaline magmatism, major structures and gold deposits: implications for greenstone belt gold metallogeny. *Econ. Geol.* **83**, 454-461.

Wyman, D.A. & Kerrich, R. 1988b. Lamprophyres a source of gold. *Nature* **332**, 209-210.

* Wyman, D.A. & Kerrich, R. 1989a. Archean lamprophyres of the Superior province, Canada: distribution, petrology and geochemical characteristics. *J.Geophys.Res.* **94 (B4)**, 4667-4696.

* Wyman, D.A. & Kerrich, R. 1989b. Archean shoshonitic lamprophyres associated with Superior Province gold mineralization: distribution, tectonic setting, noble metal abundance and significance for gold mineralization. *Econ.Geol.Monograph* **6**, 661-667.

* Yagi,K.,Ishikawa,H. & Kojima,M. 1975.Petrology of a lamprophyre sheet in Tanegashima Is.,Kagoshima prefecture,Japan. *J.Jap.Assoc.Mineral.Petrol.Econ.Geol.***70**,213-224.

* Yagi,K. & Matsumoto,H. 1966.Note on leucite-bearing rocks from the Leucite Hills,Wyoming. *J.Fac.Sci.Hokkaido Univ,Ser.IV, Geol.Mineral.* **13**,301-312.

Yen,T.P. 1985.The so-called lamprophyres of Taiwan.*Acta Geol.Taiwanica* **23**,9-18.

Yeremenko,G.K. 1968.[Essexite xenoliths in the monchiquites of the Azov region]. In Russian, Engl. summ. *Dopov.Akad.Nauk.Ukr.RSR, Ser.B*,**7**,590-593.

Yeremenko,G.K. & Shvakova,A.A. 1969.[Titanium-bearing garnet from the xenoliths of monchiquite dikes of the Azov sea area]. *Dopov.Akad.Nauk.Ukr.RSR.,Ser.B*,**11**,971-974.

Yoder,H.S. 1979.Melilite-bearing rocks and related lamprophyres. In: *Evolution of the Igneous Rocks:50th Anniversary Perspectives* (ed.H.S.Yoder),391-411.Princetown Univ.Press.

Yoder,H.S. & Tilley,C.E. 1962. Origin of basalt magmas. *J.Petrol.* **3**, 343-532.

Yoder,H.S. & Velde,D. 1976. Importance of alkali content of magma yielding melilite-bearing rocks. *Carn.Inst.Wash.Yrbk.* **75**, 580-585.

Yoshizawa,H. & Ishizaka,K. 1961.A structural and genetical study of the lamprophyre of Mt.Hiei,Kyoto, Japan. *Kyoto Univ.Coll.Sci.Mem.Ser.B,* **28(2)**,147-158.

Yur'yev,L.D. 1967a.[Kaersutite and some secondary changes in lamprophyre near village of Khlebodaravka]. In Ukranian, Engl. summ. *Dopov.Akad.Nauk.Ukr.RSR,Ser.B*,**10**, 912-916

Yur'yev,L.D. 1967b.[Saponite from the lamprophyres of the Azov region]. In Russian,Engl.summ. *Mineral.Sborn. (L'vov. Gos. Univ.),* **21(3)**,297-9.

* Zalutski,V.V. 1962.[Lamprophyres of central part of auriferous Lena region].*Geol.I geofiz.SSSR*, **5**,44-55.

Zartman,R.E.,Brock,M.R.,Heyl,A.V. & Thomas H.H. 1967.K-Ar and Rb-Sr ages of some alkalic intrusive rocks from central and eastern United States.*Am.J. Sci.* **265**,848-870.

* Zavaritsky,A.N. 1935.[The lamprophyres of Magnitnaya Mountain and related rocks]. In Russian. *Trav. Inst.Petrogr.Loewinson-Lessing,Acad.Sci.USSR*,**5**,115-126. See *Mineral. Abstr.* **6**,p.322.

Zezza,V. 1969.Filoni diabasici e lamprofirici nel granito del Biellese.*Atti Soc.Ital.Sci.Nat.Mineral.Civ.St. Nat.Milano* **109(4)**,511-538.

† Zhuykova,M.P. & Bogacheva,L.D. 1968.[Biotite lamprophyres of the Chelyabinsk-Alapayensk volcanogenic zone,E.Slope of Urals]. In Russian. In: [*Volcanism in the Urals*], 187-192. Akad.Nauk.SSSR Ural.Filial. Inst. Geol.Geokhim,Sverdlovsk.

Zimmerle,W. 1958.Die lamprophyrischen Ganggesteine des Marlsburgplutons im südwestlichen Schwarzwald. *Ber.Naturforsch.Ges. Freiburg. i.Br.***48**,175-230.

Zimmerle,W. 1977.Sphärische Texturen in Lamprophyren aus der kristallinen Umrahmung des Oberrheingrabens. *Oberrhein.Geol.Abh.***26(1-2)**,55-113.

Zujovic J. 1890. Les lamprophyres de Serbie. *Belgrade Ann.Geol.p.B. IX fasc.***2**, 76-168.

Appendix A: Abbreviations and Symbols

Explanation:

Chemical elements are referred to by standard symbols (Ba, Ce, Cr, etc.).

Measurement units also have standard symbols (cm, ha, km, etc.)

Mineral names have two letters, starting with an initial capital letter (coined to avoid duplication with chemical element symbols). The only exceptions are Cpx and Opx, used because of their already widespread adoption. Normative minerals have the same abbreviations as modal minerals, but are *italicised* and in lower-case, e.g. *ab, an, cm, qz*.

Rock-names and other terms have more than two letters, all in lower case.

Lamprophyre varieties have 2/3 letters and are in CAPITALS (e.g. AC), following Fig.1.2, App.B.

Compound names are hyphenated e.g. grt-porph = granite-porphyry, Gt-gran = garnet-granulite, Sp-lhz = spinel-lherzolite; they may include element symbols, e.g.Cr-Di = chrome-diopside.

Aa	adularia
Ab	albite
AB	bostonite (App.B)
abs	absarokite (App.B)
Ac	analcime
AC	camptonite (Fig.1.2; App.B)
AD	doleritic AL (Fig.1.3)
Ad	andalusite
Ae	aegirine(-augite)
Af	alkali feldspar
ag	agpaicity index = molar $\{Na_2O+K_2O\}/Al_2O_3$
Ah	anhydrite
Ai	almandine
ak	normative akermanite
AL	alkaline lamprophyre (App.B)
AM	monchiquite (Fig.1.2; App.B)
Am	amphibole
amph	amphibolite
An	anorthite
andes	andesite
ano	anorthosite
Ao	anorthoclase
AO	ocellus in AL
Ap	apatite
apl	aplite
App.	Appendix
ark	arkose
AS	sannaite (Fig.1.2; App.B)
As	Association (Table 2.1)
assoc.	associated with
At	actinolite
ban	banatite
bas	basalt
Bi	biotite–phlogopite
By	baryte
CA	appinite (Fig.1.2; App.B)
CAL	calc-alkaline lamp (App.B)
Cb	carbonate (Cc, Dl, etc.)
cbt	carbonatite
Cc	calcite
CC	malchite (App.B)
CE	kentallenite (Fig.1.2; App.B)
Ch	chromite
C.I.	colour index (modal mafics
Ci	cordierite
CK	kersantite (Fig.1.2; App.B)
CM	minette (Fig.1.2; App.B)
Cm	corundum
cong	conglomerate
CP	porphyry (App.B)
Cpx	clinopyroxene
CS	spessartite (Fig.1.2; App.B)
Ct	chlorite
cum	cumulate
CV	vogesite (Fig.1.2; App.B)
D.I.	Thornton-Tuttle differentiation index: *qz* + *ab* + *or* + *ne* + *lc*
Di	diopside
dio	diorite, microdiorite
Dl	dolomite (mineral)
dlr	dolerite
dun	dunite
ecl	eclogite
En	enstatite
Ep	epidote/clinozoisite
Fe_2O_3*	total Fe as Fe_2O_3
FeO*	total Fe as FeO
fels	felsite
fen	fenite
Fo	forsterite
Fp	feldspar (alkali or plagioclase)
Fs	ferrosilite
G1..*n*	garnet groups of Dawson & Stephens (1975)
Ga	billion years (or gallium)
gbr	gabbro
gdms	groundmass
ge	normative gehlenite
glm	glimmerite

gns	gneiss
Gr	grossular garnet
grdi	granodiorite
gran	granulite
grt	granite
Gt	garnet
Hb	hornblende
hfls	hornfels
HFSE	high field-strength element (e.g. P, Zr, Nb, Th)
Hm	haematite
HREE	heavy rare-earth (Er to Lu)
hrz	harzburgite
Il	ilmenite
int	intergrowth (two or more entries indicates mutuality)
intr	intrusion
K1	Group 1 (mica-poor) KIL
K2	Group 2 (mica-rich) KIL
ka	normative $KAlSiO_4$ (kalsilite)
Kf	K-feldspar
KIL	kimberlite (Fig.1.2; App.B)
ks	normative K_2SiO_3
Kt	kaersutite
KU	rock transitional between kimberlite and aillikite
la	normative Ca_2SiO_4 (larnite)
lamp (L)	lamprophyre
Lc	leucite
LC	cocite family (lamproites transition to minettes: App.B)
LF	fitzroyite lamproite (App.B)
lhz	lherzolite
LILE	large-ion lithophile element (e.g. K,Rb,Sr,Ba,Cs)
LJ	jumillite lamproite (App.B)
LL	lamproite (Fig.1.2; App.B)
LO	olivine-lamproite (App.B)
LREE	light rare-earth (La to Nd)
lst	limestone
LW	wyomingite lamproite (App.B)
Ma	million years
mac	macrocryst/megacryst
MARID	Mica-Amphibole-Rutile-Ilmenite-Diopside inclusions
Mc	monticellite
MDA	multigroup linear discriminant analysis
Me	melilite
metd	metamorphosed
mg	magnesium-number (mole % Mg/[Mg+Fe^2]: 65-80 for primary (App.B) magmas)
mg#	*mg* with molar Fe^3/[Fe^3+Fe^2] normalized to lamprophyre ratio of 55% (Table 5.1)
mig	migmatite
monz	monzonite
MORB	mid-ocean-ridge basalt
Mt	magnetite and iron-spinels
Mtn.	mountain
Ne	nepheline
Ne-sye	nepheline syenite
ns	normative Na_2SiO_3
Occ	occurrence
Og	oligoclase
OIB	ocean-island (alkali) basalt
Ol	olivine
oliv	olivinite
Opx	orthopyroxene
Or	orthoclase
peg	pegmatite
perid	peridotite
Pg	pargasite
PGE	platinum-group element (Ru,Rh,Pd,Os,Ir,Pt)
Ph	phlogopite
phen	phenocryst
phon	phonolite
Pl	plagioclase feldspar
Po	pyrope garnet
porph	porphyry (App.B)
porpht	porphyrite (App.B)
ppb	parts per billion
ppm	parts per million
Pr	prehnite
Ps	pseudoleucite
psm	psammite
publ.	published
Pv	perovskite
Px	pyroxene
pxnt	pyroxenite
Py	pyrite
qqz	quartzite
Qz	quartz
Rc	richterite
REE	rare-earth element
rhy	rhyolite
Ru	rutile
Sa	sanidine
sch	schist
Sd	sodalite, nosean, haüyne
serp	serpentine
shonk	shonkinite
shosh	shoshonite (App.B)
Sl	sillimanite
Sp	magnesian spinel
sst	sandstone
St	staurolite
sye	syenite, microsyenite
Th	thompsonite
ting	tinguaite
TNT	Ti-Nb-Ta, referring to –ve anomalies on spidergrams
Tr	tremolite-actinolite
trach	trachyte

UA	aillikite (Fig.1.2; App.B)	var	various
UD	damkjernite (Fig.1.2; App.B)	web	websterite
UL	alnöite (Fig.1.2; App.B)	weh	wehrlite
um	ultramafic	Wo	wollastonite
UML	ultramafic lamprophyre (Fig.1.2; App.B)	WR	whole-rock
		xen	xenolith/xenocryst
UO	ouachitite (Fig.1.2; App.B)	Zc	zircon
UP	polzenite (Fig.1.2; App.B)	Ze	zeolite minerals

Appendix B: Glossary of lamprophyres and ambiguous petrological terms

This glossary defines all known lamprophyric rock-names, and clarifies other terms whose definitions are obscure or controversial. Terms considered obsolete are *asterisked. Rock-names specifically approved by the IUGS Subcommission on the Systematics of Igneous Rocks (Streckeisen 1976, 1979; Le Maitre 1989) are in **BOLD CAPITALS**. Other definitions are based on the consensus in Johannsen (1938), Sørensen (1974), Tomkeieff (1983), Nixon (1987, p.761) and Bates & Jackson (1989).

Cross-references to other entries below are in **bold type.** In the main text, enclosure of a cited term in "double quotes" indicates that the usage does *not* agree with this glossary.

Absarokite. The basic member of the **shoshonite association.** A mildly potassic basaltic rock composed of plagioclase and clinopyroxene ± olivine and minor K-feldspar. Type locality Absaroka, Wyoming, USA. [Definition after Joplin 1968].

Accidental. Applies to **xenoliths** and **xenocrysts** of country wall rocks caught up by an intrusion. Opposite of **cognate** (Table 6.1).

Ach'uaine hybrid suite. A suite of highly heterogeneous, olivine-hornblende-biotite-rich rocks forming irregular, pipe-like bodies satellite to Caledonian granitoid plutons in Sutherland, Scotland. Range from ultramafic rocks (e.g. **scyelite**) to acidic rocks, with intermediate members formed by hybridisation of the two. Associated with calc-alkaline lamprophyre dykes, and closely related to the **Appinite suite.** Type locality Achu'aine, Sutherland, Scotland. The name means *green field* in Gaelic, and refers to the fertility of the soil over these bodies by comparison with the brown heath over the Moine psammite country-rocks.

Aillikite. See Table 1.1. Major variety of **ultramafic lamprophyre** consisting of combinations of forsteritic olivine, diopsidic pyroxenes, various amphiboles and phlogopite in a matrix of similar minerals with abundant primary carbonate and often minor perovskite, but little or no melilite or feldspathoids. Covers most rocks previously termed **central complex kimberlite.** Type locality Aillik Bay, Canada; named by Kranck (1939).

* **Aiounite.** An obscure local name for an augite-apatite-rich variety of perhaps **hyalomonchiquite.**

Alkaline lamprophyres. A group of **lamprophyres** consisting of the rock-types **camptonite, monchiquite, sannaite**, which correspond broadly to alkali basalts, basanites and nephelinites, but are considerably more enriched in volatile-elements (H_2O,CO_2, F, Cl) and to some extent in LILE elements (K, Rb, Ba, etc.)

Alkaline rock. An igneous rock carrying feldspathoids and/or soda pyriboles (aegirine, arfvedsonite, etc.), implying the presence of normative *ne, lc, la* and/or *ac, ns* , *ks.* [Definition after Woolley 1987].

ALNÖITE. See Table 1.1. Major variety of **ultramafic lamprophyre** consisting of combinations of forsteritic olivine, diopsidic pyroxenes, various amphiboles and phlogopite in a matrix of similar minerals with abundant melilite and often minor perovskite, primary carbonates and/or feldspathoids. Type locality Alnö, Sweden; named by Von Eckermann (1948).

Amygdale. Genetic term for a particular type of **globular structure** formed by condensation of a late-stage hydrothermal fluid into pre-existing cavities formed by gas bubbles.

Anchibasaltic rock. Termed applied by Wimmenauer (1973b) to **alkaline lamprophyres.**

Ankaratrite. A biotite-rich nephelinite, marking among the closest equivalents among volcanic rocks to **ultramafic lamprophyres** (Section 7.3.4). The type rock from Madagascar has Ol 10, Cpx 26, Bi 8, Ne 14, Il 12 (Johannsen 1938) and fits the present definition of **ouachitite**; an associated *melilite-ankaratrite* fits the definition of **alnöite.**

* **Antsohite.** An obscure local name for a lamprophyric dyke rock from Antsohy, Madagascar, consisting of phenocrysts of biotite in a groundmass of biotite, hornblende and interstitial quartz. Probably a feldspar-poor **kersantite**.

Appinite. See Section 7.1.1. Originally defined by Bailey & Maufe in 1916 as "plutonic equivalents of the hornblende vogesites and spessartites" (*see* Bailey 1960), but since used to refer to a panoply of mostly unrelated hornblende-phyric minor intrusions, ranging from gabbro to monzogabbro, diorite, syenite and even granodiorite. *Appinitic* has generally been used to imply the presence of coarse euhedral hornblendes and an association with more basic appinites. Restricted here to coarse-grained, K-rich, mafic equivalents of **vogesites** and **spessartites**, consisting essentially of abundant stumpy or prismatic hornblende ±

clinopyroxene, in a matrix of plagioclase ± alkali feldspar, etc. The type appinites of Appin, Argyll (Scotland) are predominantly gabbroic, but many contain biotite and are more K-rich than normal gabbros.

Appinite suite. See Section 7.1.1. Term used in various previous senses, but here restricted to pipes and diatremes intimately associated with CAL dyke-swarms and granitoids. Covers the rock-types **Ach'uaine hybrid, appinite, kentallenite, scyelite,** as well as various hornblende ±biotite-rich ultramafic, monzonitic, granodioritic and syenitic rocks. Excludes the older and variably metamorphosed *microdiorite suite* of D.I.Smith (1979), which in turn excludes the *West Highland appinite suite* of MacGregor & Kennedy (1931) — the latter was so named because of assumed relationships which are now known to be false (D.I.Smith 1979). See below for term **suite.**

* **Aschaffite.** An obscure local name for a xenocrystic variety of **kersantite.**

Autolith. A cognate inclusion of rock (rounded or angular) in a host of distinct composition, both being derived from a common parent magma (see Table 6.1). This is a more general definition than Nixon (1987), who uses the term specifically for *spheres* nucleated around fragments.

* **Baltorite.** An obscure local name applied to **vogesites** from the Baltoro Glacier, Karakorum, Pakistan.

Battlemented. See **castellated.**

Belt. A geographical cluster of igneous rock **occurrences**, forming part of a **province.** A **field** is a specific type. Analogous to *formation* in lithostratigraphical nomenclature. [Definition of Rock 1981].

* **Bergalite.** An obscure local name coined by Soellner (1913) for a variety of **polzenite.**

Berondrite (= foid **mafraite**). A hydrous theralite (foid gabbro), composed of kaersutite, titanaugite, plagioclase, foid (mostly nepheline) ± biotite ± minor alkali feldspar. Named after Berondra, Madagascar.

* **Bizardite.** An obscure local name coined by Stansfield (1923a) for **ultramafic lamprophyres** (mainly **aillikites**) from Ile Bizard, Montréal, Canada. The occurrence has proved controversial (cf. Raeside & Helmstaedt 1982; R.H.Mitchell 1983).

Bostonite. See Section 8.5–8.6. A useful term for often altered, mildly alkaline, porphyritic, leucocratic dyke rocks intimately associated with some **camptonite-monchiquite** dyke-suites. Corresponds in IUGS terminology to *porphyritic (biotite-hornblende) alkali feldspar trachyte* or *micro-alkali feldspar syenite.* SiO_2 content *c.* 60–65%. Type locality Boston, Mass, USA. Grades with increasing CI into **maenite.**

Calc-alkaline lamprophyres. A group of **lamprophyres** consisting of the rock-types **minette, vogesites, kersantite, spessartite, appinite** and **kentallenite**, which correspond very roughly to andesites but are considerably enriched in volatile elements (H_2O, CO_2, F, Cl), and in LILE and mafic-ultramafic elements (K, Rb, Ba, Mg, Cr, Ni, etc). More closely approximated geochemically by some **shoshonites** and **absarokites.**

CAMPTONITE. See Table 1.1. Major variety of **alkaline lamprophyre,** composed of combinations of forsteritic olivine, kaersutite, titanaugite and titanbiotite phenocrysts in a matrix of the same (minus olivine), with plagioclase and sometimes subordinate alkali feldspar and feldspathoids. Type locality Campton Falls, New Hampshire, USA; named by Hawes in 1878 and redefined by Rosenbusch.

* **Camptospessartite.** A confusing old name for **camptonites** from Lausitz, Germany, meant to indicate their occurrence in a 'Pacific' province; some are probably not lamprophyres at all.

* **Cancarixite.** An obscure local name for a diopside-phlogopite-sanidine **lamproite** from Cancarix, Spain.

* **Cascadite.** An obscure local name coined by Pirsson for a variety of lamprophyre transitional between **minette** and **lamproite**, from Cascade Creek, Highwood Mtns., Montana, USA.

Castellated. Refers to the characteristic texture of biotite-phlogopite phenocrysts in many **lamprophyres** (especially **minettes**), in which the shape resembles the battlements of a castle. See Fig.3.2d.

* **Cedricite.** A leucite-diopside **lamproite**. Type locality Mt.Cedric, W.Kimberley, Western Australia.

* **Central complex kimberlite.** See Section 1.6.6. Term introduced by Dawson (1971,1980) and formerly applied to aillikites occurring in carbonatite complexes. Recommended for abandonment by R.H.Mitchell (1970,1979,1986) and Dawson (1987,1989).

Cocite. See Fig.1.2. A local name coined by Lacroix for a olivine-diopside-phlogopite-sanidine-(leucite) rock from Coc Pia, North Vietnam, which Wagner & Velde (1986a) found to display some of the mineralogical characteristics of **lamproites** (e.g. leucite and magnetite), and some of **minettes** (e.g. no Ti-rich oxides or K-richterite). Applied here to all rocks transitional between **minettes** and **lamproites**, including those of Coc Pia, Sisco, Pendennis and Holsteinsborg (Table C4). These have low Al for minettes, and K/Al approaching unity, but do not carry the exotic mineralogy of true lamproites.

Cognate. Refers to phases which crystallized *at some stage* from the magma represented by their host rock; includes both assemblages which are in strict equilibrium with their host (**phenocrysts, autoliths**) and semi-cognate, high-*P* precipitates at greater depths. Opposite of **foreign/accidental** (Table 6.1).

Composite dyke/sheet/sill. A body formed of two or more magma injections of *distinct* compositions such as lamprophyre and porphyry (cf. **multiple**). See Table 3.1.

Coppaelite. A local term for a rock composed of pyroxene, melilite and variable amounts of phlogopite. Probably identical with alnöite. Named after Coppaeli di Sotto, Umbria, Italy.

Cortlandtite. A hornblende-rich peridotite (olivine-hornblendite) which can be regarded as an ultramafic variety of **appinite.** Type locality Cortlandt complex, New York, USA. Named by Williams (1888).

* **Cuselite.** An obscure local name coined by Rosebusch for a variety of **calc-alkaline lamprophyre** intermediate between minette, vogesite, kersantite and spessartite, originally from Cusel, Saar, Germany.

Damkjernite (damtjernite). See Table 1.1. Rare variety of **ultramafic lamprophyre** consisting of combinations of forsteritic olivine, diopsidic pyroxenes, amphiboles and phlogopite in a matrix of similar minerals with abundant primary carbonates and often minor perovskite and/or feldspathoids, together with minor alkali feldspar but no melilite. Type locality Damtjern, Fen area, Norway; named by Brøgger (1921).

Diatreme. See **vent.**

Differentiation. Used as a general term to cover any igneous process by which a melt changes composition *without external intervention*: includes fractional crystallization, liquid immiscibility, melt segregation, loss of volatiles, etc. but *not* hybridisation or contamination.

Durbachite. See Section 7.1.2. A local term coined by von Sauer (1893) for mela-syenitic rocks from Durbach, Schwarzwald, Germany composed of coarse biotite, pale green amphibole, orthoclase and prominent apatite. The type-rock forms the marginal facies of a granite pluton and grades into almost pure biotite rock, but other examples form enclaves, dykes and fairly large stocks. Probably a plutonic equivalent of **minette** or **kersantite.**

Dyke. A steeply-dipping *and* strongly discordant minor intrusion. Cf. **sheet, sill.**

* **Ehrwaldite.** An obscure local name applied by Trommsdorf (1962a) to some **monchiquites** near Ehrwald in the Northern Calcareous Alps, Austria.

Enclave. Non-genetic term for a fragment of rock (rounded or angular) in a host of distinct composition. Covers both **autholith** and **xenolith.**

* **Espichellite.** An obscure local name applied to an analcime-rich **camptonite** from Cabo Espichel near Lisbon, Portugal. Composed of Pl 48, Ac 9, Aug+Hb 8, accessories 5%.

* **Eustratite.** An obscure local name for a glassy variety of **camptonite** from Haghios Eustratios Island, Aegean Sea. Composed of Af+Gl 34, Pl 9, Bi 6, Ol 5, opaques 20%.

* **Farrisite.** An obscure local name coined by Brøgger for a variety of **alnöite** from Lake Farris, Oslo Province, Norway.

Field. A geographically distinct cluster of **pipes** and/or **vents**, forming part of a **province.** A specific type of **belt.** Analogous to *formation* in lithostratigraphical nomenclature. [Definition of Rock 1981].

Fitzroyite. A (diopside)-phlogopite-leucite **lamproite.** Type locality Fitzroy River, W.Kimberley region, Western Australia. Used here as a family term for all the lamproites of similar petrology (Fig.1.2).

* **Florinite.** An obscure local name for a melanocratic variety of **monchiquite** with phenocrysts of olivine and augite and biotite in an altered matrix.

* **Fortunite.** A phlogopite-**lamproite** often carrying enstatite (which may be xenocrystic). Type locality Fortuna, SE Spain.

* **Fourchite.** Has been applied in numerous conflicting senses: originally applied to olivine-free **monchiquite** from Fourche Mtn., Arkansas, USA (Williams 1890), but subsequently to varieties of basanites (Fenner 1938). Since the type monchiquites are often olivine-poor, the term is superfluous.

* **Fraidronite.** An obscure local name for a variety of **minette** (see Velde 1970) from Brittany.

* **Garganite.** An obscure local name for a variety of **vogesite** forming the centre of a dyke from Foggia, Italy whose margins are **kersantite.** May be a hybrid, or possibly **heteromorphic.**

* **Gaussbergite.** An obscure local name for an olivine-leucite **lamproite** from Gaussberg, Antarctica.

* **Gauteite.** An obscure local name for intermediate dyke-rocks, here termed **maenites.** Named after Gaute (Kouty), Cseske Stredehori, Czecoslovakia.

* **Giumarrite.** An obscure local name for a variety of hornblende **monchiquite** from Giumarra, Sicily.

Globular structure. Non-genetic, group term for irregular, drop-like, subrounded or circular, leucocratic to hololeucocratic bodies in mafic (e.g. lamprophyric) host-rocks. Composed of combinations of apatite, biotite, carbonate, chlorite, epidote feldspar, feldspathoid (analcime, nepheline) or quartz, occasionally with amphibole or pyroxene. Some globular structures are dominated by silicate, others by carbonate minerals. Covers **amygdales, ocelli, segregations,** spherules, **varioles.** [Definition after Phillips 1973].

Group I kimberlite. Equivalent to 'basaltic kimberlite' of Wagner (1914). Phlogopite-poor **kimberlite** consisting essentially of olivine, serpentine and calcite. [Definition after Skinner 1989].

Group II kimberlite. Equivalent to 'lamprophyric kimberlite' of Wagner (1914). Phlogopite-rich **kimberlite**. [Definition after Skinner 1989].

* **Hamrongite.** An obscure local name applied by Johannsen (1938) to a variety of quartz-bearing **kersantite** from Hamrong, Sweden.

* **Hedrumite.** An obscure local name coined by Brøgger for syenitic rocks intimately associated with camptonites in the Hedrum area, Oslo Province, Norway. **Bostonite** here covers the same compositions.

* **Heptorite.** An obscure local name for a haüyne-bearing variety of **alkaline lamprophyre** from Siebengebirge (7 hills), Germany, the name coming from the Greek word for '7'. Composed of Cpx 30, Hb 8, haüyne 8, glassy base with labradorite microlites 46%.

* **Heronite.** An obscure local name for feldspathoidal monzonitic rocks intimately associated with camptonites in the Heron Bay area of the Coldwell complex, Lake Superior, Canada. **Maenite** here covers the same compositions.

Heteromorphic. Applies to two or more rock-types which have essentially the same bulk chemistry but different modal compositions. *Local heteromorphism* indicates coexistence of the heteromorphs within a single igneous body, *global heteromorphism* a more general equivalence. Implies that a single **magma-type** can crystallize to more than one mineral assemblage under different conditions.

* **Heumite.** An obscure local name coined by Brøgger for a variety of **sannaite** from Heum, Oslo Province, Norway. Composed of Bi+Hb 45, Or 40, foids 8%.

* **Holmite.** An apochryphal local name misapplied by Flett (1900) to **monchiquites** from Orkney, in which apatite had been misidentified as melilite (Flett 1935, p.185).

Hyalomonchiquite. A **monchiquite** with a predominantly glassy base (which may, on crystallization, give rise to combinations of feldspathoids and feldspars).

Inclusion. Refers to *all* rock and crystal fragments coarser than their host rock (Table 6.1).

* **Jerseyite.** An obscure local name formerly applied to **minettes** from Jersey, Channel Is. (Velde 1970).

Jumillite. An olivine-diopside-phlogopite-leucite-sanidine **lamproite.** Type locality, Jumilla, Spain. Used here as a family term for all lamproites of similar petrology (Fig.1.2).

* **Kajanite.** An obscure local name coined by Lacroix for a rock carrying olivine, diopside, phlogopite, Cr-spinel, magnetite, leucite, nepheline and minor richterite, from Kajan River, Kalimantan, Indonesia. Probably a **cocite**: its richterite and leucite are typical of **lamproites**, but its whole-rock and mineral Al contents are more typical of **minettes**; however, the nepheline is typical of neither (Wagner 1986).

Kamafugite = **katungite** + **mafurite** + **ugandite**, Ugandan K-rich ultrabasic volcanic rocks. May be UML.

* **Kamperite.** An obscure local name coined by Brøgger for a variety of **minette** from Kamperhaug, Fen district, Norway.

Katungite. Name applied by Holmes to an extrusive **ultrapotassic,** melanocratic olivine-melilitite from Katunga volcano, SW Uganda, carrying combinations of leucite, kalsilite, nepheline and glass. Appears to be chemically similar to some **ultramafic lamprophyres** (Sections 1.6.4, 7.3.5).

Kentallenite. See Section 7.1.1. A K- and Mg-rich rock belonging to the **appinite suite**. Variously termed olivine-monzonite and shoshonitic picrite; would be termed biotite-olivine-monzogabbro on the Streckeisen (1976) nomenclature. A plutonic equivalent of **spessartite** and **vogesite.** Distinctively features two forsteritic olivine generations, the earlier mantled by pyroxene and phlogopite ± hornblende; all members of Bowen's discontinuous and continuous reaction series may thus be present. Type locality Kentallen, Appin, Argyll, Scotland; named by Teall in 1897 (Hill & Kynaston 1900).

KERSANTITE. See Table 1.1. A **calc-alkaline lamprophyre** consisting of phenocrysts of phlogopite-biotite, with or without subordinate calcic hornblende, forsteritic olivine or diopsidic clinopyroxene, in a groundmass of the same plus plagioclase and subordinate alkali feldspar. Type locality Kersanton, Brittany, France; named by Delesse in 1851.

KIMBERLITE. A group of volatile-rich potassic ultrabasic rocks, classified by the IUGS (Le Maitre 1989) as "a variety of **lamprophyric rock**", and here therefore regarded (Fig.1.2) as a family within the lamprophyre clan, which forms small subvolcanic pipes, dykes and sills. Commonly, kimberlites show a distinctive inequigranular texture reflecting the presence of **macrocrysts** in a fine-grained matrix. Macrocrysts may included rounded anhedral picroilmenite, Cr-poor titanian pyrope, forsteritic olivine, Cr-poor pyroxene, phlogopite, enstatite and Ti-poor chromite. Matrix minerals include euhedral primary olivine and/or phlogopite, together with combinations of perovskite, titanian magnesian aluminous chromite, magnesian ulvöspinel-magnetite, Al-Ti-poor diopside, monticellite, apatite, calcite and primary serpentine. See also **Group I, Group II kimberlite.** [Consensus definition between those of Clement *et al.* (1984) and R.H.Mitchell 1986,1989]. Type locality Kimberley, S.Africa; named by Lewis in 1879.

* **Kvellite.** An obscure local name coined by Brøgger for a melanocratic variety of **sannaite** from Kvelle, Oslo Province, Norway.

LAMPROITE. Term coined by Niggli (1923) for a group of often glass-rich, hypabyssal or (commonly) volcanic, lamprophyric rocks characterized geochemically by ultrabasic to intermediate levels of SiO_2, high to extremely high MgO and K_2O, with molar $K_2O/Na_2O > 4$, $K_2O/Al_2O_3 > 0.8$ (usually > 1, i.e. **perpotassic**) and *mg* (Appendix A) > 70%. Mineralogically, they contain combinations of titanian Al-poor phlogopite, Al-poor diopside, forsteritic olivine, potassium titanium richterite and/or riebeckite–arfvedsonite (*sensu* IMA 1978), Ba-Fe-rich sanidine, analcime, leucite, perovskite, armalcolite, spinel, and diagnostic rare minerals such as jeppeite, priderite, shcherbakovite and wadeite. Ilmenite is rare, and soda feldspathoids (nepheline, sodalite, nosean, haüyne), kalsilite, melanite, melilite, monticellite and primary plagioclase are all absent. Originally regarded as extrusive lamprophyres, and classified by the IUGS (Le Maitre 1989) as a "variety of **lamprophyric rock**", so here considered to be a particular branch of the lamprophyre clan. [Based on consensus of definitions in Jaques *et al.* 1984a,1986; Scott Smith & Skinner 1984b, R.H.Mitchell 1985 and Bergman 1987]. See also discussion in Section 1.3.

LAMPROPHYRE. Termed coined by Gümbel (1874) for dyke-rocks from Germany with glistening biotite phenocrysts, and derived from the Greek λαμπροσ πορπηψροσ (= glistening **porphyry**). See text!

* **Lamprophyric kimberlite.** Old name used by Wagner (1914) for **Group II kimberlite.**

LAMPROPHYRIC ROCK. Group term recommended by the IUGS (Le Maitre 1989) to cover **lamprophyres, lamproites** and **kimberlites**; adopted here in the same way, as a 'clan' name (Fig.1.2).

Lamproschist. A metamorphosed, schistose variety of **lamprophyre**, typically composed of apatite, biotite, chlorite, feldspars and sometimes relict primary phases such as clinopyroxene, hornblende or olivine. May preserve the characteristic geochemistry of lamprophyres.

* **Luhite.** An obscure local name coined by Scheumann (1913) for a haüyne-bearing variety of **polzenite** from Luh, Czechoslovakia.

Macrocryst. Non-genetic term for a macrosopically discernible crystal, significantly larger than the groundmass in which it is set. Covers **megacrysts, phenocrysts, xenocrysts.** [Definition of Clement *et al.*1984a]. See Table 6.1.

* **Madupite.** An melanocratic diopside-phlogopite **lamproite.** Name derives from an Indian word for the district in Wyoming where the rock occurs.

Madupitic lamproite. Group term used by R.H.Mitchell (1985) to refer to lamproites with poikilitic groundmass phlogopite (jumillites, madupites, wolgidites, etc.)

Maenite. A useful shorthand for intermediate, often altered, mildly alkaline porphyritic dyke-rocks, intimately associated with some camptonite-monchiquite dyke-suites. Corresponds in IUGS terminology to *porphyritic (biotite-hornblende-quartz) trachyandesite.* SiO_2 content *c.* 50–60%. Type locality Lake Maena, Oslo Province, Norway. Covers **gauteites, heronites,** etc. Grades into **bostonite** with decreasing CI.

Mafraite. A hydrous alkali gabbro, composed of brown (kaersutitic) amphibole, clinopyroxene, plagioclase ± biotite ± minor quartz. Bulk composition silica-undersaturated due to amphibole, despite possible presence of quartz, but much less so than **berondrite**. Named after Mafra, Sintra complex, Portugal.

Magma-type. Used in much the same sense as in the classic Mull Memoir (Bailey *et al.*1924) — 'Porphyritic Central magma-type', etc. — to refer to a limited compositional *range* of closely related **melts**, whose crystalline products show coherent petrographical and geochemical features, are associated together in one geological setting, and give rise to a recognizable evolutionary magma-series. Strictly, the term 'melt-type' should be used here, but was avoided because of this precedence.

Malchite. A rock long classified with the lamprophyres (Johannsen 1938), and now a useful shorthand for porphyritic dyke-rocks transitional between **calc-alkaline lamprophyres** and **porphyrites**. Composed of biotite and/or hornblende and *very minor* plagioclase phenocrysts in an andesitic biotite-hornblende-quartz-plagioclase matrix. Corresponds in IUGS terminology to *weakly plagioclase-phyric (biotite-hornblende-quartz) microdiorite/andesite.* CI < 30%, SiO_2 *c.* 57–63%. Type locality Malchen, Germany.

* **Mamilite.** A richterite-leucite **lamproite.** Type locality Mamilu Hill, W.Kimberley region, Australia.

* **Markfieldite.** Originally used for a granophyric diorite from Markfield, English Midlands, but reapplied by Kennedy & Read (1936) to a **spessartite** from SW Scotland. The original rock is a granitoid, though intimately associated with lamprophyre dykes; the second is a true lamprophyre (Macdonald *et al.* 1986).

Megacryst. A large macrocryst (usually taken to imply > 5 mm longest dimension).

Meimechite. A type of glassy, porphyritic, alkali picrite lava, regarded by some Russian petrologists as equivalent to kimberlite, but mostly now believed to be quite distinct (R.H.Mitchell 1986; Section 7.3.3).

Mesothermal. Applied to gold deposits which form at relatively substantial depths; used in contradistinction to epithermal (high-level, subvolcanic) deposits and not intended to have any age connotation, although the majority of mesothermal deposits happen to be Archaean.

Melt. Used as a non-specific term to cover the uncertain physical condition of lamprophyres when they are emplaced. Does not necessary imply a fully liquid silicate magma in the conventional sense, but may cover a hydrothermal fluid, a crystal-charged, gas-rich, fluidized phase, etc.

Mica-peridotite. A widely used but unacceptably vague term for rocks which are mostly **olivine-lamproites** or, in a few cases, **ultramafic lamprophyres.**

MINETTE. See Table 1.1. A **calc-alkaline lamprophyre** consisting of phenocrysts of phlogopite-biotite, with or without subordinate calcic or alkali amphibole, forsteritic olivine or diopsidic clinopyroxene, in a groundmass of the same plus alkali feldspar and subordinate plagioclase. The word is also used by German miners for a type of pisolitic iron ore.

* **Modlibovite.** An obscure local name coined by Scheumann (1922) for a variety of **polzenite** containing olivine and biotite phenocrysts in a groundmass of melilite and nepheline, from Modlibov, Csechoslovakia.

MONCHIQUITE. See Table 1.1. Major variety of **alkaline lamprophyre** composed of combinations of forsteritic olivine, kaersutite, titanaugite and titanbiotite phenocrysts in a matrix of the same (minus olivine), with nepheline (*nepheline-monchiquite*), analcime (*analcime-monchiquite*) and/or glass (*hyalomonchiquite*) but no modal feldspars. Type locality Serra de Monchique, Algarve, Portugal; named by Weveke in 1880, redefined by Hunter & Rosenbusch (1890).

* **Mondhaldeite.** An obscure local name for a leucite-bearing rock from Mondhalde, Kaiserstuhl, Germany which some authors have rather doubtfully called a lamprophyre. Composed of Or 38, Pl 40, Cpx+Hb 15.

* **Murambite.** A leucite basanite or absarokite from Murambe volcano, Uganda composed of olivine, augite, An_{45} plagioclase, soda anorthoclase, leucite, opaques and glass, which appears to be among the closest chemical equivalents among volcanic rocks to **kersantite**.

Multiple intrusion. A body formed of two or more magma-injections of broadly similar composition but differing macroscopic or microscopic texture (e.g. ocellar and non-ocellar lamprophyre).

Ocellus. Descriptive term for a particular type of **globular structure** which is well-rounded (hence *ocellar* = eye-like) and usually well demarcated from its host by a discontinuous or complete rim of tangential biotite or (more rarely) some other platy or acicular mineral.

Occurrence. Any local manifestation of igneous activity formed over a limited time, which can be regarded as a distinct geographical entity, and probably formed from a single parent **magma-type** or limited range of such magmas. Covers anything from single isolated dykes or plugs to volcanoes or plutons (including their associated volcanic and/or hypabyssal rocks). Grouped into larger scale **fields/belts** and **provinces.** Analogous to *member* in conventional lithostratigraphical nomenclature. [Definition of Rock 1981].

* **Odinite.** An obscure local term, named after the Odenwald, Germany, for what are here termed **malchites.** Tröger says the rock is a dense contact facies of **spessartite**, others that it is a fine-grained malchite; malchite is preferred here because the original odinite may be a hornfels.

Olivine-lamproite. Name now routinely applied to ultrabasic, Mg-rich varieties of **lamproites**, notably from the W.Kimberley province, W.Australia. Applied here (by analogy with definitions of komatiitic versus basaltic rocks) to those lamproites with whole-rock MgO > 20% (see Fig.8.2) either from the W.Kimberley province itself, or from other areas where not already given a specific name. Thus it applies to most samples from the Prairie Creek lamproite, Arkansas.

Orangite. A name half-seriously suggested by R.H.Mitchell (1989) for **Group II kimberlites**, derived from the Orange River, southern Africa.

* **Orendite.** A diopside-sanidine-phlogopite-**lamproite.** Type locality Orenda Butte, Wyoming, USA.

Ouachitite. See Table 1.1. Originally defined as a biotite-rich monchiquite but now as a rare variety of **ultramafic lamprophyre** consisting of combinations of forsteritic olivine, diopsidic pyroxenes, various amphiboles and phlogopite in a matrix of similar minerals with abundant primary carbonates and feldspathoids, often with minor perovskite, but little or no melilite. Type locality Ouachitas Mtns., Arkansas, USA; named by Williams (1890).

Panidiomorphic. Refers to the characteristic porphyritic texture of **lamprophyres,** in which nearly all the phenocrysts (olivine, pyroxene, amphibole, biotite) are strongly euhedral.

Parental melt. The melt which gives rise to a series of igneous rocks by **differentiation.** Most parental melts are also **primitive**, some are **primary.**

Peralkaline. Applies to rocks in which molecular $[K_2O + Na_2O] > Al_2O_3$, leading to the absence of plagioclase and the presence of soda pyriboles.

Perpotassic. Applies to rocks in which molecular $K_2O > Al_2O_3$, leading to the absence of plagioclase and the presence of minerals such as priderite and wadeite.

Phenocryst. Genetic term for a large **cognate** crystal which crystallised from the same melt as the groundmass in which it is set. Cf. **xenocryst.**

Phlogopite lamproite. Group term used by R.H.Mitchell (1985) to refer to lamproites with resorbed phenocrystic phlogopite (**cedricites, fitzroyites, orendites, wyomingites,** etc.)

* **Picrophyre.** An obscure local name for an olivine-bearing augite minette.

Pipe. A subvertical minor intrusion, circular to irregular in plan, often funnel-shaped in cross-section, and up to a few km in diameter, filled mostly by magmatic rather than pyroclastic rocks. Cf. **vent.**

POLZENITE. See Table 1.1. Rare variety of **ultramafic lamprophyre** consisting of combinations of forsteritic olivine, diopsidic pyroxenes, various amphiboles and phlogopite in a matrix of similar minerals with abundant melilite and feldspathoids, plus often minor perovskite and primary carbonates. Type locality Polzen area, Bohemia, Czechoslovakia.

Porphyrite. See Table 1.3. An old term useful for intermediate members of lamprophyre–porphyrite–porphyry dyke-suites. Composed of plagioclase, biotite and/or hornblende and sometimes minor quartz phenocrysts in a dacitic biotite-hornblende-quartz-feldspar matrix. Corresponds in IUGS terminology to *richly plagioclase-phyric (biotite-hornblende-quartz) microtonalite/dacite.* SiO_2 *c.* 63–70%. Considered equivalent to **porphyry** until 18th century petrographers restricted it to more basic varieties.

Porphyry. See Table 1.3. An old term useful for acidic members of lamprophyre–porphyrite–porphyry dyke-suites. Composed of combinations of plagioclase, alkali feldspar, quartz, biotite and/or hornblende phenocrysts in a rhyolitic biotite-hornblende-quartz-feldspar matrix. Corresponds in IUGS terminology to *porphyritic (biotite-hornblende-quartz) microgranite/rhyolite.* SiO_2 *c.* 67–75%. Qualified as *quartz-porphyry, feldspar-porphyry* depending on phenocrysts present. Pliny and Agricola apply the name to a purplish rock quarried in ancient Egypt; derives from a Greek work for the dyestuff Tyrian purple.

Potassic rock. An **alkaline rock** having wt.% $K_2O > [Na_2O - 2]$ (Le Bas *et al.* 1986). Leucite and sodic feldspathoids (analcime, nepheline, etc.) may coexist.

Primary melt. An undifferentiated melt which is still chemically in equilibrium with mantle source; implies an *mg* (Appendix A) ≈ 70%, coupled with high Sc, Cr, Co and Ni. [Definition of Rhodes 1981].

Primitive melt. Implies a melt which, though no longer in equilibrium with its mantle source, has undergone only slight **differentiation** and thus still retains a high *mg*, Sc, Cr, Co and Ni.

Province. A regional assemblage of igneous rocks formed at roughly the same time in the same area, and showing considerable compositional affinity. Subdivided into **fields/belts** and **occurrences** and sometimes grouped into **super-provinces.** For example, the Ellendale 9 pipe is one **occurrence** within the Miocene Ellendale **field** of the West Kimberley lamproite **province,** NW Australia. The Highwood Mountains constitute one **occurrence** within the Montana K-rich **province** of the Tertiary Cordilleran **super-province** of alkaline rocks in North America. Can encompass extended periods of magmatism as long as the products are geographically and compositionally coherent. Also covers rare occurrences separated by seas and oceans formed by subsequent plate tectonism (e.g. the Hercynian lamprophyres of continental Europe, SW England and Carolina, USA — Fig.2.5). Analogous to *group* in conventional lithostratigraphical nomenclature. [Definition of Rock 1981].

* **Prowersite (prowersose).** An obscure local name applied by Cross (1906) to a variety of **minette** or possibly **lamproite** from Two Buttes, Colorado, USA.

* **Raabsite.** An obscure local name coied by Waldemann for a variety of microcline-bearing **minette** from Raabs, Csechoslovakia.

Redwitzite. See Section 7.1.2. A local group-name for coarse-grained lamprophyric rocks characterized by large biotite and/or hornblende phenocrysts or schlieren, and varying from rare gabbroic types with relics of olivine and orthopyroxene (C.I. ≈75%), through dominant biotite-meladioritic to biotite-granodioritic rocks (C.I. ≈25%). Probably plutonic equivalents of **kersantite.** Named by Willman (1920) after a small pluton at Markredwitz, Bavaria. Also forms inclusions and schlieren in granites and gneisses.

SANNAITE. See Table 1.1. Minor variety of **alkaline lamprophyre** (AL) composed of combinations of forsteritic olivine, kaersutite, titanaugite and titanbiotite phenocrysts in a matrix of the same (minus olivine), with alkali feldspar and subordinate plagioclase and feldspathoids. Type locality Sanna, Norway; named by Brøgger (1921).

* **Scyelite.** An obscure local term for biotite-hornblende-peridotites of the **Achu'aine hybrid suite,** from Loch Scye, Sutherland, Scotland (Judd 1885). Similar to **cortlandtite**, but has biotite in addition.

Segregation. A type of **globular structure**, gradational with its host rock, and often highly irregular in shape. Segregations are much less easily distinguished from their host than **ocelli.**

* **Selagite.** A K-rich lava from the Roman province, belonging to Niggli's *lampro-sommaitic magma*, which appears to be transitional between **minette** and **lamproite** (Section 7.1.6). The name was derived in 1822 by Haüy from a Greek word meaning "to beam brightly", referring to the glistening biotite, and is thus closely related to **lamprophyre.** Type rock composed of 52% Sa, 27 Ph, 10% Di, 4% Qz.

* **Semi-lamprophyre.** Term coined by Marco (1958) for rocks here called **malchite.**

Sheet. Term used here for all minor intrusions which do not conform to the definition of either **dyke** or **sill**: e.g. concordant vertical, discordant horizontal or intermediate bodies.

Shoshonite. Intermediate (andesitic) member of the **shoshonite association,** composed of plagioclase with K-feldspar rims, clinopyroxene and sometimes minor olivine.

Shoshonite association/series. Group term for the series of **shoshonitic** rocks from basic (*absarokite*) through intermediate (*shoshonite*) to acidic (*latite*). [Definition of Joplin 1966,1968].

Shoshonitic. A group of mildly potassic rocks ($K_2O \approx Na_2O$), transitional in mineralogy and chemistry between calc-alkaline and alkaline series. They have higher K_2O for a given SiO_2 content than calc-alkaline rocks, leading to the crystallisation of alkali feldspar in more basic compositions (often as overgrowths on plagioclase). However, they lack the soda pyriboles and feldspathoids of true alkaline rocks. Includes **absarokite**, latite, some monzonites, **shoshonite,** shonkinite. [Definition of Joplin 1966,1968].

Shoshonitic lamprophyre. Synonymous with **calc-alkaline lamprophyre.**

Sill. A gently-dipping *and* (near-)concordant minor intrusion. Cf. **dyke, sheet.**

* **Sizunite.** An obscure local name applied by Cogné (1962) and others to a microcline-bearing **minette** from Finisterre (Brittany), France.

* **Soda minette.** A name applied since the late 19th century to plagioclase-free, peralkaline minettes carrying sodic pyriboles (aegirine, arfvedsonite, etc.) Abandoned here in deference to Bachinski (1986).

Sodic rock. An **alkaline rock** having $Na_2O > K_2O$. The dominant feldspathoids are analcime, cancrinite, nepheline, and/or sodalite minerals, and the dominant amphiboles are arfvedsonite, riebeckite, hastingsite, kaersutite, etc. Covers the great majority of alkaline rock varieties (nepheline syenites and phonolites, nephelinites and ijolites, alkali basalts and gabbros, theralites, etc.)

SPESSARTITE. See Table 1.1. A **calc-alkaline lamprophyre** consisting of phenocrysts of calcic hornblende, with or without subordinate phlogopite-biotite, forsteritic olivine or diopsidic clinopyroxene, in a groundmass of the same plus plagioclase and subordinate alkali feldspar. Type locality Spessart, Germany.

Suite. Term used *sensu* Bowes & McArthur (1976), Rock (1981) to indicate any assemblage of petrologically related igneous rocks that habitually occurs together in space and time. See **appinite suite.**

Super-province. Termed used to cover certain exceptionally large-scale manifestations of igneous activity which cover several examples which have been termed **provinces.** Analogous to *supergroup* in conventional lithostratigraphical nomenclature. [Definition of Rock 1981].

* **Tamaraite.** An obscure local name coined by Lacroix for a variety of feldspathoid-rich **alkaline lamprophyre** (transitional between **monchiquite** and **camptonite**) from Tamara Is., Los Archipelago, Guinea. Composed of Cpx+Hb+Bi 60, Ne 50, Fp 10%.

* **Tjosite.** An obscure local name coined by Brøgger for a melanocractic variety of **sannaite** from Tjose, Oslo Province, Norway. Composed of Cpx 49, ore 20, Ne 10, Ap 7, Or 6, Bi 5%.

* **Topsailite.** An obscure local name coined by Lacroix for a variety of **camptonite** from Topsail, Tamara Is., Los Archipelago, Guinea.

Ugandite. A potassic olivine-leucitite, composed of leucite, clinopyroxene and olivine in a glassy matrix.

Ultramafic lamprophyres. A group of **lamprophyres** consisting of the rock-types **aillikite, alnöite, damtjernite, ouachitite** and **polzenite**, which correspond broadly to melililites and melilite-nephelinites, but are considerably enriched in volatile elements (H_2O, CO_2, F, Cl) and LILE elements (K, Rb, Ba, Mg, Cr, Ni, etc).

Ultrapotassic rock. An **alkaline rock** having $K_2O \gg Na_2O$. Leucite (pseudoleucite) or kalsilite are the dominant feldspathoids, and any amphibole is usually potassium richterite. Includes several groups of rocks: *ultrapotassic fenites, kamafugites* (katungites, etc.) and **lamproites** [Definition of Foley *et al.* 1987].

Variole. A non-genetic term for a spherulitic **globular structure**, typically composed of radiating plagioclase or pyroxene. Usually applied to basic igneous rocks. [Definition after Bates & Jackson 1989].

Vaugnérite. See Section 7.1.2. A local name for texturally and compositionally variable varieties of hornblende-biotite-diorite or monzonite, with abundant accessories, apatite, allanite, magnetite, pyrite and sphene Originally defined by Drian in 1849 after Vaugnéray, near Lyons, France where they form basic enclaves and dykes in granites. The type-rocks (Michel-Levy and Lacroix 1887; Lacoix 1917; Johannsen 1938) have Qz 5, Or 1, Pl 30, Hb 42, Bi 21. Almost certainly a plutonic equivalent of **kersantite.**

Vaugnérite series. See Section 7.1.2. Informal term used here to refer collectively to **durbachite, redwitzite** and **vaugnérite**, implying petrological similarity but no spatial or temporal relationship.

Vent (= *diatreme*). A subvertical minor intrusion, circular to irregular in plan, often funnel-shaped in cross-section, and up to a few km in diameter, filled mostly by pyroclastic rocks and breccias. Cf. **pipe.**

* **VERITE.** A glassy olivine-diopside-phlogopite **lamproite**. Type locality Vera, SE Spain.

* **Vesecite.** An obscure local name coined by Scheumann (1922) for a variety of **polzenite** carrying monticellite from Vesec Svetla, Czechoslovakia.

VOGESITE. See Table 1.1. Rare variety of **calc-alkaline lamprophyre** consisting of phenocrysts of calcic hornblende, with or without subordinate phlogopite-biotite, forsteritic olivine or diopsidic clinopyroxene, in a groundmass of the same plus alkali feldspar and subordinate plagioclase. Type locality Vosges Mtns., France; named by Rosenbusch in 1887.

* **Wesselite.** An obscure local name coined by Scheumann (1922) for a variety of **polzenite** from Wesseln, Bohemia.

* **Wolgidite.** A richterite-diopside-phlogopite-leucite **lamproite.** Type locality Wolgidee Hills, W.Kimberley region, Western Australia.

Wyomingite. A diopside-phlogopite-leucite **lamproite.** Type locality Leucite Hills, Wyoming, USA. Used here as a family term for all lamproites of similar petrology (Fig.1.2).

Yamaskite. Term used here for ultramafic, hornblende-pyroxene-rich cumulate variants of **camptonites**, composed of titanaugite, kaersutite and minor calcic plagioclase. Named after rocks from Mt.Yamaska, Quebec, Canada, which are anorthite-bearing hornblende-pyroxenites.

Xenocryst. Foreign crystal normally of larger size than the groundmass in which it is set (Table 6.1).

Xenolith. Foreign fragment of rock (rounded or angular) in a host of distinct composition (Table 6.1).

Appendix C: Global compilation of lamprophyre occurrences

The 8 tables here detail occurrences of 4 groups of rocks:

(1) *Confirmed lamprophyres* (Tables C1–C6, Figs.2.1–2.13): rocks described by their original authors as lamprophyres, or under one of the variety names in Fig.1.2, which are sufficiently well documented to be seen to agree with the definitions in Chapter 1; only *kimberlites* are omitted, because they are already covered by several monographs.
(2) *Unconfirmed lamprophyres* (Table C7, distinguished mostly by unfilled symbols on Figs.2.1–2.13): as group (1), but insufficiently documented to be sure of their affinities; this covers all rocks for which few or no petrographical or geochemical data are available.
(3) *Apochryphal "lamprophyres"* (Table C8, *not* compiled on Figs.2.1–2.13): as (1), but well enough described to know they *violate* definitions in Appendix B and Chapter 1.
(4) *Additional lamprophyres* (specifically flagged, but included in Tables C1–C6): rocks not originally called "lamprophyres" but which fit the present definitions.

Entries in each table are sorted according to: (1) country; (2) petrographic province; (3) occurrence (Appendix B). Readers will note that some entries are duplicated between Tables C1 and C5 only; C5 concentrates on the appinite suite and vaugnérite series only (Sections 7.1.1–7.1.2), and Table C1 to that dealing with lamprophyres with or without these rocks.

Despite their bulk, these compilations are *preliminary*, not least because information available varies so substantially in both quality and quantity from area to area. In particular, papers concerning lamprophyres alone rarely cover all the information compiled (especially age and tectonic setting): this usually has to be gleaned from complementary papers describing associated major intrusions. Therefore, because there are commonly far more papers describing the plutonic rocks, only an incomplete compilation can be entertained here. Furthermore, a certain mixing of fact and interpretation has been unavoidable: the *As* column should be regarded with particular caution, for reasons elaborated in Chapter 8, and the *Age* column necessarily summarises multitudinous inferences from geological relationships as well as (occasionally conflicting) results from one or more direct dating methods. Wherever possible, the degree of certainty of each piece of information is carefully annotated. In the *Province* column, for example, — indicates that the occurrence is sufficiently well documented to indicate that it does *not* belong to any known province, whereas ? indicates that insufficient (or no) age data are available to indicate which of several possible provinces in the surrounding region it might belong to.

As Woolley (1987) found in a similar compilation, the matter of what should constitute a single entry also presents formidable difficulties. Complete consistency is unachievable for 2 reasons. (1) *Natural distribution:* some geographically isolated bodies (e.g. Malaita in Table C3) clearly warrant individual entries, even though their magma volumes are dwarfed by regional dyke-swarms which also receive only single entries (e.g. Bohemian Mass in Table C1). This inconsistency in the size of 'occurrences' (Appendix B) defined can nevertheless be legitimized from the analogous yet accepted variation of lithostratigraphic Formations from feather-edges to enormous thicknesses. (2) *Information availability*: in the best documented areas (e.g. North Sea province, Table C2), entries can be subdivided to cover small areas and/or distinct magmatic episodes, whereas at the other extreme (e.g. some Russian occurrences), only a single entry may be constructed for what appear to be large dyke-swarms comprising many thousands of bodies and square kilometres of ground.

Where different types of coeval lamprophyres within the same province are spatially separated, these have, as far as possible, been distributed across the appropriate tables (e.g. entries for the Ouachitas Province occur in each of Tables C1,C2,C3,C4). Table C6, by contrast, attempts to syphon off examples where the coexistence is more intimate: that is, at the *occurrence* (rather than province) level. More arbitrary choices have necessarily been involved where the degree of intimacy is unclear.

Table C1. Confirmed occurrences of calc-alkaline lamprophyres (with no other branches)

Country	Province	Occurrence	As	Age,Ma	L	Form;assoc.ig.rock	Comments	References
Algeria	— (Fig.2.12)	Oued Messelmoûn/ Oued Harbil, Cherchel	B	≈11 (K-Ar)	CK	2 dykes; associated with micromonzonite stocks & laccoliths		Lepvrier & Velde (1976)
Antarctica	— (Fig.2.1)	S.Victoria Land	A	≈470 (K-Ar, Rb-Sr)	CK CS	Granite Harbour intrusive suite; porphyry dykes		Campbell Smith (1924); McKelvey & Webb (1961); Haskell et al. (1965); Smithson et al.(1970); Manzoni & Nanni (1977); Crawford et al.(1984); Tessensohn & Roland (1987)
Argentina	— (Fig.2.1)	La Rioja province	A	lamp: 295±20 grt:348±22	CS CK	Cuesta de Miranda granitoid pluton	aplites, "filonites," "cataclasites"	Villar Favre et al. (1973)
Australia (NSW)	Tasman Fold Belt	Jenolan Caves (Figs. 2.1 & 2.8)	A	Devonian	CK	Dyke; assoc. andesite, porphyry, felsite dykes granitoid plutons		Süssmilch & Stone (1915); Joplin (1965)
Australia (NSW)	Tasman Fold Belt (Figs.2.1/8)	Mt.Woolooma	R	≈85 (K-Ar)	CM	Single dyke; part of assemblage including basalts, teschenites	Various other lamps (?AM, ?UL) of unknown age in area	Jaques & Perkin (1984); Jaques et al.(1985)
Australia (NSW)	Tasman Fold Belt (Figs.2.1/8)	New England Ranges (eg Attunga,Eumbra, Hillgrove, Nundle)	A1	Post-Carbonif-erous	CM CV CS	Dykes; probably related to Hillgrove,Tingha or other granitoid plutons	"Camptonites" may be CS or another lamp suite	Andrews et al.(1907); Benson (1913); Joplin (1965)
Australia (NSW)	Tasman Fold Belt (Figs. 2.1 & 2.8)	Mt.Dromedary	B1	94–100 (K-Ar, Rb-Sr)	C? CS	Lamp dykes; 15 small plutons of monz, pxnt, ban, shonk, sye, etc.		Joplin (1968, 1971); Jaques et al.(1985)
Australia (NSW)	Tasman Fold Belt (Figs.2.1/8)	Tumut/Happy Jacks/Adelong	A		CA CS	Belt of dykes and pipes parallel to margins of Lachlan granitoids		Joplin (1957, 1965)
Australia (NT)	— (Figs.2.1 & 2.8)	Pine Creek inlier(e.g Mt. Bundey, Tom's Gully, Woodcutters)	A	L & grt: 1812±36 (Rb-Sr)	CM	Mt.Bundey, Mt.Goyder and other granitoid & syenitoid plutons	Synchronous Au-Ag±U mineralization	Hochman (1980); Taube (1984); Jaques et al.(1985); Sheppard et al.(1990)
Australia (NT)	— (Figs.2.1 & 2.8)	Tennant Creek	A	lamp:1664 ±16; grt: ≈1850-70	CM	Tennant Creek granitoid plutons	lamps associated with younger of 2 granitoid phases	Crohn & Aldershaw (1965); Mendum & Tonkin (1976); Black (1977); Jaques et al.(1985)
Australia (SA)	— (Figs.2.1 & 2.8)	Mannum, nr.Adelaide	A	?	CS	Single dyke in small granitoid inlier among Cainozoic sediments	Transitional to malchite	Alderman (1929); Joplin (1973)
Australia (TAS)	C.Australia Mobile Belt (Figs.2.1/8)	King Island	R	≈137 (K-Ar)	CM	A few dykes cutting earlier granites		Sutherland & Corbett (1974); Jaques et al.(1985)
Australia (TAS)	C.Australia Mobile Belt (Figs.2.1/8)	W.coast swarms (Raglan Range, Queenstown, etc.)	R? A?	?Middle Devonian	All	Regional swarm of dykes, sheets; related to Varna Bay granitoids?	Remote, little-known area; may be several episodes of lamps	Sutherland (1973); Sutherland & Corbett (1974); Jaques et al.(1985)

Table C1. Confirmed occurrences of calc-alkaline lamprophyres (with no other branches)

Country	Province	Occurrence	As	Age,Ma	L	Form;assoc.ig.rock	Comments	References
Australia (TAS)	Tasman Fold Belt (Figs.2.1/8)	Blue Tier	A1	≈370 (K-Ar) ≈389 (Rb-Sr)	CS CC	Many dykes cutting various phases of Blue Tier granitoid batholith	May be contemp with batholith or Cape Portland (next entry)	Groves et al. (1977)
Australia (TAS)	Tasman Fold Belt (Figs.2.1/8)	Cape Portland	B1	91-103 (K-Ar)	All + CA	porphyrite intrusions; andesite lavas; breccias	"Vogesites" have Af+Bi and are CM on present definitions	Jennings & Sutherland (1969); Sutherland (1973); Sutherland & Corbett (1974); Jaques et al. (1985)
Australia (VIC)	Tasman Fold Belt (Figs.2.1/8)	Kiewa	A	Silurian-Devonian	CS	Pretty Valley,Big Hill, Niggerheads plutons; porph,grdi,pcg dykes	"Camptonites" may be CS or younger Tertiary rocks	Beavis (1962); Joplin (1965)
Australia (VIC)	Tasman Fold Belt (Figs.2.1/8)	Stawell/Mary-borough goldfields (Magdala mine etc.)	A1	Devonian	No info	porphyry dykes, regional granitoid plutons	Gold reefs "are closely associated with lamp dykes" and coeval	David (1950, p.191); Quick (1988)
Australia (VIC)	Tasman Fold Belt (Figs.2.1/8)	Wood's Point	A?	Middle Devonian	CM CS	U.Devonian plutons (Bow Bow, Tynong, etc) may be unrelated	Swarm covers 6,500 sq.km	Junner (1920); Hills (1952); Marsden (1988)
Australia (WA)	Pilbara	Bamboo Creek–Balfour Downs	A1	≈1800 (Pb-Pb)	CS CM	Regional dyke-swarms, plugs assoc. Shaw, Bridget etc. grt plutons	Hundreds of bodies known over several thousand sq.km	Lewis & Davy (1981); Rock & Barley (1989 & unpubl. data); Geol.Surv.W.Aust. Balfour Downs 1:250,000 sheet (1988)
Australia (WA)	Yilgarn (Figs.2.1/8)	Jimperding belt (Dilling, Hamersley Siding, York, etc.)	A	≈2700 (by analogy, cf. next entry)	CM CS CA	Various granitoid plutons	A few dykes; appinite xenoliths in granites	Simpson (1926, p.25-6); Miles (1948); Johnstone (1952); Joplin (1965); Rock et al. (1988d)
Australia (WA)	Yilgarn (Figs.2.1/8)	Norseman-Wiluna greenstone belt, Eastern Goldfields	A2	lamp:2620–2684 (U-Pb) grt: ≈2662	CK CS	"Internal" granitoids: Kambalda Granodiorite etc; porpht,porph dykes	"Camptonites" are CAL or apochryphal; oldest known lamps	Miles (1948); Hallberg (1985); Perring (1988); Perring et al.(1989a,b); Rock et al. (1987,1988d); Mueller et al. (1988)
Austria	Venetian (Fig.2.12)	Kreuzeck Mtns. & Lienz, Tauern window	R1	24–30 (K-Ar)	CM CV	Lamps, high-K basaltic andesite dykes	CM termed "SHOSH" dykes; AL dykes 30 km to E at Goldeck	Deutsch (1984)
Brazil (Rio Grande do Sul)	— (Fig.2.1)	Cerro Tupanci, São Sepé	R	?	CS	Isolated dyke associated with "large dyke of acid volcanic rocks"	Few details of geological setting available	Issler & Roîsenberg (1972)
Canada (British Columbia)	N.American Cordillera (Fig.2.9)	Bridge River mining camp	A1	43.5±3 (K-Ar)	CK	Late dykes assoc. with Bendor and Coast plutonic complexes	Dykes immediately postdate Au mineralization	Leitchl et al. (1988)
Canada (Labrador)	— (Fig.2.9)	Labrador coast (Cape Harrison, Domino, etc.)	R	≈1470 (K-Ar)	CS	Many sheets, dykes	Much younger lamps superimposed at Aillik (Table C6)	Kranck (1953); Elders & Rucklidge (1969); Currie (1976, p.115)
Canada (New Brunswick)	Caledonides	Lake George, nr. Fredericton	A1	≈412 (K-Ar)	CS	Dykes with Hawkshaw grt pluton and Qz-Fp-porphyry dykes	Lamps/porph postdate type 1 W-Mo veins but predate type 2	Seal et al.(1988)

Table C1. Confirmed occurrences of calc-alkaline lamprophyres (with no other branches)

Country	Province	Occurrence	As	Age,Ma	L	Form;assoc.ig.rock	Comments	References
Canada (New Brunswick)	Hercynides (Figs.2.1, 2.5,2.10)	Shaw's Cove, Dalhousie, Campbellton, etc.	R?	Middle–Late Devonian	CM	Sheets and dykes	Sometimes formerly misidentified as lavas	Howard (1926); Alcock (1935, p.77); Bachinski & Scott (1979,1980; Bachinski & Simpson (1984,1986); Rock (1980)
Canada (NWT)	—	Yellowknife (Con gold mine)	A?	Late Archaean	All	Single 4-30m sheet extending 2800m along strike,1100m down dip	Probably a cumulate CAL from trace chem but difficult to assign	Webb & Kerrich (1988)
Canada (Nova Scotia)	Caledonides?	Meguma zone	A1	≈370 (K-Ar Hb); grt same age	CS	"Several dykes"; grt batholiths genetically related to lamps	Dykes carry abundant lower crustal xenoliths	Giles & Chatterjee (1987); Owen et al.(1988); Kempster et al.(1989); Ruffman & Greenhough (in prep.)
Canada (NWT)	— (Fig.2.9)	Keewatin	R	≈1800 (U-Pb, K-Ar)	CM	> 750 dykes; pxnt-sye stocks; trachyandes, trach flows;felsic dykes	Assoc. syngenetic U+Th+REE, epigenetic U±Cu±Ag±Pb depos	Rimsaite (1967, p.48); Davidson (1972); Blake (1980); LeCheminant et al. (1987)
Canada (Ontario)	Superior (Fig.2.9)	Duport, Geraldton-Beardmore, Hemlo, Marathon, Kirkland	A2	lamp: 2672 ±2 (U-Pb) grt: ≈2700	CK CS	Various granitoid plutons; "albitite", porphyry dykes	"Albitites" stated to be lamps by some authors; lamps occur	McLennan(1915);Hopkins(1924);Gledhill(1927);Bruce(1935);Dyer(1936); Hurst(1936);Moore (1937);Burwash(1937);Frohberg(1937);Bartley(1938); Harcourt(1938); Thomson & Griffis(1944);Parsons(1948);Thomson et al.
Canada (Ontario)	Superior (Fig.2.9)	Lake, Birch/Pickle/ Red Lakes,Schreiber, Shebandowan,Wawa	A				widely in all Superior greenstone belts; also Rimsaite(1967,p.53)?	(1950); Abraham(1951); Satterly(1952); MacLean(1956); Prest(1957); Lawson (1959); Mackasey(1976); Pye(1976); Jensen(1978); Muir(1982a,b); McNeil & Kerrich(1986); Smith (1986); Wyman & Kerrich(1988a;1989a,b)
Canada (Ontario)	Superior? (Fig.2.9)	Lesuer township	R	?	CS	Dense dyke-swarm assoc. with 5x2 km eucrite–gabbro 'pluton'	Contemporaneous Zn-Pb-Ag mineralization	Watson (1957)
Canada (Ontario)	— (Fig.2.9)	L.Huron (Iron Bridge Ranger Lake, Massey –Espanola etc.)	A?	grt:1510±50 L:1415±40; 1530±50	CK CM	Crocker Island granitoid complex	Various conflicting K-Ar and Rb-Sr ages reported	Robertson (1966,1970); Van Schmus (1971)
Canada (Saskatchewan)	— (Figs.2.1 & 2.9)	Nevins-Forsyth Lakes area, N shore Lake Athabasca	A	2200-2500 (geological evidence)	CM	Various granitoid plutons, pegmatites	3 episodes of lamp emplacement	Blake (1955); Harper (1984,1986)
Chile	—	El Teniente	A1	Pliocene	CS	"Several lamprophyre dykes"; andesites, dacites, Qz-diorites	Dykes immediately postdate porphyry Cu deposit	Howell & Molloy (1960); Camus (1975)
China	—	Shandong Peninsula	A	L: 90–130 grt: 130–140	CS	Dyke-swarms in Linglong & Guojialng granitoid batholiths	Intimately associated gold mineralization	Liu et al.(1984); Ji Haizhang (pers.comm.1989)
Czechos -lovakia	Hercynides (Fig.2.5)	Janov-Artmanov anticlinorium	A1	Stephanian	CK CS	Various buried granitoid plutons	Distinguished from assoc. basalts by high Ba, K, etc.	Dvorak (1982)
Czechos -lovakia	Hercynides (Fig.2.5)	Prague area (Barrandian basin — Zdice etc.)	A1	290, 320, 350, 360 (K-Ar)	CM CK	"Quartz-kersantites", "minverites"; various buried plutons		Fiala (1971); Fiala & Chlupac (1973)

Table C1. Confirmed occurrences of calc-alkaline lamprophyres (with no other branches)

Country	Province	Occurrence	As	Age,Ma	L	Form;assoc.ig.rock	Comments	References
Czechos -lovakia	Hercynides? (Fig.2.5)	Tatraveporids (Malá Fatra, Nízke Tatry, Suchy, Povaszky,etc)	A	? ≈300 + ≈50	CK CS	porphyrite, granite-porphyry, diorite dykes	Dyke-swarm passes into diorites; may include Alpine dykes	Hovorka (1967); Krist (1967); Hovorka et al. (1982)
Czechos -lovakia/ Austria	Hercynides (Fig.2.5)	Bohemian Mass and extensions; (see also Table C5)	A1	lamp & grt: 350-300 (+older)	All + CC	Central Bohemian pluton; grdi/grt-porph, peg, apl, dlr, sye dykes	porph derive from CM magma; some CM transitional to LL	Hibsch (1925); Nemec (1966,1968,1970; 1971a,b; 1972a,b; 1973a,b,c; 1974; 1975a,b,c,d; 1977a,b,c; 1978a,b; 1987;1988a,b); Kesse (1971); Fiscra (1974); Schulze et al. (1985)
Eire (southern Ireland)	Caledonides	Cty.Mayo	A1	≈400 (by analogy)	CS	Termon Granite		Sutton (1970)
Eire (southern Ireland)	Caledonides (Fig.2.6)	Donegal (see also Table C5)	A1	≈400 (by analogy)	CS CA CK	Ardara, Barnesmore, Termon, Thorr grt; plugs, dykes, sills	lamps form 3 of 70 dykes at Barnesmore; grade into grdi-porph	Walker & Leedal (1954); Pitcher & Berger (1972); French (1978); Elsdon & Todd (1989)
Eire (southern Ireland)	Caledonides (Fig.2.6)	Leinster–Dublin swarm (see also Table C5)	A1	lamp & grt: Caledonian (≈400)	CK CS CA	Leinster granitoid pluton; grdi/dlr-porph dykes	Also Qz-albitites; lamps bracket grt plutons in time	McArdle (1974); McArdle & O'Connor (1987)
Finland (SW)	? (Fig.2.1)	Viljakkala, NW of Helsinki	A	?	CS	gabbro, Qz-syenite, granitoid plutons, granitic dykes		Stigzelius (1944)
Finland (SW)	Sveco-fennian (Fig.2.1)	Åland Isles	A	lamp & grt: 1800–1840 (1788 U-Pb)	All	Åva, Mosshaga, Semlinge grt stocks; younger dlr, peg	Dense radial swarms hybrid with grt; chem transitional to AL	Kaitaro (1953; 1956); Ehlers & Bergman (1984); Nurmi & Haapala (1986); Hubbard & Branigan (1987); Branigan (1989)
Finland (SW)	— (Fig.2.1)	Barösundsfjärd (Ingå, Stor Ådgrund, etc.), W of Helsinki	A	Precambrian	CS	Obbnäs granitoid pluton, migmatites, gneisses, veins	A few lamp dykes contemporaneous with grt veins and plutons	Sederholm (1926)
France	Hercynides (Fig.2.5)	Brittany (Finisterre, Rade de Brest, etc.)	A1	≈300 (by analogy)	CM CK	Trégastel & other granitoid plutons		Métais (1960,1961); Métais et al. (1962); Cogné & Giot (1961); Cogné (1962); Loy (1967); Leutwein et al. (1972); Thonon (1973)
France	Hercynides (Fig.2.5)	Hautes-Pyrénées (Neste de Saux, Montang, etc.)	A1	≈300 (by analogy)	CK	Various granitoid plutons	Contact altered by granitoids but other dykes cut granitoids	Durand-Wackenheim (1983)
France	Hercynides (Fig.2.5)	Massif Central (Puy de Dôme, Limousin, Lozère); Table C5	A1	lamp: 285± 10,312-316; grt: 312-325	CM CK CS	Millevache, Gaillard, Forez, Thiers, Ste. Sylvestre granitoids	Associated U deposits "vaugnérite" = CA (§7.1.2)	Sarcia & Sarcia (1956); Lapadu-Hargues (1959); Cantagrel et al. (1970); Sabourdy (1975); Leroy & Sonet (1976); Albarède & Weisbrod (1981); Peyronnet & Lameyre (1984); Chalier & Sabourdy (1987); Michon (1987)
France	Hercynides (Fig.2.5)	Normandy/Cherbourg Peninsula (Cape de la Hague, etc.)	A1	≈300 (by analogy)	CM CK	Trégastel, Barfleur, Flammanville granitoid plutons		Jérémine (1927)
France	Hercynides (Fig.2.5)	Pelvoux massif, Hautes Alpes	A1	lamp: ≈320 (K-Ar)	CK	Various granitoid plutons	Associated dlr occur in composite dykes, but are younger (≈197)	Bellanger & Buffet (1979)

Table C1. Confirmed occurrences of calc-alkaline lamprophyres (with no other branches)

Country	Province	Occurrence	As	Age,Ma	L	Form;assoc.ig.rock	Comments	References
France	Hercynides (Fig.2.5)	Vosges (see also Table C5)	A1	≈300 (by analogy)	?	Vic, Gerardner, Bresse granitoid plutons	lamps invade grt but grt also contains lamp xen	Choubert (1936); Métais et al. (1962); Wimmenauer & Hahn-Weinheimer (1966); Camboly et al. (1967); Zimmerle (1977); Peyronnet & Lameyre (1984); Peyronnet (1984b)
France	Hercynides? (Fig.2.5)	Languedoc (Termes, Aude,Corbières & Faugères,Hérault)	A	?	CM	Various granitoid plutons		Suvanapradip & Thiébault (1964); Bard & Gonord (1965)
France (Saône et Loire)	Hercynides (Fig.2.5)	Morvan (Saint-Aubain-Gilly sur Loir)	A?	334±1 (K-Ar)	CM	Isolated sill	Known from 4 boreholes	Bellotto & Zimmermann (1983)
Germany (DDR)	Hercynides (Fig.2.5)	Lausitz massif	A1	≈300 (by analogy)		Lausitzer granite pluton, etc.		Beger (1913,1916,1923); Tröger (1932); Grosser (1957,1966); Keiling (1961); Löffler (1961); Bautsch (1963); Hemmann & Watznauer (1964); Watznauer (1964); Rohde & Ullrich (1969); Rohde (1972); Kramer (1976c)
Germany (DDR)	Hercynides (Fig.2.5)	Thuringia (Oberhof, Saxothuringikum-Lugikum)	A1	≈300 (by analogy)		Various granitoid plutons		Kramer (1976b,1988); Beuge & Kramer (1977)
Germany (FDR)	Hercynides (Fig.2.5)	Ardennes (Aachen, etc.)	A1	≈300 (by analogy)	CV	Various granitoid plutons	Few details available	Washington (1917)
Germany (FDR)	Hercynides (Fig.2.5)	Erzgebirge /Fichtelgebirge	A1	lamp & grt: Hercynian (340-380)	CM CK CS	Various granitoid plutons	Associated dolerites; original lamp locality	Gümbel (1874); Sedlacek (1939); Acejev & Harlass (1968); Bautsch & Rohde (1970); Kramer (1976a); Kramer & Rosler (1976); Kramer & Seifert (1984); Kramer (1988); Just & Kramer (1989)
Germany (FDR)	Hercynides (Fig.2.5)	Odenwald (and adjacent areas of Rhine Graben)	A1	≈320 (K-Ar)	All	Various granitoid plutons	Gradational series from lamp to granite-porphyry	Klemm (1923, 1928); Waldmann (1935); Frenzel (1971); Zimmerle (1977); Bitschene & Mertz (1984)
Germany (FDR)	Hercynides (Fig.2.5)	Spessart	A1	≈300 (by analogy)	All		Gradational series from lamp to granite-porphyry	Mosebach (1934); Chatterjee (1959); Zimmerle (1977)
Germany (FDR) & Switzerland	Hercynides (Fig.2.5)	Schwarzwald	A1	≈300 (by analogy)	All	Marlsburg granitoid pluton; granite-porphyry dykes	Gradational series from lamp to granite-porphyry	Koch (1938); Zimmerle (1958); Kramer (1966); Wimmenauer & Hahn-Weinheimer (1966); Wimmenauer (1972); Hurrle (1976); Zimmerle (1977); Büchi et al.(1984); Müller (1984)
Germany (FDR/ DDR)	Hercynides (Fig.2.5)	Harz Mtns.	A1	≈300 (by analogy)				Gabert (1959); Schulze (1968)
Greece (Aegean Islands)	— (Fig.2.12)	Kos	A	lamp & grt: Miocene	All	Dikeos monzonite stock; apl, peg, porphyry dykes, sills	Gradational series monzonite-lamp-malchite	Wimmenauer (1976)
Greece (Akti)	— (Fig.2.1)	Mt.Athos (St.Gregory monastery)	?A	?	CK	Mt.Athos granitoid pluton	Few details available	Georgiadès (1938)

Table C1. Confirmed occurrences of calc-alkaline lamprophyres (with no other branches)

Country	Province	Occurrence	As	Age,Ma	L	Form;assoc.ig.rock	Comments	References
Greenland (E)	Caledonides (Figs.2.1 & 2.11)	Canningland & Wegener Halvø	A	Caledonian	No info	Kap Wardlaw granitoid pluton; grt, porph, Qz-dolerite dykes	Two ages lamps which pre- and post-date granitoids	Caby (1972); Escher & Watt (1976, p. 208)
Greenland (SW)	Ketilidian (Fig.2.11)	Julianehåb (Ilordleq, Brederfjord, etc.)	A1	1600–1700 (Rb-Sr, K-Ar)	CS CA	Hundreds of sheets, dykes; assoc. diorites, Hb-perids, granitoids	Some sheared; 5% of granitoid area; Opx -bearing rocks ≠ CA	Walton (1965); Watterson (1968); Escher& Watt (1976, p.138)
Hungary	Hercynides? (Fig.2.5)	Velence Hills (Sarhegy, Csala, Szekesfehervar)	A	?	CS CK	Velence granitoid plutons; aplite, grt-porph dominant	Only 4 lamps in 3 locs; all carry grt xen in margins	Embey-Isztin (1972)
India (Bihar)	?	Pat, Lohardaga district	A	?	CS	Dykes, sills assoc. with gabbros and norites	Some = microdiorites or porphyrites, but true CS may occur	Pathak (1984)
India (Bengal)	— (Figs.2.1 & 2.13)	Singhbhum	A	lamp: 1700-1000 (various)	CS	Singhbum granitoid batholith	lamp both intrude and are intruded by dlr; perid, norite, grt	Saha et al. (1973)
India (Mysore)	? (Figs.2.1, 2.13)	Yachenahalli	R	?	CS	Dykes in granulite	Strong reactions with country-rocks	Rao (1939)
India (Tamil Nadu)	? (Figs.2.1, 2.13)	Tiruppattur	S	?	CV	Single dyke	The only definite association of CAL with carbonatites	Ramasamy (1984)
Indonesia (Kali-mantan)	— (Fig.2.1)	Linhaisai, Karamu River	R	≈8 (K-Ar)	CM	shoshonite, andesite, "kajanite"	In belt of K-rich Cainozoic magmatism	Bergman et al. (1988)
Indonesia (Sumatera Barat)	— (Fig.2.1)	Natal area	A	?Tertiary	CV	Granitoid plutons of various ages	Area only reconnaissance mapped	Rock et al. (1983)
Italy	Alps/ ?Hercynides (Fig.2.12)	Adamello (Lago della Vacca, Mte.Costone, Val Nambrone, etc.)	A	Oligocene	CM CS CA	Adamello,Mte.Costone grt/Hb-dio plutons; apl,porpht,dio dykes	Appinite suite (§7.1) and Hercynian rocks may be represented	Fenoglio (1938); Hieke (1945); Callegari (1958); Monese (1969); Comin-Chiaramonti (1981); Ulmer et al.(1983)
Italy	Hercynides (Fig.2.1)	Cima d'Asta, SW Dolomites	A	≈275 (Rb-Sr)	CS	Granitoid plutons; apl, monz, peg, porph, porpht dykes	"Camptonites" = CS	Simboli (1958), Leonardi (1967, p.595-6)
Italy	Hercynides (Fig.2.6)	Biella–Locarno-Lago Maggiore area, NW Alps (also Table C5)	A1	≈280 (U-Pb Rb-Sr)	CA CS CC	Lamp-porpht-porph series slightly predate or overlap grt massifs	Assoc.Tertiary lamps (Table C6); see also Table C5; some metd.	Parker (1926); Burri & de Quervain (1934); Walter (1950); Zezza (1969); Boriani et al.(1974); Bigioggero et al.(1981); Giobbi Origoni et al. (1975;1990)
Italy	Roman/ Tuscan (Fig.2.12)	Mt.Amiata	B4	≈0.43 (K-Ar)	CM	Lamp inclusions in rhyodacite lavas		Bergen et al. (1983)

Table C1. Confirmed occurrences of calc-alkaline lamprophyres (with no other branches)

Country	Province	Occurrence	As	Age,Ma	L	Form;assoc.ig.rock	Comments	References
Italy (Sardinia)	? (Fig.2.1)	San Efisio (Nuoro)	A	?	CS CC	Nuoro granitoid pluton; aplite, porphyrite veins	dolerite, etc; CS grade into porphyrite	D'Amico & Guidicini (1961)
Japan (Honshu)	? (Fig.2.1)	Mt.Hiei, Kyoto	A	post-Palaeozoic	CS CC CP	Hiei granitoid pluton; apl; lamp may postdate acidic rocks however	Some "lamps" have Pl phen and are CC or CP; numerous dykes	Yoshizawa & Ishizaka (1961)
Japan (Honshu)	— (Fig.2.1)	Kasuga-mura, Gifu-ken	A3	grt: ≈73 (K-Ar)	CS	Kaisuki-yami granitoid pluton	Over 50 CS dykes cut granite but are cut by late veins of pluton	Suzuki & Shiraki (1980)
Japan (Shikoku)	?	Shiratori, Kagawa prefecture	A		CS	Dyke-swarm with granophyres cutting granitoid plutons	Composite lamp-granophyre dykes occur	Kawano & Kishira (1940); Hirayama (1957)
Madagascar	—	Betafo(Andrantsay, Ambatafinandrahana/ Ambatofangehana)	A	?	CK CM CS	Dykes		Lacroix (1922)
Mali	? (Fig.2.1)	Adrar das Iforas	A1 or R?	Post Pan-African orogeny	CS	lamps precede post-tectonic grt plutons; Qz-syenite, rhyolite	E-W lamps cut by later N-S swarms; microgbr dyke-swarms	Black et al.(1979)
Mexico	Mexican volcanic belt (Fig.2.1)	Colima graben	B6	Pleistocene -Recent (≈1.3)	All	Lamp lavas, dykes; high-K andes,bas,dacite ankaratrite lavas	One of few known occurrences of CAL lavas	Luhr & Carmichael (1981); Allan & Carmichael (1984); Luhr & Kyser (1989); Luhr et al.(1989)
Mexico	Mexican volcanic belt (Fig.2.1)	Mascota volcanic field	B6	Pleistocene: 3–0.5 (K-Ar)	CM	30 lava/cinder cones; associated basalt, andesite	Shows transition from calc-alkaline to CAL magmatism	Luhr et al.(1989)
Morocco	— (Fig.2.1)	Azegour, Haut Atlas	A	lamp & grt: post-Cretaceous	CM	Azegour granitoid pluton		Permingeat (1954)
Namibia	Damaraland (Fig.2.1)	Kalkfeld	S6	Jurassic	CM	Dykes; assoc. Kalkfeld carbonatite complex but may not be coeval	Petrography unknown but analysis supports description "minette"	Van Zijl (1962); Visser (1964); Heinrich (1966, p.546); Tuttle & Gittins (1966, p.740)
Norway	? (Fig.2.1)	Lyngen Peninsula, nr.Tromsø	R	?	CS	Single dyke; no known contemporaneous igneous rocks	Xenoliths comprise up to 70% of dyke	Elders (1957)
Pakistan	Himalayas (Figs.2.1 & 2.13)	Karakorum (Hindukusch, Alto Baltoro etc.)	A?	lamps: ≈3 (K-Ar); grt older?	CM CV	Keshnikhan granitoid pluton	Some lamps transitional to LL	Viterbo & Zannetin (1959); Gamerith & Kollmer (1973); Desio (1979)
Papua New Guinea	Papuan arc (Fig.2.1)	Mt.Kare, Papuan Highlands	B	?Miocene	CS	Lamp dykes, ?plugs; also porphyry; very similar to Porgera suite	Major gold prospect; poorly exposed & highly weathered	Author's unpubl.data

Table C1. Confirmed occurrences of calc-alkaline lamprophyres (with no other branches)

Country	Province	Occurrence	As	Age,Ma	L	Form;assoc.ig.rock	Comments	References
Papua New Guinea	Papuan arc (Fig.2.1)	Porgera, Papuan Highlands	B3	lamps: ≈10 (K-Ar)	CS CA CU	Lamp dykes, stocks; fels,porph,pxnt; chem. transitional to AL	Lamps formerly "andesite, Hb-dio-porph, Hb-porph" etc.	Smith & de Ferranti (1968); Davies (1983); Fleming et al. (1986); Hanley & Bradshaw (1986); Richards (1988); Rock et al. (1988b,c;1989 & unpubl.data)
Papua New Guinea	Papuan arc (Fig.2.1)	Sugarloaf volcanics, Papuan Highlands	B3	Pleistocene (0.43: K-Ar)	CS	500m lava of area 45 x 20 km + maar; shoshonitic basalt lava		Davies (1983)
Poland	Hercynides (Fig.2.5)	Intrasudetic/North Sudetic basins (Chochianów, etc.)	A1	298–346 (K-Ar) + older Caledonian	CK CM	granodiorite-porphyry dykes	Caledonian lamps as sills,pebbles in cong; Hercynian as dykes etc	Sokolowski (1970, p.490, 492); Nowakowska & Teisseyre (1971); Jermanski et al. (1974); Ryka (1974)
Poland	Hercynides (Fig.2.5)	Silesia-Cracow Uplands(Brudzowice, Zawiercie,Glasówka)	A1	≈300 (by analogy)	CM CK CC	Acidic tuffs/volcanics, porphyry, albitophyres	Mineralized; lamps immediately predate grt; some in boreholes	Sliwinski (1960); Kicula & Weiser (1970); Sokolowski (1970, p. 496); Heflik et al. (1985)
Poland	Hercynides (Fig.2.5)	Swiety Krzyz Mtns. (Iwanisk,Borzeta, Daleszyce, etc.)	A1	2 episodes: pre-/post-Devonian	CK			Pawlowska (1958); Kardymowicz (1962,1963); Rubinowski (1962); Szcepanowski (1962); Tarnowska (1967,1968,1969,1974); Pelczar (1973); Wroblewski (1974)
Poland (Silesia)	Hercynides (Fig.2.5)	Sudetes (Karkonozse block)	A1	≈300 (K-Ar, Rb-Sr)	CK CM CS	Karkonozse granitoid massif, etc.		Lis (1970); Sokolowski (1970, p.489-490)
Poland (Silesia)	Hercynides (Fig.2.5)	Sudetes (Klodzo-Zloty Stok massif)	A1	≈300 (by analogy)	CK CM CS	Klodzo-Zloty granitoid pluton; apl, peg, sye, porph dykes	Lamps extend far from mass, e.g. to Góry Bystrzyckie	Kowalski (1966); Sokolowski (1970, p.476-8); Wierzcholowski (1977, 1979); Cwodjinski (1982)
Poland (West Pomerania)	Hercynides? (Fig.2.5)	Pila boreholes, W.Pomerania	A	?Carboniferous	CK	—	Vein cutting Precambrian rocks; associated diabase	Pendias & Ryka (1974)
Poland/ Czechoslovakia	Hercynides (Fig.2.5)	Sudetes (miscellaneous areas)	A1	≈300 (by analogy)	CM CK	Kudowa, Zulova granitoid plutons	Few details available	Spangenberg (1951); Sokolowski (1970, p.470, 480)
Portugal (Alentejo)	Iberian (fig.2.12)	Sines	A6	grt/lamps: ≈72 (K-Ar, Rb-Sr)	CK CS	Sines granite-syenite pluton	Intrusion has mixed calc-alkaline/alkaline affinities	Canilho (1971); Rock (1982b)
Portugal (Minho/ Douro)	Hercynides? (Fig.2.5)	Póvoa de Varzim/ Vila do Conde/ Vilarinho das Furnas	A?	?	CV	Region of abundant Hercynian granitoid plutons (e.g. Porto)		Andrade (1950,1979); Ferreira Pinto (1968)
Portugal (Alentejo)	Hercynides (Fig.2.5)	Ossa-Morena zone (Terrugem-São Ramão area, Elvas)	A1	grt: ≈280 (Rb-Sr)	CV CS	Sto.Eulália batholith; microgrt, microdio, Qz-porphyry dykes		Mata & Munhá (1986)
Portugal (Trás-os-Montes-	Hercynides? (Fig.2.5)	Chaves	A1	?	CS CK	Santo Estevão granitoid pluton; apl, granite-porphyry dykes		Assunção & Brak-Lamy (1948)

Table C1. Confirmed occurrences of calc-alkaline lamprophyres (with no other branches)

Country	Province	Occurrence	As	Age,Ma	L	Form;assoc.ig.rock	Comments	References
S.Africa (Cape Province)	? (Fig.2.1)	Cape Town (Belville/Mooresberg Malmsbury etc.)	A	?	CS	SW Cape granitoid pluton		Walker (1949); Visser (1964)
S.Africa (Cape Province)	— (Fig.2.1)	Kuboos/Bremen, Namaqaland	A	525±60 (U-Pb)	CK CS	Kuboos granitoid batholith; Bremen granite-syenite pluton	"Camptonites" have no Am or Bi & may be dio	Biljon (1940); Middlemost (1967)
S.Africa (Cape Province)	— (Fig.2.1)	Middelplaat dyke-swarm, Namaqaland	A	lamp: 927± 182; grt:911 ±39 (Rb-Sr)	CM CK CS	Richtersweld granitoid pluton; Spektakel granite-syenite suite	Termed "lamproitic" but Al contents too high; typical CAL	De Villiers & Söhnge (1959); Reid & Barton (1983); Hunter & Reid (1987)
S.Africa (Cape Province)	— (Fig.2.1)	Philadelphia	A	?	CS	Two un-named granitoid stocks	Few details available	Beeson (1975)
S.Africa (Transvaal)	— (Fig.2.1)	Lindequesdrift/ Vereeniging-Heidel-berg, Vredefort Dome	A	≈2000 (geological evidence)	CS	Vredefort Dome grt; Roodekraal volcano; "syenodiorite", sye,peg	See also Table 3.1 for Lindequesdrift details (largest known lamp)	Rogers (1922, p.63); Willemse (1937); Nel & Jansen (1957, p.45); Bisschoff (1972)
Saudi Arabia (Asir)	— (Fig.2.1)	Abha	A	?Late Precambrian	CK CC	Abha granitoid pluton	2 dykes only	Okrusch & Jamal-Allil (1979)
Singapore	— (Fig.2.1)	Singapore Island	A1	200-220 (K-Ar)	CS CK	Central Song and P.Ubin granitoid plutons	dolerite, trachyte dykes also occur	Hutchison (1964); Chew (1970)
Spain	Hercynides? (Fig.2.5)	Segovia	A	?	CS	Sierra de Guardaramma La Granja, Novacerrada plutons;Qz-porphyry		Fúster (1951, 1955); Fúster & Ibarolla (1952)
Spain (Gerona)	? (Fig.2.12)	Costa Brava Catalane	A	?Tertiary	?	Costa Brava granitoid plutons	"green, red" lamps cut /cut by granitoids; assoc. younger AC	Arribas (1952)
Sweden	? (Fig.2.1)	Hamrånge	A?	?	CK		Possibly related to Åland Is. granitoids	Von Eckermann (1928); Johannsen (1938)
Sweden	— (Fig.2.1)	Värmland	A	lamp: ≈904 grt: ≈965 (K-Ar)	CK	Blomskog granitoid pluton		Wahlgren & Kahr (1977)
Switzerland	Hercynides (Fig.2.5)	Aar, St.Gotthard & Ticino massifs (inc. Graubunden)	A1	Hercynian (1°), Alpine (2°) ages	All		Mostly lamproschists /gneisses metamorph-osed in Alpine times	Washington (1917); Grübenmann (1919); Beger (1923); Kelterburn (1923); Bearth (1932); Baechlin (1937); Gwinner (1971, p.360); Steiner (1984); Oberhänsli (1986,1987); Giobbi Origoni et al.(1975,1990)
Taiwan	— (Fig.2.1)	General	A	Oligocene		Numerous "granitoid injections"		Yen (1985)

Table C1. Confirmed occurrences of calc-alkaline lamprophyres (with no other branches)

Country	Province	Occurrence	As	Age,Ma	L	Form;assoc.ig.rock	Comments	References
Tanzania	—	Victoria/Lupa goldfields, L.Rukwa (see also Table C5)	A2	≈2700?	CM CK CA	Dykes e.g. at New Saza and Geita mines	Associated with gold mineralization	Harris (1981)
UK (Channel Isles)	Hercynides (Fig.2.5)	Channel Is. (Jersey, Guernsey, etc.)	A1	lamp: ≈296, 254-285; grt: 297-305	CM CK	Related to Trégastel, etc. plutons—see entry France (Normandy)	apl, peg. older; dlr further E same age; some CM trans.to LL	Smith (1933,1936a,b,1939); Bishop (1964); Loy (1967); Lees (1974); Adams (1976); Wagner & Velde (1985); Turpin et al.(1988)
UK (England)	Caledonides (Fig.2.6)	Lake District (Shap, Kendal, Sedbergh, Howgill Fells, etc.)	A1	lamp & grt: ≈397, ≈438 (2 episodes)	CM CK	Shap, Threlkeld, granitoid plutons		Bonney & Houghton (1879); Harker (1892,1912); Morrison (1918); Wilson & Gilligan (1924); Smith (1930); Eastwood et al. (1968, p.130ff); Piper et al. (1978); Macdonald et al.(1985); BGS Kendal Memoir (sheet 39, p.35)
UK (England)	Caledonides (Fig.2.6)	Palaeozoic inliers (Cross Fell)	A1	Late Caledonian	CM CK	Shap granitoid pluton; porphyry dykes	Dykes radiate from pluton with more acidic dykes nearer it	Morrison(1918);Hudson(1937);Taylor et al.(1971,p.29ff);Burgess & Wadge (1974,p.130ff);Burgess & Holliday(1979,p.82ff);Arthurton & Wadge(1981, p.86ff);Macdonald et al.(1985);Gaskarth et al.(1989);BGS Brough Memoir
UK (England)	Caledonides? (Fig.2.6)	Nuneaton (Midlands) Wrekin,Malvern, etc. (Welsh Borders)	A1	Probably Caledonian	All	Mountsorrel and various buried granitoid plutons	Formerly "albite-diabases,camptonites"; widespread in bores	Brammall(1915);Allen(1968); Hains & Horton(1969, p.14); Earp & Hains (1971,p.37); Taylor & Rushton(1971); BGS Memoirs to Coventry (p.40), Shrewsbury (p.68,73,223), Atherstone (p.13); author's unpubl.data
UK (England)	Caledonides? (Fig.2.6)	Palaeozoic inliers (Cronkley,Ingleton, Haughton,Cautley)	A1	409–420 (K-Ar)	CM CK	Weardale and other buried granitoid plutons; porph dykes		Williams(1923);Moore & Ashton(1968);Taylor et al.(1971, p.29ff); Macdonald et al.(1985); Nixon et al.(1986); Gaskarth et al.(1989); BGS Memoir to Ingleborough (p.15)
UK (IOM)	Caledonides (Fig.2.6)	Isle of Man, Irish Sea	A1	Late Caledonian	CM ?	Dhoon & Foxdale granitoid plutons	Lamps not recently described	Taylor et al.(1971, p.35); BGS Isle of Man Memoir
UK (Scotland)	Caledonides (Fig.2.6)	Grampian Highlands (Aberdeen, Buchan, Tayside; see also C5)	A1	lamp: Late Caledonian; grt: 390-420	All	Monadhliath,Glen Tilt, Lochnagar, Strathspey plutons; fels, porpht	Regional dyke-swarms	Wilson (1886, p.19); Hinxman & Wilson (1902, p.42); Barrow et al. (1905, p.119; 1913,p.82); Barrow & Craig (1912,p.99); Horne (1923,p.51); Read (1923,p.164); Peacock et al. (1968, p.81)
UK (Scotland)	Caledonides (Fig.2.6)	Midland Valley (including Catacol, Isle of Arran)	A1	Late Caledonian	CK CS	Distinkhorn granitoid pluton; porphyrite, porphyry dykes	Gradational dyke-series lamp–porpht–porph-rhyolite	Eyles et al.(1949, p.43ff); Herriot (1977); Nockolds et al.(1978, p.367); Cameron & Stephenson (1985, p.39ff)
UK (Scotland)	Caledonides (Fig.2.6)	NE Highlands (Ross-shire, Sutherland; see also Table C5)	A1	lamp: Late Caledonian; grt: 390-420	All	Migdale, Rogart, Fearn granitoid plutons; fels, porpht dykes	≥2 episodes lamps, pre/postdating granitoids	Peach et al.(1912, p.121ff; 1913a, p.77; 1913b, p.87); Horne & Hinxman (1914, p.52); Read et al.(1925,p.51; 1926, p.108,166ff); Read (1931, p.183); Ramsay(1955); Smith(1979); Johnstone (1989)
UK (Scotland)	Caledonides (Fig.2.6)	NW Highlands (Assynt/Ratagain; see also Table C5)	B1/ A1	lamp & grt: 413–431 (K-Ar, Rb-Sr)	CV CS	Lamp dykes, sills; sye, borolanite, shonk,pxnt, fels intr; shosh lavas	Borralan/Ailsh sye & Ratagain grt-sye;lamp = only primitive rocks	Peach et al.(1910, p.86);Read et al. (1926, p.108ff); Sabine (1953); Sutherland (1982); Thompson & Fowler (1986); Halliday et al.(1987)
UK (Scotland)	Caledonides (Fig.2.6)	Shetland Is.	A1	lamp & grt: Late Caledonian	CS	Ronas Hill, Graven granitoid plutons; fels, porpht dykes	Also dolerite dykes	Phemister (1952); Mykura (1976, p.94ff); Mykura & Phemister (1976, p.248)
UK (Scotland)	Caledonides (Fig.2.6)	Southern Uplands (Borders, Cheviot, Galloway, etc.)	A3 or A4	lamp: 397 to >425 grt: 395-400	All inc CA	Criffel, Cockburn Law, Priestlaw, Doon grt plutons; porpht dykes	lamps bracket grt plutons in time; Read (1926);King (1937)	Gardiner & Reynolds(1932);Macgregor(1937);Phillips (1956);Lumsden et al (1967);Greig(1971);Barnes et al.(1986,1988);Leake & Cooper(1983);Rock & Rundle(1986);Macdonald et al.(1986);Rock et al.(1986a,b;1987;1988a)

Table C1. Confirmed occurrences of calc-alkaline lamprophyres (with no other branches)

Country	Province	Occurrence	As	Age,Ma	L	Form;assoc.ig.rock	Comments	References
UK (Scotland)	Caledonides (Fig.2.6)	SW Highlands (Islay Colonsay,Argyll; see also Table C5)	A1	lamp: Late Caledonian grt: 390-420	All	Arrochar, Ballachulish, Etive,Garabal, Nevis, grt; fels, porpht dykes	≥2 episodes lamps, pre/syn/postdating grt	Gunn et al.(1897,p.106); Hill et al.(1905,p.106); Kynaston & Hill (1908, p.116); Hinxman et al.(1923,p.76);Flett(1929);Anderson(1935b); Anderson & Tyrrell(1936); Bailey (1958;1960,p.226); Borradaile(1986); Rock(1988a)
UK (Scotland)	Caledonides (Fig.2.6)	W Highlands/Islands (Iona,Lochaber, Ardgour;see also C5)	A1	lamp: 406 to >420 grt: 410-420	All	Ratagain, Strontian, Ross of Mull plutons; fels, porpht dykes	≥2 episodes lamps, pre/postdating grt	Bailey & Anderson (1925, p.38); Lee & Bailey (1925, p.54); Gallagher (1963); Smith (1979); Rock & Hunter (1987); Johnstone (1989)
UK (Ulster)	Caledonides (Fig.2.6)	Ards–Newry swarm (see also Table C5)	A1	≈400 (by analogy)	All	Newry granitoid pluton; porphyry dykes	≥2 episodes lamps (foliated/unfoliated); assoc. base metals	Seymour (1900); Reynolds (1931,1938); Tomkeieff & Marshall (1940); Grant (1966); Wilson (1972, p.25); Morris et al.(1986); Murphy (1987)
UK/Ireland (British Isles)	Caledonides (Fig.2.6)	General	A1	≈400 (by analogy)	All	Numerous granitoid plutons	>5,000 dykes ranging from Shetlands to English Midlands	Richey (1939); Sutherland (1982); Rock et al.(1988a)
USA (AL)	Alaskan	Central York Mtns., Western Seward Peninsula	A	Tertiary	CK CM	Dykes associated with granitoids; age order (1) grt; (2) lamps; (3)veins	Associated Be-Sn-fluorite mineralization	Sainsbury (1969)
USA (AL)	N.American Cordillera (Fig.2.1)	Willow Creek	B1	lamp: 66±2; grt: 65,73 (K-Ar)		Dykes cutting Talkeetna Mtns. granitoid batholith	Lamps coeval with Au-Qz veins	Madden-McGuire et al.(1988)
USA (AL)	N.American Cordillera (Fig.2.9)	Salmon Bay, Prince of Wales Is.	B1	Cretaceous or Early Tertiary	CS CK	phonolites; younger basalts	Th-REE-Cb veins; "lamprophyric biotite hornblende albitite"	Houston et al.(1958); Heinrich (1966, p.369)
USA (AL)	— (Fig.2.9)	Bokan Mountain, Prince of Wales Is.	A	≈180 (K-Ar) 151±5 (Rb-Sr)	CM ?	Peralkaline granites of Bokan plutonic Complex	Lamps altered and may include CC,CP	MacKevitt (1955); Thompson et al.(1980)
USA (AZ)	N.American Cordillera (Fig.2.10)	Cerbat Range, Black Mtns., Grand Washington Cliffs	A4	Late Jurass. –Early Cretaceous	CK CM CV	granite-porphyry dykes and stocks	Important mining district	Schrader (1909); Washington (1917)
USA (CA)	N.American Cordillera (Fig.2.10)	Independence dyke-swarm	A4	lamp: ≈150 grt: 149-186 (plus ≈103)	CS	Sierra Nevada granitoid batholith	Gradational series from lamp to grdi-porph; regional lamps	Moore & Hopson (1961); Chen & Moore (1979); Tilton (1983); James (1989)
USA (CO)	N.American Cordillera	La Plata (Allard) stock	B?					Washington (1917)
USA (CO)	N.American Cordillera (Fig.2.10)	Apishapa Quadrangle	B1	Tertiary	CM CV	tinguaites, basalts	Now considered separate centre from Spanish Peaks	Cross (1914); Knopf (1936)
USA (CO)	N.American Cordillera (Fig.2.10)	Elk Mtn., Elkhead Mtns./Fortification dyke, Craig	B1	lamp: 7.6–11.1 (K-Ar)	CM	Elk Mtn.=sill of porph with CM enclaves in wide volcanic field	"Verite" of Fortification dyke = CM	Ross (1926b); Leat et al.(1988)

Table C1. Confirmed occurrences of calc-alkaline lamprophyres (with no other branches)

Country	Province	Occurrence	As	Age,Ma	L	Form;assoc.ig.rock	Comments	References
USA (CO)	N.American Cordillera (Fig.2.10)	Red Mtn.	A6	lamp & rhy: 24–33 (K-Ar)	CK	Early composite lamp dykes (Table 3.1); later rhyolite stocks, dykes	Associated porphyry-Mo systems (Climax & Urad-Henderson)	Geraghty et al.(1988)
USA (CO)	N.American Cordillera (Fig.2.10)	Ute Mtns.	B1	Early Tertiary	CS	A few lamp dykes, sills; monz, diorite, porph, gbr laccoliths	Intrudes Late Cretaceous sediments	Ekren & Houser (1965)
USA (CO)	—	Sawatch Mtns.	R	Late Precambrian	CM CV CS	Hundreds of sheets in swarm metamorphosed to amphibolite facies	Petrography and chemistry supports lamprophyric nature	Pearson (1959)
USA (MD)	Hercynides (Figs.2.1, 2.5, 2.10)	Great Falls, W of Washington DC	A1	≈360 (K-Ar)	CK	Bear Island granitic and pegmatitic plugs, dykes, veins	Localised lamp dyke-swarm	Reed & Reed(1969); Reed et al.(1970)
USA (MN, MI)	Superior (Fig.2.9)	N.Minnesota (Dead River etc), Michigan (Marquette, etc.)	A2	lamp:2600? grt: ≈2700	CK CS	Saganaga, Snowbank granitoid stocks	(in Abitibi, Opatika, Uchi, Wabigoon, Wawa sub-provinces)	Ayres & Higgins (1939); Geldon (1972); Sims & Mudrey (1972); Mudrey & Geldon (1973); Ervin & Mudrey (1975,1976)
USA (MT)	N.American Cordillera (Fig.2.10)	Highwood Mtns./ Little Belt Mtns./ Castle Mtns.	B1	48–52 (K-Ar)	CM CK dyke	Lamp,ting, trach dykes monz,shonk,fergusite, sye intr; shosh lavas	CM formerly "mafic phonolite,monchiquite camptonite" etc.	Pirsson (1905); Washington (1917); Larsen et al.(1941); Buie (1941); Burgess (1941); Witkind (1969,1970,1973); Woods (1976); O'Brien et al. (1988)
USA (NC)	Hercynides (Figs.2.1, 2.5, 2.10)	Charlotte Belt (Concord, Harrisburg)	A1	lamp: 330-350 (K-Ar); grt:300–400	CM	Concord sye,grt & other varied plutonic rocks of Charlotte belt		Mauger (1988a,b)
USA (NM)	N.American Cordillera (Fig.2.10)	Slagle Canyon, Colfax Cty.; Raton-Brilliant quadrangles	B?	Probably Quaternary	CM CV	Quaternary basalt, dacite, phonolite, trachyte lavas	Relation to Quat. volcs, nomenclature uncertain	Lee & Mertie (1922); Collins (1949); Stobbe (1949)
USA (NM)	Trans-Pecos (Fig.2.10)	Sacramento Mtns./ Sundance Quadrangle	R1	44.2±2.2 (K-Ar)	CS CA	"Numerous dykes,sills" termed "camptonite"; "diorites" = CA	Exhibit flow differentiation and late adularia cavities	Asquith (1973a,b,c)
USA (NV)	— (Fig.2.10)	Santa Rosa Mtns. (Paradise Range, Mineral County etc.)	A	?		Dykes; assoc. diorite, granodiorite, andesite porphyry, aplite	Associated brucite deposit	Wells (1937, p.32)
USA (SC)	Hercynides (Figs.2.1, 2.5, 2.10)	Winnsboro'	A1	298±11 (K-Ar)	CS	2–3 composite dykes assoc. with Rion adamellite pluton	See also Table 3.1	Vogel & Wilbrand (1978); McSween et al.(1979); Mauger (1988b)
USA (SD/WY)	N.American Cordillera (Fig.2.10)	Black Hills/ Sundance area (Squaw Creek etc.)	B1	Tertiary	CM CV	Ne-sye, Psdlc-porph, trach, phon; lamp dykes over wide area	Includes Mineral Hill ring dyke complex	Darton & O'Harra (1905); Kirchner (1979)
USA (UT)	N.American Cordillera? (Fig.2.10)	Santaquin-Nebo	?	?	CS	Dyke	Few details available	Wells (1937, p.41)

Table C1. Confirmed occurrences of calc-alkaline lamprophyres (with no other branches)

Country	Province	Occurrence	As	Age,Ma	L	Form;assoc.ig.rock	Comments	References
USA (WY)	N.American Cordillera	Granite Mtns. (Rattlesnake Hills)	B1	Middle-Late Eocene	CV	36 eruptive centres of phonolite, latite, trachyte; lamp dykes	Associated U deposits	Love (1970)
USA (WY)	N.American Cordillera (Fig.2.10)	Bear Lodge Mtns.	B	38–50 (K-Ar)	CK CS	latite, sye, Ne-sye, phonolite, Psdlc-porphyry, carbonatite	Associated Th-U deposits	Staatz (1983); Woolley (1987, p.126)
USA (WY)	N.American Cordillera (Fig.2.10)	Yellowstone Park	B1	Tertiary	CK	absarokite, shoshonite, banatite lavas; monzonite intrusions		Washington (1917); Joplin (1968)
USA (AL), Canada (Br Columbia)	N.American Cordillera (Fig.2.10)	Ketchikan–Prince Rupert–Stewart mining camp region	A? R?	?Miocene (<45); may be 2 phases	CS	Hundreds of dykes, some younger than Coast Range batholith	Several swarms over 15,000 sq.km; chem. transitional to AL	Hanson (1929,1935); Grove (1971); Smith (1973); Aldrick et al.(1987)
USA (MT) and Canada (Alberta)	N.American Cordillera (Fig.2.9/10)	Sweet Grass Hills	B1	≈48 (K-Ar)	CM	Lamp dykes, sills; dio-porph, sye-porph, tinguaite laccoliths	CM mainly in east; outlier of Highwood Mtns. magmatism	Kemp & Billingsley (1921); Williams & Dyer (1930); Currie (1976, p.172)
USA(WA/ ID)/Canada (BC)	N.American Cordillera (Fig.2.10)	49th parallel (Rossland/Coeur d'Alene,Corbaley)	A1 B1	L: 46-60 (K-Ar); Grt: 47-90	All	Central Idaho, Trail, Sheppard, Mt.Stuart plutons; Coryell intr.	lamp cuts grt but cut by grt dykes; coeval K-rich sye, monz, etc.	Barber (1904); Daly (1912); Drysdale (1915); Shannon (1920); Waters (1927); Gilbert (1948); Hopson (1957); Fyles et al.(1973)
USSR	? (Fig.2.1)	Magnitnaya	A	?		Various granitoid plutons	Most lamps are pre-granite but one chills against it	Zavaritsky (1935)
USSR	Hercynides (Asiatic) (Fig.2.1)	Chuya, Altay Sayan	A or R	lamps:av. 230 (K-Ar); grt: 240-295	CM CK	Tarkhaba granitoid plutons (Kalba complex)	Several elongate dyke-swarms invade plutons	Obolenskaya & Firsov (1966)
USSR	Kazakhstan (Fig.2.1)	Kazakshtan (central)	A	?		Various granitoid plutons		Effinova (1970)
USSR	— (Fig.2.1)	E.Transbaikal/Lena	A	Jurassic	CK CS	Amudzhikan-Shetensk, Kara-Chachinsk grt; grt/grdi-porph dykes	Also porpht, hybrids; lamps emplaced between 2 grt phases	Zalutski (1962); Solomin (1969); Solov'yeva (1973); Daniyelyants & Yakhov (1975)
USSR	— (Fig.2.1)	Talakhtakh, Anabar massif	R?	Palaeozoic	CM	Isolated vent	Termed "lamproite" but Al » K and K itself too low	Vishnevskii et al.(1986)
Yugoslavia (Macedonia Serbia,etc.)	— (Fig.2.1)	Krive Palanke, Rudnik, Boranja, Zvornik, Ripanj etc	A B	Tertiary (Neogene)	CM	Andesite-dacite-rhyolite and "bostonite, latite, shoshonite" lavas	"Kajanite" intrusives appear chemically to be typical CAL	Zujovic (1890); Dimitrijevic (1928); Tomic (1929); Karamata (1955); Knezevic (1956,1976); Trubelja & Paskvalin (1962); Majer (1965); Majer & Karamata (1983)
Yugoslavia (Slovenia)	? (Fig.2.1)	Pohorje massif	A	?	CS	Pohorje tonalite pluton	Dykes cut tonalite	Dolar-Mantuani & Klemen(1940)

Table C1. Confirmed occurrences of calc-alkaline lamprophyres (with no other branches)

Country	Province	Occurrence	As	Age,Ma	L	Form;assoc.ig.rock	Comments	References
Zimbabwe /S.Africa (Transvaal)	— (Fig.2.1)	Limpopo (central belt)	R	≈2650 (Rb-Sr, Pb-Pb)	CS	Deformed dykes formerly mapped as "amphibolites"	Chemistry somewhat transitional to AL	Watkeys & Armstrong (1985)

Table C2. Confirmed occurrences of alkaline lamprophyres (with no other branches)

Country	Province	Occurrence	As	Age,Ma	L	Form	Comments	References (plus Woolley 1987; *=refs mainly to inclusions)
Angola	Angolan (Fig.2.2)	Zenza do Itombe	M7	≈110 (assumed)	AM	Numerous dykes	Syenite-carbonatite complex	Holmes (1915); Andrade (1954); Aires Barros (1965)
Antarctica	—	Bunger Hills	R	502±10 (Rb-Sr)	AC AM	Dyke swarm	Differentiated ("trachy-basalt") associates; also older UML (Table C3)	Sheraton et al.(1990)
Antarctica	— (Fig.2.2)	Alexander Island (Uranus/Neptune glaciers)	R	≈15 (K-Ar)	AC	5 dykes	Part of basaltic province	Horne & Thompson (1967); Rex (1970)
Atlantic Ocean	Iberian (Fig.2.12)	Gorringe Bank	M6	60–66 (K-Ar)	AC AM AB	Dredge rocks assumed to be lavas	Two lava series: nephelinite-phonolite and AL-mugearite-trachyte	Cornen (1981,1982); Feraud et al. (1982)
Atlantic Ocean	Oceanic Islands (Fig.2.2)	Bermuda Seamount	R5	≈33 (K-Ar)	AC	400 sheets interfingered with lavas	Information entirely from drill-core	Aumento & Ade-Hall (1973); Aumento et al. (1974); Reynolds & Aumento (1974)
Atlantic Ocean	Oceanic Islands (Fig.2.2)	Fernando do Noronha	S5	Late Miocene	AM	Dykes, plugs, domes	Associated with basanites, phonolites	Campbell Smith & Burri (1933); De Almeida (1955)
Atlantic Ocean	Oceanic Islands (Fig.2.2)	Fuerteventura, Canary Is.	M5	Pliocene	AC	Intense dyke-swarm	Lamps form component of dominantly basaltic swarm	López Ruiz (1970a,b)
Atlantic Ocean	Oceanic Islands (Fig.2.2)	Maio, Cape Verde Is.	S5	8–16 (K-Ar)	AS	Intense swarm of sheets	Formerly "vogesites"; close assoc. alkaline plutons; 100% extension in places	Part (1950); Assunção & Canilho (1969); Mitchell-Thomé (1976); Furnes & Stillman (1985)
Atlantic Ocean	Oceanic Islands (Fig.2.2)	Trindade	S5	Tertiary	AM	Dykes, plugs		De Almeida (1961)
Australia (NSW)	Tasman Fold Belt (Fig.2.2/8)	Nandewar Mtns. (Dingo Creek, etc.)	M1	Tertiary	AM	Sills	Richly xenolithic and xenocrystic; assoc. trachyte-phonolite volcanic suite	Jensen (1907); Joplin (1965)
Australia (NSW)	Tasman Fold Belt (Fig.2.2 & 2.8)	Murrumburrah	R1	194±3 (K-Ar)	AM	Dykes	Some leucite-bearing examples may be lamproites	Harvey & Joplin (1941); Wellman et al. (1970); Jaques et al. (1985)
Australia (NSW)	Tasman Fold Belt (Fig.2.2 & 2.8)	Southern Coalfield (Bulli, Wollongong,etc.)	M	101±4 (K-Ar) plus younger	AM AC ?AD	Numerous dykes, sills	Budderoo,Bulli,Good Dog, Mt.Nebo,O'Briens, etc;publ. names vary; assoc. ting,phon	Harper (1915); Süssmilch (1922); Joplin (1965); Bowman (1971); Carr & Facer (1980); Jaques et al.(1985); Sherwin & Holmes (1986)
Australia (NSW)	Tasman Fold Belt (Fig.2.2 & 2.8)	Western Coalfield	R		AM	Dykes	Examples at Putty Creek etc.	Carne (1908); Browne (1933); Joplin (1965)

Table C2. Confirmed occurrences of alkaline lamprophyres (with no other branches)

Country	Province	Occurrence	As	Age,Ma	L	Form	Comments	References (plus Woolley 1987; *=refs mainly to inclusions)
Australia (VIC)	Lachlan Fold Belt (Fig.2.2 & 2.8)	Bendigo goldfield (also Ballarat, Gorong, etc.)	R	146-155 (K-Ar)	AM	Dykes in many mines	Though much younger, AM are guides to Au veins since they pinpoint saddle reefs	Stillwell (1912); Skeats (1914); Edwards (1934,1938); David (1950, p.195-6); McDougall & Wellman (1976); Hinde (1988); Wilkinson (1988a,b)
Australia (VIC)	Tasman Fold Belt (Fig.2.2 & 2.8)	Older Volcanics	R	Palaeocene–early Miocene	AC AM	Regional dyke-swarms a few plugs	Subordinate crinanites, bas, rare nephelinites, ting, phon; feeders to basaltic lavas	Edwards (1938); Jaques et al.(1985)
Australia (VIC)	Tasman Fold Belt (Fig.2.2 & 2.8)	S.Gippsland (Cape Paterson, Anderson Inlet)	R	Lower Oligocene	AM AD	>20 plugs, numerous dykes	Gradation between AM, AD and various basaltic/felsitic rocks; single Ol-nephelinite	Edwards (1934); Jaques et al.(1985)
Australia (WA)	C.Australian Mobile Belt (Fig.2.2/8)	Wandagee	R1	≈160 (K-Ar)	AM	14 diatremes 0.5–3.75 ha, 8 sills/ dykes	Picritic; a few diamonds; chemically transitional to KIL but mineralogically AM	Atkinson et al.(1984a); Jaques et al.(1986,1989b); *Nixon et al.(1987)
Austria	Venetian (Fig.2.12)	Goldeck Mtns.	R1	≈30 (K-Ar, Rb-Sr)	AS	Dykes	Termed "alkalibasaltic dykes" but are ocellar and rich in Kt, Bi and Af; hence = AL	Deutsch (1984)
Austria	Venetian (Fig.2.12)	N.Calcareous Alps (Ehrwald, Innsbruck, etc.)	R1	Assumed Tertiary	AC	Dykes	Formerly called "ehrwaldite"	Trommsdorf (1962a,b)
Brazil (Ceará)	Fortaleza (Fig.2.2)	General	M	≈30 (K-Ar)	AM	Dykes, plugs	AM are probably parental to associated phonolites	Sial (1987)
Brazil (Rio de Janeiro)	Brazilian (Fig.2.2)	Tíngua	M	58–66 (K-Ar)	AM	Dykes	Includes one of monchiquite type-rocks; modern dyke descriptions not available	Hunter & Rosenbusch (1890); Washington (1917)
Brazil (São Paulo)	Brazilian (Fig.2.2)	Monte de Trigo	M5	≈80 (K-Ar on Ne-sye)	AC AM AS	Numerous dykes	Part of Ne-sye/theralite intrusive complex with phonolites, shonkinites etc.	Coutinho & Melcher (1973)
Canada (British Columbia)	— (Figs.2.2 & 2.9)	Ice River	S	≈360 (K-Ar)	AM	Numerous dykes	Lamps latest phase of carbonatite complex	Heinrich (1966, p.401); Tuttle & Gittins (1966, p.526); Currie (1975,1976)
Canada (NFD)	Same age as Monteregian (Figs.2.2/9)	Notre Dame Bay (Bay of Exploits)	M6	115-144 (K-Ar)	AC	Dykes; stock at Budgells Harbour	Lamp concentrations at Twillingate,Harris; chemistry of some transitional to UML	Heyl (1936); Helwig et al.(1974); Strong & Harris (1974); Currie (1976, p.43); Lapointe (1979)
Canada (Ontario)	(Grenville subprovince of Shield)	Sturgeon Narrows	M	?	AL?	Dykes, one boudinaged	Cut sye complex, may be younger; 40-70% Ab, 25-60% Bi, 5%Cb + fluorite	Woolley (1987, p.23); Sage (1988d)
Canada (Ontario)	Coldwell	Killala Lake	M	≈1185 + ≈300 (K-Ar)	AM	Dykes cutting sye and host-rock	Several lamp generations appear to be present, some related to sye complex	Currie (1976, p.161); Woolley (1987, p.26); Sage (1988e)

Table C2. Confirmed occurrences of alkaline lamprophyres (with no other branches)

Country	Province	Occurrence	As	Age,Ma	L	Form	Comments	References (plus Woolley 1987; *=refs mainly to inclusions)
Canada (Ontario)	Kapuskasing	Seabrook Lake	S	≈1103 (K-Ar)	AL	Small dykes 3–15 cm wide	Few petrographical details available	Heinrich (1966, p.171,539); Tuttle & Gittins (19660, p.520); Currie (1976, p.92); Sage (1988c)
Canada (Ontario)	Coldwell (Figs.2.2 & 2.9)	Port Coldwell/ Heron Bay	M7	1044±6 or 1070±15 (various)	AC	Dyke-swarm close by sye pluton	Gradational from AC to "heronite" (AB) differentiates; composite Ne-sye/lamp dyke	Walker & Parsons (1927); Currie (1976, p.159; 1980); Platt & Mitchell (1982b)
Canada (Ontario)	E.Ontario/ W.Quebec (Figs.2.2/9)	Shenango	M7	1047±15 (Rb-Sr)	AL	Dykes	Associated with syenite–alkaline diorite complex	Sage (1987d)
Canada (Ontario)	E.Ontario/ W.Quebec (Figs.2.2/9)	Lackner Lake (Nemegos) /Portage	S7	1078(K-Ar) 1138 (Rb-Sr)	AS	Radial dykes, cone-sheets	Fp "clots" in gd of Cpx, Akf, Ab,Bi, Mt, Cc, Ct, Ze+Kt; might be UD	Hodder (1961, p.26); Heinrich (1966, p.393); Currie (1976, p.88); Sage (1988f)
Canada (Prince Edward Is.)	Same age as North Sea (Figs.2.2/9)	Malpeque Bay	R6	≈247 (K-Ar)	AM	Single sill	Isolated alkaline magmatism	Greenough et al. (1988)
Canada (Quebec)	Monteregian-White Mtns. (Figs.2.2/9)	Monteregian Hills (general)	M6	110-140 (K-Ar, Rb-Sr)	AC AM	Intense central dyke-swarm	E: oversat plutons+AC; centre: undersat+AM; W: carbonatite complexes+UL	Howard (1922); Poitevin & Cooke (1946); Faessler (1962); Philpotts (1974); Eby (1980,1983,1985,1987); Trzcienski & Marchildon (1989)
Canada (Quebec)	Monteregian-White Mtns. (Figs.2.2/9)	Montreal (Mts. Royal, St.Hilaire & satellites)	M6	110–120 (K-Ar)	AC AM	Dykes, sills, pipes centred on pluton	Ste.Dorothée, Ste.Monique, Brossard, etc; separate UML occur; "kersantite" = AC	Schnetzler & Philpotts (1968); Woussen (1970); Philpotts (1972,1974,1976); Currie (1976, p.46); Eby (1980)
Canada (Quebec)	Monteregian-White Mtns. (Figs.2.2/9)	Mt.Megantic	M6	≈134 (Rb-Sr)	AC AM	Intense central dyke-swarm	Gabbro-diorite-granite-quartz syenite complex	Bédard et al.(1987,1988); Bédard (1988)
China (Zheijang province)	— (Fig.2.2)	Xilong, Qu	R		AM	Dykes	Xenolithic	Cao & Zhu (1987, p.171)
Cyprus	— (Fig.2.1)	Pentadaktylos (Orga–Kormakiti)	M3	Palaeocene	AC	Dykes assoc. with basalt, trach lavas	"Lamprophyres" 10% of magmatism; Cretaceous shoshonites in same area	Baroz (1978)
Czechos-lovakia	Rhine Graben (Fig. 2.12)	Ceske Stredehori (e.g. Roztoky volcanic centre)	M6	Tertiary	AC AM	≈560 lamp dykes,mostly radial swarm	Assoc. tinguaite, bostonite, "gauteite", basalt dykes; monzodiorite, essexite stocks	Wimmenauer (1974, p.260); Ulrych (1983,1986); Jelinek et al.(1989)
Eire (southern Ireland)	North Sea (Fig.2.7)	Inishowen, Donegal	R6	Permo-Carb (by analogy)	AC	Dykes	Outlying manifestation of intense Scottish dyke-swarms	Murthy (1948); *Hunter & Upton (1987)
Finland	North Sea? (Fig.2.7)	Helsinki (Söörnainen prison)	R	?	AM	Isolated thin dyke		Eskola (1954)

Table C2. Confirmed occurrences of alkaline lamprophyres (with no other branches)

Country	Province	Occurrence	As	Age,Ma	L	Form	Comments	References (plus Woolley 1987; *=refs mainly to inclusions)
France	Iberian (Fig.2.12)	Corbières/ Pyrenees	M6	≈82 (Ar-Ar, Rb-Sr)	AM AS	≈ 20 dykes	Associated with small nepheline-syenite body at Fitou	Barrabé (1952), Azambre (1970); Wimmenauer (1974, p.241); Peyronnet (1984a)
Greenland (E)	North Sea?	Scoresby Land	R	Late Permian	AL	Dykes	Cut Permian sediments but exposed to erosion by Late Permian	Stemmerik & Sørensen (1980)
Greenland (E)	Thulean (Fig.2.11)	Kangerdlugssuaq	M6	28–55 (K-Ar)	AC	Fjord-parallel dyke-swarm	Distinct from coast-parallel dyke-swarm (next entry)	Brooks & Platt (1975); Brooks & Printzlau (1978); Nielsen (1987a,b)
Greenland (E)	Thulean (Fig.2.11)	Coast-parallel dyke-swarm	R6	30–53 (Fission track)	AC AD	Regional dyke-swarm	Distinct from fjord-parallel swarm; associated with tholeiitic dolerite dykes	Vincent (1953); Nielsen (1987a,b); Rucklidge et al.(1980); *Nixon (1987, p.14)
Greenland (E)	Thulean (Fig.2.11)	Hold with Hope	R6	Tertiary (≈50)	AM	"Probably a lava"	Termed "olivine augitite" but is ocellar and has Am, Bi so definitely a lamprophyre	Upton et al.(1984)
Greenland (E)	Thulean (Fig.2.11)	Theresabjerg	M6	Tertiary	AC ?AS	Dykes, sills	"Minette, kersantite, augite /hornblende-spessartite" are rather basic and may be AL	Kapp (1960, p.160ff)
Greenland (E)	Thulean (Fig.2.11)	Werner Bjerge	M6	≈30 (various)	AC	Dykes	Some formerly called "kersantite" or "spessartite" but are too basic	Bearth (1959)
Greenland (E)	Thulean (Fig.2.11)	Bordtinderne/ Wiedemanns Fjord	M6	47 (Fission track), 52 (K-Ar)	AC	Dozens of dykes, one xenolithic	Appear to be central dykes related to syenitic plutons rather than coastal swarm	Brooks & Rucklidge (1973); Brooks & Printzlau (1978); Brown et al.(1978); Nielsen (1987a,b); *Scott Smith (1987)
Greenland (S)	Gardar (Fig.2.11)	General	M6	≈1200 (Rb-Sr)	AC	Dykes, sills	Lamps form component of dominantly basaltic swarms	Upton (1965,1974); Watt (1966); Mitchell (1971); Nielsen (1987b); Upton & Emeleus (1987)
Greenland (S)	Gardar (Fig.2.11)	Illimaussaq	M6	1168±21 (Rb-Sr)	AS? AB?	A few thin dykes	Postdate hydrothermal veins; rather felsic to be true lamps, more likely differentiates	Sørensen et al.(1969)
Greenland (S)	Gardar (Fig.2.11)	Ivigtut/ Qaqssiarssut	M6	≈1200 (by analogy)	AC?	Dykes		Ayrton (1963); Ayrton & Burri (1967); Upton (1974)
Greenland (W)	Thulean (sensu lato) (Fig.2.11)	Ubekendt Ejland	R6	≈33 (K-Ar)	AC AM	≈100 dykes	"Kersantite" = AC; many dykes xenolithic	Clarke & Pedersen (1976, p.379); Larsen (1981,1982); Clarke et al.(1983); *Scott Smith (1987); Nielsen (1987b)
Guinea	— (Fig.2.2)	Los Is.	M					Lazarenkov (1976)

Table C2. Confirmed occurrences of alkaline lamprophyres (with no other branches)

Country	Province	Occurrence	As	Age,Ma	L	Form	Comments	References (plus Woolley 1987; *=refs mainly to inclusions)
India (Orissa)	—	Khariar, Kalahandi district	M	Pre-cambrian	AC	Body in large alkaline gneiss belt	31% Pl, 25 Cpx, 26 Am, 10 Ne; chemistry OK for AL; assoc. Malig, Ne-sye, Essex	Madhavan & Khurram (1989)
India (Andhra Pradesh)	Eastern Ghat (Fig.2.13)	Elchuru, Prakasam district	S	1242±33 (Rb-Sr)	AM AS	>100 lamps in dyke-swarm	Formerly termed "minette", "biotite-lamprophyre"; latest phase of cbt-ijolite complex	Bose & Nag (1980); Nag (1983); Nag et al.(1983); Paul & Sarkar (1984); Madhavan & Leelanandam (1988); Madhavan et al.(1989); Rao et al.(1989); Ratnakar & Leelanandam (1989)
India (Andhra Pradesh)	Eastern Ghat (Fig.2.13)	Purimetla & Settupalle, Prakasam district	M	Pre-cambrian (?≈1200)	AC AS	Ocellar dykes	Gabbro-Syenite-Nepheline syenite-Quartz Syenite Malignite plutons	Leelanandam & Ratnakar (1980); Leelanandam & Srinivasan (1986); Ratnakar & Leelanandam (1989)
India (Gujarat)	Deccan (Fig.2.13)	Amba Dongar /Phenai Mata, Panwad Kawant	S7	37.5±2.5 (K-Ar)	AC	Abundant dykes	AC occur ≈30 km NW of cbt complex, may not be directly related; "kersantite" = AC	Heinrich (1966, p.553); Sukheswela & Aavasia (1972); Sadashivaiah & Durgadmath (1976); Durgadmath (1981); Paul & Sarkar (1984); Rock & Paul (1989)
India (Gujarat)	Deccan (Fig.2.13)	Mt.Girnar	M6	≈ 57 (Rb-Sr)	AC	Central dykes —lamps also as xen in sye	Ne-sye/lamp composites; "vogesite, augite- kersantite", "spessartite"=AL too basic	Evans (1901); Krishnan (1925); Mathur et al.(1926); Desai (1964); Rao (1964,1968); Wakhaloo (1967); Sathe & Desai (1968); Chatterjee (1970,1974); Bose(1973); Sathe & Oka (1975); Paul et al.(1977); Paul & Sarkar (1984); Rock & Paul (1989)
India (Megha -laya)	— (Fig.2.13)	E.Garo/ W.Khasi Hills	S	Phanero-zoic	AC AM	"Profuse" dykes, veins	Some "alkali lamprophyres" are apochryphal (Table C8); assoc. carbonatite, ijolite	Rao (1973); Paul & Sarkar (1984); Nambiar (1987,1988,1989)
India (Rajasthan)	— (Fig.2.13)	Kishangarh	M	Pre-Cambrian	AC	Dykes		Chatterjee (1974)
Italy	Venetian (Fig.2.12)	Corvara in Badia	R1	70±3 (Rb-Sr)	AM	3 sills	Formerly considered to be part of Predazzo-Monzoni (Triassic) magmatism	Lucchini et al.(1983)
Italy	— (Fig.2.2)	Pietre Nere, Foggia	?	Post-Triassic	AC	Single dyke recorded	Termed "kersantite" but too basic; Or+Pl+Cpx+Am+Bi; transitional to UML	Restaino (1934)
Italy	— (Fig.2.12)	Predazzo-Monzoni, Dolomites	M	≈230 (K-Ar, Rb-Sr)	AC	Numerous dykes	Triassic & Cenozoic rocks occur; lamps may belong to both; also aplite	Vardabasso (1928); Alietti (1955); Leonardi (1967, p.601,611,617, etc.); Bondi et al. (1968); Lucchini et al.(1969); Lucchini & Mezzetti (1969)
Japan	Japanese Tertiary (Fig.2.2)	Tanegashima Is., Kagoshima, S of Kyushu	R3	≈16 (K-Ar)	AB	Single sheet 10m thick, 20 km long	One of largest single lamprophyre bodies known	Yagi et al.(1975)
Japan (Hokkaido)	Japanese Tertiary (Fig.2.2)	Urakawa	R3	≈17.7 (K-Ar)	AS	Numerous dykes		Kubo & Satoh (1984)
Japan (Honshu)	Japanese Tertiary (Fig.2.2)	Daido-Jima, Shizuoka	R3	Tertiary	AC	Vein in basalt lava	29% Pl, 17%Cpx, 4%Am, 2%Bi	Ohashi (1980)

Table C2. Confirmed occurrences of alkaline lamprophyres (with no other branches)

Country	Province	Occurrence	As	Age,Ma	L	Form	Comments	References (plus Woolley 1987; *=refs mainly to inclusions)
Japan (Shikoku)	Japanese Tertiary (Fig.2.2)	Shingu, Ehime	R3	≈19 (K-Ar)	AC	Two dykes	Alkaline rocks occurring on TRENCH side of Japanese arc	Goto & Arai (1987); Uto et al.(1987)
Kenya	E.African Rift (Fig.2.2)	Jombo/Mrima Hills, S. of Mombasa	S6	≈70 (K-Ar)	AC AM AS	A few dykes	"Vogesites" are AS; lamps bracket carbonatite in time	Gregory (1900); Tyrrell & Neilson (1938); Baker (1953); Heinrich (1966, p.58 & 479); Tuttle & Gittins (1966, p.421); Rock (1976a); A.Kaka (pers.comm.of new analytical data, 1989)
Madagascar	— (Fig.2.2)	Ankaramy, Antsohanina, Berondra, etc.	M		AM AC	Abundant dykes, veins		Lacroix (1922, p.637,653)
Malawi	Chilwa (Fig.2.2)	Kirk Range/ Lisungwe/Panda/ Sengeri/Zomba	M6	Jurassic (≈ 130)	AC AM AS	Dykes	"Vogesites" are mostly AC and AS	Bloomfield (1965); Bloomfield & Garson (1965); Garson (1965a)
Morocco	Iberian (Fig.2.12)	Taourirt	R6	≈57 (K-Ar)	AC	Numerous dykes, sills, vents	Xenocrystic	Mokhtari et al.(1985); Mokhtari & Velde (1987,1988)
Morocco	Iberian? (Fig.2.12)	Tamazert (Bou Agrao)	S6 ?	≈42 (K-Ar)	AC	Central dyke-swarm	"Alnöite" reported by earlier authors not mentioned in recent accounts	Jérémine(1954); Heinrich (1966, p.551); Tuttle & Gittins (1966, p.435); Bouabdli et al.(1988)
Namibia	? (Fig.2.2)	Aiais, Warmbad district	M	Pre-Dwyka Series	AC AM	Dykes, plugs	Assoc. "bostonites, diabases, porphyries" and Pb-Ag mineralization	Haughton & Frommurze (1930)
Namibia	Damaraland (Fig.2.2)	Messum	M	≈123 (K-Ar)	AC AM	Numerous late dykes	"Nephelinites" have 12–13% Bi, "Ol-tephrites" have >20% Bi+Am; both = AL	Kaiser (1923); Korn & Martin (1954); Mathias (1956)
Namibia	Damaraland (Fig.2.2)	Okonjeje	M6	≈164 (Rb-Sr)	AC	Dykes	Age is aberrantly high compared to other Damaraland intrusives	Simpson (1954)
Namibia	Lüderitz (Fig.2.2)	Pomona	M6	≈130 (K-Ar)	AM		Few details available	Hall (1937)
New Zealand	— (Figs.2.2 & 2.8)	Kaikoura Mtns. /Cape Palliser	M3	U.Jurassic-Lower Cretaceous	AC	Stock at Bluc Mtn.; dykes, sills	Gabbroic rocks are plutonic/ cumulate AL equivalents; assoc. Ne-sye, aplite	Mason (1958); Challis (1960,1963); Grapes (1975)
Norway (S)	North Sea (Fig.2.7)	Hellesund, Kristiansand	R6	Permian (by analogy)	AC AS	Dykes, sheets	"Vogesites" = AS; older (?Pre- Cambrian) lamps also present	Barth (1944); Carstens (1958,1959,1982)
Norway (S)	North Sea (Fig.2.7)	Oslo Province (Gran, Stange, Vestby, etc.)	M6	≈280 (Rb-Sr)	AC AB	Intense regional dyke-swarm	Includes sannaite type locality; some chemically transitional to UML	Brøgger (1894,1931,1933,1934); Dons (1952); Hjelle (1962); Hasan (1971); Brynhi & Dons (1975); Nystuen (1975); Russell & Smythe (1978); Scott (1980); Scott & Middleton (1983)

Table C2. Confirmed occurrences of alkaline lamprophyres (with no other branches)

Country	Province	Occurrence	As	Age,Ma	L	Form	Comments	References (plus Woolley 1987; *=refs mainly to inclusions)
Norway (SW)	North Sea (Fig.2.7)	Sunnhordland/ Hustadvika	R6	160, 220, 275 (K-Ar)	AC	> 60 dykes swarm over 1000 sq.km	Probably 3 episodes of dyke emplacement	Faerseth et al.(1976); Faerseth (1978)
Oman	—	Haybi corridor	M	Jurassic-Cretaceous	AC	Dykes plus related gabbros	Intrude tholeiitic pillow lavas in ophiolite zone	Searle (1984)
Pacific Ocean (France)	Oceanic Islands (Fig.2.2)	Tahiti	M5	Tertiary	AC		Few details available	Washington (1917); Lacroix (1928)
Pakistan	— (Fig.2.13)	Koga, Chamla, Swat	S	≈300 (Rb-Sr)		Dykes		Siddiqui et al.(1969); Le Bas et al.(1987)
Papua New Guinea	— (Fig.2.2)	Fu, Papuan Highlands	R3	≈75 (K-Ar)	AC	Numerous dykes, rare stocks	Intense carbonate alteration; includes rare stocks	Finlayson et al.(1988)
Paraguay	Brazilian (Fig.2.2)	Sapukai (and possibly other alkaline intr.)	M	98–178 (K-Ar)	AC	Dyke-swarms	Termed "minettes" and "kersantites" but too basic and carry foids	Palmieri & Arribas (1975); Druecker (1985); Woolley (1987, p.208)
Portugal (Algarve)	Iberian (Fig.2.12)	Monchique/ Algarve coast	M6	≈72 (Rb-Sr, K-Ar)	AC AM AS	Central dyke-swarm (>100 dykes)	Monchiquite type locality; many "basanites,limburgites" etc. of coast are AL	De Sousa (1926); Assunção (1940); Rock (1976a,1978,1979,1982b,c;1983a)
Portugal (Beira Alta)	Iberian Trias (Fig.2.12)	Beira Alta (Guarda area)	R	205–235 (K-Ar)	AC	Dykes, breccias, pipes	Associated "theralites, (alkali) dolerites, gabbros"	Teixeira & Assunção (1958); Ferreira & Macedo (1979)
Portugal (Estrema-dura)	Iberian (Fig.2.12)	Sintra	M6	≈82 (K-Ar, Rb-Sr)	AC	A few central dykes around sye complex	Ti-aug+Kt phen in gdms of Cpx+Pl+opaques	Matos Alves (1964); Sparks & Wadge (1975); Rock (1982c)
Portugal (Estrema-dura)	Iberian (Fig.2.12)	Lisbon area (Cabo Espichel etc.)	M6	End-Cretaceous (assumed)	AC	A few dykes		Jérémine & Sandréa (1955)
Roumania	— (Fig.2.12)	Ditro and adjacent areas	M	≈160 (K-Ar)	AC	Central dyke -swarm	Associated with Ditro Ne-sye complex; "spessartites" = AC	Vendl (1926); Atanasiu (1928); Streckeisen (1954); Streckeisen & Hunziker (1974)
Roumania	— (Fig.2.12)	Paiona-Rusca (Ogradina, etc.)	M?				Few details available	Codarcea (1936), Codarcea et al.(1965,1967)
S.Africa	"Pilanesberg" (Franspoort Line,Fig.2.2)	Leeuwfontein	M	1420±70 (Rb-Sr)	AC AM	Dykes	Part of syenitic complex; some AM chemically transitional to UML	Shand (1923); Hall (1937); Ferguson (1973)

Table C2. Confirmed occurrences of alkaline lamprophyres (with no other branches)

Country	Province	Occurrence	As	Age,Ma	L	Form	Comments	References (plus Woolley 1987; *=refs mainly to inclusions)
Spain (Balearic Isles)	Iberian? (Fig.2.12)	Ibiza (Atalaya de San Vicente)	R	Post-Kimmer-idgian	AC	Isolated sill or lava	Sometimes formerly termed "amphibolitic andesite"	Beauseigneur & Rangheard (1967)
Spain (Gérone)	Iberian? (Fig.2.12)	Costa Brava (Aigua Blava, San Feliu, etc)	R	Post-Hercynian; ?60-80	AC	Dykes, sills	Dated by analogy with Iberian province (Cretaceous) dykes of Pyrenees	Burri & Parga-Pondal (1935); Arribas (1952); Velde & Tournon (1970)
Spain (Madrid area)	Iberian Trias (Fig.2.12)	Sistema Central Español (Avila, S.de Gredos)	R	?Triassic; pre-Jurassic post-Perm	AC	N-S dykes, diatreme breccias	Rich in ocelli; E-W dykes of Hercynian CAL also occur	Nuez et al.(1981); Villaseca et al.(1983); Villaseca & Nuez (1986); Bea & Corretge (1986); Pascual & Bea (1986)
Sweden	North Sea (Fig.2.7)	Lund area	R6	≈280 (K-Ar)	AC	A few dykes	Associated syenite (single dyke) and dolerite dykes	Hjelmquist (1930,1939); Anon (1988)
Syria	— (Fig.2.2)	Tamima (Baër-Bassit)	M	≈122(K-Ar) Jurassic–Cretaceous	AM AC AS	c.10 m of pillow lavas & breccias	Stratigraphic formation of lamp, bas, phon, limestone, chert (≈100 m); Ne-sye plugs	Parrot (1974)
Tanzania	(Fig.2.2)	Lemagrut	M	Tertiary				Harpum (1959)
Tanzania	E.African Rift (Fig.2.2)	Rungwe volcanics	M	Quaternary –Recent	AC		Minor AC in mainly basaltic eruptions (last eruptions in 1800 AD)	Harkin (1960); Heinrich (1966, p.488)
UK (Scotland)	North Sea (Fig.2.7)	Ardgour/Lismore /Grampian & SW Highlands	R6 ?M	291±5 (K-Ar)	AC AM	Hundreds of dykes; 5 pipes/vents	Ocelli and trend distinguish AL from coexisting Tertiary basalts and Caledonian CAL	Kynaston & Hill (1908, p.124); McCallien (1927); Anderson (1937b); Bailey (1960, p.226,261); Gallagher (1963,1964); Johnstone (1989,1990); Speight & Mitchell (1979); Rock (1983b); Morrison et al.(1987)
UK (Scotland)	North Sea (Fig.2.7)	Caithness/ Sutherland (Thurso, etc.)	R6	249–268 (K-Ar)	AC AM	A few dozen dykes; two vents	Caledonian palaeomag ages (Storetvedt et al.1978) are untenable (Rock 1988b)	Crampton & Carruthers (1914, p.114); Read et al.(1926, p.197ff); Chapman (1975a,b); Rock (1983b); Baxter & Mitchell (1984); Storetvedt et al.(1978); Johnstone (1989)
UK (Scotland)	North Sea (Fig.2.7)	Inner Hebrides (Mull,Tiree, Colonsay,Iona)	R6	≈275 (K-Ar)	AC AM	A few dozen dykes and sheets	Locally difficult to distinguish from Caledonian CAL	Peach et al.(1909,p.90); Craig et al.(1911,p.90); Jehu (1922); Bailey et al.(1924, p.377ff); Bailey & Anderson (1925,p.82); Richey & Thomas (1930,p.359); Urry & Holmes(1941); Beckinsale & Obradovich(1973); Rock(1983b); Rock & Hunter(1987)
UK (Scotland)	North Sea (Fig.2.7)	Monar/Eil-Arkaig,Lochaber NW Highlands	R6	326±8 (K-Ar)	AC AM	Hundreds of dykes, 3 small vents	Dense dyke-swarms with crustal extensions reaching several per cent	Hartley & Leedal (1951); Leedal (1951); Walker & Ross (1955); Johnstone & Wright (1951); *Praegel (1981); Rock (1983b); *Upton et al.(1983); Baxter & Mitchell (1984); *Hunter & Upton (1987)
UK (Scotland)	North Sea (Fig.2.7)	Monar/Killilan Forest, NW Highlands	R6	323±9 (K-Ar)	AC AM	Hundreds of dykes; two vents	Overlaps with Caledonian CAL, sometimes intruding the same fractures	Peach et al.(1910, p.88; 1913a, p.78); Ramsay (1955); Morrison et al.(1980); Rock (1982a)
UK (Scotland)	North Sea (Fig.2.7)	Orkney Is.	R6	252±10; 288 (K-Ar)	AC AM AB	Regional dyke-swarm; ≈13 vents	AC and AM have distinct chemistry and orientation; K-rich bostonite at Swona	Flett (1900,1935); Brown (1975); Mykura (1976, p.96ff); Rock (1983b); *Upton et al.(1983); *Hunter & Upton (1987)

Table C2. Confirmed occurrences of alkaline lamprophyres (with no other branches)

Country	Province	Occurrence	As	Age,Ma	L	Form	Comments	References (plus Woolley 1987; *=refs mainly to inclusions)
UK (Scotland)	North Sea (Fig.2.7)	Southern Uplands (Lauderdale, etc.)	R6	Permo-Carb (by analogy)	AM	Isolated dykes; a few vents	Lauderdale dyke erroneously shown on BGS 1:50,000 map as "felsite"	Walker (1921); Gardiner & Reynolds (1937); Rock & Rundle (1986)
UK (Scotland)	North Sea? (Fig.2.7)	Outer Hebrides (mainly on Lewis & Barra)	R6	?Permian or Tertiary (47: K-Ar)	AC AM	A few dozen isolated dykes	K-Ar/geol. data for L.Roag dyke conflict; other K-Ar dyke ages prove erroneous	Jehu & Craig (1923–7); Rock (1983b); Jackson (1984); *Hunter & Upton (1987); Menzies et al.(*1987b, 1989); Fettes et al.(1989)
UK (Wales)	North Sea (Fig.2.7)	Great House, Usk, Gwent	R6	Permo-Carb (by analogy)	AM	Two isolated pipes	No other Permo-Carb magmatism known in England or Wales	Boulton (1911); Eyles & Blundell (1957); *Upton et al.(1983); *Hunter & Upton (1987)
UK/Eire	North Sea (Fig.2.7)	General	R6	≈235–340 in several episodes	AC, AM	9 swarms, >3000 dykes; ≈50 pipes	See Fig.2.7; range from Orkney in N to Usk in S; crustal extensions several %	*Eyles (1924); *MacGregor (1939,1955); Richey (1939); McIntyre (1954); Upton (1982); Rock (1983b); *Upton et al. (1983); Baxter & Mitchell (1984); Baxter (1987); *Hunter & Upton (1987); Storetvedt (1987); Rock (1988b)
USA (AL)	Alaskan	General	A? B?	97–108 (K-Ar)	AC	Dykes	Lamps may relate to high-K alkaline rocks or to coeval granitoid plutons	Miller (1972)
USA (AR)	Ouachitas (Fig.2.10)	Magnet Cove/ Potash Sulphur Spr/Fourche Mtn	S6	≈97 (K-Ar, Rb-Sr)	?AC AM AS	20 dykes	"Syenite lamp, Ne-sye lamp, dio lamp, gbr lamp, fourchite" etc.; all = AL	Tolman & Landes (1939); Erickson & Blade (1963); Heinrich (1966, p.334,337); Powell et al.(1966); Zartman et al.(1967)
USA (AZ)	— (Fig.2.10)	Boulder Dam/ Hoover Dam	R	Late Pleistocene –Recent	AC	Dykes, tuffs, lava flows	Dykes cut largely unconsolidated sediments; dykes rich in Am megacrysts	Campbell & Schenk (1950); Foland et al.(1980)
USA (CO)	N.American Cordillera (Fig.2.10)	South Park	R	Tertiary	AC	Dykes and sill-like bodies	Termed "analcite diabase" in some papers, "lamprophyre" in others	Jahns (1938); Stark et al. (1949)
USA (CO)	Wet Mtns. (Fig.2.10)	McClure Mtn. /Iron Mtn.	S	≈520 (K-Ar, Rb-Sr)	AC	Dykes	Previous terminology confused; some examples are not lamprophyres	Heinrich (1966, p.339); Heinrich & Dahlem (1967,1969); Parker & Sharp (1970)
USA (MA,ME, NY,VT)	Monteregian-White Mtns. (Fig.2.2/9)	White Mtns. (general)	M6 R6	100–200 (K-Ar, Rb-Sr)	AC AM	>1000 dykes over 75,000 sq.km	Includes camptonite type-locality at Campton Falls; several emplacement episodes	Billings & Fowler-Billings (1975); McHone (1978,1984); *Leavy & Hermes (1979); McHone & Corneille (1980); McHone & Butler (1984); Eby (1985); *Nixon (1987, p.16,45); McHone et al.(1987); *Eggler et al.(1987); also various abstracts
USA (NM)	N.American Cordillera (Fig.2.10)	Sandia Mtns.	R1	Post-Cretaceous	AC	20 dykes;	Leucocratic (Hb-sye) dykelets show differentiation; called "spessartite",chemically =AL	Woodward (1970)
USA (NM)	Trans-Pecos	N.Delawar Basin (Yeso Hills, Kerr-McGee, etc)	M	33 (K-Ar)	AC	Dyke-swarm	"Biotite basalt" = AC; dykes mainly encountered in potash mines	Calzia & Hiss (1978); Brookins (1980)
USA (OR)	N.American Cordillera (Fig.2.10)	Coast Range (Cougar Mtn., Siletz R., etc.)	M1	Eocene-Middle Oligocene	AC	One stock, numerous dykes, sills	Associated gabbroic sills, Ne-sye at Blodgett Peak, Table Mtn., etc.	Snaveley & Wagner (1961)

Table C2. Confirmed occurrences of alkaline lamprophyres (with no other branches)

Country	Province	Occurrence	As	Age,Ma	L	Form	Comments	References (plus Woolley 1987; *=refs mainly to inclusions)
USA (VT)	Monteregian-White Mtns. (Fig.2.10)	Mt.Monadnock	M6	≈120 (by analogy)	AC	A few dykes	Latest magmatism in syenite-essexite complex	Wolff (1929)
USA (VT/ME)	Monteregian-White Mtns. (Fig.2.10)	Lake Champlain (inc.Mt.Jo/Burke & adjacent areas)	M6	≈136 (K-Ar)	AC	Intense regional dyke-swarm		Kemp & Marsters (1893); Jaffe (1952); Woodland (1962); Zartman et al.(1967); McHone & Trygstad (1981); also various Geol.Soc.Am. abstracts
USSR	(Fig.2.2)	Dzhungaria (Karatal River)	R?	?	AC			Mushkin (1966); Dobretsov et al.(1969); Dobretsov & Dobretsova (1969)
USSR	(Fig.2.2)	Vladivostok, S.Primorye	R?	?	AC	Dykes, breccias		Koghan (1976)
USSR	— (Fig.2.2)	N.Timans (Cape Bol'shoy, etc.)	M	Cambrian (500-550)	AC AM	Dyke complex in Ne-sye body	Dykes include essexite-porphyries, alkali gabbroids, pyroxenites	Mal'kov (1967); Juskova & Mal'kov (1969)
USSR (Kola)	Kola Peninsula	Lovozero	M		AC AM	Dykes (phase IV of syenite complex)	Minor tinguaites,Ne-basalts; AL are the ONLY primitive rock-types in complex	Gerasimovksy et al.(1968, Vol.II, p.171-4); Vlasov et al.(1966, p.7,10);
USSR (Kola)	Kola Peninsula (Fig.2.2)	Khibina/Kanda dyke belt	M6	≈365 or ≈300? (various)	AM	Linear dyke belt 25x2 km	AM also associated with carbonatite on E margin of Khibina syenite complex	Lupanova (1934); Gerasimovsky et al.(1974, p.214); Borodin & Pyateko (1978); Grigor'yeva & Savitskiy (1979,1980); Kogarko (1987, p.535)
USSR (Uzbekh-istan	— (Fig.2.2)	Bukantan	R?	255±5 (K-Ar)	AC AM	"Closely-spaced dykes and swarms	Xenolithic; dykes traceable for up to 1 km	Mushkin & Kotunov (1976)
USSR (Georgia)	(Fig.2.2)	Kutais/ Tkvibuli/Urueh	M	Tertiary	AC AM	"Vein rocks"	Associated with teschenites, essexites, syenites, alkali gabbros, etc.	Belyankin et al.(1938); Belyankin & Petrov (1940a,b); Skhirtladze (1942)
USSR (Karelia)	— (Fig.2.2)	White Sea coast (Onega, Gulf of Kandalaksh)	R6	?Pre-Cambrian, Palaeozoic	AM	20 pipes and vents; many dykes	More lamps on extension of graben into Kandalaksh Gulf; "alnöites" unconfirmed	Gon'shakova et al.(1967, p.64); Kaminsky (1976); Milashev (1988, p.142)
USSR (Mongul-Tuva)	— (Fig.2.2)	Agardag, Sangilen Highlands	R	Post-Middle Devonian	All	Dykes, breccias + differentiates	"Vogesite,monchiquite"=AS AD (Pl phen); "kersantite" = AC;possibly coexisting CAL	Nemtsovich (1976); Kepezhinskas et al.(1984)
USSR (Tadzikh -istan)	— (Fig.2.2)	Pamirs/S.Tien Shan/S.Gissar/ Zeravshan etc.	M	199–215 (K-Ar) + Pliocene?	AC AM	Numerous dykes, ≈ 30 pipes	Several episodes/groups; Late Carboniferous CAL assoc. with granitoids in same areas	Chetverikov (1943); Gapeeva (1949); Baratov et al.(1970); Mogarovskiy et al. (1973); Gusev & Steblova (1974); Klimov (1986); Mogarovskiy (1986); Milashev (1988, p.146)
USSR (Ukraine)	— (Fig.2.2)	Azov/Donetz/ Donbas, Black Sea region	R	≈2000 + 370 (K-Ar) + Triassic	AC AM	Regional dyke-swarms pipes,sheets	Several dyke phases ?diamond; Bayrakov (1964); Gon'shakova et al.(1967);	Yur'yev(1967a,b); Yeremenko(1968); Yeremenko & Shvakova(1969); Buturlinov & Latysh(1970);Egorov & Karmazin(1972);Sviridov & Makhajeva(1972);Valter & Yere menko(1973,1974);Klitchenko & Suprichov(1974);Shatalov(1981);Berkovsky(1987)

Table C3. Confirmed occurrences of ultramafic lamprophyres (with no other branches)

Country	Province	Occurrence	As	Age,Ma	L	Form	Comments	References (+Rock 1986; * = refs mainly to inclusions)
Antarctica	—	Bunger Hills	R	≈1140 Ma	UML	Dykes	Younger associated AL (Table C2)	Sheraton et al.(1990)
Antarctica	— (Fig.2.3)	Patuxent Range, Pensacola Mtn	R	≈230 (K-Ar)	UL	Dykes		Schmidt & Ford (1969); Boyd (1986)
Argentina	— (Fig.2.3)	Los Alisos	R	≈303 (K-Ar)	UL	Long dyke 20–80m x 10km long	No associated igneous rocks known, but tinguaite occurs 125 km to north	Mendez & Villar (1977); Villar et al.(1981); Meyer & Villar (1984)
Australia (NSW)	Tasman Fold Belt (Figs. 2.3, 2.8)	Gloucester; also Mt.Ross (Inverell area)?	R	Post-Carbon-iferous	UA	2 elongate pipes	Gloucester "supposed kimberlite" has UML chem (Fig.5.5): Mt.Ross "lamprophyric pipes" recently announced as source of NSW diamonds	MacNevin (1977, p.99); Ferguson et al.(1979); Ferguson & Sheraton (1979); Stracke et al.(1979), A.J.A.Janse & C.B.Smith (pers.comm. 1989); Australian Business (Dec.1989, p.46)
Australia (NSW)	Tasman Fold Belt (Figs. 2.3, 2.8)	Jugiong	R	< 17.2	UA? UD?	8 pipes	2 pipes cut 17.2 Ma basalt lava; formerly called "kimberlitic" but carry groundmass aegirine, richterite, orthoclase, etc.	Ferguson et al.(1979); Ferguson & Sheraton (1979); Stracke et al. (1979), A.J.A.Janse & C.B.Smith (pers.comm.1989)
Australia (NSW)	White Cliffs (Fig.2.8)	White Cliffs (Kayrunnera)	R	≈260	UA	Numerous pipes	Formerly termed "kimberlites"	Ferguson et al.(1979); Ferguson & Sheraton (1979); Stracke et al. (1979), A.J.A.Janse & C.B.Smith (pers.comm.,1989)
Australia (WA)	? (Fig.2.8)	Maude Creek, Duck Creek, Devil's Elbow	R	?	UA	7 or more dykes	Called "micaceous perid with lamp affinities, kimberlitic dykes"; WR chemistry = UML and spinels "outside field for kimberlitic spinels"	Jaques et al.(1986a)
Australia (WA)	W.Australia (Figs. 2.3, 2.8)	Bow Hill, E. Kimberley	S7/ R7	≈815 (K-Ar, Rb-Sr)	UA	En echelon swarm of ≈20 dykes	Related to Cummins Range cbt, 350 km to S?; assoc. fenitization and UML-type mineralogy though WR chemistry is kimberlitic (§5.1.2)	Atkinson et al.(1984a); Jaques et al.(1986); Fielding & Jaques (1989)
Australia (WA)	W.Australia (Figs. 2.3, 2.8)	Norseman area, Yilgarn Block	R7	849±9 (Rb-Sr)	UA	About 4 dykes over wide area	Dykes occur in gold mines, mostly known from drillcore; no other contemporaneous igneous rocks known in area	Robey et al.(1989); Western Mining Co. and CRA Exploration Pty., unpubl. data
Australia (WA)	—	Melrose, Yilgarn Block	R	?	UML	Dyke-swarm		Hamilton & Rock (in prep.)
Australia (WA)	— (Figs. 2.3, 2.8)	Nabberu (Bulljah Pool, etc.)	R7	305±7	UA UD	≥ 4 sills, pipe-like bodies	No other contemporaneous igneous rocks known in area; rare diamond in concentrates; rich in mantle inclusions	Hamilton & Rock (1990); A.J.A.Janse (pers.comm., 1989); unpubl. data of various mining companies
Australia (WA)	— (Figs. 2.3, 2.8)	Shaw batholith area, Pilbara Block	R	Protero-zoic	UA	Deformed dykes	Amphibolite-like meta-lamprophyres, some schistose; no contemporaneous igneous rocks known in area	Bettenay & Rock (in prep.)
Canada (British Columbia)	N.American Cordillera (Figs.2.3/9)	Mt.Copeland	S	≈45 (K-Ar)	UA	Dykes in gneiss complex	"Nephelinitic lamprophyres"; 45 Ma may be age of gneiss formation rather than emplacement	Currie (1976, p.175)

Table C3. Confirmed occurrences of ultramafic lamprophyres (with no other branches)

Country	Province	Occurrence	As	Age,Ma	L	Form	Comments	References (+Rock 1986; * = refs mainly to inclusions)
Canada (British Columbia)	N.American Cordillera (Figs.2.3/9)	Semlin	R	Post-Triassic; ?Miocene	UL	Single dyke	May relate to Miocene alkali basalts in same region	Drysdale (1914); Currie (1976, p.117)
Canada (Labrador)	—	Saglek & Big Island	R	135-180	UA?	Dykes + 5 vents up to 200m diam	"Lamprophyric-carbonatitic" dykelets cut breccia	King & McMillan (1975); Brummer (1978); K.Collerson (unpubl.)
Canada (Ontario)	?Hudson Bay Lowlands (Fig.2.9)	Coral Rapids/ Sextant Rapids	R	128±18 (K-Ar)	UA	4 dykes, one sill	Formerly "lamprophyre" or "kimberlitic" but "clearly not kimberlite" (Currie 1976); contain melilite+Ti-Cpx; now called alnöite, melilitite	Bennett et al.(1967); Brown et al.(1967); Sandford & Norris (1975); Currie (1976, p.117); Brummer (1978); Mitchell (1979); Butler & Edgar (1988); Butler et al.(1988); A.D.Edgar (pers.comm.,1990)
Canada (Ontario)	E.Ontario/ W.Quebec (Fig.2.9)	Borden	S	≈1000 (by analogy)	UA	2 dykes	Complex unexposed; UML described as "pyroxene-lamprophyre, olivine-lamprophyre"	Sage (1987a)
Canada (Ontario)	E.Ontario/ W.Quebec (Fig.2.9)	Firesand River	S	≈1048 (K-Ar)	UA	Radial series of dykes	Associated with carbonatite complex; formerly termed "monchiquites" but carry ≈25% melanite garnet, 15% carbonate	Gledhill (1928); Frohberg (1937); Heinrich (1966, p.397); Currie (1976, p.94); Woolley (1987, p.30); Sage (1988b)
Canada (Ontario)	Kapuskasing (Fig.2.9)	McKellar Harbour (Marathon)	R	≈1650 (K-Ar)	UA	Numerous dykes	May be related to Wawa/Opapamiska dykes	Platt & Mitchell (1979, 1982a); Platt et al.(1983); *Mitchell (1987)
Canada (Ontario)	Kapuskasing ?(Fig.2.9)	Caribou Lake	S?	Protero-zoic	UA	Numerous dykes	"Ultramafic...biotite-olivine lamprophyres and carbonate-rich lamprophyres"; also "alkaline dykes" (?AL) with Am,Pl,Ol,Ap phen	Sutcliffe (1981)
Canada (Ontario)	— (Fig.2.9)	Hecla-Kilmer	S	?450 or Mesozoic	?UA	37m sill encountered in borehole	60% Ph, 20–30% Cb, 10–15% opaques; cuts carbonatite complex and may be cogenetic or substantially younger (?Mesozoic)	Sage (1988a)
Canada (Ontario)	— (Fig.2.9)	Opapamiska (Van Horne Property)	R	Probably Protero-zoic	UA	Dykes in several drill cores	Possibly related to Wawa UML dykes but age poorly constrained; not overprinted by Archaean deformation	Wyman & Kerrich (1989a,b)
Canada (Ontario)	Kapuskasing (Fig.2.9)	Wawa	R7	Protero-zoic (≈1700)	UA	Local dyke-swarm	Previously termed "monchiquites" but too ultrabasic; spatially associated with Archaean CAL of Superior Province greenstone belts	Mitchell & Janse (1982); Wyman & Kerrich (1989a,b)
Canada (Ontario)	Nipissing (Fig.2.9)	Brent Crater	S6	≈576 (K-Ar)	UL	Dykes cutting fenites	Impact crater; "lamprophyre dykes" (also termed "biotite alnöite") have phen of Ol,Cpx, Ph,Ct in gdms of Cpx,Bi,Ac,Cc; grade into cb	Currie & Shafiqullah (1967); Shafiqullah et al.(1968); Currie (1976, p. 84); Woolley (1987, p.38)
Canada (Ontario)	Ottawa-St.Lawrence (Fig.2.9)	Arvida (Saguenay River Valley)	S6	≈564 (K-Ar)	UA	Several dykes < 1 m thick	Called "kimberlitic" but contains carbonatitic differentiate; Sp composition and abundance of Px and Am typical of UML	Gittins et al.(1975); Brummer (1978)
Canada (Ontario)	— (Fig.2.9)	Hudson Bay Lowlands	S	≈154-180 (K-Ar, geology)	UL	At least 4 plugs	45 intrusions intersected by drilling beneath glacial deposits: 34 ?UML-related breccia vents and 7 carbonatites as well as definite UML	Brummer (1978); Janse et al.(1979)

Table C3. Confirmed occurrences of ultramafic lamprophyres (with no other branches)

Country	Province	Occurrence	As	Age,Ma	L	Form	Comments	References (+Rock 1986; * = refs mainly to inclusions)
Canada (Ontario)	— (Fig.2.9)	Keith township	R	?Protero -zoic	UA	Single dyke	Called "kimberlitic" but contains abundant Cpx; Il, Sp and WR compositions typical of UML (Mitchell 1979,1986)	Watson et al.(1978)
Canada (Ontario)	— (Fig.2.9)	Picton/Varty Lakes	R	≈170 (K-Ar)	UA	Two dykes	Formerly termed "kimberlite, kimberlitic rock, lamprophyre"; whole-rock geochem. and mineralogy unequivocally indicate UML	Barnett et al.(1984); Arima & Kerrich (1988)
Canada (Quebec)	E.Ontario/ W.Quebec (Fig.2.9)	Bachelor Lake–Mattawa (NW Quebec dykes)	R	1110 (K-Ar)	UA	Dykes and sills over wide area	Formerly termed "kimberlite, lamprophyre, lamprophyric nephelinite"; carry euhedral Ol,Ph phen; "NOT kimberlites" (Currie 1976)	Watson (1955,1967b); Currie (1976, p.117); Brummer (1978)
Canada (Quebec)	Monteregian -White Mtns (Fig.2.9)	Montreal (Iles Bizard,Cadieux Como, etc.)	S6	≈120 (assumed)	UA, UL	Dykes, pipes, breccias	Related to Oka cbt complex; of 3 Ile Bizard diatremes,Pain de Sucre may be diamondiferous (Brummer 1978); Ac-phonolite also occurs	Harvie (1910); Bowen (1922); Stansfield (1923b); Gold (1967,1972); Gold & Marchand (1969); Philpotts (1974); Eby (1985,1987); Raeside & Helmstaedt (1982);Mitchell (1983,*1987); Alibert & Albarede(1988)
Canada (Quebec)	Monteregian -White Mtns (Fig.2.9)	Oka complex and satellites	S6	99-120 (K-Ar); 114(RbSr)	UL ?AM	Dykes, pipes	"Alnöites" well documented, but "fourchites, kersantites,minettes,monchiquites" in previous papers unconfirmed;lamps=last intr in complex	Gold (1966); Heinrich (1966, p.373-4); Tuttle & Gittins (1966, p.517); Girault (1968); Philpotts (1974); Eby (1975); Currie (1976, p.105); McMahon & Haggerty (1979)
Canada (Quebec)	— (Figs.2.3 & 2.9)	Castignon Lake/Labrador Trough	S	≈1873 (K-Ar)	UA	Dykes, diatremes	Previously called "meimechites, kimberlitic rocks, lamprophyric alnöites"; assoc. with cbt; Labrador Trough dykes assumed to be same age	Dimroth (1970); Dressler (1975); Currie (1976, p.111); Woolley (1987, p.59)
China (PRC)	—	W.Qingling Mtns. (Hua-Ti, etc.)	R	Tertiary	UML	>10 pipes dykes, sills, vents	Mantle inclusion-bearing	Yu Xue-Hui (China University of Geosciences, pers.comm.,1989)
Czechos-lovakia (& GDR	Rhine Graben (Fig.2.12)	Polzen river (= Ploucnice)/ Hinterhermdorf	S6	Tertiary: K-Ar ≈60; Sm-Nd ≈7	UP	Dykes, pipes over 50x40 km	Assoc. "melilite-ankaratrite", trachybasalt, phonolite, Ol-nephelinite, U mineralization; subvolcanic cone-sheet complex at Osecna	Scheumann (1913,1922); Seifert & Tröger (1937); Wimmenauer (1974, p.262); Shrbeny & Machacek (1974); Novak et al.(1981); Pivec et al.(1986); Ulrych et al.(1986,1988a,b)
Finland	Kola (Fig.2.3)	Sokli	S	≈350 (K-Ar)	UA	Dykes	Associated with large carbonatite, fenite complex (among world's largest carbonatites); lamps are the ONLY primitive rock-types	Vartiainen et al.(1978)
Finland	— (Fig.2.3)	Iivaara (Kuusamo)	S	≈430 METHOD ?	UL	Dykes?	Few details available	Lokka (1934)
France (Saône-et-Loire)	Rhine Graben (Fig.2.12)	Saint-Micaud	R	Neogene	UO	Lava?	"Picrite-ankaratrite lamprophyrique"	Brousse & Nativel (1963)
Greenland (S)	Gardar (Figs.2.3, 2.11)	Narssaq, Qagssiarssuk, Igdlutalik, etc.	S	≈1300 (Rb-Sr)	UA UL	Dykes, pipes, vents	Some related to Gronnedal-Ika carbonatite complex; others intimately associated with melilite rock, pxnt, cbt; "monchiquites"=UA	Emeleus (1964); Walton (1965); Stewart (1970); Upton (1974); Upton & Thomas (1973); Emeleus & Upton (1976, p.157); Upton & Emeleus (1987); *Scott Smith (1987, p.31); J. Craven (pers. comm.)
Greenland (SW)	Ottawa-St.Lawrence (Figs.2.3/11)	Sarfartôq	S6	≈600 (K-Ar)	UA	Dykes	Related to Sarfartôq carbonatite complex; some misnamed "kimberlites"; K-Ar ages on actual dykes geologically erroneous due to excess Ar	Larsen (1980); Larsen et al.(1983); *Scott Smith (1987)

Table C3. Confirmed occurrences of ultramafic lamprophyres (with no other branches)

Country	Province	Occurrence	As	Age,Ma	L	Form	Comments	References (+Rock 1986; * = refs mainly to inclusions)
Greenland (SW)	— (Figs. 2.3 & 2.11)	Frederikshåb area (Oqúmiaq, Iluilârssuk,etc.)	M	≈162 (K-Ar max.)	UA	Dyke -swarm	"Lamprophyre, lamprophyric carbonatite" dykes; approx. contemporaneous with slightly alkaline dolerite (so-called "TD") dykes	Walton (1966); Walton & Arnold (1970); Rivalenti & Sighinolfi (1970); Larsen et al.(1983); *Scott Smith (1987)
Greenland (SW)	— (Figs. 2.3 & 2.11)	Holsteinsborg	R? L?	?	UD	A few dykes and lenses	"Anomalous lamprophyre dykes"; relations with spatially associated ≈587 Ma kimberlites and ≈1220 lamproites (Table C4) unknown	Scott (1979,1981), *Scott Smith (1987)
Greenland (SW)	— (Figs. 2.3 & 2.11)	Ravns Storø/ Frederikshåbs Isblink	S	119-141 (K-Ar)	UA, ?UL	Dykes	Assoc. cbt breccia, phon; terminology ("lamps carbonatititic lamps, nephelinites" etc.) highly ambiguous, but some appear to be true UML	Hansen & Larsen (1974), Hansen (1980,1981); Larsen et al.(1983)
Greenland (SW)	— (Figs. 2.3 & 2.11)	Sukkertoppen (Alanguarssuk, Qaqarssuk,etc.)	S	≈170 (K-Ar)	UA	Numerous dykes	Lamprophyric dykes are scattered throughout Suppertokken region; some are demonstrably related to Qaqarssuk carbonatite complex	Larsen et al.(1983); *Scott Smith (1987)
Italy	Venetian (Fig.2.12)	Val Fiscalina	R	≈34 (Rb-Sr)	?UA	A few dykes	Termed "lamprophyres", "leucite-lamproites" o "kamafugites" but chemically UML; formerly considered to be part of Predazzo magmatism	Casati & Bigioggero (1981); Lucchini et al.(1983)
Italy	— (Fig.2.12)	Mt.Queglia, Abruzzo, Apennines	R	Post-Eocene	UA	A few dykes in 2 outcrops	Emplaced along fault separating Eocene and Upper Miocene sequences	Durazzo et al.(1984)
Kenya	E.African Rift (Fig.2.3)	Homa Bay (Homa Mtn., Kisingiri, etc.)	S6	Miocene (3 & 11–14: K-Ar)	UA UL	Late dykes, tuff sheet, agglom	Composite intrusions occur; lamps = 11% of dykes around Kisingiri stratovolcano; "alnöitic agglomerate" at Got Oloo, Homa Mtn.	McCall (1958); Heinrich (1966, p.481); Tuttle & Gittins (1966, p.425); Rock (1976a); Le Bas (1977, p.93-4 & 246)
Kenya/ Uganda	E.African Rift (Fig.2.3)	Mt.Elgon	S6	Miocene (≈18)	UP	A few dykes	Formerly termed "bergalites"	Ödman (1930); Davies (1952)
Namibia	Luderitz? (Fig.2.3)	Gross Brukkaros/ Blue Hills	R	≈84 (K-Ar)	UL	Volcano: numerous sills	Mo-Me-Phl rock "of alnöitic character" (former "kimberlite-carbonatite");assoc. fenites, scale o magmatism and low MgO not like kimberlite	Janse (1969, 1971); Ferguson et al.(1975); Mitchell (1986, p.36)
Norway	Fenno-Scandian (Fig.2.3)	Fen, Telemark	S6	560–580 (K-Ar, Rb-Sr)	UD	Dykes, pipes, sills	≈7% of cbt complex outcrop, but geophysics suggests UML body extends to ≈15km depth; 50 UML sheets cover 1500 sq.km around Fen	Brøgger(1921);Saether(1957);Barth & Ramberg(1966);Heinrich (1966); Mitchell & Crocket(1972); *Griffin(1973); Mitchell & Brunfelt(1975); Griffin & Taylor(1975);Mitchell(1980,1986);*Griffin & Kresten(1987)
Norway	North Sea? (Figs.2.3 & 2.7)	Ytterøy, Trondheims-fjord	R6	≈363 (Ar-Ar); 248 (Rb-Sr)	UA	Single dyke	Possibly related to Oslo province, but may be Caledonian as suggested by palaeomagnetic and isotopic data	Carstens (1962); Storetvedt (1967); Priem et al.(1968); Mitchell & Roberts (1986)
Pacific Ocean	Oceanic Islands (Fig.2.3)	Malaita, Solomon Is.	R5	≈34 (K-Ar)	UL, UO	>2 pipes, breccias	"Ankaratrites" may also be UML. Allen & Deans (1965); Bradshaw (1968); Dawson et al. (1978); Nixon & Coleman (1978);*Delaney et	al.(1979); *Nixon & Boyd (1979); Nixon et al.(1980); Nixon (1980, *1987);*Bielski-Zyskind et al.(1984); *Neal & Nixon (1985); *Nixon & Neal (1987); *Schulze (1987); *Neal(1988); Neal & Davidson(1989)
S.Africa (Cape Province)	Cape (Fig.2.3)	Sutherland (Salpetrekop, etc.)	R	Jurassic/ Cretac -eous	UA, UL	Dykes, pyroclasts	Various previous names given; abundant Ap, purple Cpx, Ba-Ti-Ph, Ti-Gt and Il comp suggest many UML; associated with cbt, trach	Duncan et al.(1978); McIver & Ferguson (1979); Boctor & Yoder (1986)

Table C3. Confirmed occurrences of ultramafic lamprophyres (with no other branches)

Country	Province	Occurrence	As	Age,Ma	L	Form	Comments	References (+Rock 1986; * = refs mainly to inclusions)
S.Africa (Cape Province)	— (Fig.2.3)	Kimberley (De Beers & Wesselton)	R	1910±60 (Rb-Sr)	UA	5 dykes predating KIL pipes	"Mica augite peridotites"; have euhedral Ol, too Di-rich,macrocryst-poor to be KIL; range from UA to anhydrous types;sparsely diamondiferous	Clement et al.(1979)
S.Africa (Orange Free State)	—	New Elands/ Star mines	R	?	UML /KU	Dyke swarm with KIL	Carry Ti-Zr garnet,K-V-Ba-titanates,Cpx; called "micaceous kimberlites" but "unreasonable to continuing referring to them as kimberlites"!	Mitchell & Meyer (1989)
Sierra Leone	— (Fig.2.2)	Maho (Bagbe), Gola Forest	M		UD	Dykes	Termed "camptonite" but chemically = UML; associated with Ne-sye/fenite complex	Wilson (1965)
Sweden	Fenno-Scandian (Fig.2.3)	Alnö	S6	≈560 (K-Ar, Rb-Sr)	UA, UL	Hundreds of dykes, pipes	Former "kimberlites" and "kimberlitic alnöites" = UA; 70 lamps (25% of known dykes) are latest igneous activity in complex	Von Eckermann(1948,1958,1963,1966a,b;1967,1968); Howie(1963); Heinrich(1966);Kresten(*1976,1980);Kresten & Persson(1975);Kresten & Nairis (1982); Brueckner & Rex (1980); *Griffin & Kresten (1987)
Sweden	— (Fig.2.3)	Kalix-Luleå	S	≈1150 (K-Ar)	UA	≈100 dykes	Formerly "kimberlites"; associated cbt dykes; boulder of "ouachitite" from Åland Is. (Finland) may be from here;	Geijer (1928); Larsson (1943); Tuttle & Gittins (1966, p.508); Kresten & Edelman (1975); Kresten & Brunfelt (1979); Kresten et al.(1981); *Griffin & Kresten (1987)
USA (AR)	Ouachitas (Fig.2.10)	Dare Mine Knob, etc.	S6	Cretac-eous	UO, UP	Dykes, diatremes	Associated on province-scale with carbonatites, AL (AM) etc. of Magnet Cove (Table C2), LL of Prairie Creek (Table C4), etc.	Steele & Wagner (1979); Morris (1987)
USA (CA)	N.American Cordillera (Fig.2.10)	Coyote Peak	R4	≈28 (K-Ar, Ar-Ar)	UP	Single pipe 260x500 m	Formerly termed "modlibovite"	Czamanske et al.(1978,1979,1980,1981); Erd & Czamanske (1983); Czamanske & Atkin (1985); Morgan et al.(1985)
USA (IL)	N.American Cordillera (Fig.2.10)	Omaha, Rosiclare, etc, Gallatin Cty.	R	≈257 (K-Ar)	UL UA	Sills, dykes	Some formerly termed "mica peridotite, kimberlite"; associated sphalerite-galena-fluorite mineralization	English & Grogan (1948); Clegg (1955); Clegg & Bradbury (1956); Zartman et al.(1967); Meyer (1976)
USA (KY)	— (Fig.2.10)	Claylick Creek Fannery etc., Crittenden Cty.	R	Probably Permian	UA	> 50 dykes	Formerly termed "mica peridotite, kimberlite, lamprophyre", now "pseudokimberlite" (=UML); same magmatism as previous entry?	Koenig (1956); Warren (1956); Watson (1967b); Hunt & Engelhardt (1973); Meyer (1976); Mitchell (1979; 1986, p.21 & 286); Dawson (1989)
USA (MO)	— (Fig.2.10)	Avon area, St. Francois & St. Genevieve Cty	R	≈390 (K-Ar, Rb-Sr)	UL	>80 vents+ dykes over 200 sq.km.	Carbonation/brecciation around most vents; 390 = max. age?; some termed "kimberlite" but most "lamprophyre, alnöite"; autolithic	Weller & St.Clair (1928); Singewald & Milton (1930); Tarr & Keller (1933); Rust(1937); Tolman & Landes(1939); Kidwell(1947); Heinrich (1966, p.337-8); Zartman et al.(1967); Meyer (1976)
USA (MT)	N.American Cordillera (Fig.2.10)	Haystack Butte, High -wood Mtns.	S1	Tertiary	UL	Single plug	Associated with "monticellite-peridotite"; part of same magmatism as Missouri River Breaks pipes and Winnett Sill, but spatially separate	Buie (1941); Powell & Bell (1970); Wendlandt (1977); Hearn (1989)
USA (MT)	N.American Cordillera (Fig.2.10)	Missouri River Breaks	M1	≈50 (K-Ar)	UL, UP	Chonoliths plugs, pipes	Spatially associated with "monticellite-peridotite" and shoshonitic suites of Highwood Little Rocky & Bearpaw Mtns.	Powell & Bell (1970); Hearn (1968,1989); *Hearn & McGee (1983); *Eggler et al.(1987)
USA (MT)	N.American Cordillera (Fig.2.10)	Winnett	S1	≈50 (K-Ar)	UL	Single sill	Part of same magmatism as Missouri River Breaks pipes and Haystack Butte, but spatially separate	Ross (1926a); Hearn (1989)

Table C3. Confirmed occurrences of ultramafic lamprophyres (with no other branches)

Country	Province	Occurrence	As	Age,Ma	L	Form	Comments	References (+Rock 1986; * = refs mainly to inclusions)
USA (MT)	N.American Cordillera (Fig.2.10)	Yogo Gulch	R1	≈50 (K-Ar)	UD	Single dyke	Formerly termed "minette, ouachitite" etc; contains gem quality sapphire deposit	Brownlow & Komorowski (1988); Meyer & Mitchell (1988)
USSR	— (Fig.2.3)	Ferghana (Middle)	R		UL	Dykes	Associated with ?carbonatites but little other available information	Mushkin & Larin (1979)
USSR (Siberia)	E.Siberian platform (Fig.2.3)	Anabar (NE Siberian platform)	S7	Cut Palaeozoic strata	UL ?UA	Dykes, pipes, stocks	Many of "kimberlites" are closely associated with UL and are probably UA; carbonatites and carbonatite breccias also occur	Gon'shakova et al.(1967); Nikishov et al.(1972, 1979)

Table C4. Confirmed occurrences of lamproites (with no other lamprophyres)

Country	Occurrence	As	Age,Ma	Form	Comments	Refs(+Sahama 1974;Mitchell 1985;Bergman 1987;*=refs to inclusions)
Algeria (Fig.2.4)	Azzaba, Hammam-Meskoutine	R1	Tertiary	Isolated dyke	Possibly same suite as Murcia lamproites (Spain); formerly "potassic olivine trachytes" but appear to be LC	Vila et al.(1974)
Antarctica (Fig.2.4)	Gaussberg, Wilhelm II land	R7	Pleistocene -Recent (<0.025)	Lavas, tephra (volcano)	Formerly termed "gaussbergite", "leucitite" etc.; Lc-Ol-lamproite	Vyalov & Sobolev (1959); Sheraton & Cundari (1980); Collerson & McCulloch (1982); Tingey et al.(1983)
Antarctica (Fig.2.4)	Mt.Bayliss & Pristley Peak	R7	413-430 (K-Ar); Palaeozoic	Isolated dykes	Formerly termed "melasyenite"; Mt.Bayliss is in Prince Charles Mtns., Priestley Peak in Enderby Land; Am-lamproite ± Sa ± Ph	Sheraton & England (1980)
Australia (QLD;Fig 2.4,2.8)	Marraba, Cloncurry (Mt.Isa inlier)	R	?	Isolated dyke	Formerly termed "minette"	Derrick (1980); Jaques et al.(1985)
Australia (WA;Figs 2.4,2.8)	Argyle (AK1), E.Kimberley	R2	≈1100 (K-Ar, Rb-Sr)	Pipe, dykes	Formerly termed "kimberlitic"; associated en echelon dykes at Lissadell Road; produces >30% of world's diamonds by weight	Atkinson et al.(1984a,b); Jaques et al.(1985,1986,1989a,c); Skinner et al.(1985); Nixon et al.(1987); Boxer et al.(1989)
Australia (WA;Figs 2.4,2.8)	W.Kimberley	R2	18-25 (K-Ar)	Over 100 pipes	Variably diamondiferous. Three major fields at Calwynyardah, Ellendale, Noonkambah; chemically cover KIL/UML as well as LL;	Prider (1939,1960,1965,1982); Wade & Prider (1940); Prider & Cole (1942); Norrish (1951); Kaplan et al.(1967); Derrick & Gellatly (1969); Wellman (1973); Gupta & Yagi (1980, p.55); Atkinson et al.(1984a); Mitchell (1981); Arima & Edgar (1983);
Australia (WA;Figs 2.4,2.8)	W.Kimberley (contd.)	R2			older (Jurassic) ages apochryphal; only leucite-lamproites outcrop,olivine-lamproites (formerly "kimberlites) known from drilling	McCulloch et al.(1983); Nixon et al.(1984); Thompson et al.(1984); Jaques et al.(1984a,b;1985,1986); Pryce et al.(1984); Allsopp et al.(1985); Fraser et al.(1985); Jaques & Foley (1985); Nelson et al.(1986); Sobolev et al.(1989)
Bulgaria (Fig.2.4)	Svidnja, nr.Sofia	R	Palaeozoic	Small stocks	Appear to be LC, transitional between minettes and lamproites; WR chemistry classifies as CM on discriminant functions	Stefanova (1966); Stefanova & Boyadzhiyeva (1974,1975); Stefanova & Pavlova (1974); Grozdanov (1979)
France (Corsica; Fig.2.4-5)	Sisco	R1	≈14 (K-Ar)	Thin sills (1–4 m)	Formerly termed "peralkaline minette"; LC transitional in character between minettes and lamproites; Ph-Rc-lamproite	Velde (1967,1968a)
Greenland (W; Figs. 2.4, 2.11)	Holsteinsborg	L? R?	≈1220 (K-Ar)	Dykes	Formerly "potassic lamprophyres"; spatially associated with ≈587 Ma KIL and "anomalous lamprophyres" (UML:Table C3)	Brooks et al.(1978); Scott (1979,1981); Piper (1981a,b); *Scott Smith (1987); Thy (1982); Thy et al.(1987); Nelson (1989)
India (Figs. 2.4,2.13)	Chelima (Andhra Pradesh)	R7	1140,1225, 1371±45 (K-Ar)	Dyke-swarm,7 km long	Formerly termed "minette, phlogopitite, biotite-lamprophyre, calcite-rich kimberlite, kimberlite-carbonatite"; diamondiferous	Appavadhanulu (1966); Rao et al.(1966); Sen & Rao (1970,1971); Anon (1971); Crawford & Compston (1973); Chatterjee (1974); Sarma (1983); Gupta et al.(1986); Murty et al.(1987); Rock & Paul (1989); Scott Smith (1989)
India (Madhya Pradesh)	Majhgawan & Hinota, Panna	R	840-1170 (K-Ar)	2 small pipes	Formerly "kimberlite", but petrographically more like lamproite, e.g. in presence of glass and juvenile lapilli; diamondiferous	Mathur & Singh (1963,1971); Anon (1971); Paul et al.(1975); Balasubrahmanyam et al. (1978); Alexander & Shrivastava (1984); Gupta et al.(1986); Scott Smith (1989)
Italy (Figs.2.4, 2.12)	Orciatico & Montecatini, Pisa	R3	1–4 (K-Ar)	Lavas, possible plugs	Formerly termed "glimmersyenit, trap, selagite", etc.; most classify as CM on WR MDA and best termed "cocite" (see §7.2.6)	Stefanini (1934); Johannsen (1938); Barberi & Innocenti (1967); Wagner & Velde (1986b)
S.Africa, Transvaal (Fig.2.4)	Swartruggens (Helam), Pneil, Postmasburg	L7	Mesozoic	'Male' dyke, vents	Formerly "leucite lamprophyre" but few details available; assoc. with Group II KIL; apparently transitional LL/KIL	Skinner & Scott (1979); Dawson (1987)

Table C4. Confirmed occurrences of lamproites (with no other lamprophyres)

Country	Occurrence	As	Age,Ma	Form	Comments	Refs(+Sahama 1974;Mitchell 1985;Bergman 1987;*=refs to inclusions)
Spain (Fig.2.4)	Murcia (Aljorra, Mazarron, Vera outcrops)	R1	7–8 (K-Ar)	Plugs, ?lavas	Plugs at Cabeza Maria, etc.; many rocks are LC, transitional to CM	Fúster & Pedro(1953); Pellicer (1973); Larouzière & Bordet (1983); Gupta & Yagi (1980, p.78); Wagner & Velde (1987)
Spain (Fig.2.4)	Murcia (Cancarix Calasparra, Mona -grillo, Jumilla)	R1	7–8 (K-Ar)	Plugs, ?lavas	Plugs at Sierra de las Cabras, etc.; many rocks are LC, transitional to CM	Jérémine & Fallot (1929); Hernández Pacheco (1965); Bellon et al.(1981); Gupta & Yagi (1980, p.78)
Spain (Fig.2.4)	Murcia (Fortuna, Mula, Barqueros, Zeneta outcrops)	R1	7–8 (K-Ar)	Plugs, ?lavas	Many rocks are LC, transitional to CM	San Miguel (1935); San Miguel & Pedro (1945); San Miguel et al.(1951); Fúster & Gastesi (1964); Fernández & Hernández-Pacheco (1972); Gupta & Yagi (1980, p.78); Wagner & Velde (1987)
Spain (Fig.2.4)	Murcia (general)	R1	7.2–8.6 (K-Ar)	Plugs, ?lavas	Many rocks are LC, transitional to CM; slight differences in age may exist between the outcrops (Escorza & López Ruiz 1988)	Fallot & Jérémine (1932); Parga Pondal (1935); Borley (1967); Fermoso (1967a,b); Fúster et al.(1967); Velde (1969); López Ruiz & Badiola (1980); Nobel et al.(1981); Nixon et al.(1984); Venturelli et al.(1984a,1988); *Ancochea & Nixon (1987)
USA (AR;Figs. 2.4,2.10)	Prairie Creek/ Twin Knobs #1, etc.,Pike Cty.	L1	≈106 (K-Ar)	Pipes, dykes, pyroclast	Formerly "kimberlite" etc.; assoc. Ouachitas AL/UML (Table C2-3); only former US diamond mine; also American, Black Lick,etc	Miser & Ross(1923a);Miser & Purdue(1929);Meyer(1976);Bolivar (1982a,b;1987);Scott Smith & Skinner(1984a);Gogineni et al.(1978);Bolivar & Brookins(1979);Mitchell & Lewis (1983); Morris (1987); Waldman et al.(1987); Alibert & Albarede (1988)
USA (KS;Figs. 2.4,2.10)	Rose/Silver City (Hills Pond), Woodson Cty.	R7	≈90 (K-Ar)	Sills, dykes, stocks	Formerly termed "peridotites" or "kimberlitic rocks"; occur in Palaeozoic platform over Archaean craton	Tolman & Landes (1939); Franks (1959); Zartman et al.(1967); Bickford et al.(1971); Franks et al.(1971); Cullers et al.(1985); Coopersmith & Mitchell (1989)
USA (MT;Figs. 2.4,2.10)	Smoky Butte, Jordan	R7	≈27 (K-Ar)	Dykes, plugs, volcanics	Includes olivine-lamproite, phlogopite-lamproite, leucite-lamproite; also intrusive breccias, tuffs/agglomerates	Velde (1975); Fraser et al.(1985); Mitchell et al.(1987)
USA (SC;Figs. 2.4, 2.10)	Enoree vermiculite district	R1	Late Proterozoic -Palaeozoic	Small stocks	Highly altered "biotitite" considered to be lamproitic by Bergman (1987, p.109); occurs in inner piedmont of Appalachian orogen	Bergman (1987)
USA (UT;Figs. 2.4,2.10)	Kamas/Moon canyons/Whites Creek,Scott Cty.	R7	13-40 (K-Ar)	Plugs, dykes, lavas	Formerly termed "melteigites, trachytes, wyomingites, orendites, biotite olivinites", etc.; occur on Colorado Plateau margin	Best et al.(1968)
USA (WY;Figs 2.4,2.10)	Leucite Hills/ Pilot Butte, Sweetwater Cty.	R7	1.1–1.25 (K-Ar)	Dykes, sheets, volcanics	Cross (1897); Kemp (1897);Kemp & Knight (1903);Schulz & Cross(1912);Yagi & Matsu moto(1966);Carmichael(1967);Powell & Bell	(1970);Kay & Gast (1973);Sobolev et al.(1975);Barton & Wood(1976);Barton(1979); Barton & Hamilton(1978,1979,1982);Barton & Bergen(1981);Kuehner et al.(1981); Thompson et al.(1984);Vollmer et al.(1984);Nixon (1987);Nelson & McCulloch(1989)
USSR	Aldan Shield (Yakokut, Murun, etc.)	B7	Post-Middle Jurassic	Pipes, dykes	Small, highly eroded volcanic fields associated with ultrapotassic (syenitic) plutons of Aldan Shield alkaline province	Vladykin (1985); Bogatikov et al.(1986)
USSR (Kirgizia)	Chatkal-Kuraminsk, Central Asia	R5	Post-Carbon-iferous	Breccia pipes	Diamondiferous, altered glassy rocks intruding deep fault system; fragments in breccias include tuffs, granite, etc.	Bogatikov et al.(1986); Novgorodova et al.(1987)
Vietnam (Fig.2.4)	Coc-Pia	R7	Mesozoic	Dykes, lavas	LC of intermediate character between minettes and lamproites; most analyses classify on discriminant functions as CM	Lacroix (1933a,b); Wagner & Velde (1986a)
Zambia (Fig.2.4)	Kapamba, Luangwa Valley	R5	< 250 (K-Ar)	Dykes, pipes	Extension of W margin of E.African rift; wide range encompassing LL, CAL, KIL and	*Nixon (1987, p.87,192); Scott Smith et al.(1989)

Table C5. Additional documentation for the appinite suite and vaugnérite series (§7.1)

Country	Province	Occurrence	Age,Ma	Form	Ass.pluton	Comments	References (+Bowes & McArthur 1976;Wright & Bowes 1979)
Australia (NSW)	Tasman Belt (Fig.2.8)	Fifield–Bourke (Tout, etc.)	397±16 (Rb-Sr)	≈20 pipes in 20km-wide belt extending >320 km	Gobondery granitoid	Formerly termed "Alaskan-type"; PGE/Au mineralization	Raggatt (1937); Joplin (1959); Scheibner (1976); Bowman et al.(1979); Suppel & Barron (1986); Suppel et al.(1986); Rock et al.(1988c)
Australia (WA)	Yilgarn	Jimperding belt*	≈2700	Dykes and enclaves in granitoids	Various	Described under various names in other literature	Johnstone(1952)
Bulgaria	Laramide (Fig.2.1)	Vetren	Cret-aceous	≈20 gabbroic stocks up to 1.2 km across	Granites and gabbros	Hornblende-phyric gabbros very similar to type appinites	Boyadzhiyev et al.(1979)
Canada (NS)	— (Fig.2.1/10)	Greendale (Antigonish)	Camb-rian?	Single stock ≈5 km across	Greendale complex	Calc-alkaline suite of appinitic and hornblendic gbr, dio	Murphy et al.(1990)
Czechos-lovakia	Hercynides (Fig.2.5)	Bohemian Mass*	≈400 (1°) ≈300 (2°)	20sq.km redwitzite plug also enclaves,sheets	Bohemian granitoid	Weiden Rossbach,Woldviertel; Pecarady gabbro is appinitic	Willmann (1920); Hegemann (1932); Troll (1968); Palivcová & Knotek (1975); Vejnar (1975); Boyadzhiyev et al.(1979); Holl et al.(1989)
Czechos-lovakia	Hercynides (Fig.2.5)	Moldanubian plutons	≈300 (by analogy)	Enclaves in Hb-Bi-Qz-diorite and gneisses	Two lineaments	W: Zelnava to Milevsko mass E: Trebic to Rastenberg mass	Arthaud & Matte (1977); Holub (1977,1978); Bouska et al.(1984); Hopgood & Bowes (1987)
Czechos-lovakia	— (Fig.2.1)	Mutênín	Protero-zoic	Ring-intrusion with syenites	Mutênín ferrodiorite	K-rich hornblende-dioritic rocks of appinitic affinity	Tonika (1979)
Eire (S.Ireland)	Caledonides (Fig.2.6)	Donegal*	≈400 (by analogy)	> 40 pipe-complexes (Kilkenney, etc.)	Ardara granitoid	Lamps form dykes radiating from breccia bodies	Pitcher & Read(1952);French & Pitcher(1959);French(1966,1976,1977); Hall (1967); Pitcher & Berger (1972); Kennan (1979); Elsdon & Todd(1989)
Eire (S.Ireland)	Caledonides (Fig.2.6)	Leinster-Dublin*	≈400 (by analogy)	> 50 bosses in several clusters; sheets, dykes	Leinster granitoid	Probably deeper intrusion level than type appinites	Brindley (1957,1970); Connor (1974); O'Connor (1974); Brindley et al. (1976); Brück (1976); Kennan (1979)
Eire (S.Ireland)	Caledonides (Fig.2.6)	Ox Mtns.	496±8 (Rb-Sr)		Ox Mtns. granitoid	Appinites both pre- and post-date granitoids	Taylor (1968); Pankhurst et al.(1976); Long & Max (1977); Andrews et al.(1978)
France	Hercynides (Fig.2.5)	Massif Cental* (Cévennes, etc)	≈300 (by analogy)	Small intrusive bodies, enclaves in granitoids	Margeride, Mt.Lozère	Also in anatectic Velay grt; type locality at Vaugnéray	Palm(1954);Didier (1964);Couturie(1977);Sabatier (1978,1980); Albarède & Weisbrod(1981); Montel & Weisbrod(1986); Michon (1987); Montel (1988)
France	Hercynides (Fig.2.5)	Vosges*	≈300 (by analogy)	Sill-like body parallel to gneissose banding	Crêtes granitoid	Durbachites, lamps, vaugnérites	Gagny (1978); André (1981)
Greenland (E)	Caledonides (Fig.2.6/11)	Batjberg	≈445 (K-Ar,RbSr)	Rounded blocks in volcanic vents	Batbjerg pxnt–ijolite	Very similar to Scottish Caledonian examples	Brooks et al.(1981)
Italy	Hercynides (Fig.2.5)	Biella (Ivrea-Verbano zone)*	≈280 (various)	Appinite dykes, stocks; "lamp" dyke-swarms	Various granitoids	Appinites/lamps bracket granitoids; deep-seated, high-T	Buffière (1964); Boriani & Sacchi (1973); Boriani et al.(1974); Giobbi Origoni et al.(1975,1990)
Norway	Caledonides (Fig.2.1)	Bergen	430±10	Oldest part of complex sequence of gbr and dio	Sunnhordland granitoid	Numerous granitoids postdate gbr, dio, possible "appinites"	Andersen & Jansen (1987)
Norway	Caledonides (Fig.2.1)	Bindal	399±10 (K-Ar)	Single kentallenite, numerous appinite intr.	Bindal granitoid	Transitional rocks, xenoliths, assimilation with granitoids	Nordgulen & Mitchell (1988)
Norway	Caledonides (Fig.2.1)	Smola-Hitra	Ordo-vician	Small irregular intrusions, dykes	Smøla-Hitra granitoid	Hb-phyric ultramafic to dio rocks; assoc. intrusive breccias	Gautneb (1987)
Spain (Galicia)	Hercynides (Fig.2.5)	Finisterre	≈300 (by analogy)	Enclaves (a few metres long) in 2-mica granites	Finisterre granitoid	Vaugnérites, similar to those of Velay (France)	Gil Ibarguchi (1981); Gil Ibarguchi et al.(1984)
Tanzania	Dodoman (Fig.2.1)	Dodoma, Singida areas	Archaean (≈2600)	Plugs with intrusion breccias in craton	Late orogenic granitoids	Assoc.granitoids and sye-Ne-sye intrusions	Eades & Reeves (1938); Wade & Oates (1938)
Tanzania	Dodoman (Fig.2.1)	Lupa goldfield, Lake Rukwa	Archaean (≈2600)	Stocks, sills, enclaves in granitoids of craton	Chunza, Nsamya grt	Gradational to lamps; closely assoc.late-orogenic granitoids	Teal et al.(1935); Skerl & Oates (1938); Harris (1981); quarter degree sheets 208-11, 227-9, 244-5

Note: Table C1 covers rocks accompanying CAL; entries marked * appear in both C1 and C5

Table C5. Additional documentation for the appinite suite and vaugnérite series (§7.1)

Country	Province	Occurrence	Age,Ma	Form	Ass.pluton	Comments	References (+Bowes & McArthur 1976;Wright & Bowes 1979)
Tanzania	Konse (Fig.2.1)	Kinyanguku, Mtandu, etc.	< 1800	Plugs, sills; intrusive breccias in craton	Nyami syenite	Differentiated sill, kentallenite at base; appinite stocks	Whittingham (1959); Meinhold (1970); Gabert (1973); cf. Cahen et al.(1984)
Tanzania	Nyanzian (Fig.2.1)	Musoma field, Speke Gulf	Archaean (≈2600)	Small stocks with breccias	Late grt of Dodoman	Associated with spessartites and vogesites	Stockley (1936)
Tanzania	Usagaran (Fig.2.1)	Msowero	Protero-zoic	2 large kentallenite plugs (>3 km across)	Msowero syenitoid	Syenite-peridotite-appinitic complex	Fozzard (1965); Gabert (1973)
UK (Scotland)	Caledonides (Fig.2.6)	Shetland Is.* (mainland)	≈400 (by analogy)	Small complex plutons	Brae granitoid	Several other granitoids accompanied by appinitic rocks	Mykura (1976); Flinn (1988)
UK (Scotland)	Caledonides (Fig.2.6)	Grampian Highlands*	≈400 (by analogy)	Enclaves in plutons; separate sheets, sills	Glen Banvie Tilt,Foyers	CA enclosed within finer-grained diorites	Deer (1950,1953); Holgate (1950); Harry (1952); Marston (1971)
UK (Scotland)	Caledonides (Fig.2.6)	NE Highlands*	≈400 (by analogy)	Numerous pipes in several clusters	Migdale, Rogart grt	Ach'uaine hybrids: ultramafic–acid bodies cut by lamp dykes	Peach et al.(1912,p.128); Read et al.(1925,p.45;1926,p.154ff); Read (1931, p.165);Ma(1948);Smith(1979);Brown(1983);Fowler(1988);Johnstone(1989)
UK (Scotland)	Caledonides (Fig.2.6)	NW Highlands*	≈400 (by analogy)	Enclaves in diorite; satellite pipes	Ratagain, Ben Loyal	Isolated bodies and masses associated with granitoids	Nicholls (1950); Alderton (1988); Moorhouse & Moorhouse (1988)
UK (Scotland)	Caledonides (Fig.2.6)	SW Highlands (Colonsay)*	400 or 600?	Pipes at Balnahard, Kiloran, Scalasaig;dykes	?	Kiloran pipe shows complex appinite-syenite relationships	Reynolds (1936); Bentley et al.(1988); Durrance & Kearey (1988); Rock (1989b)
UK (Scotland)	Caledonides (Fig.2.6)	SW Highlands (Cowal)*	≈400 (by analogy)	Enclaves in plutons, margins of satellites	Arrochar, Garabal Hill	CS are fine-grained equivalents of appinites	Anderson (1935a,1937a); Nockolds (1941)
UK (Scotland)	Caledonides (Fig.2.6)	SW Highlands (Etive)*	≈400 (by analogy)	Isolated pipes at Glen Chàrnan, Glen Orchy	?Glen Etive granitoid	Syenites with appinites and breccias	Kynaston & Hill (1908); Bowes et al.(1963); Condliffe (1976)
UK (Scotland)	Caledonides (Fig.2.6)	SW Highlands (Lochaber)*	≈400 (by analogy)	Enclaves near contacts of granitoid phases	Ben Nevis, Glen Coe	Some skarn-like enclaves may not be magmatic	Haslam (1970); Platten (1982a)
UK (Scotland)	Caledonides (Fig.2.6)	W Highlands (Ardgour)*	≈400 (by analogy)	Enclaves in all main phases of pluton	Strontian granitoid	Spessartite dykes also as enclaves in granitoid pluton	MacGregor & Kennedy (1931); Sabine (1963)
UK (Scotland)	Caledonides (Fig.7.1)	SW Highlands (Appin)*	≈400 (K-Ar)	>100 pipe-complexes (Arsheal, Duror etc.)	Ballachulish, Etive grt	Appinite/kentallenite type-locs; lamps form margins to pipes	Walker(1927);Bowes & Wright(1961,1965,1967);Bowes(1962);Bowes et al.(1964);Westoll & Miller(1969);Rice & Davies(1979);Platten(1982b,1984)
UK (Ulster)	Caledonides (Fig.2.6)	Ards–Newry*	≈400 (by analogy)	Marginal Bi-pxnt & other appinitic rocks	Newry granitoid	No true appinites occur	Kennan (1979); Meighan & Neeson (1979)
Zambia	— (Fig.2.1)	Abercorn/ Kasama	?		Various granitoids	"matched almost exactly among the Scottish appinites"	Deans (1938)

Table C6. Confirmed mixed or transitional occurrences (covering two or more lamprophyre branches)

Country	Province	Occ.	As	Age,Ma	L	Form	Comments	References (* = mainly to inclusions)
Antarctica	— (Figs.2.2 & 2.3)	Radok Lake (Beaver), Pr. Charles Mtns	R	≈110 (K-Ar)	AM UP UD	Sills, thin dykes,	"Monchiquites" classify as UD on MDA and mineralogy; UP sills, UD dykes/pipes; richly xenolithic	Walker & Mond (1971); Ravich et al.(1985, p.206ff)
Antarctica	— (Figs.2.2 & 2.3)	Vestfold Hills	R	?	AS UD	# dykes	Associated carbonate-rich breccias; formerly "alkaline dykes"; many assoc. tholeiitic dykes of var. ages	Delor et al.(1990); Collerson & Sheraton (1986); Sheraton et al.(1987)
Australia (NSW)	Tasman Fold Belt (Fig.2.8)	Kellys Point	R	Permian (paleomag) or Tertiary?	AC/ CS	Local dyke-swarm	WR chem, mineralogy in every respect transitional to CAL; richly xenolithic; invade grt plutons	Chappell & White (1976); unpubl.data
Australia (SA)	Orroroo (Figs.2.3 & 2.8)	Eurelia/ Terowie/ Walloway	S	≈170(K-Ar)	K2? UA? KU	Belt of dykes, plugs	Formerly "lamprophyric cbt, kimberlite",etc.; some may be true kimberlites, others UML, cbt	Tucker & Collerson (1972); Colchester (1973); Ferguson et al.(1979); Ferguson & Sheraton (1979); Stracke et al.(1979); A.J.A.Janse & C.B.Smith (pers.comm. 1989)
Australia (WA — offshore)	C.Australia Mobile Belt (Fig.2.8)	Edel No.1 borehole	R1	Mesozoic (probably ≈160)	CM CV	No information	Assoc.phonolites, trachytes; WR chem, mineral and association in every respect transitional to AL	Le Maitre (1975)
Brazil (MG)	Brazilian (Figs.2.2 & 2.3)	Alto Paranaiba (Canas, etc.)	L or S	80-87 (K-Ar)	AM K1 UA	Dykes, pipes	Poorly known; cbt, "monchiquitic lamprophyre, lamprophyric-type rock" as well as true kimberlites	Svisero et al.(1983); Tompkins (1987)
Canada (Labrador)	Ottawa-St.Lawrence (Fig.2.9)	Aillik Bay	S	≈570 (K-Ar)	AS UA UD	Central dyke-swarm	Many rocks transitional AS/UD; assoc. older "vogesites, minettes, odinites" ≈1470 Ma (Table C2)	Kranck (1939,1953); Heinrich (1966, p.320); King & McMillan (1975); Currie (1976, p.115); Foley (1984); Malpas et al.(1986); *Mitchell (1987,p.40); cf. Bergman (1987, p.128)
Canada (Ontario)	E.Ontario/ W.Quebec (Fig.2.9)	Nemego-senda	S	≈1010 (K-Ar, Rb-Sr)	AM UA		Lamps represent penultimate phase of complex (only calcite-chlorite veins are later)	Heinrich (1966, p.396); Currie (1976, p.87); Sage (1987b)
Canada (Ontario)	Nipissing (Figs.2.2 & 2.9)	Callander Bay	S	≈570 (K-Ar)	AS AM UO	Many dykes	Lamps parental to fenite–carbonatite-trachyte complex; some "monchiquites" = UO	Currie & Ferguson (1970); Ferguson & Currie (1971,1972); Currie (1976, p. 72); Cullers & Medaris (1977); Eby (1980)
Egypt	? (Fig.2.1)	Asswan & Eastern Desert	A	?	AC CS CK	> 38 dykes, 14 locs	CAL may relate to granitoids, AL to alkaline ring-complexes, i.e. the two may be totally unrelated	Lehmann (1977); Basta et al.(1985); Eyal & Eyal (1985)
Finland (Middle)	— (Figs.2.1 & 2.2)	Haukivesi, Pielavesi, Nilsiä, etc.	R	≈1830 (U-Pb in zircon)	AC CK CM	42 dykes	"Camptonites" = CS; mineralogy /age not well known; may be 2 CAL/AL groups of different ages	Hackmann (1914); Huhma (1981); Laukkanen & Mäkipää (1983); Laukkanen (1987)
Germany (FDR)	Rhine Graben (Fig.2.12)	Kaiserstuhl	S6	Oligocene -Miocene	AM UP	Dykes, Decken -tuff	"Bergalites" =UP; assoc. with melilitites,cbt,phon,tephrite,etc.; AM predate but UP postdate cbt	Soellner (1913,1939); Heinrich (1966, p. 57,436); Wimmenauer (1966;1973b,p.31;1974); Hubberten et al.(1988)
Hungary	— (Fig.2.12)	Buda/Velence Hills, Trans-danube Mtns	S	69-77 (K-Ar)	AM AC UO	Dyke-swarms	Data mostly from 11 boreholes; only 2 outcrops; transition series between UML/AL; associated	Horváth & Odor (1984); Kubovics et al.(1985,1989); Szabó (1985,1986); Dobosi & Horváth (1988)

NB. Coincidental juxtapositions of lamprophyres known to be of different ages are NOT covered, but some may be included for lack of data

Table C6. Confirmed mixed or transitional occurrences (covering two or more lamprophyre branches)

Country	Province	Occ.	As	Age,Ma	L	Form	Comments	References (* = mainly to inclusions)
India (Andhra Pradesh)	— (Fig.2.13)	Wajrakarur/ Lattavaram/ Murigiripalle	R	900–1000 (K-Ar)	KU UA LL?	8 pipes in belt	Termed "kimberlites, olivine-lamproites" etc.; geochemistry & mineralogy transitional (Fig.5.5b)	Rao & Phadtre(1966);Anand(1971);Anon(1971); Murthy (1977);Balasubrahmanyan et al.(1978); Akella et al.(1979);Nagabhushanam(1985);Gupta et al.(1986);Gupta Sarma et al.(1986); *Nehru & Reddy(1989);Reddy(1986-7);Ganguly & Bhattacharya(1987);Scott Smith(1989)
India (Assam)	Indian Gondwanas (Fig.2.13)	E.Himalayas coalfield, Darjeeling	L	≈121 (K-Ar)	CM LL	Sills, dykes	dominantly CM, but many transitional rocks and some unequivocal LL; most of the	Acharyya (1969); Sarkar et al.(1980); Paul & Potts (1981); Paul & Sarkar (1984)
India (Bihar)	Indian Gondwanas (Fig.2.13)	Bokaro coalfield	L	Cretaceous (by analogy)	CM LL	Sills, dykes	So-called "mica-peridotites" are best described as LO, though transitional to KIL in chemistry	Mukherjee (1961); Gupta & Yagi (1980, p.78); Sarkar et al.(1980); Paul & Potts (1981); Paul & Sarkar (1984); Middlemost et al.(1988); Rock & Paul (1989); Scott Smith (1989)
India (Bihar)	Indian Gondwanas (Fig.2.13)	Dhalbhum/ Giridih coalfield	L	Cretaceous (by analogy)	CM LL	Dykes	Extensions of main coalfields lamps	Biswas & Ray (1952); Subramanyam & Nandan (1961); Paul & Sarkar (1984)
India (Bihar)	Indian Gondwanas (Fig.2.13)	Jharia coalfield	L	105-113 (K-Ar)	CM LL	Sills, dykes	See previous entries; outliers occur at Chhipia, Mirzapur district, etc.	Fox (1930); Sharma (1939); Ghose (1949); Sanyal (1961,1964); Pareek (1966); Sarkar et al. (1980); Paul & Potts (1981); Singh (1981); Gupta & Yagi (1980); Gupta et al.(1983); Paul & Sarkar(1984); Middlemost et al.(1988); Rock & Paul(1989);Rock et al.(1990)
India (Maha-rashtra)	Deccan (Fig.2.13)	Bombay/W coast (Parol, Murud, etc.)	R6	Tertiary (≈ 60)	AC AM UD	Major dyke-swarm	Some misclassified (Table C8); transitional AL/UML; assoc tephriphonolite, nephelinite plugs	Auden (1949); Sukheswela & Poldevaart (1948); Sukheswela & Sethna (1962); Deshpande & Chakranarayan (1973); Chatterjee (1974); Dessai (1985,1987); Mahoney et al.(1985); Kochhar (1987); Rock & Paul (1989); Sethna (1989); Dessai et al.(1990)
India (West Bengal)	Indian Gondwanas (Fig.2.13)	Raniganj coalfield	L	110–112 (K-Ar)	CM LL	Sills, dykes	Formerly termed "glimmerite, lamprophyre, leucitite, mica-peridotite, monchiquite", etc.;	Gee (1932); Sharma & Subramanyam (1952); Banerjee (1953); Ganju & Pant (1962); Gupta & Yagi (1980, p.78); Sarkar et al.(1980); Paul & Potts (1981); Paul & Sarkar (1984); Middlemost et al.(1988); Rock & Paul (1989); Scott Smith (1989)
Italy	Venetian (Fig.2.12)	Altipiano di Tonezza/ Posina Vall.	R	Post-Oligocene	AC AM UO	Dykes	Most "ankaratrites" have essential Am and are better termed UO; assoc. alkali basalts	Vecchi (1966); Pieri et al.(1969)
Italy (Valle d'Aosta)	Venetian (Fig.2.4 & 2.12)	NW Alps (Biella, Plan d'Albard,etc.)	B1/ A1	29–33 (K-Ar)	CM CV LC	Dykes, ?rare lavas	Au veins coeval with lamps; CM transitional to LL; ands, high-K ands lavas; sye, monz, grt stocks	Marco (1958); Dal Piaz et al.(1979); Venturelli et al.(1984b); Wagner & Velde (1985); Diamond & Wiedenbeck (1986)
Ivory Coast (Fig.2.4)	—	Bobi, Seguela	R	1150–1430 (K-Ar)	KU LO UL	Dykes, some sheared	Formerly "metakimberlite", "mica-peridotite", "fitzroyite", "alnöite"; chemically transitional (Fig.5.5b)	Bardet & Vachette (1966); Knopf (1970); Bardet (1974); Mitchell (1986, p.28)
Kenya	E.African Rift (Fig.2.3)	Homa Bay (Bondo & Yala)	S6	Tertiary	KU	Cluster of pipes	Termed "kimberlites" (Ito), UML (Dawson); Am abundant; chemically transitional (Fig.5.5b)	Ito (1986); Dawson (1989)
Malawi	Chilwa (Figs.2.2 & 2.3)	Chilwa Is./ Tundulu	S	Jurassic (≈130)	AC UA UL	Central dyke-swarm	"Beforsitic lamprophyres" = UA; one late plug AC at Chilwa Is; lamps = latest magmatism in area	Garson (1962,1965a,b;1966); Garson & Campbell Smith (1965); Heinrich (1966, p.522); Woolley & Jones (1987, p.339,348-9)
New Zealand	— (Figs.2.2, 2.3,2.8)	Hohunu/ Buller (inc. Reefton)	R3	84-90 (field data, K-Ar, palaeomag)	AC CM CV	Major dyke-swarms	Trachytes, rhyolites, etc; CAL/AL swarms may be of different ages; associated Au mineralization	Henderson (1917); Washington (1917); Nathan (1968); Pirajno (1982); Tulloch (1984); Cooper et al.(1987)

Table C6. Confirmed mixed or transitional occurrences (covering two or more lamprophyre branches)

Country	Province	Occ.	As	Age,Ma	L	Form	Comments	References (* = mainly to inclusions)
New Zealand (Otago)	— (Figs.2.2, 2.3,2.8)	Alpine swarm, Haast /Burke River	S4	20-25 (K-Ar, Rb-Sr)	AC AM UO	Dykes, a few vents	Intense regional dyke-swarm with assoc. cbt,ting, breccias; previous (Cretaceous) ages incorrect	Turner (1932); Hutton (1940); Mason (1961); Grindley (1963); Wellman & Cooper (1971); Wallace (1975); Barreiro (1983); Cooper (1971,1979,1986); Barreiro & Cooper (1987); Cooper et al.(1987); Brodie & Cooper (1989)
Norway	Fenno-Scandian? (Fig.2.2/3)	Lerkehaug, (Vive, Bruem),	R6	< 688 (Ar-Ar)	AM UA	≈10 dykes	Metamorphosed; may be similar age to Fen, etc; analyses define two groups of AL and UML	Andreasson et al.(1979); Mitchell & Roberts (1986); D.Roberts (pers.comm.)
Portugal (Alentejo)	Iberian (Fig.2.12)	Sines	M6	≈72 (Rb-Sr)	AC ?CK	Central dykes	"Spessartites" have normative Ne, probably AL; "kersantites may be genuine; oversat. sye complex	Canilho (1971); Rock (1982b)
Roumania	—	Fagaras Mtns.	M?	Post-Liassic	AC AM CS	Dykes	May be two separate suites, one possibly related to Ditro nepheline syenite complex (Table C2)	Anton & Constantinescu (1977)
S.Africa	Swart-ruggens, Pneil, etc.	'Male' dyke, Helam mine; Baster's mine	R?	Mesozoic	K2; UA LO?	Dykes, vents	Variously termed "Ol-Ph-Lc-lamproite; Ph-lamp, Sa-Lc-lamp"; assoc. definite diamondiferous K2	Skinner & Scott (1979); Bergman (1987); Dawson (1980,1989); A.J.A.Janse (pers.comm., 1989)
S.Africa (Cape Province)	Cape (Figs.2.2 & 2.3)	Bushmanland /Bitterfontein /Pofadder	L	50–75 (various methods)	AM CK? UA	Belt of >270 pipes	Formerly "kimberlites" etc; rocks have Mg-Il,Cr-Po,Cr-Di but no gdms Ol; assoc.cbt, Ol-melilitites	Visser (1964); McIver (1981); Moore & Verwoerd (1985)
S.Africa (Bechua-naland)	Kuruman (Fig.2.3)	Dundrum/ Riries	R	≈1600 (K-Ar, Rb-Sr)	K1 UA	2 dykes	Intimately associated with oldest known kimberlites	Shee et al.(1989)
S.Africa (Trans-vaal)	"Pilanes-berg" (Figs 2.2 & 2.3)	General (Derdepoorte Goudini,etc.)	M S	146–1420 (mainly 1195-1420)	AC AM UL	Dykes, plugs	Terminology confused; several magmatic phases but data too sparse to relate lamps to each one	Wagner (1912); Hall (1937); Visser (1964); Ferguson (1973)
UK (England)	Hercynides (Figs.2.4 & 2.5)	Cornish dykes, Cornwall	A1	lamp: ≈279; grt:≈282	CM LL	Dykes	Some CM transitional to LL; true LL at Pendennis; assoc. granitoids of Cornubian batholith; Qz-porph	Reid (1906,p.58ff;1910,1911,1912); Smith (1916,1929); Flett & Hill (1946); Edmonds et al. (1969, p.47); Exley & Stone (1982); Hall (1982); Jones & Smith (1985); Leat et al.(1987); Bergman (1987)
UK (England)	Hercynides (Figs.2.4 & 2.5)	Exeter traps, Devon	A1	lamp: ≈279 grt:≈282 (K- Ar,	CM LL	Dykes, ?lavas	Some CM transitional to LL; true LL at Holmead, etc.; assoc. shosh abs,bas,lavas,Cornubian batholith	Tidmarsh (1932); Edmonds et al.(1968, p.134ff; 1969, p.55); Knill (1969,1982); Velde (1971c); Cosgrove (1972); Jones & Smith (1985); Thorpe et al.(1986); Leat et al.(1987)
USA (AR)	Ouachitas (Fig.2.10)	"V" intr/ Perryville/ Benton, etc	M R S	Cretaceous (≈100)	AM LO UA	Dykes, pipes, sills	Terminology confused; see also Prairie Creek in Table C4	Williams (1890); Miser & Ross (1923b); Croneis & Billings (1929); Tolman & Landes (1939); Gordon et al.(1958); Steele & Wagner(1979); Janse & Sheahan (1987); Morris(1987)
USA (AZ)	N.American Cordillera (Fig.2.10)	Hopi Buttes (inc. outliers nr. Cameron)	R1	Middle Pliocene	AM UO	300 vents; maars	Assoc. U deposits; includes Tuba dyke,etc.;"Ne-monchiquite" = UO; unusual AL lavas and pyroclastics	Shoemaker (1956); Barrington & Kerr (1961,1962); Shoemaker et al.(1962); McBirney (1963); Heinrich (1966, p.352-3); Watson (1967a); Suda et al.(1982); Alibert et al.(1986)
USA (AZ/CO/ NM/UT)	N.American Cordillera (Fig.2.10)	Navajo (Colorado Plateau)	L	20-25 (K-Ar)	CM UL UP	Sills, dykes, pipes,	> 30 CM bodies; "katungites"= UML; "kimberlites" (now termed "serpentinite microbreccias") rel.	[1]: Williams (1936); Allen & Balk (1954); Shoemaker et al.(1962); Watson (1967a); McGetchin & Silver (1970); Naeser (1971); Schmitt et al.(1974); Meyer (1976); Roden & Smith (1979); Roden et al.(1979)

Table C6. Confirmed mixed or transitional occurrences (covering two or more lamprophyre branches)

Country	Province	Occ.	As	Age,Ma	L	Form	Comments	References (* = mainly to inclusions)
USA (AZ/CO/ NM/UT)	N.American Cordillera (Fig.2.10)	Navajo (continued 2)				vents	to CM (Mitchell 1986,p.36,151) form 6 diatremes at Buell Park, Cane Valley, Garnet Ridge, Green	[2]: Gupta & Yagi (1980, p.63); Aoki (1981); Delaney & Pollard (1981); Roden (1981); Rogers et al.(1982); Jones & Smith (1983); Laughlin et al.(1985-6,1989); Vaniman et al.(1985); McDowell et al.(1986); Esperança & Holloway (1987); Alibert et al.(1986)
USA (AZ/CO/ NM/UT)	N.American Cordillera (Fig.2.10)	Navajo (continued 3)					Knobs, Mule Ear, Moses Rock; refs listed cover: [1] serpentinites; [2] minettes; *[3] inclusions	[3]:McGetchin & Silver(1972);Helmstaedt & Doig(1975);Helmstaedt & Schulze(1979); Ehrenberg(1979,1982ab);Ehrenberg & Griffin(1979);Smith(1979,1987);Padovani & Tracy (1981);Smith & Ehrenberg(1984);Broadhurst(1986);Menzies et al.(1987a);Wilshire et al.(1989)
USA (CO)	N.American Cordillera (Fig.2.10)	Cripple Creek	M1	28-34 (K-Ar)	AM CV UO	Dykes	Associated with phonolitic subvolcanic complex; goldfield; Fp-free "monchiquite" = UO	Lindgren & Ransome (1906); recent unpubl.data
USA (CO)	N.American Cordillera (Fig.2.10)	Spanish Peaks/Two Buttes	A1 B1	20–27 (fission track)	AC CM	930 sq. km swarm	Terminology very confused; dykes up to 1.65 km long; 2 sets lamps (ENE cut by radial dykes)	Cross (1906); Knopf (1936); Johnson (1961,1964,1968); Joplin (1966); Podwysocki & Dutcher (1971); Jahn (1973); Jahn et al.(1979); Smith (1987)
USA (MT)	N.American Cordillera (Fig.2.10)	Castle Mtns.	A1	lamp & grt: Tertiary	AM CM CV	Dykes	lamp supposedly intrusive equivalents of basaltic flows	Weed & Pirsson (1896)
USA (NY)	— (Fig.2.10)	Ithaca area (Glenwood, Six Mile,	R	≈155 (K-Ar, Rb-Sr)	KU	Many dykes	Formerly "kimberlite, kimberlitic rock, alnöite" etc.; chemically transitional (Fig.5.5b); ??melilite	Martens (1924); Watson (1967b); Zartman et al.(1967); Meyer (1976); Mitchell (1979,1986); *Kay et al.(1983); Basu et al.(1984); *Eggler et al.(1987)
USA (NY)	— (Figs.2.2, 2.3 & 2.10)	Cortlandt–Beemerville complexes	M S	430-460 (K-Ar,Rb-Sr,Sm-Nd)	AC CK CS	Belt of dykes	AL rel. Beemerville Ne-sye, CAL to Cortlandt complex;"spessartite" =AC; Rosetown, Stony Pt.,etc.	Aurousseau & Washington (1922); Maxey (1976); Ratcliffe (1981); Domenick & Basu (1982); Ratcliffe et al.(1983); Bender et al.(1984)
USSR	Kola (Figs.2.2 & 2.3)	Cap Turij/ Kovdozero/ Afrikanda	S7	≈365	AM UA UL	Dyke-swarm	Associated cbt, turjaite, ijolite; Dawson (1989) reinterprets some previously noted dykes as UML	Kranck (1928); Heinrich (1966, p.451); Tuttle & Gittins (1966, p.505); Popov (1967); Mal'kov (1970); Kapustin (1974); Borodin & Pyateko (1978); Kogarko (1987)
USSR	Aldan Shield	(Yakokut, Murun, etc.)	Post-Mi	Pipes, dykes	CM LL	Dykes, pipes	Small, eroded volcanic fields associated with ultrapotassic (syenitic) plutons of Aldan Shield	Vladykin (1985); Bogatikov et al.(1986); Vishnevskii et al.(1986)
USSR (Siberia)	Meimecha-Kotui (Figs. 2.2 & 2.3)	General	S	Permo-Triassic	AM UL	Dykes	Major province of strongly alkaline complexes	Heinrich (1966, p.457); Egorov (1970)

Table C7. Unconfirmed occurrences of "lamprophyres"

Country	Province	Occurrence	Age,Ma	Known description(s)	Lamps	*	References
Argentina		Chajan	66–75 (K-Ar)	"Nepheline basalt" has Bi+Ol+Px+Ne+Pv+Mt+Ac	UO?	2	Woolley (1987,p.164)
Argentina		Las Chacras	83–85 (K-Ar)	"Basalts" and "nephelinites" have Bi+Ac etc	AM?	2	Woolley (1987,p.164)
Argentina		San Miguel	Late Cretac.	"Biotite-analcime monchiquites"	AM?	1	Quartino (1961);Quartino & Llambias(1964);Woolley(1987)
Argentina		Sierra Queupuniyeu	?Cretaceous	"Basalt,basanites" but phen include Am, gd Bi	AC?	2	Woolley (1987,p.164)
Argentina	—	Patagonia	?	"Camptonite"	AC?	3	Washington (1917)
Atlantic Ocean	?	Falkland Is. (UK)	?	"Camptonite"	AC?	3	Washington (1917)
Austria	Hercynides?	Tyrol	?300	"Kersantite"	CK?	3	Washington (1917)
Brazil (BA)	Bahia	Itabuna	556±20 (K-Ar)	Called "nephelinites" but have Kt+Ne+Pl+Cb	AC?	2	Woolley (1987,p.175)
Brazil (GO)		Agua Emendada	?Cretaceous	"Alnöite/monchiquite type rocks"	UL/AM?	2	Woolley (1987,p.179)
Brazil (GO)		Montes Claros	≈89	"Monchiquites"	AM?	2	Woolley (1987,p.177)
Brazil (GO)		S.Antonio da Barra	85–130 (var.)	"Fourchites and monchiquites hypabyssal..lavas"	AM?	2	Woolley (1987,p.179)
Brazil (MG)		Bambui	?Cretaceous	"Soda minette": Af+Ag+Am+Bi+Ne+Ac+access.	CM?	2	Woolley (1987,p.181)
Brazil (MG)		Bebedouro do Salitre	?	"Minette"	CM?	2	Heinrich (1966, p.412)
Brazil (MG)	Brazilian	Araxá	87±4 (K-Ar)	"Lamprophyric types"	AL/UML?	2	Woolley (1987,p.181)
Brazil (RJ)		Rio Bonito	68–70 (K-Ar)	"Minette" dykes	AL/UML?	2	Woolley (1987,p.184)
Brazil (RJ)		Sãorinho	?Cretaceous	"Lamprophyre"	AL/UML?	2	Woolley (1987,p.184)
Brazil (RJ)	Brazilian	Cabo Frio	48–70 (K-Ar)	"Lamprophyres"	AM?	2	Woolley (1987,p.183)
Brazil (SP)	Brazilian	Jacupiranga	125–161(K-Ar)	Conflicting:"many...monchiquites"; 6km "vogesite"	AM?	1	Tuttle & Gittins(1966,p.171,539); Heinrich(1966,p.58,407)
Canada (NWT)	?	Ellesmere Is.	?	"Kersantite"	CK?	3	Washington (1917)
Canada (ONT)		Lake Nosbonsing		"Lamprophyric dykes" associated with fenitization	AL/UML?	2	Woolley (1987,p.37)
Canada (ONT)		Martison Lake		"Serpentinized Ol-Ph-Cc "kimberlitic rocks"	UA?	2	Woolley (1987,p.27)
Canada (ONT)	?	Chipman Lake	?	Single dyke of "lamprophyre" or "alkali Ol diabase"	AL?	1	Sage (1985)
Canada (ONT)	?	Squaw Lake	?	"1.2m wide minette-type (biotitic) lamprophyre"	AL/CAL?	2	Trowell (1983); Sage (1988d)
Canada (ONT)	Kapuskasing	Big Beaverhouse	≈1005 (K-Ar)	Dyke of "highly altered biotite lamprophyre"	AL/UML?	1	Sage (1987c)
Canada (ONT)	Kapuskasing	Prairie Lake	≈1112 (K-Ar)	"Mica and pyroxene lamprophyres"	AL/UML?	2	Currie (1976, p.97); Woolley (1987,p.24)
Canada (ONT)	Nipissing	Burritt Is.	≈570(assumed)	"Lamprophyre dykes"	AL/UML?	2	Lumbers (1971); Currie (1976,p.82); Woolley (1987,p.36)
Canada (ONT)	Nipissing	Iron Is.	≈570(assumed)	"Various late lamprophyric and basic dykes"	AL/UML?	2	Currie (1976, p.82); Woolley (1987,p.35)
Canada (ONT)	Nipissing	Manitou Is.	≈570(assumed)	"Numerous lamprophyric...dykes"	AL/UML?	2	Heinrich(1966,p.389);Currie(1976,p.80);Woolley(1987,p.36)
Canada (QBC)		Mutton Bay	562–577 (var.)	"Carbonate-rich lamprophyres"	UA?	2	Woolley (1987,p.63)
Chile	?	Coquimbo	?	"Kersantite"	CK?	3	Washington (1917)
China (PRC)	?	Kaushu	?	"Vogesite"	CV?	3	Washington (1917)
Cuba	—	Sierra de Escambray	?	"Appinite series of basic rocks"	?CA	2	Palivcová & Knotek (1975)
Egypt	?	Eustratos	?	Called "camptonite" (eustratite) but chemically CAL	CS?	4	Johannsen (1938)
Ethiopia	?	Mulazenai/Saganati	?	"Kersantite"	CK?	3	Washington (1917); Mohr (1970)
Greenland (E)	?	Arnold Eschers Land		"Nephelinites, basanites" have Kt xen,gd Bi	AM?	2	Woolley (1987,p.74)
Greenland (E)	Caledonides?	Forsblads Fjord	?Archaean	"Swarm of Bi-Hb dykes" called "older lamprophyres"	CAL?	2	Escher & Watt (1976, p.228)
Greenland (E)	Thulean	Kap Dalton	Tertiary	"Monchiquite" pebbles in Eocene sediments	AM?	1	Deer (1976, p.408); Woolley (1987,p.76)
Greenland (E)	Thulean?	Liverpool Land	?	"Alnöite"	AL?	3	Washington (1917)

* 1=no further details available; 2=no further details obtained; 3=chemistry lamprophyric but mineralogical details lacking; 4 = descriptions unconvincing

Table C7. Unconfirmed occurrences of "lamprophyres"

Country	Province	Occurrence	Age,Ma	Known description(s)	Lamps	*	References
Greenland (N)	—	Roosevelt Fjelde	Cretaceous?	"Lamprophyric" dykes "with megascopic mica"	AL/UML?	1	Dawes & Soper (1970); Dawes (1976, p.290)
Greenland (S)	—	Julianehåb	217±5 (K-Ar)	"Camptonite sill"	AC?	1	Hansen et al.(1983)
Greenland (W)	—	Søndre Strømfjord	≈1800 (K-Ar)	"Lamprophyre dykes"; ≈600 Ma dykes in same area	UML?	1	Hansen et al.(1983)
Guinea	—	Gbenko–Tissinkoro	Cut 95Ma KIL	"Ultrabasic..alkaline lamps"; Phl,Ti-Am,Ol,Di,Mt	AL/UML?	1	Rombouts (1987)
Guyana	—	W.Guyana Shield	L.Proterozoic	"Lamprophyres"	?	2	Gibbs (1985)
Italy	Roman?	Monte Vulture	Quaternary	"Lamprophyric rocks, polzenite, missourite–alnöite"	UML?	4	Amodio & Hieke Merlin (1966)
Italy (Sardinia)	?	Ogliastra	?	Schistose "spessartite" dyke with Mt concentrations	CS?	4	Piepoli & Collari (1939)
Japan	?	Hokkaido	?Tertiary	"Monchiquitic rocks"	AM?	1	Sanbonsugi (1938)
Japan	?	Kozaki,Bungo	?	"Camptonite"	AC?	3	Washington (1917)
Kenya	E.African Rift	Lengetet	Tertiary	"Alnöite" dykes		2	Heinrich (1966, p.487)
Malawi	Chilwa	Kangankunde		"possibly biotite alnöite"	UL?	2	Tuttle & Gittins (1966, p.428)
Malawi	Chilwa	Nkalonje-Matopon		"dykes and sheets of...lamprophyre"	UML?	2	Tuttle & Gittins (1966, p.431)
Malawi	Chilwa	Nsengwa		Dykes of "lamprophyre", "monchiquite"	AM?	2	Heinrich (1966, p.530); Tuttle & Gittins (1966, p.429)
Mexico	N.Am.Cord.	Sonora	≈50 (K-Ar)	"Lamprophyre dykes" in several mines	CAL?	2	Mead et al.(1988)
Mozambique	Chilwa	Salambidwe		Dykes of "camptonite"	AC?	2	Heinrich (1966, p.531)
Pacific Ocean	Oceanic Is.	Ponape		"Monchiquite"	AM?	1	Yagi (1960)
Paraguay		Cerro Guazu	117±4 (K-Ar)	"Radial swarm of lamprophyre dykes"	AL/UML?	2	Woolley (1987,p.207)
Paraguay		Soto-Rugua	132±10 (K-Ar)	"Kersantite dyke"	CK?	2	Woolley (1987,p.207)
Roumania	?	Bihor massif	?		?	2	Manea (1983)
S.Africa		Magnet Heights		"Elongate body of lamprophyre"	AL/UML?	2	Heinrich (1966, p.539)
S.Africa	Angola border	Swaartbooisdrif		"dykes of porphyritic biotite-rich lamprophyric rock"	UML?	2	Tuttle & Gittins (1966, p.464); Mathias (1974, p.192)
Tanzania	E.African Rift	Essimingor	Tertiary	"dykes of monchiquite and fourchite"	AM?	2	Tuttle & Gittins (1966, p.475)
Tanzania	E.African Rift	Mbeya		"Carbonatized alnöite dykes"	UL?	2	Heinrich (1966, p.499)
Tanzania	E.African Rift	Ruhuru Valley	Post-Triassic	"Alnöitic"; "carbonatite-kimberlite"	UML?	2	Heinrich (1966, p.505); Mitchell (1986, p.116)
Uganda	E.African Rift	Tororo		"Lamprophyre dykes"	AL/UML?	2	Heinrich (1966, p.470,473)
USA (AR)		Raton-Clayton	Olig–Recent	"Vogesite,minette,kersantite...lamps...kimberlitic"	CAL?	2	Woolley (1987,p.134)
USA (AR)	Ouachitas	Granite Mtn.	86–90 (Rb-Sr)	"Fourchites...are probably monchiquites"	AM?	2	Woolley (1987,p.145)
USA (AR)	Ouachitas	Little Rock	Cretaceous	"Fourchite-monchiquite type"	AM?	2	Woolley (1987,p.145)
USA (AR)	Ouachitas	Perry/Conway Cty.	Cretaceous	"Ouachitite"	AM/UO?	2	Woolley (1987,p.143)
USA (AR)	Ouachitas	Saline Cty.	Cretaceous	"Monchiquitic dykes of several varieties"	AM?	2	Woolley (1987,p.146)
USA (CO)		Goldie	?	"Carbonated lamprophyre"	UA?	2	Woolley (1987,p.132)
USA (CO)		McCoy Gulch	?	"Lamprophyre"	UA?	2	Woolley (1987,p.132)
USA (MT)	N.American	Crazy Mtns.	Tertiary	Very altered "monchiquite"	AL?	4	Wolff (1938)
USA (NM)	Cordillera	Kerr-McGee mine	32–35 (K-Ar)	"Biotite-basalt, lamprophyre dike"	?	2	Calzie & Hiss (1978); Brookins (1980)
USA (VA)		Augusta Cty.	114,145–153	"Camptonites"	AC?	2	Woolley (1987,p.154)
USA (VI)	?	Shenandoah	?Triassic	"Camptonite" associated with teschenite,syenite	AC?	2	Johnson & Milton (1955)
USA (WI)	—	Wausau	≈1450 (Rb-Sr)	"Lamprophyres" as veins, segregations, dykes	CAL/AL?	4	Emmons et al.(1953)
USA (WY)		Mineral Hill	?	"Lamprophyres"	CAL?	2	Woolley (1987,p.127)

Table C7. Unconfirmed occurrences of "lamprophyres"

Country	Province	Occurrence	Age,Ma	Known description(s)	Lamps	*	References
USSR	?	Baikal	?	"Monchiquite"	AM?	2	Butakova (1974, p.178)
USSR	?	Chelyabinsk	?	"Biotite lamprophyres"	?	2	Zhuykova & Bogacheva (1968)
USSR	?	Chukotka(Chukchi)	Cretaceous	"lamprophyre dykes"	CAL?	2	Makeyev & Yefimov (1972); Ivanyuk (1986)
USSR	?	Elbrus	?	"Spessartites"	CS?	2	Kumar (1968)
USSR	?	Goray-Altay	?	"Lamprophyres"	?	2	Michalewa (1962)
USSR	?	Jenissel,Siberia	?	"Camptonite"	AC?	3	Washington (1917)
USSR	?	Kadainskoe	?	"Lamprophyres"	?	2	Kudryavtseva et al.(1967)
USSR	?	Maykhurin,Tadzh.	?	"Lamprophyre dykes"	?	2	Rosseykin & Razhmanov (1971)
USSR	?	N.Nuratau Mtns.	?	"Lamprophyre"	?	2	Kushmuradov (1970)
USSR	?	Obitochnaya River	?	"Lamprophyre dykes"	?	2	Ivanushko (1973)
USSR	?	Rudny Altai	?	"Lamprophyres"	CAL?	2	Kuzebny & Mar'in (1973)
USSR	?	Shaktaminskiy	?	Lamp–dio porph–grt porph dykes	CAL?	2	Skuridin et al.(1972)
USSR	?	Tayun	?	"Kersantite"	CK?	2	Gapeeva (1950)
USSR	?	Tchakyl-Kalyan	?	"Lamprophyres"	?	2	Kuchukova (1973)
USSR	?	Tchiatour	?	"Lamprophyre dykes"	?	2	Kanchaveli (1964)
USSR	?	Varzob-Bolo	?	"Lamprophyres"	?	2	Babakhodzhaev (1956)
USSR	?	Volhynia	?	"Dioritic lamprophyres"	CAL?	2	Lichak & Piskorskaya (1968)
USSR	Aldan Shield	Konderskii	?	"Lamprophyres" among dyke-rocks	?	2	Heinrich (1966, p.462)
USSR	Azov	Shakhty,Donbass	?	"Lamprophyres,kersantites,spessartites"	CS,CK?	1	Ignatev (1936); Buturlinov (1959); Razdorozhryi (1974)
USSR	Baltic Shield	Yelet'zero	Proterozoic	"Spessartite" and "vogesite" dykes	AL/CAL?	2	Kogarko (1987, p.531,533)
Yugoslavia	?	Cer Mountain	?	"Lamprophyres"	?	2	Knezevic (1976)
Yugoslavia	?	Klisura,Selecevica	?	"Lamprophyre dykes"	?	2	Simic et al.(1965)

Table C8. Apochryphal occurrences of "lamprophyres"

Country	Occurrence	Published name	Comments/petrology	Preferred name	References
Antarctica	Raymond Fosdick Mtns.	"Olivine fourchite"	Ol + Cpx in gdms	Basanite	Fenner (1938)
Argentina	Cerro Tandileofu	"Spessartite"	Cataclased, with Pl phen	Meta-microdiorite?	Cortellezzi & Rabassa (1976)
Australia (NSW)	Wellington district	"Lamprophyre"	Metd./sheared but with ophitic texture, Pl phen, little Hb/Bi	Porphyrite, dolerite, etc.	Basnett (1942)
Australia (TAS)	Little Mt. Horror	"Lamprophyre"	80% Ab, 15% Ol, 4% Am, 1% Py	Malchite?	Baillie (1973)
Australia (WA)	Yilgarn Block	"Camptonite"	Variable metamorphosed rocks with Ol,Cpx,Am, etc.	High-Mg basalt, amphibolite, etc	Miles (1948); cf. Rock et al. (1988d)
France	Pelvoux	"Lamprophyre"	True lamps do occur in same area (Table C1)	Intrusive spilites	Tane (1963,1974)
France	Saint Bresson	"Lamprophyric rock"	Contact facies of microgranite intrusion; too acid to be lamp	Malchite	Guintrand et al.(1963)
Germany	Steinhügel, Münchberg	"Augite-spessartite"	60% Pl, 20% Opx, 11% Cpx, 4% Bi, 3% Qz, 3% accessories	Basaltic andesite?	Eigenfeld (1960)
Greece	Cozani	"Lamprophyres"	Schistose At+Bi+Na-Pl dykes in ultrabasic mass	Metamorphosed ultramafites?	Netelbeck (1959)
Greenland (SW)	Narssarssuaq	"Appinitic rocks"	Not Hb-rich, carry Opx; too low in MgO and CaO for CA	Norite, gabbro, etc.	Walton (1965)
Hungary	Komló-173 borehole	"Lamprophyric dyke...augitite"	40% Cpx + Pl,Me,Ol,Ae,Mt,Ap,Ac	Possibly a lime-contaminated rock	Szilágyi(1981)
India (Gujarat)	Bassein, Bombay	"Camptonite"	Cpx phen in gdms of Pl,Cpx, Ac/Gl — no Am or Bi	Analcite- basanite	Sukheswela & Sethna (1962); true AL do occur in same area
India (Me -ghalaya)	Garo Hills	"Alkali lamprophyres"	30-50% Ae, 1-16% Or, 9-22% Cc, 1-18% Bi, 9-40% Ne	Shonkinite or ijolite	Rao (1973); true AL do occur in same area (Table C2)
India (Mysore)	Kushalnagar, Mysore	"Camptonitic lamprophyre"	Carries 56% En	Norite?	Chatterjee (1974)
Italy	Alpe Lusia, Buddoso	"Lamprophyre", "Spessartite"	Have Pl phen, little or no Hb, ± Opx ± pigeonite	Andesite, porphyrite?	Emiliani (1958), Scribano (1976)
Italy	Val d'Ultimo (Alto Adige)	"Kersantite", "camptonite"	"Kersantite" has Pl phen; Qz in norm/mode of "camptonite"	Porphyrite, ?andesite	Minguzzi (1940)
Italy (Sardinia)	Bitti-Onani	"Lamprophyre of diabase tendency"	Carries Pl,Ch,Cpx,Il and 2° products; no 1° Am or Bi	Microdiorite or diabase	D'Amico (1960)
Japan	Shodo-Shima	"Spessartite"	6% Pl phen; 6% altered Cpx; 6% At;gdms: Pl+Bi+At+Il+Py	Altered basalt?	Ujike (1979)
Japan	Yatsugataka	"Kersantite"	Carries Opx with low SiO2 (32%) & high Mt (23%)	Mela-andesite?	Oji & Takeshita (1970)
Norway	Jotunheim	"Spessartite"	Opx-Pl-Di, mineralogically quite unlike spessartite	Microdiorite?	Dietrichson (1955)
Norway	Langesund-fjord	"Soda-minette"	Strongly Fp-phyric and show no lamp textures	'Rhomb-porphyry' (trachyte)	Hasan (1969)
S.Africa	Roberts Victor	"Equivalent to lamprophyre"	Carries Pl, Ac phen	Ne-bostonite or similar	Windom & Boettcher (1980)
Spain	Avila, Rio Espinaredo	"Lamprofido (mangerítico)"	Fp-phyric, related to acid porphyry dykes nearby	Porphyrite, microdiorite?	Mulas Sanchez (1963a,b)
Tanzania	Njombe district	"Appinite"	Metamorphic rocks derived from calcareous shales	Skarn, etc.	Stockley (1948, p.15)
UK (Channel)	Jersey, Guernsey	"Appinites, appinitic diorites"	Textural (mainly, pegmatitic) variants of normal diorites	Diorite varieties	Wells & Bishop (1955); Key (1977)
UK (England)	Torquay	"Augite lamprophyre"	No Am or Bi	Microdiorite?	Shannon (1924)
UK (Scotland)	Midland Valley	"Monchiquite"	Ol,Cpx,Ac,Ne,Gl±Pl(Cameron & Stephenson 1985, p.114)	Ol-analcimite, Ol-nephelinite	Tyrrell (1912,1928); Forsyth & Chisholm (1978); Upton (1973)
UK (Scotland)	NW Highlands	"Metamorphosed lamprophyre"	Foliated/schistose with Hb,Bi,Ct,Qz,Pl	Microdiorite (Smith 1979)	Johnson & Dalziel (1966)
UK (Scotland)	Outer Hebrides	Laxfordian "lampro-phyric dykes"	Hornblendic intermediate dykes predating Scourie dykes	Metabasalt or microdiorite	Watson (1975, p.18 & pers.comm.,1983)
USA (CO)	Crazy Mtns., La Plata Mtns	"Spessartite, diorite lamprophyre", etc.	Fp-phyric; Am, Bi absent or not primary; some have Opx	Microdiorite, andesite, etc.	Rogers & Longshore (1960); Gross & Heinrich (1966)
USA (MI)	Northern Gulf coastal plain	"Monchiquite"	No Am or Bi	Nephelinite	Moody (1949); Woolley (1987, p.147)
USA (MT)	Crazy Mtns.	"Camptonite"	Fp phen, gdms Qz, etc.	Altered basalt, etc.	Wolff (1938)
Zimbabwe	Nuanetsi province	"Kersantite dykes"	Oligoclase phenocrysts up to 13 mm long	Porphyrite or porphyry	Cox et al.(1965, p.148)

Appendix D: The computerized lamprophyre database LAMPDA

All the summaries in this book are derived from a large computer database built up over some 17 years. The database currently occupies some 8.5 megabytes, and contains whole-rock data for over 5,000 rock specimens, plus a similar number of constituent mineral grains. Computing, bibliographic, data capture and validation aspects relevant to *all* large petrological databases can be consulted in Le Maitre (1982) and Rock (1987b). Since the construction and validation of LAMPDA required major effort, a brief discussion of critical aspects is warranted below; a full description can be consulted in Rock & Wheatley (1989).

D1 Data capture

LAMPDA incorporates data from ≈1,500 references, and its construction was intimately linked to the Bibliography. Data capture proceeded by an expansion of the process outlined by Métais & Chayes (1963) but, unlike many existing geological databases, did *not* rely on literature descriptions alone; thus LAMPDA is *not* merely a compilation of data for rocks *called* "lamprophyre", etc., but a *carefully screened* collection of data for rocks which conform to the definitions in Appendix B/Chapter 1. LAMPDA contains the following data:

(1) Published (and much unpublished) data for rocks described as one of the lamprophyre varieties (Fig.1.2), *and* which also comply with the present definitions (Chapter 1).

(2) *Some* data for rocks which comply with the present definition of lamprophyre but which were originally given other names (e.g. "biotite-nephelinite, diabase, K-rich diorite, mafic porphyry, mica-peridotite"). Here, it is impossible to judge the proportion of available data captured, since locating it can only be by manual search, personal communication, etc. The references concerned are however carefully annotated in the Bibliography, and justification appears under the appropriate entries in Appendix C.

(3) A moderate but not so exhaustive amount of data for kimberlites and for intermediate-felsic and cumulate rocks *intimately* associated with lamprophyres (felsic ocelli, phonolites, porphyrites, porphyries, syenites, tinguaites, trachytes, felsic veins, etc.)

Two arguments suggest that LAMPDA contains the *vast majority* of published data:

— Dr.S.C.Bergman kindly supplied a printout from his own (1987) independently compiled database, but this proved to contain very few analyses not already incorporated in LAMPDA (and nearly all of these were from unpublished US theses);

— the rate of ongoing *retrospective* data capture has decreased exponentially to the point where only a handful of analyses (other than newly published ones) are now located each year; this suggests that the total of available data has been closely approached.

In all three above categories, however, coverage of literature in European languages (e.g. Czech, English, French, German, Italian, Norwegian, Polish, Portuguese, Serbo-Croat, Spanish) is strongest, that in Soviet languages weaker, and that in Chinese minimal.

The following general exclusions apply to all data compiled in LAMPDA:

— Analyses regarded as inferior in quality by Washington (1917).

— All other pre-1920 data not compiled by Washington (1917).

— Analyses of so-called "lamprophyres" which do not conform to the present definitions.

— Partial analyses with figures for less than 10 of the 14 major/minor oxides in the case of whole-rock analyses, or less than 4 critical oxides in the case of mineral analyses.
— Analyses with unacceptable major element totals (see below for validation procedures).
— Analyses of "altered" rocks, breccias and pyroclastic rocks where alternative analyses of fresh, magmatic rocks from the same occurrence are available.
— Analyses of confirmed xenoliths and xenocrysts (Types **E** & **F** of Chapters 4 and 6).

D2 Database structure

LAMPDA is implemented on a Macintosh® IIX microcomputer under the database quasi-relational management system *Acius 4th Dimension™*. As shown in Fig.D1, it consists of 20 computerized information tables (*relations*), which are linked via an identification 'rock number' (*primary key*) which is unique to each lamprophyre specimen in the database. The 8 relations containing actual geological and analytical data are as follows:

— *ROCK* , the 'core' of LAMPDA, carries geological data for each analyzed specimen.
— *MAJ* carries data for 14 major and minor elements (in wt.%).
— *TRA* carries data for 29 trace elements (in ppm).
— *REE* carries data for 14 rare-earth elements (in ppm).
— *PGE* carries data for 7 precious metals (PGEs and Au, in ppb).
— *MIN* carries oxide data for analyzed mineral grains or bulk mineral separates.
— *ISO* carries data for 8 isotopic ratios.
— *Occurrence* encodes the geology of known lamprophyre occurrences (Appendix C).

Much of the geological and petrographical information is in coded rather than free-text form, and further 6 tables elucidate the codes used. The remaining 6 tables contain validation parameters for the analytical data, described in Section D4 of this Appendix.

D3 Data assignation and rock classification

Data were assigned for consistency using an iterative, five-pass approach, as follows:
(1) Analyses were first assigned according to authors' original published names.
(2) The resulting datafile was used to erect consistent definitions of the 5 branches (Fig.1.2; Appendix B). Wherever authors' original names proved to be *blatantly* inconsistent with these definitions, at the first-pass, analyses were reclassified at this stage (e.g. "alkali lamprophyres" of Vartiainen *et al.*(1981) are plainly UA; "camptonites" of Asquith (1973c) are CS; "peralkaline minettes" of Velde (1967) and Hall (1982) are LL, and so on). Each reference where such reclassifications were made is annotated by ¶ in the Bibliography, and explanations are given under the relevant entries in Appendix C.
(3) Major oxide data were recalculated to 100% free of H_2O, CO_2 and MnO, and with total Fe as Fe_2O_3, to allow for variations in analytical totals, oxidation state, and secondary alteration. Analyses with missing values for TiO_2 or P_2O_5 were rejected. Two-group discriminant analysis was then performed, using the 9 remaining major oxides (SiO_2, Al_2O_3, $Fe_2O_{3(t)}$, MgO, CaO, Na_2O, K_2O, TiO_2, P_2O_5) as variables, on the UML and KIL groups *alone*. These are the only lamprophyres which cannot be distinguished on their simple modal mineralogy (Table 1.1), and chemical criteria therefore assume greater importance in their classification. Based on Fig.5.5, analyses were reclassified wherever the whole-rock geochemistry clarified previously conflicting classifications of the particular occurrence and allowed an unequivocal assignation (e.g. Gloucester and

Pilanesberg occurrences on Fig.5.5c were confidently assigned to UML; Homa Bay, Wajrakarur to 'transitional KIL/UML'. Careful account was taken in this step of the statistical probabilities of group membership for *all* analyses from a particular occurrence. Take, for example, the following hypothetical MDA results:

Rock	Occurrence	prob(A)	prob(B)	Rock	Occurrence	prob(A)	prob(B)
1	Timbuktu	0.05	0.95	5	Blongovia	0.55	0.45
2	Timbuktu	0.01	0.99	6	Blongovia	0.57	0.43
3	Timbuktu	0.51	0.49	7	Blongovia	0.44	0.56
4	Timbuktu	0.25	0.75	8	Blongovia	0.41	0.59

Statistical decision rules would classify rocks 1,2,4 as group A and rock 3 as B, but in a *geological* context, rock 3 can be regarded as aberrant, not only because the other three rocks are so consistent, but also because the probabilities of group membership to both A and B for no.3 are subequal. The 'Timbuktu' occurrence would thus tend to be classified as B; if it was originally classified as A, reclassification would occur provided this were consistent with the geological context, and would be more likely, the greater the number of analyses assigned by MDA to B. By contrast, the 'Blongovia' occurrence clearly straddles the A/B boundary. If it were originally classified as A, it would be left as such in LAMPDA. If it were doubtful, it would be classified as 'transitional'.

(4) Five-group multiple discriminant analysis (MDA) was performed on the third-pass data-set to produce a further set of discriminant functions, again based on major oxides only, but this time for the complete range of analyses in LAMPDA. Several hundred misclassified examples were then re-examined. Using the same criteria as in step (3), misclassified analyses were reclassified into a group which the MDA showed to be much more likely, *wherever this was consistent with available mineralogical data*, in other words, where whole-rock and mineralogical criteria, applied not only to the particular rock but to the whole occurrence in question, gave a consistent and geologically reasonable answer. In accordance with IUGS recommendations, mineralogical data were given most weight in these considerations. In nearly all cases, reclassification only took place for occurrences which have already received conflicting classifications in the literature. As discussed in Chapter 5, the groups of analyses giving lowest classification efficiencies were olivine-lamproites and kimberlites (because these are probably the same thing), cumulates (including appinites) and fractionated rocks (including bostonites and porphyrites), because accumulation tends to generate AL-like chemistry in CAL, fractionation to generate CAL-like chemistry in AL.

(5) The discriminant functions from step (4) were therefore further refined by excluding the obvious cumulate and fractionated rocks (essentially, all rocks with SiO_2 > 60%, all appinites and kentallenites, and all felsic veins or ocelli). This allowed the MDA to home in on the more subtle distinctions between the main lamprophyre types, and gave the final classification efficiency shown in Table 5.2 and illustrated in Figs.5.1–5.4.

The hierarchical classification in Fig.1.2 allows all cases to be classified as precisely as available data warrant — no more and no less. Where assignation to individual rock-types (CM, etc.) was not possible, analyses were assigned to lamprophyre branches (CAL, AL, UML, LL, KIL), and, in the most doubtful cases of all, to the general category lamprophyric rock (coded '??'). Analyses given compound names in the literature (e.g. "minette-kersantite") were classified under the first part of the name if sufficient modal data were provided to fit them into the present definitions, otherwise subsidiary categories were erected (e.g. CB = calc-alkaline biotite-lamprophyre covers "minette-kersantite", and also allows for the many biotite-lamprophyres whose dominant feldspar is not known).

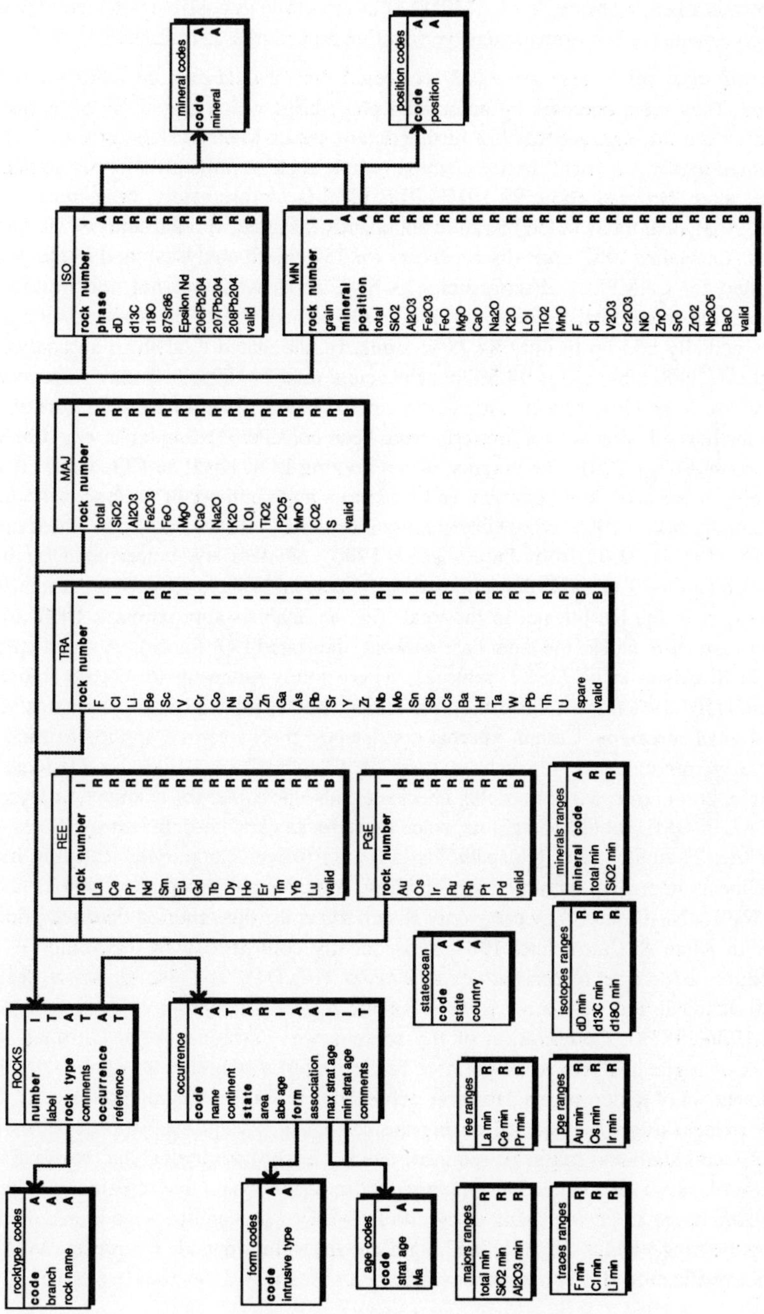

Fig.D1 Structure of the database LAMPDA. Arrows indicate relational links between tables.

D4 Data validation

Great care was taken to ensure that LAMPDA is as error-free as possible, via a combination of exhaustive manual and automatic data verification procedures, as follows:

(1) *Duplicate analyses* : these are easily collected during data capture from secondary sources. They were detected by automated procedures which sorted on ≥2 variables, then checked adjacent records (for further details see Le Maitre 1982, Rock 1987b).

(2) *Analytical totals*. An 'ideal' major element (whole-rock or mineral) analysis should, of course, total between about 99-101%. This ideal is, unfortunately, little more than a dream. Analytical totals in the literature abundantly disagree with the sums of the quoted figures (Le Maitre 1982 finds discrepancies for 15% of all analyses), and in the papers consulted for LAMPDA, discrepancies as high as 15 wt.% are not uncommon: for example, the stated total for analysis G14 in Oberhänsli (1986, p.336) is 99.79 but the oxides actually add up to only 84.78%; similarly, the stated total from an analysis in Milashev (1988, table 18) is 98.5% but the actual total in 109.2%. Today's literature is even worse in another respect, namely that analyses are blithely published with *stated* totals far beyond what would formerly have been considered acceptable: e.g. 106.09% in Morris (1987, p.220). The practice of not quoting LOI, $H_2O\pm$ or CO_2 values is also becoming more and more common, and leads to a huge number of analyses with totals substantially below 100% whose correctness is extremely difficult to judge (the extreme in LAMPDA is 72.65 from Jaques *et al.* 1986). Matters are exacerbated by some authors including LOI with the other major oxides, and others (e.g. Jahn *et al.* 1979) including it in the list but not in the total (i.e. an analysis approximates 100% in the former case *with* and in the latter case *without* the stated LOI figure). A worst-case is seen in Morrison *et al.* (1987, table 2), where totals range up to 121.06 if LOI is included (107.19% if not); it is difficult to know whether to take such data seriously!

(3) *Petrological screening*. Certain internal consistency checks can be applied to rock and especially mineral analyses. An analysis of a "spinel" with high Na_2O must, for example, be in error. More complex checks are also possible: for example, analyses of fresh AL or UML should contain *ne*, since these rocks carry Si-deficient pyriboles (e.g. kaersutite, Ti-augite) and feldspathoids; a *qz*-normative "camptonite" carrying modal nepheline is internally inconsistent. Similarly, minettes and lamproites should always have $K_2O > Na_2O$ since they carry only K-rich minerals (this criterion detected printers' errors in Allan & Carmichael 1984, subsequently confirmed with the authors). This procedure led to a set of *acceptable range tables* (Fig.D1), and also to a more detailed set of chemical and normative screens for the different lamprophyre varieties (Rock 1984,1986,1987a). Compilation of the screens was combined with elimination of suspect or rogue analyses in an iterative fashion, until a stable set of screens and self-consistent set of lamprophyre data were achieved. For some rocks, one set of data (e.g. major oxides) might be rejected during the screening process even though other data (e.g. mineral analyses) might be retained: this is clearly both logical and reasonable in the case of, say, a carbonated lamprophyre whose phenocrysts have remained unaltered.

(4) *Automatic range screening*: After entry into LAMPDA, procedures were invoked which call up the *range* tables (Fig.D1), and highlight any values outside the preset ranges for manual verification. This procedure complements the manual methods in step (b).

INDEX

Citations to the body of the text are in normal type; where appropriate, **bold** indicates the most significant of many citations for one entry; page-numbers referring to figures are in *italics* and those referring to tables are underlined. Indexing of geographical localities is at the level of countries (nations) or oceanic islands, and not down to that of states, regions, individual provinces or occurrences. Hence to find 'Argyle' look under 'Australia'. However, a very few exceptionally well-documented provinces are indexed separately (e.g. Caledonides, Hercynides, Navajo). Similarly, individual mineral species are indexed under mineral groups (e.g. biotite under *Mica*, nepheline under *Feldspathoid*).